Systems Analysis and Design with UML

AN OBJECT-ORIENTED APPROACH

Third Edition

International Student Version

Systems Analysis and Design with UML

AN OBJECT-ORIENTED APPROACH

Third Edition

International Student Version

Alan Dennis
Indiana University

Barbara Haley Wixom
University of Virginia

David Tegarden
Virginia Tech

John Wiley & Sons, Inc.

ISBN: 978-0-470-40030-2

Printed in Asia

10 9 8 7 6 5 4 3 2

BRIEF CONTENTS

CONTENTS

PREFACE

PURPOSE OF THIS BOOK

Systems Analysis and Design (SAD) is an exciting, active field in which analysts continually learn new techniques and approaches to develop systems more effectively and efficiently. However there is a core set of skills that all analysts need to know—no matter what approach or methodology is used. All information systems projects move through the four phases of planning, analysis, design, and implementation; all projects require analysts to gather requirements, model the business needs, and create blueprints for how the system should be built; and all projects require an understanding of organizational behavior concepts like change management and team building. Today, the cost of developing modern software is composed primarily of the cost associated with the developers themselves and not the computers. As such, object-oriented approaches to developing information systems hold much promise in controlling these costs.

Today, the most exciting change to systems analysis and design is the move to object-oriented techniques, which view a system as a collection of self-contained objects that have both data and processes. This change has been accelerated through the creation of the Unified Modeling Language (UML). UML provides a common vocabulary of object-oriented terms and diagramming techniques that is rich enough to model any systems development project from analysis through implementation.

This book captures the dynamic aspects of the field by keeping students focused on doing SAD while presenting the core set of skills that we feel every systems analyst needs to know today and in the future. This book builds on our professional experience as systems analysts and on our experience in teaching SAD in the classroom.

This book will be of particular interest to instructors who have students do a major project as part of their course. Each chapter describes one part of the process, provides clear explanations on how to do it, gives a detailed example, and then has exercises for the students to practice. In this way, students can leave the course with experience that will form a rich foundation for further work as a systems analyst.

OUTSTANDING FEATURES

A Focus on Doing SAD

The goal of this book is to enable students to do SAD—not just read about it, but understand the issues so they can actually analyze and design systems. The book introduces each major technique, explains what it is, explains how to do it, presents an example, and provides opportunities for students to practice before they do it for real in a project. After reading each chapter, the student will be able to perform that step in the system development life cycle (SDLC) process.

Rich Examples of Success and Failure

The book includes a running case about a fictitious company called CD Selections. Each chapter shows how the concepts are applied in situations at CD Selections. Unlike running cases in other books, we have tried to focus these examples on planning, managing, and executing the activities described in the chapter, rather than on detailed dialogue between fictious actors. In this way, the running case serves as a template that students can apply to their own work. Each chapter also includes numerous Concepts in Action boxes, many of which were written by Dr. Bruce White from Quinnipiac University, that describe how real companies succeeded—and failed—in performing the activities in the chapter.

Real World Focus

The skills that students learn in a systems analysis and design course should mirror the work that they ultimately will do in real organizations. We have tried to make this book as "real" as possible by building extensively on our experience as professional systems analysts for organizations such as Arthur Andersen, IBM, the U.S. Department of Defense, and the Australian Army. We have also worked with a diverse industry advisory board of IS professionals and consultants in developing the book and have incorporated their stories, feedback, and advice throughout. Many students who use this book will eventually use the skills on the job in a business environment, and we believe they will have a competitive edge in understanding what successful practitioners feel is relevant in the real world.

Project Approach

We have presented the topics in this book in the SDLC order in which an analyst encounters them in a typical project. Although the presentation is necessarily linear (because students have to learn concepts in the way in which they build on each other), we emphasize the iterative, complex nature of SAD as the book unfolds. The presentation of the material should align well with courses that encourage students to work on projects because it presents topics as students need to apply them.

WHAT'S NEW IN THIS EDITION

In this edition, we have increased the coverage of and better organized the text around the enhanced Unified Process; provided a greater focus on nonfunctional requirements; provided a greater emphasis on the iterative and incremental development associated with object-oriented analysis and design; added figures and examples, along with additional explanatory text that addresses some of the more difficult concepts to learn; better aligned the CD selections case material; and did some minor reorganization. However, the biggest changes that have been included address the issues surrounding information systems development and the so-called flat world. The global economy has brought up the need for much greater understanding of cultural issues, regulatory issues, and the need for testing. The third edition covers this type of material throughout the text. Details of the major changes are as follows:

1. To better align the text with a Unified Process–based methodology, all references to the MOOSAD methodology have been removed, the object-oriented systems analysis and design material and the short overview of UML 2.0 have been moved to Chapter 1, and "Basic Characteristics of Object-Oriented Systems" section is now included in an optional appendix. This last change was driven by the desire of instructors to have the flexibility to cover this material at the most appropriate time for the students based on the students' backgrounds. In some cases, students

may have had multiple object-oriented programming courses, whereas in other cases, students may not have had any programming courses at all. Either way, the material is not really necessary to understand until the functional modeling or the class and method design material is covered. Finally, we have tied the "Evolutionary Work Breakdown Structures and Iterative Workplans" section of the Project Management chapter to the enhanced Unified Process. This allowed us better to apply the iterative and incremental development characteristics of object-oriented systems development to the project management material.

2. With regards to the requirements determination, we introduced the idea that in today's world, there are additional nonfunctional requirements, such as Sarbanes-Oxley, COBIT, ISO 9000, and CMM, being added to the set of non-functional requirements that the analyst must address. We also have introduced new requirements-gathering techniques, for example, throwaway prototyping, role-playing CRC cards, and mind/concept mapping. And, we have introduced the system proposal idea as a separate topic instead of only having it embedded in the CD Selections case. Furthermore, in the design chapters (data management layer design, human–computer interaction layer design, and physical architecture layer design), more emphasis has been placed on the impact of nonfunctional requirements on the design.

3. The material included within the functional, structural, and behavioral modeling chapters has been more tightly coupled. This is especially true with regard to the idea of iterative and incremental development. The text now emphasizes that systems must be incrementally built by iterating over each of the models and over the intersection of the models. For example, the normal flow of events contained within a use-case description is associated with the activities on an activity diagram, the operations on a class diagram, the behaviors on the CRC cards, the messages on sequence and communication diagrams, and transitions on behavioral state machines. As such, any change to any one of these most likely will force changes in the others. Furthermore, we have promoted the use of CRUD analysis up to a behavioral modeling technique in and of itself instead of having it associated only with the communications diagram.

4. A major new section has been added to the design chapter that addresses the verification and validation of the analysis models. This section goes through great detail on how to verify and validate the analysis models developed during functional, structural, and behavioral modeling. Furthermore, there had been intentional oversights or errors placed in a couple of the earlier models. During this new section, those errors are uncovered and corrected. Furthermore, to enhance the coverage of testing material, we have added use-case testing as an integration-testing approach in the construction chapter.

5. We have included material that addresses global concerns throughout the text. This includes material with regard to requirements determination, outsourcing, and class and method design. With regard to the construction chapter, a new section on cultural issues has been included with the managing programmers section. Finally, with the installation and operations chapter a new major section has been added to address cultural issues and information technology. This new section is based on the work of Geert Hofstede.

6. Additional figures and explanatory material have been added throughout the text. However, special attention was paid to the material contained in the structural modeling, behavioral modeling, and the class and method design chapters.

7. The CD Selections case has now been more closely aligned with each chapter. Each chapter simply has a section of the case associated with it. For example, the introduction to the case is now associated with Chapter 1. The case has been slightly modified to ensure that the case itself is more cohesive. Furthermore, in some situations, the CD Selections case had introduced new content material. We have moved any new content material into the content of the corresponding chapter.

ORGANIZATION OF THIS BOOK

This book is organized by the phases of the Systems Development Life Cycle (SDLC). Each chapter has been written to teach students specific tasks that analysts need to accomplish over the course of a project, and the deliverables that will be produced from the tasks. As students complete the book, tasks will be "checked off" and deliverables will be completed. Along the way, students will be reminded of their progress using roadmaps that indicate where their current task fits into the larger context of SAD.

Chapter 1 introduces the SDLC, systems development methodologies, Object-oriented systems analysis, the Unified Process, UML 2.0, and describes the roles and skills needed for a project team.

Part One contains material that in many ways goes across the phases of the traditional SDLC. Chapter 2 presents project initiation, with a focus on project identification, system request, feasibility analysis, and project selection. In Chapter 3, students learn about project management, with emphasis on the workplan, staffing plan, project charter, and risk assessment that are used to help manage and control the project. Chapter 4 introduces students to an assortment of analysis techniques to help with business automation, business improvement, and business process reengineering, a variety of requirements-gathering techniques that are used to determine the functional and nonfunctional requirements of the system, and to a system proposal.

Part Two focuses on creating analysis models. Chapter 5 focuses on constructing functional models, Chapter 6 addresses producing structural models, and Chapter 7 tackles creating behavioral models.

Part Three addresses design modeling. In Chapter 8, students learn how to verify and validate the analysis models created during analysis modeling and to evolve the analysis models into design models via the use of factoring, partitions, and layers. The students also learn to create an alternative matrix that can be used to compare custom, packaged, and outsourcing alternatives. Chapter 9 concentrates on designing the individual classes and their respective methods through the use of contracts and method specifications. Chapter 10 presents the issues involved in designing persistence for objects. These issues include the different storage formats that can be used for object persistence, how to map an object-oriented design into the chosen storage format, and how to design a set of data access and manipulation classes that act as a translator between the classes in the application and the object persistence. This chapter also focuses on the nonfunctional requirements that impact the data management layer. Chapter 11 presents the design of the human–computer interaction layer, where students learn how to design user interfaces using use scenarios, windows navigation diagrams, storyboards, Windows layout diagrams, HTML prototypes, real use cases, interface standards, and user interface templates, to perform user interface evaluations using heuristic evaluation, walkthrough evaluation, interactive evaluation, and formal usability testing, and to address nonfunctional requirements such as user interface layout, content awareness, aesthetics, user experience, and consistency. Chapter 12 focuses on the physical architecture and infrastructure design, which includes deployment diagrams and hardware/software specification. This chapter, like the previous design chapters, covers the impact that nonfunctional requirements can have on the physical architecture layer.

Part Four provides material that is related to the construction, installation, and operations of the system. Chapter 13 focuses on system construction, where students learn how to build, test, and document the system. Installation and operations are covered in Chapter 14, where students learn about the conversion plan, change management plan, support plan, and project assessment. Additionally, these chapters address the issues related to developing systems in a flat world, where developers and users are distributed throughout the world.

SUPPLEMENTS http://www.wiley.com/go/global/dennis

Instructor's Resources Web Site

- PowerPoint slides, which instructors can tailor to their classroom needs and that students can use to guide their reading and studying activities
- The Instructor's Manual provides
  Short experiential exercises,
  short real world stories,
  additional minicases for every chapter, and
  solutions to end of chapter questions and exercises.

Student Website

- Relevant Web links, including career resources Web site.
- Web quizzes help students prepare for class tests.

ACKNOWLEDGMENTS

For the third edition, we would like to thank the students of the ACIS 3515: Information Systems Development I and ACIS 3516: Information Systems Development II classes at Virginia Tech for giving many suggestions that drove most of the changes from the second edition to the third edition. Their feedback was invaluable in improving the text and examples.

We would like to thank the following reviewers for their helpful and insightful comments on the third edition: Evans Adams, Fort Lewis College; Murugan Anandarajan, Drexel University; Rob Anson, Boise State University; Ravi Krovi, University of Akron; Leo Legorreta, California State University Sacramento; Diane Lending, James Madison University; Major Fernando Maymi, West Point University; J. Drew Procaccino, Rider University; Bill Watson, Indiana University–Purdue University Indianapolis; and Amy B. Woszczynski, Kennesaw State University.

We also thank the following reviewers from the first and second edition: Evans Adams, Fort Lewis College; Noushin Ashrafi, University of Massachusetts, Boston; Dirk Baldwin, University of Wisconsin-Parkside; Qing Cao, University of Missouri–Kansas City; Ahmad Ghafarian, North Georgia College & State University; Daniel V. Goulet, University of Wisconsin–Stevens Point; Harvey Hayashi, Loyalist College of Applied Arts and Technology; Jean-Piere Kuilboer, University of Massachusetts, Boston; Daniel Mittleman, DePaul University; Fred Niederman, Saint Louis University; H. Robert Pajkowski, DeVry Institute of Technology, Scarborough, Ontario; June S. Park, University of Iowa; Tom Pettay, DeVry Institute of Technology, Columbus, Ohio; Neil Ramiller, Portland State University; Eliot Rich, University at Albany, State University of New York; Carl Scott, University of Houston; Keng Siau, University of Nebraska–Lincoln; Jonathan Trower, Baylor University; Anna Wachholz, Sheridan College; Randy S. Weinberg, Carnegie Mellon University; Eli J. Weissman, DeVry Institute of Technology, Long Island City, NY; Heinz Roland Weistroffer, Virginia Commonwealth University; Amy Wilson, DeVry Institute of Technology, Decatur, GA; Vincent C. Yen, Wright State University; Murugan Anandarajon, Drexel University; Ron Anson, Boise State University; Noushin Ashrafi, University of Massachusetts Boston; Dirk Baldwin, University of Wisconsin; Robert Barker, University of Louisville; Terry Fox, Baylor University; Donald Golden, Cleveland State University; Cleotilde Gonzalez, Carnegie Melon University; Scott James, Saginaw Valley State University; Rajiv Kishore, State University of New York–Buffalo; Ravindra Krovi, University of Akron; Fernando Maymi, United States Military Academy at West Point; Fred Niederman, Saint Louis University; Graham Peace, West Virginia University; J. Drew Procaccino, Rider University; Marcus Rothenberger,University of Wisconsin–Milwaukee; June Verner, Drexel University; Heinz Roland Weistroffer, Virginia Commonwealth University; and Amy Woszczynski, Kennesaw State University.

CHAPTER 1

INTRODUCTION

Chapter 1 first introduces the systems development life cycle (SDLC), the fundamental four-phase model (planning, analysis, design, and implementation) common to all information system development projects. Second, it describes the evolution of system development methodologies. Third, the chapter overviews object-oriented systems analysis and design and describes the Unified Process and its extensions. Finally, the chapter closes with a discussion of the roles and skills necessary within the project team.

OBJECTIVES

- Understand the fundamental systems development life cycle and its four phases.
- Understand the evolution of systems development methodologies.
- Be familiar with the Unified Process and its extensions.
- Be familiar with the different roles on the project team.

CHAPTER OUTLINE

Introduction
The Systems Development Life Cycle
   Planning
   Analysis
   Design
   Implementation
Systems Development Methodologies
   Structured Design
   Rapid Application Development (RAD)
   Agile Development
   Selecting the Appropriate Development
   Methodology
Object-Oriented Systems Analysis and
  Design (OOSAD)
   Use-Case Driven
   Architecture Centric

   Iterative and Incremental
   Benefits of Object-Oriented Systems
    Analysis and Design
The Unified Process
   Phases
   Workflows
   Extensions to the Unified Process
The Unified Modeling Language
Project Team Roles and Skills
   Business Analyst
   Systems Analyst
   Infrastructure Analyst
   Change Management Analyst
   Project Manager
Applying the Concepts at CD Selections
Summary

INTRODUCTION

The *systems development life cycle (SDLC)* is the process of understanding how an information system (IS) can support business needs by designing a system, building it, and delivering it to users. If you have taken a programming class or have programmed on your own, this probably sounds pretty simple. Unfortunately, it is not. A 1996 survey by the Standish Group found that 42 percent of all corporate IS projects were abandoned before completion. A similar study done in 1996 by the General Accounting Office found 53 percent of all U.S. government IS projects were abandoned. Unfortunately, many of the systems that aren't abandoned are delivered to the users significantly late, cost far more than planned, and have fewer features than originally planned.

Most of us would like to think that these problems only occur to "other" people or "other" organizations, but they happen in most companies. Even Microsoft has a history of failures and overdue projects (e.g., Windows 1.0, Windows 95).[1]

Although we would like to promote this book as a "silver bullet" that will keep you from IS failures, we readily admit that a silver bullet that guarantees IS development success simply does not exist. Instead, this book will provide you with several fundamental concepts and many practical techniques that you can use to improve the probability of success.

The key person in the SDLC is the systems analyst, who analyzes the business situation, identifies opportunities for improvements, and designs an information system to implement them. Being a systems analyst is one of the most interesting, exciting, and challenging jobs around. Systems analysts work with a variety of people and learn how they conduct business. Specifically, they work with a team of systems analysts, programmers, and others on a common mission. Systems analysts feel the satisfaction of seeing systems that they designed and developed make a significant business impact, knowing that they contributed unique skills to make that happen.

However, the primary objective of a systems analyst is not to create a wonderful system; instead, it is to create value for the organization, which for most companies means increasing profits (government agencies and not-for-profit organizations measure value differently). Many failed systems have been abandoned because the analysts tried to build a wonderful system without clearly understanding how the system would fit with an organization's goals, current business processes, and other information systems to provide value. An investment in an information system is like any other investment, such as a new machine tool. The goal is not to acquire the tool, because the tool is simply a means to an end; the goal is to enable the organization to perform work better so it can earn greater profits or serve its constituents more effectively.

This book introduces the fundamental skills a systems analyst needs. This pragmatic book discusses best practices in systems development; it does not present a general survey of systems development that presents everything about the topic. By definition, systems analysts *do things* and challenge the current way that organizations work. To get the most out of this book, you will need actively to apply the ideas and concepts in the examples and in the "Your Turn" exercises that are presented throughout to your own systems development project. This book guides you through all the steps for delivering a successful information system. Also, it illustrates how one organization (called CD Selections) applies the steps in one project (developing a Web-based CD sales system). By the time you finish the book, you won't be an expert analyst, but you will be ready to start building systems for real.

---

[1] For more information on the problem, see Capers Jones, *Patterns of Software System Failure and Success* (London: International Thompson Computer Press, 1996); Capers Jones, *Assessment and Control of Software Project Risks* (Englewood Cliffs, NJ: Yourdon Press, 1994); Julia King, "IS Reins in Runaway Projects," *Computerworld* (February 24, 1997).

This chapter first introduces the basic SDLC that IS projects follow. This life cycle is common to all projects, although the focus and approach to each phase of the life cycle may differ. The next section describes three fundamentally different types of systems development methodologies: structured design, rapid application development, and agile development.

The next three sections introduce the fundamental characteristics of object-oriented systems analysis and design, a specific object-oriented systems development methodology (The Unified Process), and a specific object-oriented systems development graphical notation (The Unified Modeling Language). Finally, the chapter discusses one of the most challenging aspects of systems development, the depth and breadth of skills required of systems analysts. Today, most organizations use project teams that contain members with unique, but complementary, skills. The chapter closes with a discussion of the key roles played by members of the systems development team.

CONCEPTS

IN ACTION

1–A   An Expensive False Start

A real-estate group in the federal government cosponsored a data warehouse with the IT department. In the formal proposal written by IT, costs were estimated at $800,000, the project duration was estimated to be eight months, and the responsibility for funding was defined as the business unit's. The IT department proceeded with the project before it even knew if the project had been accepted.

The project actually lasted two years because requirements gathering took nine months instead of one and a half, the planned user base grew from 200 to 2,500, and the approval process to buy technology for the project took a year. Three weeks prior to technical delivery, the IT director canceled the project. This failed endeavor cost the organization and taxpayers $2.5 million.

Source: Hugh J. Watson et al., "Data Warehousing Failure: Case Studies and Findings," *The Journal of Data Warehousing* 4, (no. 1) (1999): 44–54.

Questions

1. Why did this system fail?
2. Why would a company spend money and time on a project and then cancel it?
3. What could have been done to prevent this?

THE SYSTEMS DEVELOPMENT LIFE CYCLE

In many ways, building an information system is similar to building a house. First, the house (or the information system) starts with a basic idea. Second, this idea is transformed into a simple drawing that is shown to the customer and refined (often through several drawings, each improving on the last) until the customer agrees that the picture depicts what he or she wants. Third, a set of blueprints is designed that presents much more detailed information about the house (e.g., the type of water faucets, where the telephone jacks will be placed). Finally, the house is built following the blueprints, often with some changes directed by the customer as the house is erected.

The SDLC has a similar set of four fundamental *phases*: planning, analysis, design, and implementation. Different projects may emphasize different parts of the SDLC or approach the SDLC phases in different ways, but all projects have elements of these four phases. Each *phase* is itself composed of a series of *steps,* which rely upon *techniques* that produce *deliverables* (specific documents and files that provide understanding about the project).

For example, when you apply for admission to a university, all students go through the same phases: information gathering, applying, and accepting. Each of these phases has steps—information gathering includes steps such as searching for schools, requesting information, and reading brochures. Students then use techniques (e.g., Internet searching) that can be applied to steps (e.g., requesting information) to create *deliverables* (e.g., evaluations of different aspects of universities).

In many projects, the SDLC phases and steps proceed in a logical path from start to finish. In other projects, the project teams move through the steps consecutively, incrementally, iteratively, or in other patterns. In this section, we describe the phases, actions, and some of the techniques that are used to accomplish the steps at a very high level. Not all organizations follow the SDLC in exactly the same way. As we shall shortly see, there are many variations on the overall SDLC.

For now, there are two important points to understand about the SDLC. First, you should get a general sense of the phases and steps through which IS projects move and some of the techniques that produce certain deliverables. Second, it is important to understand that the SDLC is a process of *gradual refinement*. The deliverables produced in the analysis phase provide a general idea of the shape of the new system. These deliverables are used as input to the design phase, which then refines them to produce a set of deliverables that describes in much more detailed terms exactly how the system will be built. These deliverables, in turn, are used in the implementation phase to produce the actual system. Each phase refines and elaborates on the work done previously.

Planning

The *planning phase* is the fundamental process of understanding *why* an information system should be built and determining how the project team will go about building it. It has two steps:

1. During *project initiation,* the system's business value to the organization is identified: how will it lower costs or increase revenues? Most ideas for new systems come from outside the IS area (from the marketing department, accounting department, etc.) in the form of a *system request.* A system request presents a brief summary of a business need, and it explains how a system that supports the need will create business value. The IS department works together with the person or department that generated the request (called the *project sponsor*) to conduct a *feasibility analysis.* The feasibility analysis examines key aspects of the proposed project:

   - The idea's technical feasibility (Can we build it?)
   - The economic feasibility (Will it provide business value?)
   - The organizational feasibility (If we build it, will it be used?)

   The system request and feasibility analysis are presented to an information systems *approval committee* (sometimes called a steering committee), which decides whether the project should be undertaken.

2. Once the project is approved, it enters *project management.* During project management, the *project manager* creates a *workplan,* staffs the project, and puts techniques in place to help the project team control and direct the project through the entire SDLC. The deliverable for project management is a *project plan,* which describes how the project team will go about developing the system.

Analysis

The *analysis phase* answers the questions of *who* will use the system, *what* the system will do, and *where* and *when* it will be used. During this phase, the project team investigates any current system(s), identifies improvement opportunities, and develops a concept for the new system.

This phase has three steps:

1. An *analysis strategy* is developed to guide the project team's efforts. Such a strategy usually includes an analysis of the current system (called the *as-is system*) and its problems, and then ways to design a new system (called the *to-be system*).

2. The next step is *requirements gathering* (e.g., through interviews or questionnaires). The analysis of this information—in conjunction with input from project sponsor and many other people—leads to the development of a concept for a new system. The system concept is then used as a basis to develop a set of business *analysis models,* which describe how the business will operate if the new system is developed. The set of models typically includes models that represent the data and processes necessary to support the underlying business process.

3. The analyses, system concept, and models are combined into a document called the *system proposal,* which is presented to the project sponsor and other key decision makers (e.g., members of the approval committee) who decide whether the project should continue to move forward.

The system proposal is the initial deliverable that describes what business requirements the new system should meet. Because it is really the first step in the design of the new system, some experts argue that it is inappropriate to use the term analysis as the name for this phase; some argue a better name would be analysis and initial design. Most organizations continue use to the name *analysis* for this phase, however, so we use it in this book as well. Just keep in mind that the deliverable from the analysis phase is both an analysis and a high-level initial design for the new system.

Design

The *design phase* decides *how* the system will operate, in terms of the hardware, software, and network infrastructure; the user interface, forms and reports; and the specific programs, databases, and files that will be needed. Although most of the strategic decisions about the system were made in the development of the system concept during the analysis phase, the steps in the design phase determine exactly how the system will operate. The design phase has four steps:

1. The *design strategy* is first developed. It clarifies whether the system will be developed by the company's own programmers, whether the system will be outsourced to another firm (usually a consulting firm), or whether the company will buy an existing software package.

2. This leads to the development of the basic *architecture design* for the system, which describes the hardware, software, and network infrastructure to be used. In most cases, the system will add or change the infrastructure that already exists in the organization. The *interface design* specifies how the users will move through the system (e.g., navigation methods such as menus and on-screen buttons) and the forms and reports that the system will use.

3. The *database and file specifications* are developed. These define exactly what data will be stored and where they will be stored.

4. The analyst team develops the *program design,* which defines the programs that need to be written and exactly what each program will do.

This collection of deliverables (architecture design, interface design, database and file specifications, and program design) is the *system specification* that is handed to the programming team for implementation. At the end of the design phase, the feasibility analysis and project plan are reexamined and revised, and another decision is made by the project sponsor and approval committee about whether to terminate the project or continue.

Implementation

The final phase in the SDLC is the *implementation phase,* during which the system is actually built (or purchased, in the case of a packaged software design). This is the phase that usually gets the most attention, because for most systems it is the longest and most expensive single part of the development process. This phase has three steps:

1. System *construction* is the first step. The system is built and tested to ensure it performs as designed. Because the cost of bugs can be immense, testing is one of the most critical steps in implementation. Most organizations give more time and attention to testing than to writing the programs in the first place.

2. The system is installed. *Installation* is the process by which the old system is turned off and the new one is turned on. It may include a direct cutover approach (in which the new system immediately replaces the old system), a parallel conversion approach (in which both the old and new systems are operated for a month or two until it is clear that there are no bugs in the new system), or a phased conversion strategy (in which the new system is installed in one part of the organization as an initial trial and then gradually installed in others). One of the most important aspects of conversion is the development of a *training plan* to teach users how to use the new system and help manage the changes caused by the new system.

3. The analyst team establishes a *support plan* for the system. This plan usually includes a formal or informal post-implementation review as well as a systematic way for identifying major and minor changes needed for the system.

CONCEPTS

IN ACTION

1–B Keeping Up with Consumer Electronics

Consumer electronics is a very competitive business. What might be the success story of the year one year is a forgotten item two years later. Rapid product commoditization makes the consumer electronic marketplace very competitive. Getting the right products to market at the right time with the right components is an ongoing challenge for telecommunications and consumer electronic goods companies.

Questions

1. What external data analysis should a consumer electronics company use to determine marketplace needs and its abilities to compete effectively in a marketplace?

2. Staying one step ahead of competitors requires a corporate strategy and the support of information systems. How can information systems and systems analysts contribute to an aggressive corporate strategy?

SYSTEMS DEVELOPMENT METHODOLOGIES

A *methodology* is a formalized approach to implementing the SDLC (i.e., it is a list of steps and deliverables). There are many different systems development methodologies, and each one is unique, based on the order and focus it places on each SDLC phase. Some methodologies are formal standards used by government agencies, whereas others have been developed by consulting firms to sell to clients. Many organizations have internal methodologies that have been honed over the years, and they explain exactly how each phase of the SDLC is to be performed in that company.

There are many ways to categorize methodologies. One way is by looking at whether they focus on business processes or the data that support the business. A *process-centered*

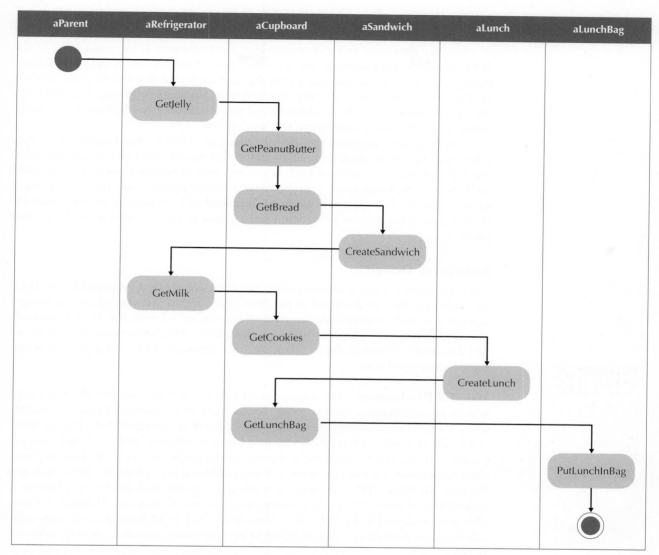

aParent	aRefrigerator	aCupboard	aSandwich	aLunch	aLunchBag

FIGURE 1-1   A Simple Behavioral Model for Making Lunch

methodology emphasizes process models as the core of the system concept. In Figure 1-1, for example, process-centered methodologies would focus first on defining the processes (e.g., assemble sandwich ingredients). *Data-centered* methodologies emphasize data models as the core of the system concept. In Figure 1-1 data-centered methodologies would focus first on defining the contents of the storage areas (e.g., refrigerator) and how the contents were organized.[2] By contrast, *object-oriented methodologies* attempt to balance the focus between process and data by incorporating both into one model. In Figure 1-1, these

---

[2] The classic modern process-centered methodology is that by Edward Yourdon, *Modern Structured Analysis* (Englewood Cliffs, NJ: Yourdon Press, 1989). An example of a data-centered methodology is information engineering; see James Martin, *Information Engineering*, vols. 1–3 (Englewood Cliffs, NJ: Prentice Hall, 1989). A widely accepted standardized non–object-oriented methodology that balances processes and data is IDEF; see FIPS 183, *Integration Definition for Function Modeling,* Federal Information Processing Standards Publications, U.S. Department of Commerce, 1993.

methodologies would focus first on defining the major elements of the system (e.g., sandwiches, lunches) and look at the processes and data involved with each element.

Another important factor in categorizing methodologies is the sequencing of the SDLC phases and the amount of time and effort devoted to each.[3] In the early days of computing, programmers did not understand the need for formal and well-planned life cycle methodologies. They tended to move directly from a very simple planning phase right into the construction step of the implementation phase—in other words, from a very fuzzy, not-well-thought-out system request into writing code.

This is the same approach that you sometimes use when writing programs for a programming class. It can work for small programs that require only one programmer, but if the requirements are complex or unclear, you may miss important aspects of the problem and have to start all over again, throwing away part of the program (and the time and effort spent writing it). This approach also makes teamwork difficult because members have little idea about what needs to be accomplished and how to work together to produce a final product.

Structured Design

The first category of systems development methodologies is called *structured design*. These methodologies became dominant in the 1980s, replacing the previous, ad hoc, and undisciplined approach. Structured design methodologies adopt a formal step-by-step approach to the SDLC that moves logically from one phase to the next. Numerous process-centered and data-centered methodologies follow the basic approach of the two structured design categories outlined next.

Waterfall Development The original structured design methodology (still used today) is *waterfall development*. With waterfall development–based methodologies, the analysts and users proceed in sequence from one phase to the next (see Figure 1-2). The key deliverables for each phase are typically very long (often hundreds of pages in length) and are presented to the project sponsor for approval as the project moves from phase to phase. Once the sponsor approves the work that was conducted for a phase, the phase ends and the next one begins. This methodology is referred to as waterfall development because it moves forward from phase to phase in the same manner as a waterfall. Although it is possible to go backward in the SDLC (e.g., from design back to analysis), it is extremely difficult (imagine yourself as a salmon trying to swim upstream against a waterfall, as shown in Figure 1-2).

Structured design also introduced the use of formal modeling or diagramming techniques to describe the basic business processes and the data that support them. Traditional structured design uses one set of diagrams to represent the processes and a separate set of diagrams to represent data. Because two sets of diagrams are used, the systems analyst must decide which set to develop first and use as the core of the system—process-model diagrams or data-model diagrams. There is much debate over which should come first, the processes or the data, because both are important to the system. As a result, several different structured design methodologies have evolved that follow the basic steps of the waterfall model but use different modeling approaches at different times. Those that attempt to emphasize process-model diagrams as the core of the system are process centered, whereas those that emphasize data-model diagrams as the core of the system concept are data centered.

[3] A good reference for comparing systems development methodologies is Steve McConnell, *Rapid Development* (Redmond, WA: Microsoft Press, 1996).

FIGURE 1-2
A Waterfall
Development–based
Methodology

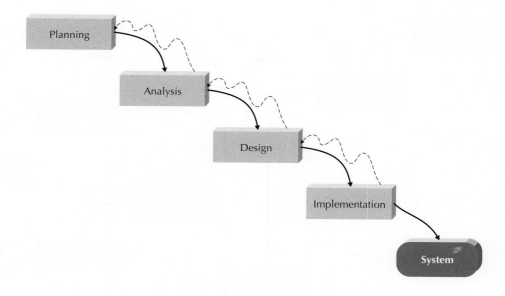

The two key advantages of the structured design waterfall approach are that it identifies system requirements long before programming begins and it minimizes changes to the requirements as the project proceeds. The two key disadvantages are that the design must be completely specified before programming begins and that a long time elapses between the completion of the system proposal in the analysis phase and the delivery of the system (usually many months or years). Lengthy deliverables often result in poor communication; the result is that important requirements can be overlooked in the voluminous documentation. Users are rarely prepared for their introduction to the new system, which occurs long after the initial idea for the system was introduced. If the project team misses important requirements, expensive post-implementation programming may be needed (imagine yourself trying to design a car on paper; how likely would you be to remember interior lights that come on when the doors open or to specify the right number of valves on the engine?).

A system may also require significant rework because the business environment has changed from the time that the analysis phase occurred. When changes do occur, it means going back to the initial phases and following the change through each of the subsequent phases in turn.

Parallel Development   *Parallel development* methodology attempts to address the problem of long delays between the analysis phase and the delivery of the system. Instead of doing design and implementation in sequence, it performs a general design for the whole system and then divides the project into a series of distinct subprojects that can be designed and implemented in parallel. Once all subprojects are complete, there is a final integration of the separate pieces, and the system is delivered (see Figure 1-3).

The primary advantage of this methodology is that it can reduce the schedule time to deliver a system; thus, there is less chance of changes in the business environment causing rework. However, the approach still suffers from problems caused by paper documents. It also adds a new problem: Sometimes the subprojects are not completely independent; design decisions made in one subproject may affect another, and the end of the project may require significant integration efforts.

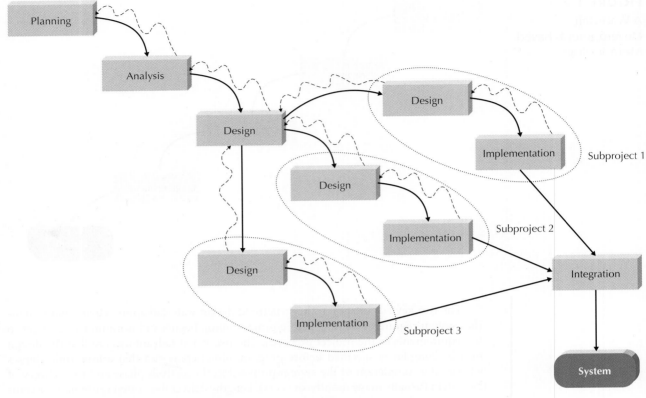

FIGURE 1-3 A Parallel Development–based Methodology

Rapid Application Development (RAD)

A second category of methodologies includes *rapid application development (RAD)*–based methodologies. These are a newer class of systems development methodologies that emerged in the 1990s. RAD-based methodologies attempt to address both weaknesses of structured design methodologies by adjusting the SDLC phases to get some part of the system developed quickly and into the hands of the users. In this way, the users can better understand the system and suggest revisions that bring the system closer to what is needed.[4]

Most RAD-based methodologies recommend that analysts use special techniques and computer tools to speed up the analysis, design, and implementation phases, such as CASE tools, joint application design (JAD) sessions, fourth-generation/visual programming languages that simplify and speed up programming (e.g., Visual Basic), and code generators that automatically produce programs from design specifications. The combination of the changed SDLC phases and the use of these tools and techniques improves the speed and quality of systems development. However, there is one possible subtle problem with RAD-based methodologies: managing user expectations. Due to the use of the tools and techniques that can improve the speed and quality of systems development, user expectations of what is possible may dramatically change. As a user better

[4] One of the best RAD books is Steve McConnell, *Rapid Development* (Redmond, WA: Microsoft Press, 1996).

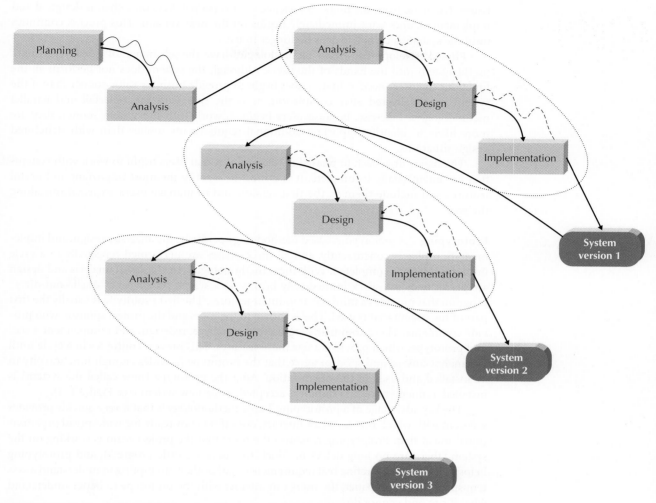

FIGURE 1-4   A Phased Development–based Methodology

understands the information technology, the systems requirements tend to expand. This was less of a problem when using methodologies that spent a lot of time thoroughly documenting requirements. Process-centered, data-centered, and object-oriented methodologies that follow the basic approaches of the three RAD categories are described in the following sections.

Phased Development   A *phased development*–based methodology breaks an overall system into a series of *versions,* which are developed sequentially. The analysis phase identifies the overall system concept, and the project team, users, and system sponsor then categorize the requirements into a series of versions. The most important and fundamental requirements are bundled into the first version of the system. The analysis phase then leads into design and implementation—but only with the set of requirements identified for version 1 (see Figure 1-4).

Once version 1 is implemented, work begins on version 2. Additional analysis is performed based on the previously identified requirements and combined with new ideas and issues that arose from the users' experience with version 1. Version 2 then is designed and implemented, and work immediately begins on the next version. This process continues until the system is complete or is no longer in use.

Phased development–based methodologies have the advantage of quickly getting a useful system into the hands of the users. Although the system does not perform all the functions the users need at first, it does begin to provide business value sooner than if the system were delivered after completion, as is the case with the waterfall and parallel methodologies. Likewise, because users begin to work with the system sooner, they are more likely to identify important additional requirements sooner than with structured design situations.

The major drawback to phased development is that users begin to work with systems that are intentionally incomplete. It is critical to identify the most important and useful features and include them in the first version and to manage users' expectations along the way.

Prototyping   A *prototyping*-based methodology performs the analysis, design, and implementation phases concurrently, and all three phases are performed repeatedly in a cycle until the system is completed. With these methodologies, the basics of analysis and design are performed, and work immediately begins on a *system prototype,* a "quick-and-dirty" program that provides a minimal amount of features. The first prototype is usually the first part of the system that is used. This is shown to the users and the project sponsor, who provide comments. These comments are used to reanalyze, redesign, and reimplement a second prototype, which provides a few more features. This process continues in a cycle until the analysts, users, and sponsor agree that the prototype provides enough functionality to be installed and used in the organization. After the prototype (now called the system) is installed, refinement occurs until it is accepted as the new system (see Figure 1-5).

The key advantage of a prototyping-based methodology is that it *very* quickly provides a system with which the users can interact, even if it is not ready for widespread organizational use at first. Prototyping reassures the users that the project team is working on the system (there are no long delays in which the users see little progress), and prototyping helps to more quickly refine real requirements. Rather than attempting to understand a system specification on paper, the users can interact with the prototype to better understand what it can and cannot do.

The major problem with prototyping is that its fast-paced system releases challenge attempts to conduct careful, methodical analysis. Often the prototype undergoes such significant changes that many initial design decisions become poor ones. This can cause

FIGURE 1-5
A Prototyping-based
Methodology

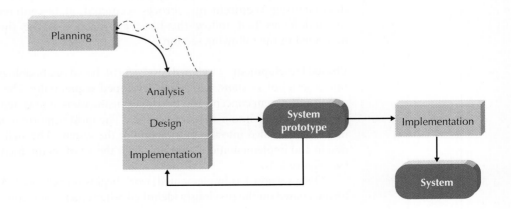

problems in the development of complex systems because fundamental issues and problems are not recognized until well into the development process. Imagine building a car and discovering late in the prototyping process that you have to take the whole engine out to change the oil (because no one thought about the need to change the oil until after it had been driven 10,000 miles).

Throwaway Prototyping   *Throwaway prototyping*–based methodologies are similar to prototyping-based methodologies in that they include the development of prototypes; however, throwaway prototypes are done at a different point in the SDLC. These prototypes are used for a very different purpose than those previously discussed, and they have a very different appearance (see Figure 1-6).

The throwaway prototyping–based methodologies have a relatively thorough analysis phase that is used to gather information and to develop ideas for the system concept. However, users may not completely understand many of the features they suggest, and there may be challenging technical issues to be solved. Each of these issues is examined by analyzing, designing, and building a *design prototype*. A design prototype is not a working system; it is a product that represents a part of the system that needs additional refinement, and it contains only enough detail to enable users to understand the issues under consideration. For example, suppose users are not completely clear on how an order entry system should work. The analyst team might build a series of HTML pages viewed using a Web browser to help the users visualize such a system. In this case, a series of mock-up screens *appear* to be a system, but they really do nothing. Or, suppose that the project team needs to develop a sophisticated graphics program in Java. The team could write a portion of the program with pretend data to ensure that they could do a full-blown program successfully.

A system developed using this type of methodology probably relies on several design prototypes during the analysis and design phases. Each of the prototypes is used to minimize the risk associated with the system by confirming that important issues are understood before the real system is built. Once the issues are resolved, the project moves into design and implementation. At this point, the design prototypes are thrown away, which is an important difference between these methodologies and prototyping methodologies, in which the prototypes evolve into the final system.

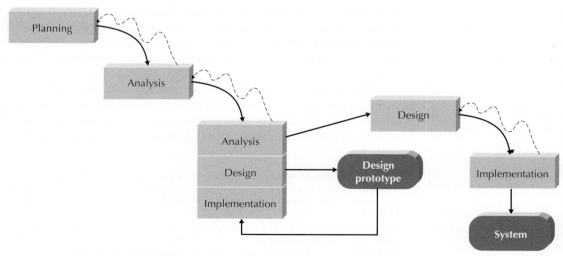

FIGURE 1-6   A Throwaway Prototyping–based Methodology

FIGURE 1-7
The Extreme
Programming
Methodology

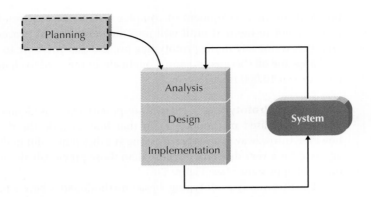

Throwaway prototyping–based methodologies balance the benefits of well-thought-out analysis and design phases with the advantages of using prototypes to refine key issues before a system is built. It may take longer to deliver the final system as compared to prototyping-based methodologies (because the prototypes do not become the final system), but this type of methodology usually produces more stable and reliable systems.

Agile Development[5]

A third category of systems development methodologies is still emerging today: *agile development*. These programming-centric methodologies have few rules and practices, all of which are fairly easy to follow. They focus on streamlining the SDLC by eliminating much of the modeling and documentation overhead and the time spent on those tasks. Instead, projects emphasize simple, iterative application development. Examples of agile development methodologies include extreme programming, Scrum, and the Dynamic Systems Development Method (DSDM). The agile development approach, as described next, typically is used in conjunction with object-oriented methodologies.

Extreme Programming[6] *Extreme programming* (*XP*) is founded on four core values: communication, simplicity, feedback, and courage. These four values provide a foundation that XP developers use to create any system. First, the developers must provide rapid feedback to the end users on a continuous basis. Second, XP requires developers to follow the KISS principle.[7] Third, developers must make incremental changes to grow the system, and they must not only accept change, they must embrace change. Fourth, developers must have a quality-first mentality. XP also supports team members in developing their own skills.

Three of the key principles that XP uses to create successful systems are continuous testing, simple coding performed by pairs of developers, and close interactions with end users to build systems very quickly. After a superficial planning process, projects perform analysis, design, and implementation phases iteratively (see Figure 1-7).

---

[5] Two good sources of information on agile development and object-oriented systems is S. W. Ambler, *Agile Modeling: Effective Practices for Extreme Programming and The Unified Process* (New York: Wiley, 2002), and R. C. Martin, *Agile Software Development: Principles, Patterns, and Practices* (Upper Saddle River, NJ: Prentice Hall, 2003).

[6] For more information, see K. Beck, *eXtreme Programming Explained: Embrace Change* (Reading, MA: Addison-Wesley, 2000), M. Lippert, S. Roock, and H. Wolf, *eXtreme Programming in Action: Practical Experiences from Real World Projects* (New York: Wiley, 2002), or www.extremeprogramming.com.

[7] Keep It Simple, Stupid.

Testing and efficient coding practices are core to XP. In fact, code is tested each day and is placed into an integrative testing environment. If bugs exist, the code is backed out until it is completely free of errors. XP relies heavily on refactoring, which is a disciplined way to restructure code to keep it simple.

An XP project begins with user stories that describe what the system needs to do. Then, programmers code in small, simple modules and test to meet those needs. Users are required to be available to clear up questions and issues as they arise. Standards are very important to minimize confusion, so XP teams use a common set of names, descriptions, and coding practices. XP projects deliver results sooner than even the RAD approaches, and they rarely get bogged down in gathering requirements for the system.

For small projects with highly motivated, cohesive, stable, and experienced teams, XP should work just fine. However, if the project is not small or the teams aren't jelled,[8] then the success of an XP development effort is doubtful. This tends to throw the whole idea of bringing outside contractors into an existing team environment using XP into doubt.[9] The chance of outsiders "jelling" with insiders may simply be too optimistic. XP requires a great deal of discipline; otherwise projects will become unfocused and chaotic. Furthermore, it is recommended only for small groups of developers—no more than ten developers, and it is not advised for large mission-critical applications. Due to the lack of analysis and design documentation, there is only code documentation associated with XP, so maintaining large systems built with XP may be impossible. And because mission-critical business information systems tend to exist for a long time, the utility of XP as a business information system development methodology is in doubt. Finally, the methodology needs a lot of on-site user input, something to which many business units cannot commit.[10]

Selecting the Appropriate Development Methodology

Because there are many methodologies, the first challenge faced by analysts is to select which methodology to use. Choosing a methodology is not simple, because no one methodology is always best. (If it were, we'd simply use it everywhere!) Many organizations have standards and policies to guide the choice of methodology. You will find that organizations range from having one "approved" methodology to having several methodology options to having no formal policies at all.

Figure 1-8 summarizes some important methodology selection criteria. One important item not discussed in this figure is the degree of experience of the analyst team. Many of the RAD-based methodologies require the use of "new" tools and techniques that have a significant learning curve. Often these tools and techniques increase the complexity of the project and require extra time for learning. However, once they are adopted and the team becomes experienced, the tools and techniques can significantly increase the speed in which the methodology can deliver a final system.

Clarity of User Requirements   When the user requirements for a system are unclear, it is difficult to understand them by talking about them and explaining them with written reports.

---

[8] A *jelled team* is one that has low turnover, a strong sense of identity, a sense of eliteness, a feeling that they jointly own the product being developed, and enjoyment in working together. For more information regarding jelled teams, see T. DeMarco and T. Lister. *Peopleware: Productive Projects and Teams* (New York: Dorset/House, 1987).

[9] Considering the tendency for offshore outsourcing, this is a major obstacle for XP to overcome. For more information on offshore outsourcing, see P. Thibodeau, "ITAA Panel Debates Outsourcing Pros, Cons," *Computerworld Morning Update* (September 25, 2003), and S. W. Ambler, "Chicken Little Was Right," *Software Development* (October 2003).

[10] Many of the observations described on the utility of XP as a development approach were based on conversations with Brian Henderson-Sellers.

Ability to Develop Systems	Structured Methodologies			RAD Methodologies		Agile Methodologies
	Waterfall	Parallel	Phased	Prototyping	Throwaway Prototyping	XP
With Unclear User Requirements	Poor	Poor	Good	Excellent	Excellent	Excellent
With Unfamiliar Technology	Poor	Poor	Good	Poor	Excellent	Poor
That Are Complex	Good	Good	Good	Poor	Excellent	Poor
That Are Reliable	Good	Good	Good	Poor	Excellent	Good
With a Short Time Schedule	Poor	Good	Excellent	Excellent	Good	Excellent
With Schedule Visibility	Poor	Poor	Excellent	Excellent	Good	Good

FIGURE 1-8   Criteria for Selecting a Methodology

Users normally need to interact with technology to really understand what a new system can do and how to best apply it to their needs. Prototyping- and throwaway prototyping–based RAD methodologies are usually more appropriate when user requirements are unclear because they provide prototypes for users to interact with early in the SDLC.

Familiarity with Technology   When the system will use new technology with which the analysts and programmers are not familiar (e.g., the first Web development project with Java), early application of the new technology in the methodology will improve the chance of success. If the system is designed without some familiarity with the base technology, risks increase because the tools might not be capable of doing what is needed. Throwaway prototyping–based methodologies are particularly appropriate if users lack familiarity with technology because they explicitly encourage the developers to develop design prototypes for areas with high risks. Phased development–based methodologies are good as well, because they create opportunities to investigate the technology in some depth before the design is complete. Although you might think prototyping-based methodologies are also appropriate, they are much less so because the early prototypes that are built usually only scratch the surface of the new technology. It is generally only after several prototypes and several months that the developers discover weaknesses or problems in the new technology.

System Complexity   Complex systems require careful and detailed analysis and design. Throwaway prototyping–based methodologies are particularly well suited to such detailed analysis and design, as opposed to prototyping-based methodologies, which are not. The traditional structured design–based methodologies can handle complex systems, but without the ability to get the system or prototypes into the users' hands early on, some key issues may be overlooked. Although phased development–based methodologies enable users to interact with the system early in the process, we have observed that project teams who follow these tend to devote less attention to the analysis of the complete problem domain than they might using other methodologies.

System Reliability   System reliability is usually an important factor in system development—after all, who wants an unreliable system? However, reliability is just one factor among several. For some applications reliability is truly critical (e.g., medical equipment, missile control systems), whereas for other applications (e.g., games, Internet video) it is merely important. Throwaway prototyping methodologies are the most appropriate when system reliability is a high priority, because it combines detailed analysis and design phases with the ability for the project team to test many different approaches through design prototypes before completing the design. Prototyping methodologies are generally not a good choice when reliability is critical because it lacks the careful analysis and design phases that are essential for dependable systems.

Short Time Schedules   Projects that have short time schedules are well suited for RAD-based methodologies. This is due to them being designed to increase the speed of development. Prototyping and phased development–based methodologies are excellent choices when timelines are short because they best enable the project team to adjust the functionality in the system based on a specific delivery date, and if the project schedule starts to slip, it can be readjusted by removing functionality from the version or prototype under development. Waterfall-based methodologies are the worst choice when time is at a premium because they do not allow for easy schedule changes.

Schedule Visibility   One of the greatest challenges in systems development is determining whether a project is on schedule. This is particularly true of the structured design methodologies because design and implementation occur at the end of the project. The RAD-based methodologies move many of the critical design decisions earlier in the project to help project managers recognize and address risk factors and keep expectations in check.

YOUR TURN

1-1 Selecting a Methodology

*S*uppose you are an analyst for the Roanoke Software Consulting Company (RSCC), a large consulting firm with offices around the world. The company wants to build a new knowledge management system that can identify and track the expertise of individual consultants anywhere in the world based on their education and the various consulting projects on which they have worked. Assume that this is a new idea that has never before been attempted in RSCC or elsewhere. RSCC has an international network, but the offices in each country may use somewhat different hardware and software. RSCC management wants the system up and running within a year.

Question

1. What type of methodology would you recommend RSCC use? Why?

OBJECT-ORIENTED SYSTEMS ANALYSIS AND DESIGN (OOSAD)

Object-oriented approaches to developing information systems, technically speaking, can use any of the traditional methodologies (waterfall development, parallel development, phased development, prototyping, and throwaway prototyping). However, the object-oriented approaches are most associated with a phased development RAD methodology. The primary difference between a traditional approach like structured design and an object-oriented approach is how a problem is decomposed. In traditional approaches, the problem decomposition process is either process centric or data centric. However, processes and data are so closely related that it is difficult to pick one or the other as the primary focus. Based on this lack of congruence with the real world, new *object-oriented methodologies* have emerged that use the RAD-based sequence of SDLC phases but attempt to balance the emphasis between process and data by focusing the decomposition of problems on objects that contain both data and processes. Both approaches are valid approaches to developing information systems. In this book, we focus only on object-oriented approaches.[11]

According to the creators of the Unified Modeling Language (UML), Grady Booch, Ivar Jacobson, and James Rumbaugh,[12] any modern object-oriented approach to developing information systems must be (1) use-case driven, (2) architecture-centric, and (3) iterative and incremental.

Use-Case Driven

Use-case driven means that *use cases* are the primary modeling tools defining the behavior of the system. A use case describes how the user interacts with the system to perform some activity, such as placing an order, making a reservation, or searching for information. The use cases are used to identify and to communicate the requirements for the system to the programmers who must write the system.

Use cases are inherently simple because they focus on only one activity at a time. In contrast, the process model diagrams used by traditional structured and RAD methodologies are far more complex because they require the system analyst and user to develop models of the entire system. With traditional methodologies, each business activity is decomposed into a set of subprocesses, which are, in turn, decomposed into further subprocesses, and so on. This goes on until no further process decomposition makes sense, and it often requires dozens of pages of interlocking diagrams. In contrast, use cases focus on only one activity at a time, so developing models is much simpler.[13]

Architecture Centric

Any modern approach to systems analysis and design should be architecture centric. *Architecture centric* means that the underlying software architecture of the evolving system specification drives the specification, construction, and documentation of the system. Modern object-oriented systems analysis and design approaches should support at least three separate but interrelated architectural views of a system: functional, static, and dynamic. The *functional,* or *external, view* describes the behavior of the system from the perspective of the user. The *structural,* or *static, view* describes the system in terms of attributes, methods, classes, and relationships. The *behavioral,* or *dynamic, view* describes the behavior of the system in terms of messages passed among objects and state changes within an object.

Iterative and Incremental

Modern object-oriented systems analysis and design approaches emphasize *iterative* and *incremental* development that undergoes continuous testing and refinement throughout the life of the project. This implies that the systems analysts develop their understanding of a user's problem by building up the three architectural views little by little. The systems analyst does this by

---

[11] See Alan Dennis, Barbara Haley Wixom, and Roberta M. Roth, *Systems Analysis and Design: An Applied Approach, 3rd ed.* (New York: Wiley, 2006) for a description of the traditional approaches.

[12] Grady Booch, Ivar Jacobson, and James Rumbaugh, *The Unified Modeling Language User Guide* (Reading, MA: Addison-Wesley, 1999).

[13] For those of you that have experience with traditional structured analysis and design, this will be one of the most unusual aspects of object-oriented analysis and design using UML. Structured approaches emphasize the decomposition of the complete business process into subprocesses and sub-subprocesses. Object-oriented approaches stress focusing on just one use-case activity at a time and distributing that single use case over a set of communicating and collaborating objects. Therefore, use-case modeling may seem initially unsettling or counterintuitive, but in the long run this single focus does make analysis and design simpler.

FIGURE 1-9
Iterative and
Incremental
Development

Object-Oriented Analysis

Functional
view

Structural
view

Behavioral
view

working with the user to create a functional representation of the system under study. Next, the analyst attempts to build a structural representation of the evolving system. Using the structural representation of the system, the analyst distributes the functionality of the system over the evolving structure to create a behavioral representation of the evolving system.

As an analyst works with the user in developing the three architectural views of the evolving system, the analyst will iterate over each of and among the views. That is, as the analyst better understands the structural and behavioral views, the analyst will uncover missing requirements or misrepresentations in the functional view. This, in turn, can cause changes to be cascaded back through the structural and behavioral views. All three architectural views of the system are interlinked and dependent on each other (see Figure 1-9). As each increment and iteration is completed, a more complete representation of the user's real functional requirements are uncovered.

Benefits of Object-Oriented Systems Analysis and Design

Concepts in the object-oriented approach enable analysts to break a complex system into smaller, more manageable modules, work on the modules individually, and easily piece the modules back together to form an information system. This modularity makes system development easier to grasp, easier to share among members of a project team, and easier to communicate to users, who are needed to provide requirements and confirm how well the system meets the requirements throughout the SDLC. By modularizing system development, the project team actually is creating reusable pieces that can be plugged into other systems efforts or used as starting points for other projects. Ultimately, this can save time because new projects don't have to start completely from scratch.

Many people argue that "object-think" is a much more realistic way to think about the real world. Users typically do not think in terms of data or process; instead, they see their business as a collection of logical units that contain both—so communicating in terms of objects improves the interaction between a user and an analyst or developer.

THE UNIFIED PROCESS

The Unified Process is a specific methodology that maps out when and how to use the various UML techniques for object-oriented analysis and design. The primary contributors were Grady Booch, Ivar Jacobsen, and James Rumbaugh of Rational. Whereas the UML provides structural support for developing the structure and behavior of an information system, the Unified Process provides the behavioral support. The Unified Process, of course, is use-case driven, architecture centric, and iterative and incremental.

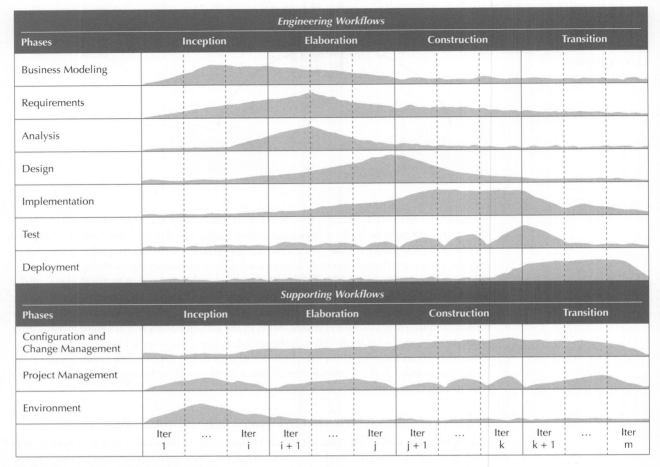

FIGURE 1-10 The Unified Process

The Unified Process is a two-dimensional systems development process described by a set of phases and workflows. The phases are inception, elaboration, construction, and transition. The workflows include business modeling, requirements, analysis, design, implementation, test, deployment, project management, configuration and change management, and environment. In the remainder of this section, we describe the phases and workflows of the Unified Process.[14] Figure 1-10 depicts the Unified Process.

Phases

The *phases* of the Unified Process support an analyst in developing information systems in an iterative and incremental manner. The phases describe how an information system evolves through time. Depending on which development phase the evolving system is

[14] The material in this section is based on Khawar Zaman Ahmed and Cary E. Umrysh, *Developing Enterprise Java Applications with J2EE and UML* (Boston, MA: Addison-Wesley, 2002); Jim Arlow and Ila Neustadt, *UML and The Unified Process: Practical Object-Oriented Analysis & Design* (Boston, MA: Addison-Wesley, 2002); Peter Eeles, Kelli Houston, Wojtek Kozacynski, *Building J2EE Applications with the Rational Unified Process*, (Boston, MA: Addison-Wesley, 2003); Ivar Jacobson, Grady Booch, and James Rumbaugh, *The Unified Software Development Process* (Reading, MA: Addison-Wesley, 1999); Phillipe Krutchten, *The Rational Unified Process: An Introduction, 2nd ed.* (Boston, MA: Addison-Wesley, 2000).

currently in, the level of activity will vary over the *workflows.* The curve in Figure 1-10 associated with each workflow approximates the amount of activity that takes place during the specific phase. For example, the inception phase primarily involves the business modeling and requirements workflows, while practically ignoring the test and deployment workflows. Each phase contains a set of iterations, and each iteration uses the various workflows to create an incremental version of the evolving information system. As the system evolves through the phases, it improves and becomes more complete. Each phase has objectives, a focus of activity over the workflows, and incremental deliverables. Each of the phases is described next.

Inception   In many ways, the *inception phase* is very similar to the planning phase of a traditional SDLC approach. In this phase, a business case is made for the proposed system. This includes feasibility analysis that should answer questions such as the following:

Do we have the technical capability to build it (technical feasibility)?

If we build it, will it provide business value (economic feasibility)?

If we build it, will it be used by the organization (organizational feasibility)?

To answer these questions, the development team performs work related primarily to the business modeling, requirements, and analysis workflows. In some cases, depending on the technical difficulties that could be encountered during the development of the system, a throwaway prototype is developed. This implies that the design, implementation, and test workflows could also be involved. The project management and environment supporting workflows are very relevant to this phase. The primary deliverables from the inception phase are (1) a vision document that sets the scope of the project, identifies the primary requirements and constraints, sets up an initial project plan, and describes the feasibility of and risks associated with the project, and (2) the adoption of the necessary environment to develop the system.

Elaboration   When we typically think about object-oriented systems analysis and design, the activities related to the *elaboration phase* of the Unified Process are the most relevant. The *analysis* and *design workflows* are the primary focus during this phase. The elaboration phase continues with developing the vision document, including finalizing the business case, revising the risk assessment, and completing a project plan in sufficient detail to allow the stakeholders to be able to agree with constructing the actual final system. It deals with gathering the requirements, building the UML structural and behavioral models of the problem domain, and detailing how the problem domain models fit into the evolving system architecture. Developers are involved with all but the *deployment engineering workflow* in this phase. As the developers iterate over the workflows, the importance of addressing configuration and change management becomes apparent. Also, the development tools acquired during the inception phase become critical to the success of the project during this phase.[15] The primary deliverables of this phase include (1) the UML structure and behavior diagrams and (2) an executable of a baseline version of the evolving information system. The baseline version serves as the foundation for all later iterations. By providing a solid foundation at this point in time, the developers have a basis for completing the system in the construction and transition phases.

---

[15] With UML comprising fourteen different, related diagramming techniques, keeping the diagrams coordinated and the different versions of the evolving system synchronized is typically beyond the capabilities of a mere mortal systems developer. These tools typically include project management and CASE (Computer-Aided Software Engineering) tools. We describe the use of these tools in Chapter 3.

Construction   The *construction phase* focuses heavily on programming the evolving information system. As such, it is primarily concerned with the *implementation workflow*. However, the *requirements workflow* and the analysis and design workflows also are involved with this phase. It is during this phase that missing requirements are uncovered, and the analysis and design models are finally completed. Typically, there are iterations of the workflows during this phase, and during the last iteration, the deployment workflow kicks into high gear. The *configuration and change management workflow*, with its version control activities, becomes extremely important during the construction phase. At times, an iteration may have to be rolled back. Without good version controls, rolling back to a previous version (incremental implementation) of the system is nearly impossible. The primary deliverable of this phase is an implementation of the system that can be released for beta and acceptance testing.

Transition   Like the construction phase, the *transition phase* addresses aspects typically associated with the implementation phase of a traditional SDLC approach. Its primary focus is on the testing and deployment workflows. Essentially, the business modeling, requirements, and analysis workflows should have been completed in earlier iterations of the evolving information system. Depending on the results from the testing workflow, it is possible that some redesign and programming activities on the design and implementation workflows could be necessary, but they should be minimal at this point in time. From a managerial perspective, the project management, configuration and change management, and environment are involved. Some of the activities that take place are beta and acceptance testing, fine-tuning the design and implementation, user training, and the actual rolling out of the final product onto a production platform. Obviously, the primary deliverable is the actual executable information system. The other deliverables include user manuals, a plan to support the users, and a plan for upgrading the information system in the future.

Workflows

The workflows describe the tasks or activities that a developer performs to evolve an information system over time. The workflows of the Unified Process are grouped into two broad categories: engineering and supporting.

Engineering Workflows   Engineering workflows include business modeling, requirements, analysis, design, implementation, test, and deployment workflows. The engineering workflows deal with the activities that produce the technical product (i.e., the information system).

Business Modeling Workflow   The *business modeling workflow* uncovers problems and identifies potential projects within a user organization. This workflow aids management in understanding the scope of the projects that can improve the efficiency and effectiveness of a user organization. The primary purpose of business modeling is to ensure that both developer and user organizations understand where and how the to-be-developed information system fits into the business processes of the user organization. This workflow is primarily executed during the inception phase to ensure that we develop information systems that make business sense. The activities that take place on this workflow are most closely associated with the planning phase of the traditional SDLC; however, requirements gathering and use-case and business process modeling techniques also help to understand the business situation.

Requirements Workflow   In the Unified Process, the requirements workflow includes eliciting both functional and nonfunctional requirements. Typically, requirements are gathered from project stakeholders, such as end users, managers within the end user organization, and even customers. There are many different ways to capture requirements, including interviews, observation techniques, joint application development, document analysis, and questionnaires. The requirements workflow is utilized the most during the inception and elaboration phases. The identified requirements are very helpful for developing the vision document and the use cases used throughout the development process. Additional requirements tend to be discovered throughout the development process. In fact, only the transition phase tends to have few, if any, additional requirements identified.

Analysis Workflow   The analysis workflow primarily addresses the creation of an analysis model of the problem domain. In the Unified Process, the analyst begins designing the architecture associated with the problem domain; using the UML, the analyst creates structural and behavioral diagrams that depict a description of the problem domain classes and their interactions. The primary purpose of the analysis workflow is to ensure that both the developer and user organizations understand the underlying problem and its domain without overanalyzing. If they are not careful, analysts can create *analysis paralysis,* which occurs when the project becomes so bogged down with analysis that the system is never actually designed or implemented. A second purpose of the analysis workflow is to identify useful reusable classes for class libraries. By reusing predefined classes, the analyst can avoid "reinventing the wheel" when creating the structural and behavioral diagrams. The analysis workflow is predominantly associated with the elaboration phase, but like the requirements workflow, it is possible that additional analysis will be required throughout the development process.

Design Workflow   The design workflow transitions the analysis model into a form that can be used to implement the system: the *design model.* Whereas the analysis workflow concentrated on understanding the problem domain, the design workflow focuses on developing a solution that will execute in a specific environment. Basically, the design workflow simply enhances the description of the evolving information system by adding classes that address the environment of the information system to the evolving analysis model. As such, the design workflow uses activities such as user interface design, database design, physical architecture design, detailed problem domain class design, and the optimization of the evolving information system. The design workflow is associated primarily with the elaboration and construction phases of the Unified Process.

Implementation Workflow   The primary purpose of the implementation workflow is to create an executable solution based on the design model (i.e., programming). This includes not only writing new classes but also incorporating reusable classes from executable class libraries into the evolving solution. As with any programming activity, testing of the new classes and their interactions with the incorporated reusable classes must occur. Finally, in the case of multiple groups performing the implementation of the information system, the implementers also must integrate the separate, individually tested modules to create an executable version of the system. The implementation workflow is associated primarily with the elaboration and construction phases.

Testing Workflow   The primary purpose of the *testing workflow* is to increase the quality of the evolving system. As such, testing goes beyond the simple unit testing associated with the implementation workflow. In this case, testing also includes testing the integration of all modules used to implement the system, user acceptance testing, and the actual alpha testing of the software. Practically speaking, testing should go on throughout the development of the system; testing of the analysis and design models occurs during the elaboration and construction phases, whereas implementation testing is performed primarily during the construction and, to some degree, transition phases. Basically, at the end of each iteration during the development of the information system, some type of test should be performed.

Deployment Workflow   The deployment workflow is most associated with the transition phase of the Unified Process. The deployment workflow includes activities, such as software packaging, distribution, installation, and beta testing. When actually deploying the new information system into a user organization, the developers may have to convert the current data, interface the new software with the existing software, and provide end user training on the use of the new system.

Supporting Workflows   The supporting workflows include the project management, configuration and change management, and the environment workflows. The supporting workflows focus on the managerial aspects of information system development.

Project Management Workflow   Whereas the other workflows associated with the Unified Process are technically active during all four phases, the *project management workflow* is the only truly cross-phase workflow. The development process supports incremental and iterative development, so information systems tend to grow or evolve over time. At the end of each iteration, a new incremental version of the system is ready for delivery. The project management workflow is quite important due to the complexity of the two-dimensional development model of the Unified Process (workflows and phases). This workflow's activities include risk identification and management, scope management, estimating the time to complete each iteration and the entire project, estimating the cost of the individual iteration and the whole project, and tracking the progress being made toward the final version of the evolving information system.

Configuration and Change Management Workflow   The primary purpose of the configuration and change management workflow is to keep track of the state of the evolving system. In a nutshell, the evolving information system comprises a set of artifacts, including, for example, diagrams, source code, and executables. During the development process, these artifacts are modified. A substantial amount of work—and, hence, dollars—is involved in the development of the artifacts. As such, the artifacts themselves should be handled as any expensive asset would be handled—access controls must be put into place to safeguard the artifacts from being stolen or destroyed. Furthermore, because the artifacts are modified on a regular, if not continuous, basis, good version control mechanisms should be established. Finally, a good deal of project management information needs to be captured (e.g., author, time, and location of each modification). The configuration and change management workflow is associated mostly with the construction and transition phases.

Environment Workflow   During the development of an information system, the development team needs to use different tools and processes. The *environment workflow* addresses these needs. For example, a computer-aided software engineering tool that supports the development of an object-oriented information system via the UML could be required. Other tools necessary include programming environments, project management tools, and configuration management tools. The environment workflow involves acquiring and installing these tools. Even though this workflow can be active during all of the phases of the Unified Process, it should be involved primarily with the inception phase.

Extensions to the Unified Process

As large and as complex as the Unified Process is, many authors have pointed out a set of critical weaknesses. First, the Unified Process does not address staffing, budgeting, or contract management issues. These activities were explicitly left out of the Unified Process. Second, the Unified Process does not address issues relating to maintenance, operations, or support of the product once it has been delivered. As such, it is not a

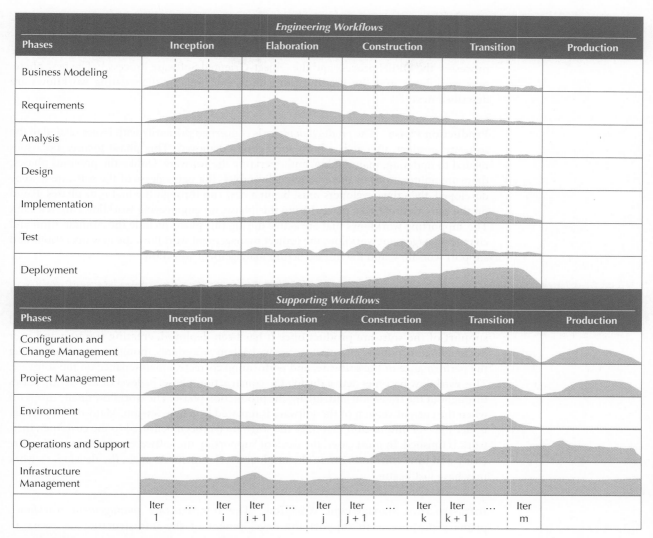

	Engineering Workflows				
Phases	**Inception**	**Elaboration**	**Construction**	**Transition**	**Production**
Business Modeling					
Requirements					
Analysis					
Design					
Implementation					
Test					
Deployment					

	Supporting Workflows				
Phases	**Inception**	**Elaboration**	**Construction**	**Transition**	**Production**
Configuration and Change Management					
Project Management					
Environment					
Operations and Support					
Infrastructure Management					
	Iter 1 ... Iter i	Iter i + 1 ... Iter j	Iter j + 1 ... Iter k	Iter k + 1 ... Iter m	

FIGURE 1-11   The Enhanced Unified Process

complete software process; it is only a development process. Third, the Unified Process does not address cross- or interproject issues. Considering the importance of reuse in object-oriented systems development and the fact that in many organizations employees work on many different projects at the same time, leaving out interproject issues is a major omission.

To address these omissions, Ambler and Constantine suggest the addition of a production phase and two workflows: the operations and support workflow and the infrastructure management workflow (see Figure 1-11).[16] In addition to these new workflows,

[16] S. W. Ambler and L. L. Constantine, *The Unified Process Inception Phase: Best Practices in Implementing the UP* (Lawrence, KS: CMP Books, 2000); S. W. Ambler and L. L. Constantine, *The Unified Process Elaboration Phase: Best Practices in Implementing the UP* (Lawrence, KS: CMP Books, 2000); S. W. Ambler and L. L. Constantine, *The Unified Process Construction Phase: Best Practices in Implementing the UP* (Lawrence, KS: CMP Books, 2000); S. W. Ambler and L. L. Constantine, *The Unified Process Transition and Production Phases: Best Practices in Implementing the UP* (Lawrence, KS: CMP Books, 2002).

the test, deployment, and environment workflows are modified, and the project management and configuration and change management workflows are extended into the production phase. These extensions are based on alternative object-oriented software processes: the OPEN process and the Object-Oriented Software Process.[17] The new phase, new workflows, and the modifications and extensions to the existing workflows are described next.

Production Phase   The *production phase* is concerned primarily with issues related to the software product after it has been successfully deployed. This phase focuses on issues related to updating, maintaining, and operating the software. Unlike the previous phases, there are no iterations or incremental deliverables. If a new release of the software is to be developed, then the developers must begin a new run through the first four phases. Based on the activities that take place during this phase, no engineering workflows are relevant. The supporting workflows that are active during this phase include the configuration and change management workflow, the project management workflow, the new operations and support workflow, and the infrastructure management workflow.

Operations and Support Workflow   The *operations and support workflow*, as you might guess, addresses issues related to supporting the current version of the software and operating the software on a daily basis. Activities include creating plans for the operation and support of the software product once it has been deployed, creating training and user documentation, putting into place necessary backup procedures, monitoring and optimizing the performance of the software, and performing corrective maintenance on the software. This workflow becomes active during the construction phase; its level of activity increases throughout the transition and, finally, the production phase. The workflow finally drops off when the current version of the software is replaced by a new version. Many developers are under the false impression that once the software has been delivered to the customer, their work is finished. In most cases, the work of supporting the software product is much more costly and time consuming than the original development. As such, the developer's work may have just begun.

Infrastructure Management Workflow   The *infrastructure management workflow's* primary purpose is to support the development of the infrastructure necessary to develop object-oriented systems. Activities such as development and modification of libraries, standards, and enterprise models are very important. When the development and maintenance of a problem domain architecture model goes beyond the scope of a single project and reuse is going to occur, the infrastructure management workflow is essential. Another very important set of cross-project activities is the improvement of the software development process. Because the activities on this workflow tend to affect many projects and the Unified Process focuses only on a specific project, the Unified Process tends to ignore these activities (i.e., they are simply beyond the scope and purpose of the Unified Process).

---

[17] S. W. Ambler, *Process Patterns—Building Large-Scale Systems Using Object Technology* (Cambridge, UK: SIGS Books/Cambridge University Press, 1998); S. W. Ambler, *More Process Patterns—Delivering Large-Scale Systems Using Object Technology* (Cambridge, UK: SIGS Books/Cambridge University Press, 1999); I. Graham, B. Henderson-Sellers, and H. Younessi, *The OPEN Process Specification* (Harlow, UK: Addison-Wesley, 1997); B. Henderson-Sellers and B. Unhelkar, *OPEN Modeling with UML* (Harlow, UK: Addison-Wesley, 2000).

Existing Workflow Modifications and Extensions   In addition to the workflows that were added to address deficiencies contained in the Unified Process, existing workflows had to be modified and/or extended into the production phase. These workflows include the test, deployment, environment, project management, and configuration and change management workflows.

Test Workflow   For high-quality information systems to be developed, testing should be done on every deliverable, including those created during the inception phase. Otherwise, less than quality systems will be delivered to the customer.

Deployment Workflow   Legacy systems exist in most corporations today, and these systems have databases associated with them that must be converted to interact with the new systems. Due to the complexity of deploying new systems, the conversion requires significant planning. As such, the activities on the deployment workflow need to begin in the inception phase instead of waiting until the end of the construction phase, as suggested by the Unified Process.

Environment Workflow   The environment workflow needed to be modified to include activities related to setting up the operations and production environment. The actual work performed is similar to the work related to setting up the development environment that was performed during the inception phase. In this case, the additional work is performed during the transition phase.

Project Management Workflow   Even though the project management workflow does not include staffing the project, managing the contracts among the customers and vendors, and managing the project's budget, these activities are crucial to the success of any software development project. As such, we suggest extending project management to include these activities. Furthermore, this workflow should additionally occur in the production phase to address issues such as training, staff management, and client relationship management.

Configuration and Change Management Workflow   The configuration and change management workflow is extended into the new production phase. Activities performed during the production phase include identifying potential improvements to the operational system and assessing the potential impact of the proposed changes. Once developers have identified these changes and understood their impact, they can schedule the changes to be made and deployed with future releases.

Figure 1-12 shows the chapters in which the Enhanced Unified Process's phases and workflows are covered. Given the offshore outsourcing and automation of information technology,[18] in this textbook, we focus primarily on the elaboration phase and the business modeling, requirements, analysis, design, and project management workflows of the Enhanced Unified Process. However, as Figure 1-12 shows, the other phases and workflows are covered. In many object-oriented systems development environments today, code generation is supported. Thus, from a business perspective, we believe the activities associated with these workflows are the most important.

---

[18] See Thomas L. Friedman, *The World Is Flat: A Brief History of the Twenty-First Century, Updated and Expanded Edition* (New York: Farrar, Straus, and Giroux, 2006); and Daniel H. Pink, *A Whole New Mind: Why Right-Brainers Will Rule the Future* (New York: Riverhead Books, 2006).

Enhanced UP Phases	Chapters
Inception	2–5
Elaboration	4–12
Construction	9, 13
Transition	13, 14
Production	14
Enhanced UP Engineering Workflows	**Chapters**
Business Modeling	2, 4–6
Requirements	4–6, 11
Analysis	5–7
Design	8–12
Implementation	10, 13
Test	8, 13
Deployment	14
Enhanced UP Supporting Workflows	**Chapters**
Project Management	2, 3, 5, 14
Configuration and Change Management	2, 14
Environment	3
Operations and Support	14
Infrastructure Management	3

FIGURE 1-12   The Enhanced Unified Process and the Textbook Organization

YOUR TURN

1-2 OO Systems Analysis and Design Methodology

Review Figures 1-10, 1-11, and 1-12. Based on your understanding of the UP and the EUP, suggest a set of steps for an alternative object-oriented systems development method. Be sure that the steps are capable of delivering an executable and maintainable system.

THE UNIFIED MODELING LANGUAGE

Until 1995, object concepts were popular but implemented in many different ways by different developers. Each developer had his or her own methodology and notation (e.g., Booch, Coad, Moses, OMT, OOSE, SOMA.)[19] Then in 1995, Rational Software brought three industry leaders together to create a single approach to object-oriented systems development. Grady Booch, Ivar Jacobson, and James Rumbaugh worked with others to create a standard set of diagramming techniques known as the *Unified Modeling Language (UML)*. The objective of UML was to provide a common vocabulary of object-oriented terms and diagramming techniques rich enough to model any systems development project from analysis through implementation. In November 1997, the *Object Management Group (OMG)* formally accepted UML as the standard for all object developers. During the following years, the UML has gone through multiple minor revisions. The current version of UML, Version 2.0, was accepted by the members of the OMG during their spring and summer meetings of 2003.

Version 2.0 of the UML defines a set of fourteen diagramming techniques used to model a system. The diagrams are broken into two major groupings: one for modeling structure of a system and one for modeling behavior. *Structure diagrams* provide a way to represent the data and static relationships in an information system. The structure diagrams include class, object, package, deployment, component, and composite structure diagrams. *Behavior diagrams* provide the analyst with a way to depict the dynamic relationships among the instances or objects that represent the business information system. They also allow the modeling of the dynamic behavior of individual objects throughout their lifetime. The behavior diagrams support the analyst in modeling the functional requirements of an evolving information system. The behavior modeling diagrams include activity, sequence, communication, interaction overview, timing, behavior state machine, protocol state machine, and use-case diagrams.[20] Figure 1-13 provides an overview of these diagrams.

Depending on where in the development process the system is, different diagrams play a more important role. In some cases, the same diagramming technique is used throughout the development process. In that case, the diagrams start off very conceptual and abstract. As the system is developed, the diagrams evolve to include details that ultimately lead to code generation and development. In other words, the diagrams move from documenting the requirements to laying out the design. Overall, the consistent notation, integration among the diagramming techniques, and application of the diagrams across the entire development process makes the UML a powerful and flexible language for analysts and developers. Later chapters provide more detail on using a subset of the UML in object-oriented systems analysis and design. In particular, these chapters describe activity, use-case, class, object, sequence, communication, and package diagrams and the behavioral state machines.

---

[19] See Grady Booch, *Object-Oriented Analysis and Design with Applications*, 2nd ed. (Redwood City, CA: Benjamin/Cummings, 1994); Peter Coad and Edward Yourdon, *Object-Oriented Analysis*, 2nd ed. (Englewood Cliffs, NJ: Yourdon Press, 1991); Peter Coad and Edward Yourdon, *Object-Oriented Design* (Englewood Cliffs, NJ: Yourdon Press, 1991); Brian Henderson-Sellers and Julian Edwards, *Book Two of Object-Oriented Knowledge: The Working Object* (Sydney, Australia: Prentice Hall, 1994); James Rumbaugh, Michael Blaha, William Premerlani, Frederick Eddy, and William Lorensen, *Object-Oriented Modeling and Design* (Englewood Cliffs, NJ: Prentice Hall, 1991); Ivar Jacobson, Magnus Christerson, Patrik Jonsson, and Gunnar Overgaard, *Object-Oriented Software Engineering: A Use Case Approach* (Wokingham, England: Addison-Wesley, 1992); Ian Graham, *Migrating to Object Technology* (Wokingham, England: Addison-Wesley, 1994).

[20] The material contained in this section is based on the *Unified Modeling Language: Superstructure Version 2.0, ptc/03-08-02* (www.uml.org). Additional useful references include Michael Jesse Chonoles and James A. Schardt, *UML 2 for Dummies* (Indianapolis, IN: Wiley, 2003); Hans-Erik Eriksson, Magnus Penker, Brian Lyons, and David Fado, *UML 2 Toolkit* (Indianapolis, IN: Wiley, 2004); and Kendall Scott, *Fast Track UML 2.0* (Berkeley, CA: Apress, 2004). For a complete description of all diagrams, see www.uml.org.

Diagram Name	Used to...	Primary Phase
Structure Diagrams		
Class	Illustrate the relationships between classes modeled in the system.	Analysis, Design
Object	Illustrate the relationships between objects modeled in the system. Used when actual instances of the classes will better communicate the model.	Analysis, Design
Package	Group other UML elements together to form higher-level constructs.	Analysis, Design, Implementation
Deployment	Show the physical architecture of the system. Can also be used to show software components being deployed onto the physical architecture.	Physical Design, Implementation
Component	Illustrate the physical relationships among the software components.	Physical Design, Implementation
Composite Structure	Illustrate the internal structure of a class, i.e., the relationships among the parts of a class.	Analysis, Design
Behavioral Diagrams		
Activity	Illustrate business workflows independent of classes, the flow of activities in a use case, or detailed design of a method.	Analysis, Design
Sequence	Model the behavior of objects within a use case. Focuses on the time-based ordering of an activity.	Analysis, Design
Communication	Model the behavior of objects within a use case. Focuses on the communication among a set of collaborating objects of an activity.	Analysis, Design
Interaction Overview	Illustrate an overview of the flow of control of a process.	Analysis, Design
Timing	Illustrate the interaction that takes place among a set of objects and the state changes in which they go through along a time axis.	Analysis, Design
Behavioral State Machine	Examine the behavior of one class.	Analysis, Design
Protocol State Machine	Illustrates the dependencies among the different interfaces of a class.	Analysis, Design
Use-Case	Capture business requirements for the system and to illustrate the interaction between the system and its environment.	Analysis

FIGURE 1-13   UML 2.0 Diagram Summary

PROJECT TEAM ROLES AND SKILLS

It is clear from the various phases and steps performed during the SDLC that the project team needs a variety of skills. Project members are *change agents* who identify ways to improve an organization, build an information system to support them, and train and motivate others to use the system. Leading a successful organizational change effort is one of the most difficult jobs that someone can do. Understanding what to change and how to change it—and convincing others of the need for change—requires a wide range of skills. These skills can be broken down into six major categories: technical, business, analytical, interpersonal, management, and ethical.

Role	Responsibilities
Business analyst	Analyzing the key business aspects of the system Identifying how the system will provide business value Designing the new business processes and policies
Systems analyst	Identifying how technology can improve business processes Designing the new business processes Designing the information system Ensuring that the system conforms to information systems standards
Infrastructure analyst	Ensuring the system conforms to infrastructure standards Identifying infrastructure changes needed to support the system
Change management analyst	Developing and executing a change management plan Developing and executing a user training plan
Project manager	Managing the team of analysts, programmers, technical writers, and other specialists Developing and monitoring the project plan Assigning resources Serving as the primary point of contact for the project

FIGURE 1-14
Project Team Roles

Analysts must have the technical skills to understand the organization's existing technical environment, the technology that will comprise the new system, and the way in which both can be fit into an integrated technical solution. Business skills are required to understand how IT can be applied to business situations and to ensure that the IT delivers real business value. Analysts are continuous problem solvers at both the project and the organizational level, and they put their analytical skills to the test regularly.

Analysts often need to communicate effectively one-on-one with users and business managers (who often have little experience with technology) and with programmers (who often have more technical expertise than the analyst). They must be able to give presentations to large and small groups and write reports. Not only do they need to have strong interpersonal abilities, but they also need to manage people with whom they work and they need to manage the pressure and risks associated with unclear situations.

Finally, analysts must deal fairly, honestly, and ethically with other project team members, managers, and system users. Analysts often deal with confidential information or information that, if shared with others, could cause harm (e.g., dissent among employees); it is important to maintain confidence and trust with all people.

In addition to these six general skill sets, analysts require many specific skills associated with roles performed on a project. In the early days of systems development, most organizations expected one person, the analyst, to have all the specific skills needed to conduct a systems development project. Some small organizations still expect one person to perform many roles, but because organizations and technology have become more complex, most large organizations now build project teams containing several individuals with clearly defined responsibilities. Different organizations divide the roles differently, but Figure 1-14 presents one commonly used set of project team roles. Most IS teams include many other individuals, such as the *programmers,* who actually write the programs that make up the system, and *technical writers,* who prepare the help screens and other documentation (e.g., users manuals and systems manuals).

Business Analyst

A *business analyst* focuses on the business issues surrounding the system. These issues include identifying the business value that the system will create, developing ideas and suggestions for how the business processes can be improved, and designing the new processes and policies in conjunction with the systems analyst. This individual will likely have business experience and some type of professional training (e.g., the business analyst for accounting systems will likely be a CPA [in the United States] or a CA [in Canada]). He or she represents the interests of the project sponsor and the ultimate users of the system. A business analyst assists in the planning and design phases but is most active in the analysis phase.

Systems Analyst

A *systems analyst* focuses on the IS issues surrounding the system. This person develops ideas and suggestions for how information technology can improve business processes, designs the new business processes with help from the business analyst, designs the new information system, and ensures that all IS standards are maintained. A systems analyst will likely have significant training and experience in analysis and design, programming, and even areas of the business. He or she represents the interests of the IS department and works intensively through the project but perhaps less so during the implementation phase.

Infrastructure Analyst

An *infrastructure analyst* focuses on the technical issues surrounding how the system will interact with the organization's technical infrastructure (e.g., hardware, software, networks, and databases). An infrastructure analyst's tasks include ensuring that the new information system conforms to organizational standards and identifying infrastructure changes needed to support the system. This individual will probably have significant training and experience in networking, database administration, and various hardware and software products. He or she represents the interests of the organization and IS group that will ultimately have to operate and support the new system once it has been installed. An infrastructure analyst works throughout the project but perhaps less so during planning and analysis phases.

Change Management Analyst

A *change management analyst* focuses on the people and management issues surrounding the system installation. The roles of this person include ensuring that the adequate documentation and support are available to users, providing user training on the new system, and developing strategies to overcome resistance to change. This individual should have significant training and experience in organizational behavior in general and change management in particular. He or she represents the interests of the project sponsor and users for whom the system is being designed. A change management analyst works most actively during the implementation phase but begins laying the groundwork for change during the analysis and design phases.

Project Manager

A project manager is responsible for ensuring that the project is completed on time and within budget and that the system delivers all benefits intended by the project sponsor. The role of the project manager includes managing the team members, developing the project plan, assigning resources, and being the primary point of contact when people outside the team have questions about the project. This individual will likely have significant experience in project management and has probably worked for many years as a systems analyst beforehand. He or she represents the interests of the IS department and the project sponsor. The project manager works intensely during all phases of the project.

YOUR
TURN

1-3 Being an Analyst

*S*uppose you decide to become an analyst after you graduate. Decide what type of analyst you would most prefer to be and what type of courses you should take before you graduate. Then decide the type of summer job or internship you should seek.

Question

Develop a short plan that describes how you will prepare for your career as an analyst.

APPLYING THE CONCEPTS AT CD SELECTIONS

Throughout this book, many new concepts about object-oriented systems analysis and design are introduced. As a way to make these new concepts more relevant, we apply them to a fictitious company called CD Selections. CD Selections is a chain of fifty music stores located in California, with headquarters in Los Angeles. Annual sales last year were $50 million, and they have been growing at about 3 to 5 percent per year for the past few years. However, the firm has been interested in expanding their presence beyond California. Margaret Mooney, vice president of marketing, has recently become both excited by and concerned with the rise of Internet sites selling CDs. She believes that the Internet has great potential, but she wants to use it in the right way. Rushing into e-commerce without considering things such as its effect on existing brick-and-mortar stores and the implications on existing systems at CD Selections could cause more harm than good. Currently, CD Selections has a Web site that provides basic information about the company and about each of its stores (e.g., map, operating hours, phone number, etc.). The Web site was developed by an Internet consulting firm and is hosted by a prominent local Internet service provider (ISP) in Los Angeles. The IT department at CD Selections has become experienced with Internet technology as it has worked with the ISP to maintain the site; however, it still has a lot to learn when it comes to conducting business over the Web. As such, Margaret is interested in investigating the possibility of creating an e-commerce site that will work with the current systems used by CD Selections. In future chapters, we revisit CD Selections to see how the concepts introduced in the individual chapters impact Margaret and CD Selections.

SUMMARY

The Systems Development Life Cycle

All systems development projects follow essentially the same fundamental process, called the system development life cycle (SDLC). SDLC starts with a planning phase in which the project team identifies the business value of the system, conducts a feasibility analysis, and plans the project. The second phase is the analysis phase, in which the team develops an analysis strategy, gathers information, and builds a set of analysis models. In the next phase, the design phase, the team develops the physical design, architecture design, interface design, database and file specifications, and program design. In the final phase, implementation, the system is built, installed, and maintained.

The Evolution of Systems Development Methodologies

System development methodologies are formalized approaches to implementing an SDLC. System development methodologies have evolved over the decades. Structured design methodologies, such as waterfall and parallel development, emphasize decomposition of a problem by either focusing on process decomposition (process-centric methodologies) or data decomposition (data decomposition). They produce a solid, well-thought-out system but can overlook requirements because users must specify them early in the design process before seeing the actual system. RAD-based methodologies attempt to speed up development and make it easier for users to specify requirements by having parts of the system developed sooner either by producing different versions (phased development) or by using prototypes (prototyping, throwaway prototyping) through the use of CASE tools and fourth-generation/visual programming languages. However, RAD-based methodologies still tend to be either process-centric or data-centric. Agile development methodologies, such as XP, focus on streamlining the SDLC by eliminating many of the tasks and time associated with requirements definition and documentation. Several factors influence the choice of a methodology: clarity of the user requirements; familiarity with the base technology; system complexity; need for system reliability; time pressures; and the need to see progress on the time schedule.

Object-Oriented Systems Analysis and Design

Object-oriented systems analysis and design (OOSAD) is most associated with a phased-development RAD-based methodology, where the time spent in each phase is very short. OOSAD uses a use-case-driven, architecture-centric, iterative, and incremental information systems development approach. It supports three different views of the evolving system: functional, static, and dynamic. OOSAD allows the analyst to decompose complex problems into smaller, more manageable components using a commonly accepted set of notations. Also, many people believe that users do not think in terms of data or processes but instead think in terms of a collection of collaborating objects. As such, object-oriented systems analysis and design allows the analyst to interact with the user with objects from the user's environment instead of a set of separate processes and data.

One of the most popular approaches to object-oriented systems analysis and design is the Unified Process. The Unified Process is a two-dimensional systems development process described with a set of phases and workflows. The phases consist of the inception, elaboration, construction, and transition phases. The workflows are organized into two subcategories: engineering and supporting. The engineering workflows include business modeling, requirements, analysis, design, implementation, test, and deployment workflows, and the supporting workflows comprise the project management, configuration and change management, and environment workflows. Depending on which development phase the evolving system is currently in, the level of activity will vary over the workflows.

The Unified Modeling Language

The Unified Modeling Language, or UML, is a standard set of diagramming techniques that provide a graphical representation rich enough to model any systems development project from analysis through implementation. Today most object-oriented systems analysis and design approaches use the UML to depict an evolving system. The UML uses a set of different diagrams to portray the various views of the evolving system. The diagrams are grouped into two broad classifications: structure and behavior. The structure diagrams include class, object, package, deployment, component, and composite structure diagrams. The behavior diagrams include activity, sequence, communication, interaction overview, timing, behavior state machine, protocol state machine, and use case diagrams.

Project Team Roles and Skills

The project team needs a variety of skills. All analysts need to have general skills, such as change management, ethics, communications, and technical. However, different kinds of analysts require specific skills in addition to these. Business analysts usually have business skills that help them to understand the business issues surrounding the system, whereas systems analysts also have significant experience in analysis and design and programming. The infrastructure analyst focuses on technical issues surrounding how the system will interact with the organization's technical infrastructure, and the change management analyst focuses on people and management issues surrounding the system installation. In addition to analysts, project teams will include a project manager, programmers, technical writers, and other specialists.

KEY TERMS

Agile development
Analysis model
Analysis paralysis
Analysis phase
Analysis strategy
Analysis workflow
Approval committee
Architecture centric
Architecture design
As-is system
Behavior diagrams
Behavioral view
Business analyst
Business modeling workflow
Change agent
Change management analyst
Configuration and change
   management workflow
Construction
Construction phase
Database and file specification
Data-centered methodology
Deliverable
Deployment workflow
Design model
Design phase
Design prototype
Design strategy
Design workflow
Dynamic view
Elaboration phase

Engineering workflow
Environment workflow
External view
Extreme programming (XP)
Feasibility analysis
Functional view
Gradual refinement
Implementation phase
Implementation workflow
Inception phase
Incremental
Infrastructure analyst
Infrastructure management workflow
Interface design
Iterative
Methodology
Object management group (OMG)
Object-oriented methodologies
Operations and support workflow
Parallel development
Phased development
Phases
Planning phase
Process-centered methodology
Production phase
Program design
Programmer
Project management
Project management workflow
Project manager
Project plan

Project sponsor
Prototyping
Rapid application development
   (RAD)
Requirements gathering
Requirements workflow
Static view
Structural view
Structure diagrams
Structured design
Support plan
Systems development life cycle
   (SDLC)
System proposal
System prototype
System request
System specification
Systems analyst
Technical writer
Testing workflow
Throwaway prototyping
Training plan
Transition phase
Unified Modeling Language (UML)
Use case
Use-case driven
Version
Waterfall development
Workflows
Workplan

QUESTIONS

1. Compare and contrast the role of a systems analyst, business analyst, and infrastructure analyst.
2. Compare and contrast phases, steps, techniques, and deliverables.
3. Describe the major phases in the SDLC.
4. Which phase in the SDLC is most important? Why?
5. What does *gradual refinement* mean in the context of SDLC?
6. What are the key factors in selecting a methodology?
7. Compare and contrast process-centered methodologies with data-centered methodologies.
8. Compare and contrast structured design-based methodologies in general to RAD-based methodologies in general.
9. Describe the major elements and issues with prototyping.
10. Describe the major elements and issues with throwaway prototyping.
11. Compare and contrast extreme programming and throwaway prototyping.
12. Describe the principal steps in the planning phase. What are the major deliverables?
13. Describe the principal steps in the analysis phase. What are the major deliverables?
14. Describe the principal steps in the design phase. What are the major deliverables?
15. Describe the principal steps in the implementation phase. What are the major deliverables?
16. What are the roles of a project sponsor and the approval committee?
17. Describe the major elements and issues with waterfall development.
18. Describe the major elements and issues with phased development.
19. Describe the major elements and issues with parallel development.
20. Describe the major elements and issues with an object-oriented approach to developing information systems.
21. What are the phases and workflows of the Unified Process?
22. Compare the phases of the Unified Process with the phases of the waterfall model.
23. Who is the Object Management Group?
24. What is the Unified Modeling Language?
25. What does it mean for an OOSAD approach to be incremental and iterative?
26. Why is it important for an OOSAD approach to be architecture centric?
27. What is a use case?
28. What is the primary purpose of structure diagrams? Give some examples of structure diagrams.
29. For what are behavior diagrams used? Give some examples of behavior diagrams.

EXERCISES

A. The basic types of methodologies discussed in this chapter can be combined and integrated to form new hybrid methodologies. Suppose you were to combine throwaway prototyping with the use of waterfall development. What would the methodology look like? Draw a picture (similar to those in Figures 1–2 through 1–7). How would this new methodology compare to the others?

B. Suppose you are a project manager using a waterfall development–based methodology on a large and complex project. Your manager has just read the latest article in *Computerworld* that advocates replacing this methodology with prototyping and comes to your office requesting that you switch. What would you say?

C. Suppose you are a project manager who typically has been using a waterfall development–based methodology on a large and complex project. Your manager has just read the latest article in *Computerworld* that advocates replacing this methodology with the Unified Process and comes to your office requesting you to switch. What do you say?

D. Investigate IBM's Rational Unified Process (RUP) on the Web. RUP is a commercial version that extends aspects of the Unified Process. Write a brief memo describing how it is related to the Unified Process as described in this chapter. (*Hint:* A good Web site with which to begin is www.306.ibm.com/software/awdtools/rup/.)

E. Suppose you are an analyst working for a small company to develop an accounting system. Would you use the Unified Process to develop the system, or would you prefer one of the traditional approaches? Why?

F. Suppose you are an analyst developing a new information system to automate the sales transactions and manage inventory for each retail store in a large chain. The system would be installed at each store and exchange data with a mainframe computer at the company's head office. Would you use the Unified Process to develop the system or would you prefer one of the traditional approaches? Why?

G. Suppose you are an analyst developing a new executive information system intended to provide key strategic information from existing corporate databases to senior executives to help in their decision making. What type of methodology would you use? Why?

H. Suppose you are an analyst working for a small company to develop an accounting system. What type of methodology would you use? Why?

I. Investigate the Unified Modeling Language on the Web. Write a paragraph news brief describing the current state of the UML. (*Hint:* A good Web site with which to begin is www.uml.org.)

J. Investigate the Object Management Group (OMG) on the Web. Write a report describing the purpose of the OMG and what it is involved with besides the UML. (*Hint:* A good Web site with which to begin is www.omg.org.)

K. Using the Web, find a set of CASE tools that support the UML. A couple of examples include Rational Rose and Visual Paradigm. Find at least two more. Write a short report describing how well they support the UML, and make a recommendation as to which one you believe would be best for a project team to use in developing an object-oriented information system using the UML.

L. Think about your ideal analyst position. Write a newspaper ad to hire someone for that position. What requirements would the job have? What skills and experience would be required? How would an applicant be able to demonstrate having the appropriate skills and experience?

M. Look in the classified section of your local newspaper. What kinds of job opportunities are available for people who want analyst positions? Compare and contrast the skills that the ads ask for to the skills that we presented in this chapter.

MINICASES

1. Joe Brown, the president of Roanoke Manufacturing, requested that Jack Jones, the MIS department manager, investigate the viability of selling their products over the Web. Currently, the MIS department is still using an IBM mainframe as their primary deployment environment. As a first step, Jack contacted his friends at IBM to see if they had any suggestions as to how Roanoke Manufacturing could move toward supporting sales in an electronic commerce environment while keeping their mainframe as their main system. His friends explained that IBM (www.ibm.com) now supports Java and Linux on their mainframes. Furthermore, Jack learned that IBM owns Rational (www.rational.com), the creator of the UML and the Unified Process. As such, they suggested that Jack investigate using object-oriented systems as a basis for developing the new system. They also suggested that using the Rational Unified Process (RUP), Java, and virtual Linux machines on his current mainframe as a way to support the movement toward a distributed electronic commerce system would protect his current investment in his legacy systems while allowing the new system to be developed in a more modern manner.

Even though Jack's IBM friends were very persuasive, Jack is still a little wary about moving his operation from a structured systems approach to this new object-oriented approach. Assuming that you are one of Jack's IBM friends, how would you convince him to move toward using an object-oriented systems development method, such as RUP, and using Java and Linux as a basis for developing and deploying the new system on Roanoke Manufacturing's current mainframe.

2. Maria Carrasquillo is the logistics manager for El Corte de Mejias, a Spanish company that makes and distributes sporting goods throughout Europe. El Corte de Mejias started as a small specialty company, but has grown rapidly over the past three years. However, its information systems (IS) have not kept pace. Señora Carrasquillo requested that the IS department develop a logistics system that would ensure that the right products are in-stock in El Corte de Mejias stores. Due to the massive backlog of work faced by the IS department, her request was given low priority. After six months of inaction by the department, Maria decided to take matters into her own hands. Based on her friends' advice , Maria purchased a simple database

software package and created a logistics system on her own.

Although Maria's system has been installed for about six weeks, it still does not work correctly, and is error-prone. Maria's assistant is so mistrustful of the system that she has secretly gone back to using her old paper-based system, since it is much more reliable.

Over dinner one evening, Maria complained to a systems analyst friend, "I don't know what went wrong with this project. It seemed pretty simple to me. Those IS guys wanted me to follow their elaborate set of steps and tasks, but I didn't think all that would really apply to a PC-based system. I just thought I could build this system and tweak it around without all the fuss and bother of the methodology the IS guys were pushing for. I mean, that applies only to big, expensive systems, right?"

Assuming you are Maria's systems analyst friend, how would you respond to her complaint?

3. Marcus Weber, IS project manager at ICAN Mutual Insurance Co., is reviewing the staffing arrangements for his next major project, the development of an expert system-based underwriters assistant. This new system will involve a whole new way for the underwriters to perform their tasks. The underwriters assistant system will function as sort of an underwriting supervisor, reviewing key elements of each application, checking for consistency in the underwriter's decisions, and ensuring that no critical factors have been overlooked. The goal of the new system is to improve the quality of the underwriters' decisions and to improve underwriter productivity. It is expected that the new system will substantially change the way the underwriting staff do their jobs.

Marcus is dismayed to learn that due to budget constraints, he must choose between one of two available staff members. Barry Filmore has had considerable experience and training in individual and organizational behavior. Barry has worked on several other projects in which the end users had to make significant adjustments to the new system, and Barry seems to have a knack for anticipating problems and smoothing the transition to a new work environment. Marcus had hoped to have Barry's involvement in this project.

Marcus's other potential staff member is Kim Danville. Prior to joining ICAN Mutual, Kim had considerable work experience with the expert system technologies that ICAN has chosen for this expert system project. Marcus was counting on Kim to help integrate the new expert system technology into ICAN's systems environment, and also to provide on-the-job training and insights to the other developers on this team.

Given that Marcus's budget will only permit him to add Barry or Kim to this project team, but not both what choice do you recommend for him? Justify your answer.

PART ONE

INITIATION

The Planning Phase is the fundamental process of understanding why an information system should be built, and determining how the project team will build it. The deliverables are combined into a system request which is presented to the project sponsor and approval committee. They decide whether it is advisable to proceed with the system. If the request is approved, detailed workplans, staffing plans, risk assessments, and a project charter is created. Finally, the detailed requirements are identified and a system proposal is created. The activities described in this section are continuously revisited throughout the lifetime of the system.

System
Request

Feasibility
Analysis

Workplan

Staffing
Plan

Risk
Assessment

Project
Charter

Requirements
Definition

System
Proposal

CHAPTER 2

PROJECT IDENTIFICATION AND SELECTION

This chapter describes Project Initiation, the point at which an organization creates and assesses the original goals and expectations for a new system. The first step in the process is to identify a project that will deliver value to the business and to create a system request that provides basic information about the proposed system. Next, the analysts perform a feasibility analysis to determine the technical, economic, and organizational feasibility of the system; if appropriate, the system is selected and the development project begins.

OBJECTIVES

- Understand the importance of linking the information system to business needs.
- Be able to create a system request.
- Understand how to assess technical, economic, and organizational feasibility.
- Be able to perform a feasibility analysis.
- Understand how projects are selected in some organizations.

CHAPTER OUTLINE

INTRODUCTION

The first step in any new development project is for someone—a manager, staff member, sales representative, or systems analyst—to see an opportunity to improve the business. New systems start first and foremost from a business need or opportunity. Many ideas for new systems or improvements to existing ones arise from the application of a new technology, but an understanding of technology is usually secondary to a solid understanding of the business and its objectives.

This idea may sound like common sense, but, unfortunately, many projects are started without a clear understanding of how the system will improve the business. The IS field is filled with thousands of buzzwords, fads, and trends (e.g., customer relationship management (CRM), mobile computing, data mining). The promise of these innovations can appear so attractive that organizations begin projects even if they are not sure what value they offer because they believe that the technologies are somehow important in their own right. A 1996 survey by the Standish Group found that 42 percent of all corporate IS projects were abandoned before completion; a similar 1996 study by the General Accounting Office found 53 percent of all U.S. government IS projects were abandoned. Problems can usually be traced back to the very beginning of the SDLC, where too little attention was given to the identifying business value and understanding the risks associated with the project.

This does not mean that technical people should not recommend new systems projects. In fact, the ideal situation is for both IT people (i.e., the experts in systems) and the business people (i.e., the experts in business) to work closely to find ways for technology to support business needs. In this way, organizations can leverage the exciting technologies that are available while ensuring that projects are based upon real business objectives, such as increasing sales, improving customer service, and decreasing operating expenses. Ultimately, information systems need to affect the organization's bottom line (in a positive way!).

In general, a *project* is a set of activities with a starting point and an ending point meant to create a system that brings value to the business. *Project initiation* begins when someone (or some group) in the organization (called the *project sponsor*) identifies some business value that can be gained from using information technology. The proposed project is described briefly using a technique called the *system request,* which is submitted to an *approval committee* for consideration. The approval committee reviews the system request and makes an initial determination, based on the information provided, of whether to investigate the proposal or not. If so, the next step is the feasibility analysis.

A *feasibility analysis* plays an important role in deciding whether to proceed with an information systems development project. It examines the technical, economic, and organizational pros and cons of developing the system, and it gives the organization a slightly more detailed picture of the advantages of investing in the system as well as any obstacles that could arise. In most cases, the project sponsor works together with an analyst (or analyst team) to develop the feasibility analysis for the approval committee.

Once the feasibility analysis has been completed, it is submitted to the approval committee, along with a revised system request. The committee then decides whether to approve the project, decline the project, or table it until additional information is available. Projects are selected by weighing risks and return and by making trade-offs at the organizational level.

CONCEPTS

IN ACTION

2-A   Interview with Lyn McDermid, CIO, Dominion Virginia Power

A CIO needs to have a global view when identifying and selecting projects for her organization. I would get lost in the trees if I were to manage on a project-by-project basis. Given this, I categorize my projects according to my three roles as a CIO, and the mix of my project portfolio changes depending on the current business environment.

My primary role is to **keep the business running**. That means every day when each person comes to work, he or she can perform his or her job efficiently. I measure this using various service level, cost, and productivity measures. Projects that keep the business running could have a high priority if the business were in the middle of a merger or a low priority if things were running smoothly and it were "business as usual."

My second role is to push **innovation that creates value for the business**. I manage this by looking at our lines of business and asking which lines of business create the most value for the company. These are the areas for which I should be providing the most value. For example, if we had a highly innovative marketing strategy, I would push for innovation there. If operations were running smoothly, I would push less for innovation in that area.

My third role is strategic, to look beyond today and find **new opportunities** for both IT and the business of providing energy. This may include investigating process systems, such as automated meter reading or looking into the possibilities of wireless technologies.

—*Lyn McDermid*

PROJECT IDENTIFICATION

A project is identified when someone in the organization identifies a *business need* to build a system. This could occur within a business unit or IT, come from a steering committee charged with identifying business opportunities, or evolve from a recommendation made by external consultants. Examples of business needs include supporting a new marketing campaign, reaching out to a new type of customer, or improving interactions with suppliers. Sometimes, needs arise from some kind of "pain" within the organization, such as a drop in market share, poor customer service levels, or increased competition. Other times, new business initiatives and strategies are created, and a system is required to enable them.

Business needs also can surface when the organization identifies unique and competitive ways of using IT. Many organizations keep an eye on *emerging technology,* which is technology that is still being developed and is not yet viable for widespread business use. For example, if companies stay abreast of technology such as the Internet, smart cards, and scent-technology in their earliest stages, they can develop business strategies that leverage the capabilities of these technologies and introduce them into the marketplace as a *first mover.* Ideally, they can take advantage of this first-mover advantage by making money and continuing to innovate while competitors trail behind.

The project sponsor is someone who recognizes the strong business need for a system and has an interest in seeing the system succeed. He or she will work throughout the SDLC to make sure that the project is moving in the right direction from the perspective of the business. The project sponsor serves as the primary point of contact for the system. Usually, the sponsor of the project is from a business function, such as marketing, accounting, or finance; however, members of the IT area also can sponsor or cosponsor a project.

The size or scope of a project determines the kind of sponsor needed. A small, departmental system may require sponsorship from only a single manager; however, a large, organizational initiative may need support from the entire senior management team and even the CEO. If a project is purely technical in nature (e.g., improvements to the existing IT infrastructure or research into the viability of an emerging technology), then sponsorship from IT is appropriate. When projects have great importance to the business yet are technically complex, joint sponsorship by both the business and IT may be necessary.

The business need drives the high-level *business requirements* for the system. Requirements are what the information system will do, or the *functionality* it will contain. They need to be explained at a high level so that the approval committee and, ultimately, the project team understand what the business expects from the final product. Business requirements are the features and capabilities the information system will have to include, such as the ability to collect customer orders online or the ability for suppliers to receive inventory information as orders are placed and sales are made.

The project sponsor also should have an idea of the *business value* to be gained from the system, both in tangible and intangible ways. *Tangible value* can be quantified and measured easily (e.g., 2 percent reduction in operating costs). An *intangible value* results from an intuitive belief that the system provides important, but hard-to-measure, benefits to the organization (e.g., improved customer service or a better competitive position).

Once the project sponsor identifies a project that meets an important business need and he or she can identify the system's business requirements and value, it is time to formally initiate the project. In most organizations, project initiation begins with a technique called a system request.

2-1 Identify Tangible and Intangible Value

*D*ominion Virginia Power is one of the nation's ten largest investor-owned electric utilities. The company delivers power to more than two million homes and businesses in Virginia and North Carolina. In 1997, the company over-hauled some of its core processes and technology. The goal was to improve customer service and cut operations costs by developing a new workflow and geographic information system. When the project was finished, ser-vice engineers who had sifted through thousands of paper maps could use computerized searches to pinpoint the locations of electricity poles. The project helped the utility improve management of all its facilities, records, maps, scheduling, and human resources. That, in turn, helped increase employee productivity, improve customer response times, and reduce the costs of operating crews.

Source: Computerworld (November 11, 1997).

Questions

1. What kinds of things does Dominion Virginia Power do that require it to know power pole locations? How often does it do these things? Who benefits if the company can locate power poles faster?

2. Based on your answers to question 1, describe three tangible benefits that the company can receive from its new computer system. How can these be quantified?

3. Based on your answers to question 1, describe three intangible benefits that the company can receive from its new computer system. How can these be quantified?

System Request

A system request is a document that describes the business reasons for building a sys-tem and the value that the system is expected to provide. The project sponsor usually completes this form as part of a formal system project selection process within the orga-nization. Most system requests include five elements: project sponsor, business need, business requirements, business value, and *special issues* (see Figure 2-1). The sponsor describes the person who will serve as the primary contact for the project, and the busi-ness need presents the reasons prompting the project. The business requirements of the project refer to the business capabilities that the system will need to have, and the busi-ness value describes the benefits that the organization should expect from the system. Special issues are included on the document as a catch-all for other information that should be considered in assessing the project. For example, the project may need to be completed by a specific deadline. Project teams need to be aware of any special circum-stances that could affect the outcome of the system. Figure 2-2 shows a template for a system request.

The completed system request is submitted to the approval committee for consideration. This approval committee could be a company steering committee that meets regularly to make information systems decisions, a senior executive who has control of organizational resources, or any other decision-making body that governs the use of business investments. The committee reviews the system request and makes an initial determination, based on the information provided, of whether to investigate the proposal or not. If so, the next step is to conduct a feasibility analysis.

Feasibility Analysis

Once the need for the system and its business requirements have been defined, it is time to create a more detailed business case to better understand the opportunities and limitations

Element	Description	Examples
Project Sponsor	The person who initiates the project and who serves as the primary point of contact for the project on the business side.	Several members of the Finance department Vice President of Marketing IT Manager Steering committee CIO CEO
Business Need	The business-related reason for initiating the system.	Increase sales Improve market share Improve access to information Improve customer service Decrease product defects Streamline supply acquisition Processes
Business Requirements	The business capabilities that the system will provide.	Provide online access to information Capture customer demographic information Include product search capabilities Produce management reports Include online user support
Business Value	The benefits that the system will create for the organization.	3 percent increase in sales 1 percent increase in market share Reduction in headcount by 5 FTEs* $200,000 cost savings from decreased supply costs $150,000 savings from removal of existing system
Special Issues or Constraints	Issues that are relevant to the implementation of the system and decisions made by the committee about the project.	Government-mandated deadline for May 30 System needed in time for the Christmas holiday season Top-level security clearance needed by project team to work with data

* = Full-time equivalent

FIGURE 2-1
Elements of the
System Request Form

associated with the proposed project. Feasibility analysis guides the organization in determining whether or not to proceed with a project. Feasibility analysis also identifies the important *risks* associated with the project that must be addressed if the project is approved. As with the system request, each organization has its own process and format for the feasibility analysis, but most include three techniques: technical feasibility,

System Request—Name of Project	
Project Sponsor:	Name of project sponsor
Business Need:	Short description of business need
Business Requirements:	Description of business requirements
Business Value:	Expected value that the system will provide
Special Issues or Constraints:	Any additional information that may be relevant to the stakeholders

FIGURE 2-2
System Request
Template

CONCEPTS   **2-B Interview with Don Hallacy, President, Technology Services, Sprint Corporation**

IN ACTION

At Sprint, network projects originate from two vantage points—IT and the business units. IT projects usually address infrastructure and support needs. The business unit projects typically begin after a business need is identified locally, and a business group informally collaborates with IT regarding how a solution can be delivered to meet customer expectations.

Once an idea is developed, a more formal request process begins, and an analysis team is assigned to investigate and validate the opportunity. This team includes members from the user community and IT, and they scope out at a high level what the project will do; create estimates for technology, training, and development costs; and create a business case. This business case contains the economic value-add and the net present value of the project.

Of course, not all projects undergo this rigorous process. The larger the project, the more time is allocated to the analysis team. It is important to remain flexible and not let the process consume the organization. At the beginning of each budgetary year, specific capital expenditures are allocated for operational improvements and maintenance. Moreover, this money is set aside to fund quick projects that deliver immediate value without going through the traditional approval process.

—Don Hallacy

economic feasibility, and organizational feasibility. The results of these techniques are combined into a *feasibility study* deliverable, which is given to the approval committee at the end of project initiation (see Figure 2-3).

Although we now discuss feasibility analysis within the context of project initiation, most project teams will revise their feasibility study throughout the SDLC and revisit its contents at various checkpoints during the project. If at any point the project's risks and limitations outweigh its benefits, the project team may decide to cancel the project or make necessary improvements.

Technical Feasibility

The first technique in the feasibility analysis is to assess the *technical feasibility* of the project, the extent to which the system can be successfully designed, developed, and installed by the IT group. Technical feasibility analysis is in essence a *technical risk analysis* that strives to answer this question: *Can* we build it?[1]

YOUR   **2–2 Create a System Request**

TURN

Think about your own university or college, and choose an idea that could improve student satisfaction with the course enrollment process. Currently can students enroll for classes from anywhere? How long does it take? Are directions simple to follow? Is online help available?

Next, think about how technology can help support your idea. Would you need completely new technology? Can the current system be changed?

Question

Create a system request that you could give to the administration that explains the sponsor, business need, business requirements, and potential value of the project. Include any constraints or issues that should be considered.

---

[1] We use *build it* in the broadest sense. Organizations can also choose to buy a commercial software package and install it, in which case, the question might be, can we select the right package and successfully install it?

Technical Feasibility: Can We Build It?

- Familiarity with Functional area: Less familiarity generates more risk
- Familiarity with Technology: Less familiarity generates more risk
- Project Size: Large projects have more risk
- Compatibility: The harder it is to integrate the system with the company's existing technology, the higher the risk

Economic Feasibility: Should We Build It?

- Development costs
- Annual operating costs
- Annual benefits (cost savings and revenues)
- Intangible costs and benefits

Organizational Feasibility: If We Build It, Will They Come?

- Project champion(s)
- Senior management
- Users
- Other stakeholders
- Is the project strategically aligned with the business?

FIGURE 2-3
Feasibility Analysis
Assessment Factors

Many risks can endanger the successful completion of a project. First and foremost is the users' *and* analysts' lack of *familiarity with the functional area*. When analysts are unfamiliar with the business functional area, they have a greater chance of misunderstanding the users or of missing opportunities for improvement. The risk increases dramatically when the users themselves are less familiar with an application, such as with the development of a system to support a business innovation (e.g., Microsoft starting up a new Internet dating service). In general, the development of new systems is riskier than extensions to an existing system because existing systems tend to be better understood.

Familiarity with the technology is another important source of technical risk. When a system uses technology that has not been used before *within the organization*, there is a greater chance that problems will occur and delays will be incurred because of the need to learn how to use the technology. Risk increases dramatically when the technology itself is new (e.g., a new Java development toolkit).

Project size is an important consideration, whether measured as the number of people on the development team, the length of time it will take to complete the project, or the number of distinct features in the system. Larger projects present more risk, both because they are more complicated to manage and because there is a greater chance that important system requirements will be overlooked or misunderstood. The extent to which the project is highly integrated with other systems (which is typical of large systems) can cause problems because complexity increases when many systems must work together.

Finally, project teams need to consider the *compatibility* of the new system with the technology that already exists in the organization. Systems are rarely built in a vacuum—they are built in organizations that already have numerous systems in place. New technology and applications need to be able be integrated with the existing environment for many reasons. They may rely on data from existing systems, they may produce data that feed other applications, and they may have to use the company's existing communications infrastructure. A new CRM system, for example, has little value if it does not use customer data found across the organization in existing sales systems, marketing applications, and customer service systems.

The assessment of a project's technical feasibility is not cut and dried because in many cases, some interpretation of the underlying conditions is needed (e.g., how large a project needs to grow before it becomes less feasible). One approach is to compare the project under consideration with prior projects undertaken by the organization. Another option is to consult with experienced IT professionals in the organization or external IT consultants; often they will be able to judge whether a project is feasible from a technical perspective.

CONCEPTS **2-C Caring for Grandpa and Grandma**
IN ACTION

*H*ealth care is a big industry in the United States. And with the baby boomers born in the late 1940s and 1950s (after World War II) starting to retire, there will be huge demands for senior health care. The desire is for better technologies to allow grandpa and grandma to live independently in their own homes or apartments longer—and not to use the more expensive options of nursing homes and assisted living centers. Some technologies include vital sign monitoring and reporting; motion detectors that sense if somebody has fallen; sensors to turn off the stove that may have been left on; and internet portals so that family members can check on the health of their loved ones.

Questions

1. How can technology assist with keeping retirees healthy?
2. How can technology help keep retirees out of expensive nursing homes and centers?

Economic Feasibility

The second element of a feasibility analysis is to perform an *economic feasibility* analysis (also called a *cost–benefit analysis*), which identifies the financial risk associated with the project. It attempts to answer this question: *Should* we build the system? Economic feasibility is determined by identifying costs and benefits associated with the system, assigning values to them, and then calculating the cash flow and return on investment for the project. The more expensive the project, the more rigorous and detailed the analysis should be. Figure 2-4 lists the steps in performing a cost benefit analysis; each step is described in the following sections.

Identifying Costs and Benefits   The first task when developing an economic feasibility analysis is to identify the kinds of costs and benefits the system will have and list them along the left-hand column of a spreadsheet. Figure 2-5 lists examples of costs and benefits that may be included.

Costs and benefits can be broken down into four categories: (1) development costs, (2) operational costs, (3) tangible benefits, and (4) intangibles. *Development costs* are those tangible expenses incurred during the construction of the system, such as salaries for the project team, hardware and software expenses, consultant fees, training, and office space and equipment. Development costs are usually thought of as one-time costs. *Operational costs* are those tangible costs required to operate the system, such the salaries for operations staff, software licensing fees, equipment upgrades, and communications charges. Operational costs are usually thought of as ongoing costs.

1. Identifying Costs and Benefits	List the tangible costs and benefits for the project. Include both one-time and recurring costs.
2. Assigning Values to Costs and Benefits	Work with business users and IT professionals to create numbers for each of the costs and benefits. Even intangibles should be valued if at all possible.
3. Determining Cash Flow	Project what the costs and benefits will be over a period of time, usually three to five years. Apply a growth rate to the numbers, if necessary.
4. Determining Net Present Value	Calculate what the value of future costs and benefits are if measured by today's standards. You will need to select a rate of growth to apply the NPV formula.
5. Determining Return on Investment	Calculate how much money the organization will receive in return for the investment it will make using the ROI formula.
6. Determining the Break-Even Point	Find the first year in which the system has greater benefits than costs. Apply the break-even formula using figures from that year. This will help you understand how long it will take before the system creates real value for the organization.
7. Graphing the Break-Even Point	Plot the yearly costs and benefits on a line graph. The point at which the lines cross is the break-even point.

FIGURE 2-4
Steps for Conducting Economic Feasibility

Revenues and cost savings are the *tangible benefits* the system enables the organization to collect or the tangible expenses the system enables the organization to avoid. Tangible benefits may include increased sales, reductions in staff, and reductions in inventory.

Development Costs	Operational Costs
Development Team Salaries	Software Upgrades
Consultant Fees	Software Licensing Fees
Development Training	Hardware Repairs
Hardware and Software	Hardware Upgrades
Vendor Installation	Operational Team Salaries
Office Space and Equipment	Communications Charges
Data Conversion Costs	User Training

Tangible Benefits	Intangible Benefits
Increased Sales	Increased Market Share
Reductions in Staff	Increased Brand Recognition
Reductions in Inventory	Higher Quality Products
Reductions in IT Costs	Improved Customer Service
Better Supplier Prices	Better Supplier Relations

FIGURE 2-5
Example Costs and Benefits for Economic Feasibility

Of course, a project also can affect the organization's bottom line by reaping *intangible benefits* or incurring *intangible costs*. Intangible costs and benefits are more difficult to incorporate into the economic feasibility because they are based on intuition and belief rather than "hard numbers." Nonetheless, they should be listed in the spreadsheet along with the tangible items.

Assigning Values to Costs and Benefits   Once the types of costs and benefits have been identified, analysts assign specific dollar values to them. This may seem impossible; how can someone quantify costs and benefits that haven't happened yet? And how can those predictions be realistic? Although this task is very difficult, they have to do the best they can to come up with reasonable numbers for all the costs and benefits. Only then can the approval committee make an educated decision about whether or not to move ahead with the project.

The best strategy for estimating costs and benefits is to rely on the people who have the clearest understanding of them. For example, costs and benefits related to the technology or the project itself can be provided by the company's IT group or external consultants, and business users can develop the numbers associated with the business (e.g., sales projections, order levels). Analysts can also consider past projects, industry reports, and vendor information, although these approaches probably will be a bit less accurate. All the estimates will probably be revised as the project proceeds.

Sometimes, it is acceptable for analysts to list intangible benefits, such as improved customer service, without assigning a dollar value; whereas other times they have to make estimates regarding the value of an intangible benefit. If at all possible, they should quantify intangible costs or benefits. Otherwise, it will not be apparent whether they have been realized. Consider a system that is supposed to improve customer service. This is intangible, but assume that the greater customer service will decrease the number of customer complaints by 10 percent each year over three years and that $200,000 is spent on phone charges and phone operators who handle complaint calls. Suddenly there are some very tangible numbers with which to set goals and measure the original intangible benefit.

CONCEPTS

IN ACTION

2-D Intangible Value at Carlson Hospitality

I conducted a case study at Carlson Hospitality, a global leader in hospitality services, encompassing more than 1,300 hotel, resort, restaurant, and cruise ship operations in seventy-nine countries. One of its brands, Radisson Hotels & Resorts, researched guest stay information and guest satisfaction surveys. The company was able to quantify how much of a guest's lifetime value can be attributed to his or her perception of the stay experience. As a result, Radisson knows how much of the collective future value of the enterprise is at stake given the perceived quality of stay experience. Using this model, Radisson can confidently show that a 10 percent increase in customer satisfaction among the 10 percent of highest-quality customers will capture a one-point market share for the brand. Each point in market share for the Radisson brand is worth $20 million in additional revenue.

— *Barbara Wixom*

Question

How can a project team use this information to help determine the economic feasibility of a system?

Benefits[a]	
Increased sales	500,000
Improved customer service[b]	70,000
Reduced inventory costs	68,000
Total benefits	**638,000**
Development costs	
2 servers @ $125,000	250,000
Printer	100,000
Software licenses	34,825
Server software	10,945
Development labor	1,236,525
Total development costs	**1,632,295**
Operational costs	
Hardware	54,000
Software	20,000
Operational labor	111,788
Total operational costs	**185,788**
Total costs	**1,818,083**

[a] An important yet intangible benefit will be the ability to offer services that our competitors currently offer.

[b] Customer service numbers have been based on reduced costs for customer complaint phone calls.

FIGURE 2-6

Assigning Values to Costs and Benefits

Figure 2-6 shows costs and benefits along with assigned dollar values. Notice that the customer service intangible benefit has been quantified based on fewer customer complaint phone calls. The intangible benefit of being able to offer services that competitors currently offer was not quantified, but it was listed so that the approval committee will consider the benefit when assessing the system's economic feasibility.

Determining Cash Flow   A formal cost benefit analysis usually contains costs and benefits over a selected number of years (usually three to five years) to show cash flow over time (see Figure 2-7). When using this *cash flow method,* the years are listed across the top of the spreadsheet to represent the time period for analysis, and numeric values are entered in the appropriate cells within the spreadsheet's body. Sometimes fixed amounts are entered into the columns. For example, Figure 2-7 lists the same amount for customer complaint calls and inventory costs for all five years. Usually amounts are augmented by some rate of growth to adjust for inflation or business improvements, as shown by the 6 percent increase that is added to the sales numbers in the sample spreadsheet. Finally, totals are added to determine what the overall benefits will be; the higher the overall total, the more feasible the solution is in terms of its economic feasibility.

Determining Net Present Value and Return on Investment   There are several problems with the cash flow method because it does not consider the time value of money (i.e., a dollar today is *not* worth a dollar tomorrow), and it does not show the overall "bang for the buck" that the organization is receiving from its investment. Therefore, some project teams add additional calculations to the spreadsheet to provide the approval committee with a more accurate picture of the project's worth.

	2008	2009	2010	2011	2012	Total
Increased sales	500,000	530,000	561,800	595,508	631,238	
Reduction in customer complaint calls	70,000	70,000	70,000	70,000	70,000	
Reduced inventory costs	68,000	68,000	68,000	68,000	68,000	
TOTAL BENEFITS:	638,000	668,000	699,800	733,508	769,238	
PV OF BENEFITS:	**619,417**	**629,654**	**640,416**	**651,712**	**663,552**	**3,204,752**
PV OF ALL BENEFITS:	**619,417**	**1,249,072**	**1,889,488**	**2,541,200**	**3,204,752**	
2 Servers @ $125,000	250,000	0	0	0	0	
Printer	100,000	0	0	0	0	
Software licenses	34,825	0	0	0	0	
Server software	10,945	0	0	0	0	
Development labor	1,236,525	0	0	0	0	
TOTAL DEVELOPMENT COSTS:	1,632,295	0	0	0	0	
Hardware	54,000	81,261	81,261	81,261	81,261	
Software	20,000	20,000	20,000	20,000	20,000	
Operational labor	111,788	116,260	120,910	125,746	130,776	
TOTAL OPERATIONAL COSTS:	185,788	217,521	222,171	227,007	232,037	
TOTAL COSTS:	1,818,083	217,521	222,171	227,007	232,037	
PV OF COSTS:	**1,765,129**	**205,034**	**203,318**	**201,693**	**200,157**	**2,575,331**
PV OF ALL COSTS:	**1,765,129**	**1,970,163**	**2,173,481**	**2,375,174**	**2,575,331**	
TOTAL PROJECT BENEFITS − COSTS:	(1,180,083)	450,479	477,629	506,501	537,201	
YEARLY NPV:	(1,145,712)	424,620	437,098	450,019	463,395	**629,421**
CUMULATIVE NPV:	**(1,145,712)**	**(721,091)**	**(283,993)**	**166,026**	**629,421**	
RETURN ON INVESTMENT:	**24.44%**	(629,421/2,575,331)				
BREAK-EVEN POINT:	**3.63 years**	[break-even occurs in year 4; (450,019 − 166,026)/450,019 = 0.63]				
INTANGIBLE BENEFITS:	This service is currently provided by competitors Improved customer satisfaction					

FIGURE 2-7   Cost–Benefit Analysis

Net present value (NPV) is used to compare the present value of future cash flows with the investment outlay required to implement the project. Consider the table in Figure 2-8, which shows the future worth of a dollar investment today, given different numbers of years and different rates of change. If you have a friend who owes you a dollar today but instead gives you a dollar three years from now, you've been had! Given a 10 percent increase in value, you'll be receiving the equivalent of 75 cents in today's terms.

NPV can be calculated in many different ways, some of which are extremely complex. Figure 2-9 shows a basic calculation that can be used to in your cash flow analysis to get more relevant values. In Figure 2-7, the present value of the costs and benefits are calculated first (i.e., they are shown at a discounted rate). Then, net present value is calculated, and it shows the discounted rate of the combined costs and benefits.

The *return on investment* (ROI) is a calculation listed somewhere on the spreadsheet that measures the amount of money an organization receives in return for the

Number of years	6%	10%	15%
1	0.943	0.909	0.870
2	0.890	0.826	0.756
3	0.840	0.751	0.572
4	0.792	0.683	0.497

This table shows how much a dollar today is worth 1–4 years from now in today's terms using different interest rates.

FIGURE 2-8
The Value of a Future Dollar Today

money it spends. A high ROI results when benefits far outweigh costs. ROI is determined by finding the total benefits less the costs of the system and dividing that number by the total costs of the system (see Figure 2-9). ROI can be determined per year or for the entire project over a period of time. One drawback of ROI is that it considers only the end points of the investment, not the cash flow in between, so it should not be used as the sole indicator of a project's worth. The spreadsheet in Figure 2-7 show a ROI figure.

Determining the Break–Even Point   If the project team needs to perform a rigorous cost benefit analysis, it may need to include information about the length of time before the project will break even, or when the returns will match the amount invested in the project. The greater the time it takes to break even, the riskier the project. The *break–even point* is determined by looking at the cash flow over time and identifying the year in which the benefits are larger than the costs (see Figure 2-7). Then, the difference between the yearly and cumulative NPV for that year is divided by the yearly NPV to determine how far into the year the break-even point will occur. See Figure 2-9 for the break-even calculation.

Calculation	Definition	Formula
Present Value (PV)	The amount of an investment today compared to that same amount in the future, taking into account inflation and time.	$\dfrac{\text{Amount}}{(1 + \text{interest rate})^n}$ n = number of years in future
Net Present Value (NPV)	The present value of benefit less the present value of costs.	PV Benefits − PV Costs
Return on Investment (ROI)	The amount of revenues or cost savings results from a given investment.	$\dfrac{\text{Total benefits} - \text{Total costs}}{\text{Total costs}}$
Break-Even Point	The point in time at which the costs of the project equal the value it has delivered.	$\dfrac{\text{Yearly NPV* } - \text{ Cumulative NPV}}{\text{Yearly NPV*}}$

*Use the Yearly NPV amount from the first year in which the project has a positive cash flow.

Add the above amount to the year in which the project has a positive cash flow.

FIGURE 2-9 Financial Calculations Used For Cost–Benefit Analysis

CONCEPTS IN ACTION

2-E The FBI Pulls the Plug on a Project

*T*he FBI's failure to roll out an expanded computer system that would help agents investigate criminals and terrorists is the latest in a series of costly technology blunders by the government over more than a decade. Experts blame poor planning, rapid industry advances, and the massive scope of some complex projects, whose price tags can run into billions of dollars at U.S. agencies with tens of thousands of employees. "There are very few success stories," said Paul Brubaker, former deputy chief information officer at the Pentagon. "Failures are very common, and they've been common for a long time." The FBI said earlier this month it might shelve its custom-built, $170 million "Virtual Case File" project because it is inadequate and outdated. The system was intended to help agents, analysts, and others around the world share information without using paper or time-consuming scanning of documents. Officials said commercial software might accomplish some of what the FBI needs. The bureau's mess—the

subject of an investigation by the Justice Department and an upcoming congressional hearing—was the latest black eye among ambitious technology upgrades by the government since the 1990s.

Questions

Some systems like this are very complex. They must have security, and they must interface between the FBI, CIA, and other government agencies as well as state and local law enforcement groups. Such complexity can take years to build—and is almost guaranteed to fail because of newer technologies that have come along during the wait. How might you keep a complex project on track? What commercial software might work in this case (as mentioned in the case?)

Source: www.securityfocus.com/news/10383

The break-even point also can be depicted graphically, as shown in Figure 2-10. The cumulative present value of the costs and benefits for each year are plotted on a line graph; the point at which the lines cross is the break-even point.

Alternatives to Traditional Cost–Benefit Analysis   Concerns have been raised as to the appropriateness of using traditional cost–benefit analysis with NPV and ROI to determine economic feasibility of an IT project. One of the major problems of using traditional cost–benefit analysis to determine the economic feasibility of an IT investment is that traditional cost–benefit analysis is based on the assumption that the investor must either invest now or not invest at all. However, in most IT investment decisions, the decision to invest is not a now-or-never decision. In most situations, an information system is already in place. As

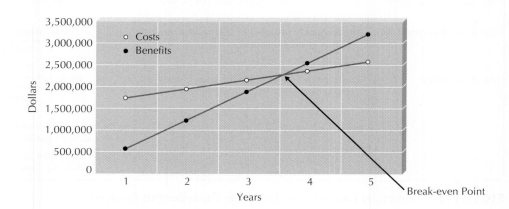

FIGURE 2-10
Break-even Graph

such, the decision to replace or upgrade the current information system can usually be delayed. Different proposals have been made to overcome some of the weaknesses in traditional cost–benefit analysis. For example, economic production models, activity-based costing, and balanced score cards have been suggested.[2] However, in this section, we describe the primary alternative that has been proposed for object-oriented systems: *option pricing models (OPMs)*.[3]

At this point in time, OPMs have had limited use in economic feasibility analysis for IT investment decisions in industry. In fact, there is some controversy as to whether an instrument created for a traded asset (stock) can be used in evaluating IT investment opportunities. However, the preliminary research results demonstrate that their use in IT investment evaluations may be warranted. OPMs have shown promise in evaluating the potential future value of an investment in IT. In many cases in which traditional cost–benefit analysis of investments in IT has predicted that the investment would be a failure, OPMs have shown that it may indeed be feasible.

With object-oriented systems, where classes are designed not only for the current application but also for use in future development efforts, an investment in developing a class or a set of classes may pay dividends well beyond the original system development effort. Furthermore, with the iterative and incremental development emphasis in object-oriented systems development approaches, an object-oriented project can be viewed as a sequence of smaller projects. As such, you might treat investments in an object-oriented project much as you would an investment in a call option in finance. A *call option* is essentially a contract that gives the right to purchase an amount of stock for a given price for a specified period of time to the purchaser of the call option. However, a call option does not create an obligation to buy the stock.

Treating an IT investment as a call option allows management, at a relevant point in the future, to determine whether additional investment into the evolving system is reasonable. This gives management the flexibility to determine the economic feasibility of a project to decide whether to continue with the development as planned, to abandon the project, to expand the scope of the project, to defer future development, or to shrink the current development effort. In many ways, treating IT investments as call options simply allows management to delay investment decisions until more information is available.

Once the decision is made to invest (i.e., the call option is exercised) the decision is considered irreversible. The idea of irreversible decisions is one of the fundamental

---

[2] See, for example, Q. Hu, R. Plant, and D. Hertz, "Software Cost Estimation Using Economic Production Models," *Journal of MIS* 15, no. 1 (Summer 1998): 143–163; G. Ooi and C. Soh, "Developing an Activity-based Approach for System Development and Implementation," *ACM Data Base for Advances in Information Systems* 34, no. 3 (Summer 2003): 54–71, and K. Milis and R. Mercken, "The Use of the Balanced Scorecard for the Evaluation of Information and Communication Technology Projects," *International Journal of Project Management* 22 (2004): 87–97.

[3] For more information regarding the use of option pricing models in evaluating economic feasibility of information systems, see M. Benaroch and R. Kauffman, "A Case for Using Real Options Pricing Analysis to Evaluate Information Technology Project Investments," *Information Systems Research* 10, no. 1 (March 1999): 70–86; M. Benaroch and R. Kauffman, "Justifying Electronic Banking Network Expansion Using Real Options Analysis," *MIS Quarterly* 24, no. 2 (June 2000): 197–225; Q. Dai, R. Kauffman, and S. March, "Analyzing Investments in Object-Oriented Middleware," *Ninth Workshop on Information Technologies and Systems* (December 1999): 45–50; A. Kambil, J. Henderson, and H. Mohsenzadeh, Strategic Management of Information Technology Investments: An Options Perspective, in R. D. Banker, R. J. Kauffman, and M. A. Mahmood (eds.), *Strategic Information Technology Management: Perspectives on Organizational Growth and Competitive Advantage* (Idea Group, 1993); A. Taudes, "Software Growth Options," *Journal of Management Information Systems* 15, no. 1 (Summer 1998): 165–185; A. Taudes, M. Feurstein, and A. Mild, "Options Analysis of Software Platform Decisions: A Case Study," *MIS Quarterly* 24, no. 2 (June 2000): pp. 227–243.

assumptions on which OPMs are based. This assumption fits quite well with modern object-oriented systems development approaches, in which, once an iteration has begun, an increment is completed before another investment decision is made.

Researchers have studied many different OPMs in terms of their applicability to IT investment.[4] However, all OPMs share a common thread: both the direct benefit of the proposed project and the indirect value (option value) must be computed to determine economic feasibility of an IT investment using an OPM. The direct benefit can be computed using the traditional NPV, whereas the value of the option can be computed using one of the OPMs in the literature. Given that the minimum expected value of an option is always zero, the minimum estimated value for investing using an OPM will be the same as the value given by the traditional approach. However, when the expected value of the option (e.g., future iterations or projects) exceeds zero, an OPM will give an estimated value greater than the traditional approach. The actual calculation of the value of an option is quite complex and, as such, is beyond the scope of this book. However, given how well the OPMs fit the object-oriented systems development approaches, it seems reasonable that OPMs should be considered as alternatives to evaluating IT investments in object-oriented systems.

Organizational Feasibility

The final technique used for feasibility analysis is to assess the *organizational feasibility* of the system, how well the system ultimately will be accepted by its users and incorporated into the ongoing operations of the organization. There are many organizational factors that can have an impact on the project, and seasoned developers know that organizational feasibility can be the most difficult feasibility dimension to assess. In essence, an organizational feasibility analysis attempts to answer this question: If we build it, will they come?

One way to assess the organizational feasibility of the project is to understand how well the goals of the project align with business objectives. *Strategic alignment* is the fit between the project and business strategy—the greater the alignment, the less risky the project will be from an organizational feasibility perspective. For example, if the marketing department has decided to become more customer focused, then a CRM project that produces integrated customer information would have strong strategic alignment with marketing's goal. Many IT projects fail when the IT department initiates them, because there is little or no alignment with business unit or organizational strategies.

A second way to assess organizational feasibility is to conduct a *stakeholder analysis*.[5] A stakeholder is a person, group, or organization that can affect (or will be affected by) a new system. In general, the most important stakeholders in the introduction of a new system are the project champion, system users, and organizational management (see Figure 2-11), but systems sometimes affect other stakeholders as well. For example, the IS department can be a stakeholder of a system because IS jobs or roles may be changed significantly after its implementation. One key stakeholder outside of the champion, users, and management in Microsoft's project that embedded Internet Explorer as a standard part of Windows was the U.S. Department of Justice.

---

[4] Two of the more important OPMs used for evaluating IT investments are the binomial OPM and the Black-Scholes OPM. For more information on these models see J. C. Hull, *Options, Futures, and Other Derivative Securities* (Englewood Cliffs, NJ: Prentice-Hall, 1993).

[5] A good book that presents a series of stakeholder analysis techniques is R. O. Mason and I. I. Mittroff, *Challenging Strategic Planning Assumptions: Theory, Cases, and Techniques* (New York: Wiley, 1981).

	Role	Techniques for improvement
Champion	A champion: • Initiates the project • Promotes the project • Allocates his or her time to project • Provides resources	• Make a presentation about the objectives of the project and the proposed benefits to those executives who will benefit directly from the system • Create a prototype of the system to demonstrate its potential value
Organizational Management	Organizational managers: • Know about the project • Budget enough money for the project • Encourage users to accept and use the system	• Make a presentation to management about the objectives of the project and the proposed benefits • Market the benefits of the system using memos and organizational newsletters • Encourage the champion to talk about the project with his or her peers
System Users	Users: • Make decisions that influence the project • Perform hands-on activities for the project • Ultimately determine whether the project is successful by using or not using the system	• Assign users official roles on the project team • Assign users specific tasks to perform with clear deadlines • Ask for feedback from users regularly (e.g., at weekly meetings)

FIGURE 2-11 Some Important Stakeholders for Organizational Feasibility

The *champion* is a high-level, noninformation systems executive who is usually, but not always, the project sponsor who created the system request. The champion supports the project with time, resources (e.g., money), and political support within the organization by communicating the importance of the system to other organizational decision makers. More than one champion is preferable because if the champion leaves the organization, the support could leave as well.

Whereas champions provide day-to-day support for the system, *organizational management* also needs to support the project. Such management support conveys to the rest of the organization the belief that the system will make a valuable contribution and that necessary resources will be made available. Ideally, management should encourage people in the organization to use the system and to accept the many changes that the system will likely create.

A third important group of stakeholders are the *system users* who will ultimately use the system once it has been installed in the organization. Too often, the project team meets with users at the beginning of a project and then disappears until after the system is created. In this situation, rarely does the final product meet the expectations and needs of those who are supposed to use it because needs change and users become savvier as the project progresses. User participation should be promoted throughout the development process to make sure that the final system will be accepted and used by getting users actively involved in the development of the system (e.g., performing tasks, providing feedback, making decisions).

Finally, the feasibility study helps organizations make wiser investments regarding information systems because it forces project teams to consider technical, economic, and organizational factors that can impact their projects. It protects IT professionals from criticism by keeping the business units educated about decisions and positioned as the leaders in the decision-making process. Remember, the feasibility study should be revised several times during the project at points where the project team makes critical decisions about the system (e.g., before the design begins). It can be used to support and explain the critical choices that are made throughout the SDLC.

PROJECT SELECTION

Once the feasibility analysis has been completed, it is submitted to the approval committee, along with a revised system request. The committee then decides whether to approve the project, decline the project, or table it until additional information is available. At the project level, the committee considers the value of the project by examining the business need (found in the system request) and the risks of building the system (presented in the feasibility analysis).

Before approving the project, however, the committee also considers the project from an organizational perspective; it has to keep in mind the company's entire portfolio of projects. This way of managing projects is called *portfolio management*. Portfolio management takes into consideration the different kinds of projects that exist in an organization—large and small, high risk and low risk, strategic and tactical. (See Figure 2-12 for the different ways of classifying projects.) A good project portfolio will have the most appropriate mix of projects for the organization's needs. The committee acts as portfolio manager with the goal of maximizing the cost/benefit performance and other important factors of the projects in their portfolio. For example, an organization may want to keep high-risk projects to less than 20 percent of its total project portfolio.

Size	What is the size? How many people are needed to work on the project?
Cost	How much will the project cost the organization?
Purpose	What is the purpose of the project? Is it meant to improve the technical infrastructure? Support a current business strategy? Improve operations? Demonstrate a new innovation?
Length	How long will the project take before completion? How much time will go by before value is delivered to the business?
Risk	How likely is it that the project will succeed or fail?
Scope	How much of the organization is affected by the system? A department? A division? The entire corporation?
Return on investment	How much money does the organization expect to receive in return for the amount the project costs?

FIGURE 2-12
Ways to Classify Projects

The approval committee must be selective about where to allocate resources, because the organization has limited funds. This involves *trade-offs*, in which the organization must give up something in return for something else to keep its portfolio well balanced. If there are three potentially high-payoff projects, yet all have very high risk, then perhaps only one of the projects will be selected. Also, there are times when a system at the project level makes good business sense, but it does not make sense at the organization level. Thus, a project may show a very strong ROI and support important business needs for a part of the company; however, it is not selected. This could happen for many reasons—because there is no money in the budget for another system, the organization is about to go through some kind of change (e.g., a merger or an implementation of a companywide system like an ERP), projects that meet the same business requirements already are underway, or the system does not align well with current or future corporate strategy.

YOUR TURN

2-4 To Select or Not to Select

It seems hard to believe that an approval committee would not select a project that meets real business needs, has a high potential ROI, and has a positive feasibility analysis. Think of a company you have worked for or know about. Describe a scenario in which a project may be very attractive at the project level but not at the organization level.

CONCEPTS IN ACTION

2-F Interview with Carl Wilson, CIO, Marriott Corporation

At Marriott, we don't have IT projects—we have business initiatives and strategies that are enabled by IT. As a result, the only time a traditional "IT project" occurs is when we have an infrastructure upgrade that will lower costs or leverage better functioning technology. In this case, IT has to make a business case for the upgrade and prove its value to the company.

The way IT is involved in business projects in the organization is twofold. First, senior IT positions are filled by people with good business understanding. Second, these people are placed on key business committees and forums where the real business happens, such as finding ways to satisfy guests. Because IT has a seat at the table, we are able to spot opportunities to support business strategy. We look for ways in which IT can enable or better support business initiatives as they arise.

Therefore, business projects are proposed, and IT is one component of them. These projects are then evaluated the same as any other business proposal, such as a new resort—by examining the return on investment and other financial measures.

At the organizational level, I think of projects as must-do's, should-do's, and nice-to-do's. The must-do's are required to achieve core business strategy, such as guest preference. The should-do's help grow the business and enhance the functionality of the enterprise. These can be somewhat untested, but good drivers of growth. The nice-to-do's are more experimental and look farther out into the future.

The organization's project portfolio should have a mix of all three kinds of projects, with a much greater proportion devoted to the must-do's.

—*Carl Wilson*

CONCEPTS   **2-G A Project That Does Not Get Selected**

IN ACTION

*H*ygeia Travel Health is a Toronto-based health insurance company whose clients are the insurers of foreign tourists to the United States and Canada. Its project selection process is relatively straightforward. The project evaluation committee, consisting of six senior executives, splits into two groups. One group includes the CIO, along with the heads of operations and research and development, and it analyzes the costs of every project. The other group consists of the two chief marketing officers and the head of business development, and they analyze the expected benefits. The groups are permanent, and to stay objective, they don't discuss a project until both sides have evaluated it. The results are then shared, both on a spreadsheet and in conversation. Projects are then approved, passed over, or tabled for future consideration.

Last year, the marketing department proposed purchasing a claims database filled with detailed information on the costs of treating different conditions at different facilities. Hygeia was to use this information to estimate how much money insurance providers were likely to owe on a given claim if a patient was treated at a certain hospital as opposed to any other. For example, a 45-year-old man suffering a heart attack may accrue $5,000 in treatment costs at hospital A but only $4,000 at hospital B. This information

would allow Hygeia to recommend the cheaper hospital to its customer. That would save the customer money and help differentiate Hygeia from its competitors.

The benefits team used the same three-meeting process to discuss all the possible benefits of implementing the claims database. Members of the team talked to customers and made a projection using Hygeia's past experience and expectations about future business trends. The verdict: The benefits team projected a revenue increase of $210,000. Client retention would rise by 2 percent. And overall, profits would increase by 0.25 percent.

The costs team, meanwhile, came up with large estimates: $250,000 annually to purchase the database and an additional $71,000 worth of internal time to make the information usable. Putting it all together, it was a financial loss of $111,000 in the first year.

The project still could have been good for marketing—maybe even good enough to make the loss acceptable. But some of Hygeia's clients were also in the claims information business and therefore were potential competitors. This, combined with the financial loss, was enough to make the company reject the project.

Source: Ben Worthen, "Two Teams are Better than One" *CIO Magazine,* (July 15, 2001).

YOUR   **2-5 Project Selection**

TURN

*I*n April 1999, one of Capital Blue Cross's health-care insurance plans had been in the field for three years but hadn't performed as well as expected. The ratio of premiums to claims payments wasn't meeting historic norms. In order to revamp the product features or pricing to boost performance, the company needed to understand why it was underperforming. The stakeholders came to the discussion already knowing they needed better extraction and analysis of usage data in order to understand product shortcomings and recommend improvements.

After listening to input from the user teams, the stakeholders proposed three options. One was to persevere with the current manual method of pulling data from flat files via ad hoc reports and retyping it into spreadsheets.

The second option was to write a program to dynamically mine the needed data from Capital's customer information control system (CICS). While the system was

processing claims, for instance, the program would pull out up-to-the-minute data at a given point in time for users to analyze.

The third alternative was to develop a decision support system to allow users to make relational queries from a data mart containing a replication of the relevant claims and customer data. Each of these alternatives was evaluated on cost, benefits, risks and intangibles.

Questions

1. What are three costs, benefits, risks, and intangibles associated with each project?
2. Based on your answer to question 1, which project would you choose?

Source: Richard Pastore, "Capital Blue Cross," *CIO Magazine* (February 15, 2000).

APPLYING THE CONCEPTS AT CD SELECTIONS

So far we have introduced the concepts of identifying a project, creating a system request, performing a feasibility analysis, and selecting a project. Now we see how CD Selections performs these tasks.

Project Identification and System Request

At CD Selections, all potential IT projects are reviewed and approved by a project steering committee that meets quarterly. The committee has representatives from IT as well as from the major areas of the business. For Margaret, the first step was to prepare a system request for the committee. Using the system request template (see Figure 2-2), Margaret prepared a system request (see Figure 2-13). Of course, the sponsor is Margaret, and the business needs are to increase sales and to better service retail customers. Notice that the need does not focus on the technology, such as the need "to upgrade our Web page." The focus is on the business aspects: sales and customer service.

System Request—Internet Order Project

Project sponsor:	Margaret Mooney, Vice President of Marketing
Business Need:	This project has been initiated to reach new Internet customers and to better serve existing customers using Internet sales support.

Business Requirements:

Using the Web, customers should be able to search for products and identify the brick-and-mortar stores that have them in stock. They should be able to put items on hold at a store location or place an order for items that are not carried or are not in stock. The functionality that the system should have is as follows:

- Search through the CD Selections inventory of products.
- Identify the retail stores that have the product in stock.
- Put a product on hold at a retail store and schedule a time to pick up the product.
- Place an order for products not currently in stock or not carried by CD Selections.
- Receive confirmation that an order can be placed and when the item will be in stock.

Business Value:

We expect that CD Selections will increase sales by reducing lost sales due to out-of-stock or nonstocked items and by reaching out to new customers through its Internet presence. We expect the improved services will reduce customer complaints, primarily because 50% of all customer complaints stem from out-of-stocks or nonstocked items. Also, CD Selections should benefit from improved customer satisfaction and increased brand recognition due to its Internet presence.

Conservative estimates of tangible value to the company include:

- $750,000 (75% of $1,000,000) in sales from new customers
- $1,875,000 (75% of $2,500,000) in sales from existing customers
- $50,000 in sales from customers not facing "out-of-stock or nonstocked" items

Special Issues or Constraints:

- The Marketing Department views this as a strategic system. This Internet system will add value to our current business model, and it also will serve as a proof-of-concept for future Internet endeavors. For example, in the future, CD Selections may want to sell products directly over the Internet.
- The system should be in place for the holiday shopping season next year.

FIGURE 2-13

System Request for CD Selections

For now, the business requirements are described at a very high level of detail. In this case, Margaret's vision for the requirements includes the ability to help brick-and-mortar stores reach out to new customers. Specifically, customers should be able to search for products over the Internet, locate a retail store that contains the product, put a product on hold for later store pickup, and order products that are not currently being stocked.

The business value describes how the requirements will affect the business. Margaret found identifying intangible business value to be fairly straightforward in this case. The Internet is a "hot" area, so she expects the Internet to improve customer recognition and satisfaction. Estimating tangible value is more difficult. She expects that Internet ordering will increase sales in the retail stores, but by how much?

Margaret decided to have her marketing group do some market research to learn how many retail customers do not complete purchases because the store does not carry the item they want. They learned that stores lose approximately 5 percent of total sales from "out-of-stocks and nonstocks." This number gave Margaret some idea of how much sales could increase from the existing customer base (i.e., about $50,000 per store), but it does not indicate how many new customers the system will generate.

Estimating how much revenue CD Selections should anticipate from new Internet customers was not simple. One approach was to use some of CD Selections' standard models for predicting sales of new stores. Retail stores average about $1 million in sales per year (after they have been open a year or two), depending upon location factors such as city population, average incomes, proximity to universities, and so on. Margaret estimated that adding the new Internet site would have an effect similar to adding a new store. This would suggest ongoing revenues of $1 million, give or take several hundred thousand dollars, after the Web site had been operating for a few years.

Together, the sales from existing customers ($2.5 million) and new customers ($1 million) totaled approximately $3.5 million. Margaret created conservative and optimistic estimates by reducing and increasing this figure by 25 percent. This created a possible range of values from $2,625,000 to $4,375,000. Margaret is conservative, so she decided to include the lower number as her sales projection. Figure 2-13 shows the completed system request.

Feasibility Analysis

Once Margaret and her marketing group completed the system request, they submitted it to the steering committee for their next meeting. When the steering committee met, they placed the Internet Order project high on its list of projects. A senior systems analyst, Alec Adams, was assigned to help Margaret conduct a feasibility analysis because of his familiarity with CD Selections' sales and distribution systems. He also was an avid user of the Web and had been offering suggestions for the improvement of CD Selections' Web site.

Alec and Margaret worked closely together over the next few weeks on the feasibility analysis. Figure 2-14 presents the executive summary page of the feasibility analysis; the report itself was about 10 pages long, and it provided additional detail and supporting documentation.

As shown in Figure 2-14, the project is somewhat risky from a technical perspective. CD Selections has minimal experience with the proposed application and the technology because the ISP has been managing most of the website technology to date. One solution may be to hire a consultant with e-commerce experience to work with the IT department and to offer guidance. Further, the new system would have to exchange order information

Internet Order Feasibility Analysis Executive Summary

Margaret Mooney and Alec Adams created the following feasibility analysis for the CD Selections Internet Order System Project. The System Proposal is attached, along with the detailed feasibility study. The highlights of the feasibility analysis are as follows:

Technical Feasibility

The Internet Order System is feasible technically, although there is some risk.

CD Selections' risk regarding familiarity with Internet order applications is high

• The Marketing Department has little experience with Internet-based marketing and sales.
• The IT Department has strong knowledge of the company's existing order systems; however, it has not worked with Web-enabled order systems.
• Hundreds of retailers that have Internet Order applications exist in the marketplace.

CD Selections' risk regarding familiarity with the technology is medium.

• The IT Department has relied on external consultants and an Information Service Provider to develop its existing Web environment.
• The IT Department has gradually learned about Web systems by maintaining the current Web site.
• Development tools and products for commercial Web application development are available in the marketplace, although the IT department has little experience with them.
• Consultants are readily available to provide help in this area.

The project size is considered medium risk.

• The project team likely will include fewer than ten people.
• Business user involvement will be required.
• The project timeframe cannot exceed a year because of the holiday season implementation deadline, and it should be much shorter.

The compatibility with CD Selections' existing technical infrastructure should be good.

• The current Order System is a client-server system built using open standards. An interface with the Web should be possible.
• Retail stores already place and maintain orders electronically.
• An Internet infrastructure already is in place at retail stores and at the corporate headquarters.
• The ISP should be able to scale their services to include a new Order System.

Economic Feasibility

A cost–benefit analysis was performed; see attached spreadsheet for details. A conservative approach shows that the Internet Order System has a good chance of adding to the bottom line of the company significantly.

ROI over 3 years: 229 percent

Total benefit after three years: $3.5 million (adjusted for present value)

Break-even occurs: after 1.32 years

Intangible Costs and Benefits

  • Improved customer satisfaction
  • Greater brand recognition

Organizational Feasibility

From an organizational perspective, this project has low risk. The objective of the system, which is to increase sales, is aligned well with the senior management's goal of increasing sales for the company. The move to the Internet also aligns with Marketing's goal to become more savvy in Internet marketing and sales.

  The project has a project champion, Margaret Mooney, Vice President of Marketing. Margaret is well positioned to sponsor this project and to educate the rest of the senior management team when necessary. To date, much of senior management is aware of and supports the initiative.

  The users of the system, Internet consumers, are expected to appreciate the benefits of CD Selections' Web presence. And, management in the retail stores should be willing to accept the system, given the possibility of increased sales at the store level.

Additional Comments

• The Marketing Department views this as a strategic system. This Internet system will add value to our current business model, and it also will serve as a proof of concept for future Internet endeavors.
• We should consider hiring a consultant with expertise in similar applications to assist with the project.
• We will need to hire new staff to operate the new system, from both the technical and business operations aspects.

TEMPLATE
can be found at
www.wiley.com
/college/dennis

FIGURE 2-14 Feasibility Analysis for CD Selections

with the company's brick-and-mortar order system. Currently, individual retail stores submit orders electronically, so receiving orders and exchanging information with the Internet systems should be possible.

The economic feasibility analysis includes refined assumptions that Margaret made in the system request. Figure 2-15 shows the summary spreadsheet that lead to the conclusions on the feasibility analysis. Development costs are expected to be about $250,000. This is a very rough estimate, because Alec has had to make some assumptions about the amount of time it will take to design and program the system. These estimates will be revised after a detailed workplan has been developed and as the project proceeds.[6] Traditionally, operating costs include the costs of the computer operations. In this case, CD Selections has had to include the costs of business staff because they are creating a new business unit, resulting in a total of about $450,000 each year. Margaret and Alec have decided to use a conservative estimate for revenues, although they note the potential for higher returns. This shows that the project can still add significant business value, even if the underlying assumptions prove to be overly optimistic. The spreadsheet was projected over three years, and the ROI and break-even point were included.

The organizational feasibility is presented in Figure 2-14. There is a strong champion, well placed in the organization to support the project. The project originated in the business or functional side of the company, not the IT department, and Margaret has carefully built up support for the project among the senior management team.

This is an unusual system in that the ultimate end users are the consumers external to CD Selections. Margaret and Alec have not done any specific market research to see how well potential customers will react to the CD Selections system, so this is a risk.

An additional stakeholder in the project is the management team responsible for the operations of the traditional stores and the store managers. They should be quite supportive, given the added service that they now can offer. However, Margaret must convince them that the Internet Sales System will not be viewed as a threat to stores' future sales. As such, Margaret and Alec need to make sure that the management team and store managers are included in the development of the system so that they can incorporate the system into their business processes.

Project Selection

The approval committee met and reviewed the Internet Sales System project along with two other projects—one calling for the implementation of a corporate Intranet and another proposing in-store kiosks that would provide customers with information about the CDs that the store carried. Unfortunately, the budget would allow for only one project to be approved, so the committee carefully examined the costs, expected benefits, risks, and strategic alignment of all three projects. Currently, a primary focus of upper management is increasing sales in the retail stores and the Internet system and kiosk project best aligned with that goal. Given that both projects had equal risk but that the Internet Order project expected a much greater return, the committee decided to fund the Internet Sales System.

---

[6] Some of the salary information may seem high. Most companies use a "full-cost" model for estimating salary cost, in which all benefits (e.g., health insurance, retirement, and payroll taxes) are included in salaries when estimating costs.

	2008	2009	2010	Total
Increased sales from new customers	0	750,000	772,500	
Increased sales from existing customers	0	1,875,000	1,931,250	
Reduction in customer complaint calls	0	50,000	50,000	
TOTAL BENEFITS:	<u>0</u>	<u>2,675,000</u>	<u>2,753,750</u>	
PV of BENEFITS:	**<u>0</u>**	**<u>2,521,444</u>**	**<u>2,520,071</u>**	**<u>5,041,515</u>**
PV of ALL BENEFITS:	**<u>0</u>**	**<u>2,521,444</u>**	**<u>5,041,515</u>**	
Labor: Analysis and Design	42,000	0	0	
Labor: Implementation	120,000	0	0	
Consultant Fees	50,000	0	0	
Training	5,000	0	0	
Office Space and Equipment	2,000	0	0	
Software	10,000	0	0	
Hardware	25,000	0	0	
TOTAL DEVELOPMENT COSTS:	254,000	0	0	
Labor: Webmaster	85,000	87,550	90,177	
Labor: Network Technician	60,000	61,800	63,654	
Labor: Computer Operations	50,000	51,500	53,045	
Labor: Business Manager	60,000	61,800	63,654	
Labor: Assistant Manager	45,000	46,350	47,741	
Labor: 3 Staff	90,000	92,700	95,481	
Software Upgrades	1,000	1,000	1,000	
Software Licenses	3,000	1,000	1,000	
Hardware Upgrades	5,000	3,000	3,000	
User Training	2,000	1,000	1,000	
Communications Charges	20,000	20,000	20,000	
Marketing Expenses	25,000	25,000	25,000	
TOTAL OPERATIONAL COSTS:	446,000	452,700	464,751	
TOTAL COSTS:	<u>700,000</u>	<u>452,700</u>	<u>464,751</u>	
PV of COSTS:	**<u>679,612</u>**	**<u>426,713</u>**	**<u>425,313</u>**	**<u>1,531,638</u>**
PV of ALL COSTS:	**679,612**	**1,106,325**	**1,531,638**	
Total Project Benefits − Costs :	**(700,000)**	**2,222,300**	**2,288,999**	
Yearly NPV:	**(679,612)**	**2,094,731**	**2,094,758**	**3,509,878**
Cumulative NPV:	**(679,612)**	**1,415,119**	**3,509,878**	
Return on Investment:	**229.16%**	(3,509,878/1,531,638)		
Break-even Point:	**<u>1.32 years</u>**	(Break-even occurs in year 2; (2,094,731 − 1,415,119)/2,094,731 = 0.32)		
Intangible Benefits:	Greater brand recognition			
	Improved customer satisfaction			

FIGURE 2-15 Economic Feasibility Analysis for CD Selections

SUMMARY

Project Initiation

Project initiation is the point at which an organization creates and assesses the original goals and expectations for a new system. The first step in the process is to identify the business value for the system by developing a system request that provides basic information about the proposed system. Next, the analysts perform a feasibility analysis to determine the technical, economic, and organizational feasibility of the system; if appropriate, the system is approved, and the development project begins.

Feasibility Analysis

A feasibility analysis is then used to provide more detail about the risks associated with the proposed system, and it includes technical, economic, and organizational feasibilities. The technical feasibility focuses on whether the system *can* be built by examining the risks associated with the users' and analysts' familiarity with the functional area, familiarity with the technology, and project size. The economic feasibility addresses whether the system *should* be built. It includes a cost–benefit analysis of development costs, operational costs, tangible benefits, and intangible costs and benefits. Finally, the organizational feasibility assesses how well the system will be accepted by its users and incorporated into the ongoing operations of the organization. The strategic alignment of the project and a stakeholder analysis can be used to assess this feasibility dimension.

Project Selection

Once the feasibility analysis has been completed, it is submitted to the approval committee, along with a revised system request. The committee then decides whether to approve the project, decline the project, or table it until additional information is available. The project selection process uses portfolio management to take into account all the projects in the organization. The approval committee weighs many factors and makes trade-offs before a project is selected.

KEY TERMS

Approval committee	Feasibility analysis	Project size
Break-even point	Feasibility study	Project sponsor
Business need	First mover	Return on investment (ROI)
Business requirement	Functionality	Risks
Business value	Intangible benefits	Special issues
Call option	Intangible costs	Stakeholder
Cash flow method	Intangible value	Stakeholder analysis
Champion	Net present value (NPV)	Strategic alignment
Compatibility	Operational costs	System request
Cost–benefit analysis	Option pricing models (OPMs)	System users
Development costs	Organizational feasibility	Tangible benefits
Economic feasibility	Organizational management	Tangible value
Emerging Technology	Portfolio management	Technical feasibility
Familiarity with the functional area	Project	Technical risk analysis
Familiarity with the technology	Project initiation	Trade-offs

QUESTIONS

1. Why should the system request be created by a businessperson as opposed to an IS professional?
2. Give three examples of business needs for a system.
3. What is the purpose of an approval committee? Who is usually on this committee?
4. What is the difference between intangible value and tangible value? Give three examples of each.
5. List two intangible benefits. Describe how these benefits can be quantified.
6. List two tangible benefits and two operational costs for a system. How would you determine the values that should be assigned to each item?
7. Describe the three techniques for feasibility analysis.
8. What are the purposes of the system request and the feasibility analysis? How are they used in the project selection process?
9. Describe two special issues that may be important to list on a system request.
10. What are the steps for assessing economic feasibility? Describe each step.
11. Describe a risky project in terms of technical feasibility. Describe a project that would *not* be considered risky.
12. What factors are use to determine project size?
13. What is stakeholder analysis? Discuss three stakeholders that would be relevant for most projects.
14. What is the break-even point for the project? How is it calculated?
15. Explain the net present value and return on investment for a cost benefit analysis. Why would these calculations be used?

EXERCISES

A. Locate a news article in an IT trade magazine (e.g., *Computerworld*) about an organization that is implementing a new computer system. Describe the tangible and intangible value that the organization is likely to realize from the new system.
B. Supposed that you are interested in buying a new computer. Create a cost-benefit analysis that illustrates the return on investment that you would receive from making this purchase. Computer-related Web sites (e.g., Dell Computers, Compaq Computers) should have real tangible costs that you can include in your analysis. Project your numbers out to include a three-year period of time and provide the net present value of the final total.
C. Car dealers have realized how profitable it can be to sell automobiles using the Web. Pretend you work for a local car dealership that is part of a large chain such as CarMax. Create a system request you might use to develop a Web-based sales system. Remember to list special issues that are relevant to the project.
D. Reread *Your Turn 2-1*. (Identify Tangible and Intangible Value). Create a list of the stakeholders that should be considered in a stakeholder analysis of this project.
E. Interview someone who works in a large organization and ask him or her to describe the approval process that exists for approving new development projects. What do they think about the process? What are the problems? What are the benefits?
F. Consider the Amazon.com Web site. The management of the company decided to extend their Web-based system to include products other than books (e.g., wine and specialty gifts). How would you have assessed the feasibility of this venture when the idea first came up? How risky would you have considered the project that implemented this idea? Why?

MINICASES

1. The Seisakusho Company operates a fleet of 20 trucks and crews that provide a variety of repair services to residential customers in Tokyo. Currently, it takes on average about six hours before a service team responds to a request. Each truck and crew averages 10 service calls per week, and the average revenue earned per service call is ¥20,000. Each truck is in service 50 weeks per year. Due to difficulty in scheduling and routing, there is considerable slack time for each truck and crew during a typical week.

   To schedule the trucks and crews more efficiently and improve their productivity, the Seisakusho management is evaluating the installation of a routing and scheduling software package. The benefits would include reduced response time to service requests and result in more productive service teams. However, the management has difficulty quantifying these benefits.

One approach is to make an estimate of how much service response time will decrease with the new system, which then can be used to project the increase in the number of service calls made each week. For example, if the system permits the average service response time to fall to four hours, then each truck will be able to make 15 service calls per week on average—an increase of five calls per week. With each truck making five additional calls per week and the average revenue per call at ¥20,000, the revenue increase per truck per week is ¥100,000 (5 × ¥20,000). With 20 trucks in service 50 weeks per year, the average annual revenue increase will be ¥100,000,000 (¥100,000 × 20 × 50).

The Seisakusho management is unsure if the new system will enable response time to fall to four hours on average. Therefore, it has compiled the following data on possible outcomes of the new system,.

New Response Time	# Calls/Truck/Week	Likelihood
3 hours	18	30%
4 hours	15	50%
5 hours	12	20%

Given these figures, prepare a spreadsheet model that computes the expected value of the annual revenues to be produced by this new system.

2. The Amberssen Specialty Company is a chain of twelve retail stores that sell a variety of imported gift items, gourmet chocolates, cheeses, and wines in the Toronto area. Amberssen has an IS staff of three people who have created a simple but effective information system of networked point-of-sale registers at the stores and a centralized accounting system at the company headquarters. Harry Hilman, the head of Amberssens IS group, has just received the following memo from Bill Amberssen, Sales Director (and son of Amberssen's founder).

Harry—it's time Amberssen Specialty launched itself on the Internet. Many of our competitors are already there, selling to customers without the expense of a retail storefront, and we should be there too. I project that we could double or triple our annual revenues by selling our products on the Internet. I'd like to have this ready by early November in time for the prime holiday gift-shopping season. Bill

After pondering this memo for several days, Harry scheduled a meeting with Bill so that he could clarify Bill's vision of this venture. Using the standard content of a system request as your guide, prepare a list of questions that Harry needs to have answered about this project.

CHAPTER 3

PROJECT MANAGEMENT

This chapter describes the important steps of project management, which begins in planning and continues throughout the SDLC. First, the project manager estimates the size of the project and identifies the tasks that need to be performed. Next, he or she staffs the project and puts several activities in place to help coordinate project activities. These steps produce important project management deliverables, including the workplan, staffing plan, and standards list.

OBJECTIVES

- Become familiar with estimation.
- Be able to create a project workplan.
- Understand why project teams use timeboxing.
- Become familiar with how to staff a project.
- Understand how computer-aided software engineering, standards, and documentation improve the efficiency of a project.
- Understand how to reduce risk on a project.

CHAPTER OUTLINE

Introduction
Identifying Project Size
  Function Point Approach
Creating and Managing the Workplan
  Identifying Tasks
  The Project Workplan
  Gantt Chart
  PERT Chart
  Refining Estimates
  Scope Management
  Timeboxing
  Evolutionary Work Breakdown
    Structures and Iterative Workplans

Staffing the Project
  Staffing Plan
  Motivation
  Handling Conflict
Coordinating Project Activities
  CASE Tools
  Standards
  Documentation
  Managing Risk
Applying the Concepts at CD Selections
  Staffing the Project
  Coordinating Project Activities
Summary

INTRODUCTION

Think about major projects that occur in people's lives, such as throwing a big party, like a wedding or graduation celebration. Months are spent in advance identifying and performing all the tasks that need to get done, such as sending out invitations and selecting a menu, and time and money are carefully allocated among them. Along the way, decisions are recorded, problems are addressed, and changes are made. The increasing popularity of the party planner, a person whose sole job is to coordinate a party, suggests how tough this job can be. In the end, the success of any party has a lot to do with the effort that went into planning along the way. System development projects can be much more complicated than the projects we encounter in our personal lives—usually, more people are involved (e.g., the organization), the costs are higher, and more tasks need to be completed. Therefore, it is not surprising that "party planners" exist for information system projects; they are called *project managers*.

Project management is the process of planning and controlling the development of a system within a specified time frame at a minimum cost with the right functionality.[1] A project manager has the primary responsibility for managing the hundreds of tasks and roles that need to be carefully coordinated. Today, project management is an actual profession, and analysts spend years working on projects prior to tackling the management of them. In a 1999 *Computerworld* survey, more than half of 103 companies polled said they now offer formal project management training for information technology (IT) project teams. There also is a variety of *project management software* available, such as Microsoft Project, PlanView, and PMOffice, which support project management activities.

Although training and software are available to help project managers, unreasonable demands set by project sponsors and business managers can make project management very difficult. Too often, the approach of the holiday season, the chance at winning a proposal with a low bid, or a funding opportunity pressures project managers to promise systems long before they are able to deliver them. These overly optimistic timetables are thought to be one of the biggest problems that projects face; instead of pushing a project forward faster, they result in delays.

Thus, a critical success factor for project management is to start with a realistic assessment of the work that needs to be accomplished and then manage the project according to that assessment. This can be achieved by carefully following the four steps presented in this chapter: identifying the project size, creating and managing the workplan, staffing the project, and coordinating project activities. The project manager ultimately creates a workplan, staffing plan, and standards list, which are used and refined throughout the entire system development lifecycle.

IDENTIFYING PROJECT SIZE

The science (or art) of project management is in making *trade-offs* among three important concepts: the size of the system (in terms of what it does), the time to complete the project (when the project will be finished), and the cost of the project. Think of these three things as interdependent levers that the project manager controls throughout the SDLC. Whenever one lever is pulled, the other two levers are affected in some way. For example, if a project manager needs to readjust a deadline to an earlier date, then the only solutions are to decrease the size of the system (by eliminating some of its functions) or to increase costs by adding more people or having them work overtime. Often, a project manager will have to work with the project sponsor to change the goals of the project, such as developing a

---

[1] The following are good books on project management for object-oriented projects: Grady Booch, *Object Solutions: Managing the Object-Oriented Project* (Menlo Park, CA: Addison-Wesley, 1996); Murray R. Cantor, *Object-Oriented Project Management with UML* (New York, NY: Wiley, 1998); Alistair Cockburn, *Surviving Object-Oriented Projects: A Manager's Guide* (Reading, MA: Addison-Wesley, 1998); and Walker Royce, *Software Project Management: A Unified Framework* (Reading, MA: Addison-Wesley, 1998). Also, the Project Management Institute (www.pmi.org) and the Information Systems Special Interest Group of the Project Management Institute (www.pmi-issig.org) have valuable resources on project management in information systems.

system with less functionality or extending the deadline for the final system, so that the project has reasonable goals that can be met.

Therefore, in the beginning of the project, the manager needs to estimate each of these levers and then continuously assess how to roll out the project in a way that meets the organization's needs. *Estimation*[2] is the process of assigning projected values for time and effort, and it can be performed manually or with the help of estimation software packages such as Costar and Construx—there are more than fifty available on the market. The estimates developed at the start of a project are usually based on a range of possible values (e.g., the design phase will take three to four months) and gradually become more specific as the project moves forward (e.g., the design phase will be completed on March 22).

The numbers used to calculate these estimates can come from several sources. They can be provided with the methodology that is used, taken from projects with similar tasks and technologies, or provided by experienced developers. Generally speaking, the numbers should be conservative. A good practice is to keep track of the actual values for time and effort during the SDLC so that numbers can be refined along the way and the next project can benefit from real data. One of the greatest strengths of systems consulting firms is the past experience they offer to a project; they have estimates and methodologies that have been developed and honed over time and applied to hundreds of projects.

There are a variety of ways to estimate the time required to build a system. For example, it is possible to use the function point approach, a task-decomposition approach using work breakdown structures, and the timeboxing approach. Each of these is described later in this chapter. However, the simplest method uses the amount of time spent in the planning phase to predict the time required for the entire project. The idea is that a simple project will require little planning and a complex project will require more planning, so using the amount of time spent in the planning phase is a reasonable way to estimate overall project time requirements.

CONCEPTS IN ACTION   3-A Trade-offs

I was once on a project to develop a system that should have taken a year to build. Instead, the business need demanded that the system be ready within five months—impossible!

On the first day of the project, the project manager drew a triangle on a white board to illustrate some trade-offs that he expected to occur over the course of the project. The corners of the triangle were labeled Functionality, Time, and Money. The manager explained, "We have too little time. We have an unlimited budget. We will not be measured by the bells and whistles that this system contains. So over the next several weeks, I want you as developers to keep this triangle in mind and do everything it takes to meet this five-month deadline."

At the end of the five months, the project was delivered on time; however, the project was incredibly over

budget, and the final product was "thrown away" after it was used because it was unfit for regular usage. Remarkably, the business users felt that the project was very successful because it met the very specific business needs for which it was built. They believed that the trade-offs that were made were worthwhile.

—*Barbara Wixom*

Questions

1. What are the risks in stressing only one corner of the triangle?
2. How would you have managed this project? Can you think of another approach that might have been more effective?

---

[2] A good book for further reading on software estimation is that by Capers Jones, *Estimating Software Costs* (New York: McGraw-Hill, 1989).

	Planning	Analysis	Design	Implementation
Typical industry standards for business applications	15%	20%	35%	30%
Estimates based on actual figures for first stages of SDLC	Actual: 4 person-months	Estimated: 5.33 person-months	Estimated: 9.33 person-months	Estimated: 8 person-months
SDLC = systems development life cycle.				

FIGURE 3-1

Estimating Project Time Using the Planning Phase Approach

With this approach, you take the time spent in (or estimated for) the planning phase and use industry standard percentages (or percentages from the organization's own experiences) to calculate estimates for the other SDLC phases. Industry standards suggest that a typical business application system spends 15 percent of its effort in the planning phase, 20 percent in the analysis phase, 35 percent in the design phase, and 30 percent in the implementation phase. This would suggest that if planning a project takes 4 months, then the rest of the project will probably take a total of 22.67 person-months $[(4 \div 0.15) - 4 = 22.67]$. These same industry percentages are then used to estimate the amount of time in each phase (Figure 3-1). The obvious limitation of this approach is that it can be difficult to take into account the specifics of your individual project, which may be simpler or more difficult than the typical project.

FIGURE 3-2

Estimating Project Time Using the Function Point Approach

Function Point Approach[3]

The *function point approach* to estimation uses a more complex—and, it is hoped, more reliable—three-step process (Figure 3-2). First, the project manager estimates the size of the project in terms of the number of lines of code the new system will require. This size estimate is then converted into the amount of effort required to develop the system in terms of a number of person-months. The estimated effort is then converted into an estimated schedule time in terms of the number of months from start to finish.

Step 1: Estimate System Size   The first step is to estimate the size of a project using function points, a concept developed in 1979 by Allen Albrecht of IBM. A *function point* is a measure of program size based on the system's number and *complexity* of inputs, outputs, queries, files, and program interfaces.

[3] Two good books that focus on function points are J. Brian Dreger, *Function Point Analysis* (Englewood Cliffs, NJ: Prentice Hall, 1989) and C. R. Symons, *Software Sizing and Estimating: MK II FPA* (New York: Wiley, 1991). Additional information on function point analysis can be found at www.ifpug.org. We introduce a variation on function points, use-case points, in Chapter 5.

To calculate the function points for a project, components are listed on a worksheet to represent the major elements of the system. For example, data entry screens are kinds of inputs, reports are outputs, and database queries are kinds of queries (see Figure 3-3). The project manager records the total number of each component that the system will include, and then he or she breaks down the number to show the number of components that have low, medium, and high complexity. In Figure 3-3, there are nineteen outputs that need to be developed for the system, four of which have low complexity, ten that have medium complexity, and five that are very complex. After each line is filled in, a total number of points is calculated per line by multiplying each number by a complexity index. The line totals are added to determine the *total unadjusted function points (TUFP)* for the project.

The complexity of the overall system is greater than the sum of its parts. Things such as the familiarity of the project team with the business area and the technology that will be used to implement the project may also influence how complex a project will be. A project that is very complex for a team with little experience might have little complexity for a team with lots of experience. To create a more realistic size for the project, a number of additional system factors, such as end user efficiency, reusability, and data communications, are assessed in terms of their effect on the project's complexity (see Figure 3-3). These assessments are totaled and placed into a formula to calculate an *adjusted project complexity (APC)* score. The TUFP value is multiplied by the APC value to determine the ultimate size of the project in terms of *total adjusted function points (TAFP)*. This number should give the project manager a reasonable idea as to how big the project will be.

Sometimes a shortcut is used to determine the complexity of the project. Instead of calculating the complexity for the fourteen factors listed in Figure 3-3, project managers choose to assign an APC value that ranges from .65 for very simple systems, to 1.00 for normal systems, to as much as 1.35 for complex systems; then, they multiply the value to the TUFP score. For example, a very simple system that has 200 unadjusted function points

---

CONCEPTS   **3-B Function Points at Nielsen**

IN ACTION

Nielsen Media used function point analysis (FPA) for an upgrade to the Global Sample Management System (GSMS) for Nielsen Media/NetRatings, which keeps track of the Internet rating sample, a group of 40,000 homes nationwide that volunteer to participate in ongoing ratings.

In late fall of 1998, Nielsen Media did an FP count based on the current GSMS. (FPA is always easier and more accurate for an existing system.) Nielsen Media had its counters—three quality assurance staff—do their FPA and then input their count into KnowledgePlan, a productivity modeling tool. In early 1999, seven programmers began writing code for the system, which they were expected to complete in ten months. As November approached, the project was adding staff to try to meet the deadline. When it became evident that the deadline would not be met, a new FP count was conducted. The GSMS had grown to

900 FPs. Besides the original 500 plus 20 percent, there were 300 FPs attributable to features and functions that had crept into the project.

How did that happen? The way it always does: The developers and users had added a button here and a new feature there, and soon the project was much larger than it was originally. But Nielsen Media had put a stake in the ground at the beginning from which they could measure growth along the way.

The best practice is to run the FPA and productivity model at the project's launch and again when there is a full list of functional requirements. Then do another analysis any time there is a major modification in the functional definition of the project.

Source: Bill Roberts, "Ratings Game," *CIO Magazine* (October 2000).

System Components:

Description	Total Number	Complexity			Total
		Low	Medium	High	
Inputs	6	3 × 3	2 × 4	1 × 6	23
Outputs	19	4 × 4	10 × 5	5 × 7	101
Queries	10	7 × 3	0 × 4	3 × 6	39
Files	15	0 × 7	15 × 10	0 × 15	150
Program Interfaces	3	1 × 5	0 × 7	2 × 10	25
Total Unadjusted Function Points (TUFP):					338

Overall System:

Data communications	3
Heavy use configuration	0
Transaction rate	0
End-user efficiency	0
Complex processing	0
Installation ease	0
Multiple sites	0
Performance	0
Distributed functions	2
Online data entry	2
Online update	0
Reusability	0
Operational ease	0
Extensibility	0
Total Processing Complexity (PC):	7

(0 = no effect on processing complexity; 5 = great effect on processing complexity)

Adjusted Processing Complexity (APC):

0.65 + (0.01 x 7) = **0.72**

Total Adjusted Function Points (TAFP):

0.72 (APC) x 338 (TUFP) = 243 **(TAFP)**

FIGURE 3-3
Function
Point Estimation
Worksheet

would have a size of 130 adjusted function points (200 * .65 = 130). However, if the system with 200 unadjusted function points were very complex, its function point size would be 270 (200 * 1.35 = 270).

In planning, the exact nature of the system has not yet been determined, so it is impossible to know *exactly* how many inputs, outputs, and so forth, will be in the system. It is up to the project manager to make an intelligent guess. Some people feel that using function points early on in a project is not practical for this reason. We believe function points can be a useful tool for understanding a project's size at any point in the SDLC. Later in the project, once more is known about the system, the project manager will revise the estimates using this better knowledge to produce more accurate results.

Once you have estimated the number of function points, you need to convert the number of function points into the lines of code that will be required to build the system. The number of lines of code depends on the programming language you decide to use. Figure 3-4 presents a very rough conversion guide for some popular languages.

For example, the system in Figure 3-3 has 243 total adjusted function points. To develop the system in C would typically require approximately 35,964 (148 * 243) lines of code to write it. Conversely, Visual Basic would typically take 12,150 (50 * 243) lines of code. Developing the system using a package such as Excel or Access would take between 11,421 (47 * 243) and 8,505 (35 * 243) lines of code, respectively. There is a great range for packages because different packages enable you to do different things, and not all systems can be built using certain packages. Sometimes lots of extra code must be written to do some simple function because the package does not have the needed capabilities.

There is also a very important message from the data in this figure. Because there is a direct relationship between lines of code and the amount of effort and time required to develop a system, the choice of development language has a significant impact on the time and cost of projects.

Language	Average Number of Lines of Code per Function Point
Access	35
ASP	69
C	148
C++	60
C#	59
COBOL	73
Excel	47
HTML	43
Java	60
Javascript	56
JSP	59
Lotus Notes	21
Smalltalk	35
SQL	39
Visual Basic	50

Source: QSM Function Point Programming Languages Table, http://www.qsm.com/FPGearing.html

FIGURE 3-4

Converting from Function Points to Lines of Code

YOUR TURN

3-1 Calculating System Size

*I*magine that job hunting has been going so well that you need to develop a system to support your efforts. The system should allow you to input information about the companies with which you interview, the interviews and office visits that you have scheduled, and the offers you receive. It should be able to produce reports, such as a company contact list, an interview schedule, and an office visit schedule, as well as produce thank-you letters to be brought into a word processor to customize. You also need the system to answer queries, such as the number of interviews by city and your average offer amount.

Questions

1. Determine the number of inputs, outputs, interfaces, files, and queries that this system requires. For each element, determine if the complexity is low, medium, or high. Record this information on a worksheet similar to the one in Figure 3-3.

2. Calculate the total function points for each line on your worksheet by multiplying the number of each element with the appropriate complexity score.

3. Sum up the total unadjusted function points.

4. Suppose the system will be built by you using Visual Basic (VB). Given your VB skills, multiply the TUFP score by the APC score that best estimates how complex the system will be for you to develop (.65 = simple, 1 = average, 1.35 = complex), and calculate a TAFP value.

5. Using the table in Figure 3-4, determine the number of lines of code that correspond to VB. Multiply this number with the TAFP to find the total lines of code that your system will require.

Step 2: Estimate Required Effort   Once an understanding is reached about the size of a system, the next step is to estimate the effort required to build it. *Effort* is a function of the system size combined with production rates (how much work someone can complete in a given time). Much research has been done on software production rates. One of the most popular algorithms, the *COCOMO*[4] *model*, was designed by Barry W. Boehm to convert a lines-of-code estimate into a person–month estimate. There are different versions of the COCOMO model, which vary based on the complexity of the software, the size of the system, the experience of the developers, and the type of software being developed (e.g., business application software such as the registration system at a university; commercial software such as Word; or system software such as Windows). For small- to moderate-sized business software projects (i.e., 100,000 lines of code and ten or fewer programmers), the model is quite simple:

$$\text{effort (in person-months)} = 1.4 * \text{thousands of lines of code}$$

For example, to develop a business software system requiring 10,000 lines of code would typically take 14 person-months to complete. If the system in Figure 3-3 were developed in C (which equates to 35,964 lines of code), it would require about 50.35 person-months of effort.

Step 3: Estimate Time Required   Once the effort is understood, the optimal schedule for the project can be estimated. Historical data or estimation software can be used as aids. One rule of thumb is to determine schedule using the following equation:

$$\text{schedule time (months)} = 3.0 * \text{person-months}^{1/3}$$

---

[4] The original COCOMO model is presented by Barry W. Boehm in *Software Engineering Economics* (Englewood Cliffs, NJ: Prentice-Hall, 1981). Since then, much additional research has been done. The latest version of COCOMO, COCOMO II, is described in B.W. Boehm, C. Abts, A.W. Brown, S. Chulani, B.K. Clark, E. Horowitz, R. Madachy, D. Reifer, and B. Steece, *Software Cost Estimation with COCOMO II* (Upper Saddle River, NJ: Prentice Hall PTR, 2000). For the latest updates on COCOMO, see http://sunset.usc.edu/csse/research/ COCOMOII/cocomo_main.html.

This equation is widely used, although the specific numbers vary (e.g., some estimators may use 3.5 or 2.5 instead of 3.0). The equation suggests that a project that has an effort of 14 person-months should be scheduled to take a little more than 7 months to complete. Continuing the example of Figure 3-3, 50.35 person-months would require very slightly more than 11 months. It is important to note that this estimate is for the analysis, design, and implementation; it does not include the planning.

CREATING AND MANAGING THE WORKPLAN

Once a project manager has a general idea of the size and approximate schedule for the project, he or she creates a *workplan*, which is a dynamic schedule that records and keeps track of all the tasks that need to be accomplished over the course of the project. The workplan lists each task, along with important information about it, such as when it needs to be completed, the person assigned to do the work, and any deliverables that will result (Figure 3-5). The level of detail and the amount of information captured by the workplan depend on the needs of the project (and the detail usually increases as the project progresses). The workplan is usually the main component of the project management software mentioned earlier.

To create a workplan, the project manager first identifies the tasks that need to be accomplished and determines how long they will take. Then the tasks are organized within a workplan and presented graphically using Gantt and PERT charts. All these techniques help a project manager understand and manage the project's progress over time.

Workplan Information	Example
Name of the task	Perform economic feasibility
Start date	Jan 05, 2010
Completion date	Jan 19, 2010
Person assigned to the task	Project sponsor: Mary Smith
Deliverable(s)	Cost–benefit analysis
Completion status	Open
Priority	High
Resources that are needed	Spreadsheet software
Estimated time	16 hours
Actual time	14.5 hours

FIGURE 3-5
Workplan Information

Identifying Tasks

The overall objectives for the system should be listed on the system request, and it is the project manager's job to identify all the tasks that need to be accomplished to meet those objectives. This sounds like a daunting task—how can someone know everything that needs to be done to build a system that has never been built before?

One approach for identifying tasks is to get a list of tasks that has already been developed and to modify it. There are standard lists of tasks, or methodologies, that are available for use as a starting point. As we stated in Chapter 1, a *methodology* is a formalized approach to implementing an SDLC (i.e., it is a list of steps and deliverables). A project manager can take an existing methodology, select the steps and deliverables that apply to the current project, and add them to the workplan. If an existing methodology is not available within the organization, methodologies can be purchased from consultants or vendors, or books such as this textbook can serve as a guide. Using an existing methodology is the most popular way to create a workplan because most organizations have a methodology they use for projects.

If a project manager prefers to begin from scratch, he or she can use a structured, top-down approach whereby high-level tasks are first defined and then broken down into subtasks. For example, Figure 3-6 shows a list of high-level tasks needed to implement a new IT training class. Some of the main steps in the process include identifying vendors, creating and administering a survey, and building new classrooms. Each step is then broken down in turn and numbered in a hierarchical fashion. There are eight subtasks (i.e., 7.1–7.8) for creating and administering a survey, and there are three subtasks (7.2.1–7.2.3) that comprise the review initial survey task. A list of tasks hierarchically numbered in this way is called a *work breakdown structure (WBS)*, and it is the backbone of the project workplan. The number of tasks and level of detail depend on the complexity and size of the project. The larger the

Task Number	Task Name	Duration (in weeks)	Dependency	Status
1	Identify vendors	2		Complete
2	Review training materials	6	1	Complete
3	Compare vendors	2	2	In Progress
4	Negotiate with vendors	3	3	Open
5	Develop communications information	4	1	In Progress
6	Disseminate information	2	5	Open
7	Create and administer survey	4	6	Open
7.1	Create initial survey	1		Open
7.2	Review initial survey	1	7.1	Open
7.2.1	Review by Director of IT Training	1		Open
7.2.2	Review by Project Sponsor	1		Open
7.2.3	Review by Representative Trainee	1		Open
7.3	Pilot test initial survey	1	7.1	Open
7.4	Incorporate survey changes	1	7.2, 7.3	Open
7.5	Create distribution list	0.5		Open
7.6	Send survey to distribution list	0.5	7.4, 7.5	Open
7.7	Send follow-up message	0.5	7.6	Open
7.8	Collect completed surveys	1	7.6	Open
8	Analyze results and choose vendor	2	4, 7	Open
9	Build new classrooms	11	1	In Progress
10	Develop course options	3	8, 9	Open

FIGURE 3-6
Work Breakdown Structure

project, the more important it becomes to define tasks at a low level of detail so that essential steps are not overlooked.

There are two basic approaches to organizing a WBS: by SDLC phase or by product. For example, if a firm decided that it needed to develop a Web site, the firm could create a WBS based on the SDLC: planning, analysis, design, and implementation. In this case, a typical task that would take place during planning would be feasibility analysis. This task would be broken down into the different types of feasibility analysis: technical, economic, and organizational. Each of these would be further broken down into a set of subtasks. Alternatively, the firm could organize the workplan along the lines of the different products to be developed. For example, in the case of a Web site, the products could include applets, application servers, database servers, the various sets of Web pages to be designed, a site map, and so on. Then these would be further decomposed into the different tasks associated with the phases of the SDLC. As described previously, Figure 3-6 depicts the tasks necessary for creating a new IT training class. Either way, once the overall structure is determined, tasks are identified and included in the WBS of the workplan. We return to the topic of WBS and their use in iterative planning later in this chapter.

The Project Workplan

The project workplan is the mechanism that is used to manage the tasks that are listed in the work breakdown structure. It is the project manager's primary tool for managing the project. Using it, the project manager can tell if the project is ahead or behind schedule, how well the project was estimated, and what changes need to be made to meet the project deadline.

Basically, the workplan is a table that lists all the tasks in the WBS, along with important task information, such as the people who are assigned to perform the tasks, the actual hours that the tasks took, and the variances between estimated and actual completion times (see Figure 3-6). At a minimum, the information should include the duration of the task, the current statuses of the tasks (i.e., open, complete), and the *task dependencies*, which occur when one task cannot be performed until another task is completed. For example, Figure 3-6 shows that incorporating changes to the survey (task 7.4) takes a week to perform, but it cannot occur until after the survey is reviewed (task 7.2) and pilot tested (task 7.3). Key *milestones*, or important dates, are also identified on the workplan. Presentations to the approval committee, the start of end user training, a company retreat, and the due date of the system prototype are the types of milestones that may be important to track.

Gantt Chart

A *Gantt chart* is a horizontal bar chart that shows the same task information as the project workplan but in a graphical way. Sometimes a picture really is worth a thousand words, and the Gantt chart can communicate the high-level status of a project much faster and easier than the workplan. Creating a Gantt chart is simple and can be done using a spreadsheet package, graphics software (e.g., Microsoft VISIO), or a project management package.

First, tasks are listed as rows in the chart, and time is listed across the top in increments based on the needs of the projects (see Figure 3-7). A short project may be divided into hours or days; whereas, a medium-sized project may be represented using weeks or months. Horizontal bars are drawn to represent the duration of each task; the bar's beginning and end mark exactly when the task will begin and end. As people work on tasks, the appropriate bars are filled in proportionately to how much of the task is finished. Too many tasks on a Gantt chart can become confusing, so it's best to limit the number of tasks to around twenty or thirty. If there are more tasks, break them down into subtasks and create Gantt charts for each level of detail.

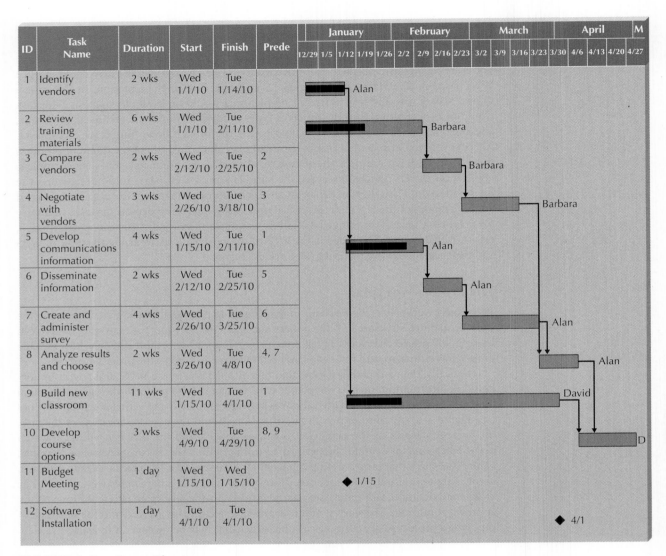

ID	Task Name	Duration	Start	Finish	Prede
1	Identify vendors	2 wks	Wed 1/1/10	Tue 1/14/10	
2	Review training materials	6 wks	Wed 1/1/10	Tue 2/11/10	
3	Compare vendors	2 wks	Wed 2/12/10	Tue 2/25/10	2
4	Negotiate with vendors	3 wks	Wed 2/26/10	Tue 3/18/10	3
5	Develop communications information	4 wks	Wed 1/15/10	Tue 2/11/10	1
6	Disseminate information	2 wks	Wed 2/12/10	Tue 2/25/10	5
7	Create and administer survey	4 wks	Wed 2/26/10	Tue 3/25/10	6
8	Analyze results and choose	2 wks	Wed 3/26/10	Tue 4/8/10	4, 7
9	Build new classroom	11 wks	Wed 1/15/10	Tue 4/1/10	1
10	Develop course options	3 wks	Wed 4/9/10	Tue 4/29/10	8, 9
11	Budget Meeting	1 day	Wed 1/15/10	Wed 1/15/10	
12	Software Installation	1 day	Tue 4/1/10	Tue 4/1/10	

FIGURE 3-7   Gantt Chart

There are many things a project manager can see by quickly looking at a Gantt chart. In addition to seeing how long tasks are and how far along they are, the project manager also can tell which tasks are sequential, which tasks occur at the same time, and which tasks overlap in some way. He or she can get a quick view of tasks that are ahead of schedule and behind schedule by drawing a vertical line on today's date. If a bar is not filled in and is to the left of the line, that task is behind schedule.

There are a few special notations that can be placed on a Gantt chart. Project milestones are shown using upside-down triangles or diamonds. Arrows are drawn between the task bars to show task dependencies. Sometimes, the names of people assigned to each task are listed next to the task bars to show what human resources have been allocated to the tasks.

Pert Chart

A second graphical way to look at project workplan information is the *PERT chart*, which lays out the project tasks in a flowchart (see Figure 3-8). PERT, which stands for program evaluation and review technique, is a network analysis technique that can be used when the individual task time estimates are fairly uncertain. Instead of simply putting a point estimate for the duration estimate, PERT uses three time estimates: optimistic, most likely, and a pessimistic. It then combines the three estimates into a single weighted average estimate using the following formula:

$$\text{PERT weighted average} = \frac{\text{optimistic estimate} + (4 * \text{most likely estimate}) + \text{pessimistic estimate}}{6}$$

The PERT chart is drawn as a node-and-arc type of graph that shows time estimates in the nodes and task dependencies on the arcs. Each *node* represents an individual task, and a line connecting two nodes represents the dependency between two tasks. Partially completed tasks are usually displayed with a diagonal line through the node, and completed tasks contain crossed lines. Milestone tasks are emphasized in some way; in Figure 3-8, for example, the milestone tasks have solid blue borders.

PERT charts are the best way to communicate task dependencies because they lay out the tasks in the order in which they need to be completed. The *critical path method* (CPM) simply allows the identification of the critical path in the network. The critical path is the longest path from the project inception to completion. The critical path shows

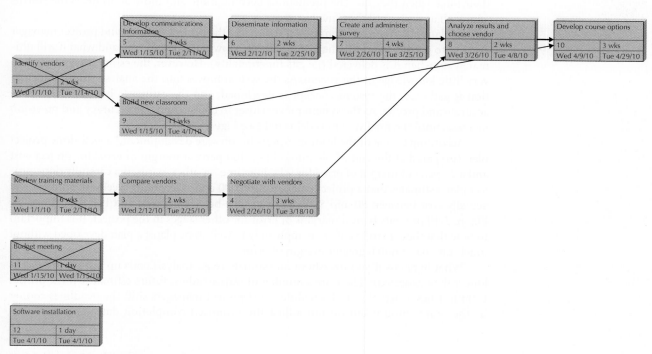

FIGURE 3-8   Pert Chart

all the tasks that must be completed on schedule for a project as a whole to finish on schedule. If any tasks on the critical path take longer than expected, the entire project will fall behind. Each task on the critical path is a *critical task*, and they are usually depicted in a unique way; in Figure 3-8 they are shown with double borders. CPM can be used with or without PERT.

The benefit of using project management software packages such as Microsoft Project is that the workplan can be input once, and then the software can display the information in many different formats. The user can toggle between the workplan, a Gantt chart, and a PERT chart, depending on project management needs.

Refining Estimates

The estimates that are produced during planning need to be refined as the project progresses. This does not mean that estimates were poorly done at the start of the project; rather, it is virtually impossible to develop an exact assessment of the project's schedule before the analysis and design phases are conducted. A project manager should expect to be satisfied with broad ranges of estimates that become more and more specific as the project's product becomes better defined.

In many respects, estimating what an IS development project will cost, how long it will take, and what the final system will actually do follows a *hurricane model*. When storms and hurricanes first appear in the Atlantic or Pacific, forecasters watch their behavior and, on the basis of minimal information about them (but armed with lots of data on previous storms), attempt to predict when and where the storms will hit and what damage they will do when they arrive. As storms move closer to North America, forecasters refine their tracks and develop better predictions about where and when they are most likely to hit and their force when they do. The predictions become more and more accurate as the storms approach a coast, until they finally arrive.

In planning, when a system is first requested, the project sponsor and project manager attempt to predict how long the SDLC will take, how much it will cost, and what it will ultimately do when it is delivered (i.e., its functionality). However, the estimates are based on very little knowledge of the system. As the system moves into the analysis, more information is gathered, the system concept is developed, and the estimates become even more accurate and precise. As the system moves closer to completion, the accuracy and precision increase, until the final system is delivered (see Figure 3-9).

According to one of the leading experts in software development,[5] a well-done project plan (prepared at the end of planning) has a 100 percent margin of error for project cost and a 25 percent margin of error for schedule time. In other words, if a carefully done project plan estimates that a project will cost $100,000 and take twenty weeks, the project will actually cost between $0 and $200,000 and take between fifteen and twenty-five weeks. Figure 3-10 presents typical margins of error for other stages in the project. It is important to note that these margins of error apply only to well-done plans; a plan developed without much care has a much greater margin of error.

What happens if you overshoot an estimate (e.g., analysis ends up lasting two weeks longer than expected)? There are a number of ways to adjust future estimates. If the project team finishes a step ahead of schedule, most project managers shift the deadlines sooner by the same amount but do not adjust the promised completion date. The challenge,

---

[5] Barry W. Boehm et al., "Cost Models for Future Software Life Cycle Processes: COCOMO 2.0," in J. D. Arthur and S. M. Henry (eds.), *Annals of Software Engineering: Special Volume on Software Process and Product Measurement* (Amsterdam: J. C. Baltzer AG Science Publishers, 1995).

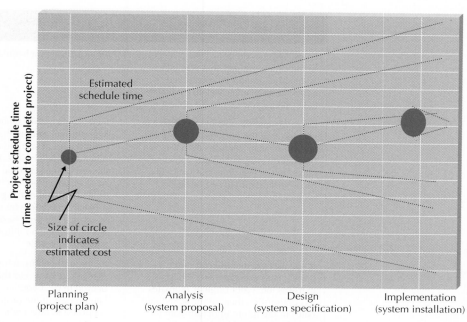

Project schedule time
(Time needed to complete project)

Estimated
schedule time

Size of circle
indicates
estimated cost

| Planning | Analysis | Design | Implementation |
| (project plan) | (system proposal) | (system specification) | (system installation) |

Stage at which estimate is made

FIGURE 3-9
Hurricane Model

however, occurs when the project team is late in meeting a scheduled date. Three possible responses to missed schedule dates are presented in Figure 3-11. If an estimate proves too optimistic early in the project, planners should not expect to make up for lost time—very few projects end up doing this. Instead, they should change future estimates to include an increase similar to the one that was experienced. For example, if the first phase was completed 10 percent over schedule, planners should increase the rest of their estimates by 10 percent.

Scope Management

An analyst may assume that a project will be safe from scheduling problems because he or she carefully estimated and planned the project up front. However, the most common reason for schedule and cost overruns—*scope creep*—occurs after the project is underway.

Scope creep happens when new requirements are added to the project after the original project scope was defined and "frozen." It can happen for many reasons: users may suddenly

| Phase | Deliverable | Typical Margins of Error for Well-Done Estimates | |
		Cost (%)	Schedule Time (%)
Planning	System request	400	60
	Project plan	100	25
Analysis	System proposal	50	15
Design	System specifications	25	10

Source: Barry W. Boehm et al., "Cost Models for Future Software Life Cycle Processes: COCOMO 2.0," in J. D. Arthur and S. M. Henry (eds.) *Annals of Software Engineering Special Volume on Software Process and Product Measurement* (Amsterdam: J. C. Baltzer AG Science Publishers, 1995).

FIGURE 3-10
Margins of Error in Cost and Time Estimates

Assumptions	Actions	Level of Risk
If you assume the rest of the project is simpler than the part that was late and is also simpler than believed when the original schedule estimates were made, you can make up lost time.	Do not change schedule.	High risk
If you assume the rest of the project is simpler than the part that was late and is no more complex than the original estimate assumed, you can't make up the lost time, but you will not lose time on the rest of the project.	Increase the entire schedule by the total amount of time that you are behind (e.g., if you missed the scheduled date by two weeks, move the rest of the schedule dates to two weeks later). If you included padded time at the end of the project in the original schedule, you may not have to change the promised system delivery date; you'll just use up the padded time.	Moderate risk
If you assume that the rest of the project is as complex as the part that was late (your original estimates too optimistic), then all the scheduled dates in the future underestimate the real time required by the same percentage as the part that was late.	Increase the entire schedule by the percentage of weeks that you are behind (e.g., if you are two weeks late on part of the project that was supposed to take eight weeks, you need to increase all remaining time estimates by 25 percent). If this moves the new delivery date beyond what is acceptable to the project sponsor, the scope of the project must be reduced.	Low risk

FIGURE 3-11

Possible Actions When a Schedule Date Is Missed

understand the potential of the new system and realize new functionality that would be useful; developers may discover interesting capabilities to which they become very attached; a senior manager may decide to let this system support a new strategy that was developed at a recent board meeting.

Unfortunately, after the project begins, it becomes increasingly difficult to address changing requirements. The ramifications of change become more extensive, the focus is removed from original goals, and there is at least some impact on cost and schedule. Therefore, the project manager plays a critical role in managing this change to keep scope creep to a minimum.

The keys are to identify the requirements as well as possible in the beginning of the project and to apply analysis techniques effectively. For example, if needs are fuzzy at the project's onset, a combination of intensive meetings with the users and prototyping would allow users to "experience" the requirements and better visualize how the system could support their needs. In fact, the use of meetings and prototyping has been found to reduce scope creep to less than 5 percent on a typical project.

Of course, some requirements may be missed no matter what precautions are taken, but several practices can help control additions to the task list. First, the project manager should allow only absolutely necessary requirements to be added after the project begins. Even at that point, members of the project team should carefully assess the ramifications of the addition and present the assessment to the users. For example, it may require two more person-months of work to create a newly defined report, which would throw off the entire project deadline by several weeks. Any change that is implemented should be carefully tracked so that an audit trail exists to measure the change's impact.

Sometimes changes cannot be incorporated into the present system even though they truly would be beneficial. In this case, these additions to scope should be recorded as future enhancements to the system. The project manager can offer to provide functionality in future releases of the system, thus getting around telling someone no.

Timeboxing

Another approach to scope management is a technique called timeboxing. Up until now, we have described task-oriented projects. In other words, we have described projects that have a schedule driven by the tasks that need to be accomplished, so the greater number of tasks and requirements, the longer the project will take. Some companies have little patience for development projects that take a long time, and these companies take a time-oriented approach that places meeting a deadline above delivering functionality.

Think about the use of word processing software. For 80 percent of the time, only 20 percent of the features, such as the spelling checker, boldfacing, and cutting and pasting, are used. Other features, such as document merging and creation of mailing labels, may be nice to have, but they are not a part of day-to-day needs. The same goes for other software applications; most users rely on only a small subset of their capabilities. Ironically, most developers agree that typically 75 percent of a system can be provided relatively quickly, with the remaining 25 percent of the functionality demanding most of the time.

To resolve this incongruency, the technique of timeboxing has become quite popular, especially when using RAD methodologies. This technique sets a fixed deadline for a project and delivers the system by that deadline no matter what, even if functionality needs to be reduced. Timeboxing ensures that project teams don't get hung up on the final finishing touches that can drag out indefinitely, and it satisfies the business by providing a product within a relatively fast time frame.

There are several steps involved in implementing timeboxing on a project (see Figure 3-12). First, set the date of delivery for the proposed goals. The deadline should not be impossible to meet, so it is best to let the project team determine a realistic due date. Next, build the core of the system to be delivered; you will find that timeboxing helps create a sense of urgency and helps keep the focus on the most important features. Because the schedule is absolutely fixed, functionality that cannot be completed needs to be postponed. It helps if the team prioritizes a list of features beforehand to keep track of what functionality the users absolutely need. Quality cannot be compromised, regardless of other constraints, so it is important that the time allocated to activities is not shortened unless the requirements are changed (e.g., don't reduce the time allocated to testing without reducing features). At the end of the time period, a high-quality system is delivered, but it is likely that future iterations will be needed to make changes and enhancements. In that case, the timeboxing approach can be used once again.

FIGURE 3-12
Steps for Timeboxing

1. Set the date for system delivery.
2. Prioritize the functionality that needs to be included in the system.
3. Build the core of the system (the functionality ranked as most important).
4. Postpone functionality that cannot be provided within the time frame.
5. Deliver the system with core functionality.
6. Repeat steps 3 through 5 to add refinements and enhancements.

3-C Faster Products to Market—with IT

*T*ravelers Insurance Company of Hartford Connecticut has adopted agile development methodologies. The insurance field can be competitive, and Travelers wanted to have the shortest "time to implement" in the field. Travelers set up development teams of six people: two systems analysts, two representatives from the user group (such as claim services), a project manager, and a clerical support person. In the agile approach, the users are physically assigned to the development team for the project. Although at first it might seem that the users might just be sitting around drinking coffee and watching the developers come up with appropriate software solutions, that is not the case. The rapport that is developed within the team allows for instant communication. The interaction is very deep and profound. The resulting software product is delivered quickly—and generally with all the features and nuances that the users wanted.

Questions

1. Could this be done differently, such as by using JAD sessions or having the users review the program on a weekly basis rather than taking the users away from their real job to work on development?

2. What mind-set does an analyst need to work on such an approach?

Evolutionary Work Breakdown Structures and Iterative Workplans

Because object-oriented systems approaches to systems analysis and design support incremental and iterative development, any project planning approach for object-oriented systems development also requires an incremental and iterative process. In the description of the enhanced Unified Process in Chapter 1, the development process was organized around iterations, phases, and workflows. In many ways, a workplan for an incremental and iterative development process is organized in a similar manner. For each iteration, there are different tasks executed on each workflow. This section describes an incremental and iterative process using evolutionary WBSs for project planning that can be used with objects-oriented systems development.

According to Royce,[6] most approaches to developing conventional WBSs tend to have three underlying problems:

1. *They tend to be focused on the design of the information system being developed.* As such, the creation of the WBS forces the premature decomposition of the system design and the tasks associated with creating the design of the system. Where the problem domain is well understood, tying the structure of the workplan to the product to be created makes sense. However, in cases where the problem domain is not well understood, the analyst must commit to the architecture of the system being developed before the requirements of the system are fully understood.

2. *They tend to force too many levels of detail very early on in the SDLC for large projects or they tend to allow too few levels of detail for small projects.* Because the primary purpose of a WBS is to allow cost estimation and scheduling to take place, in conventional approaches to planning, the WBS must be done correctly and completely at the beginning of the SDLC. To say the least, this is

---

[6] Walker Royce, *Software Project Management: A Unified Framework* (Reading, MA: Addison-Wesley, 1998).

a very difficult task to accomplish with any degree of validity. In such cases, it is no wonder that cost and schedule estimation for many information systems development projects tend to be wildly inaccurate.

3. *Because they are project specific, they are very difficult to compare across projects.* This leads to ineffective learning across the organization. Without some standard approach to create WBSs, it is difficult for project managers to learn from previous projects managed by others. This tends to encourage the "reinventing of the wheel" and allows managers to make the same mistakes that previous managers have made.

Evolutionary WBSs allow the analyst to address all three problems by allowing the development of an *iterative workplan.* First, evolutionary WBSs are organized in a standard manner across all projects: by workflows, phases, and then tasks. This decouples the structure of an evolutionary WBS from the structure of the design of the product. This prevents prematurely committing to a specific architecture of a new system. Second, evolutionary WBSs are created in an incremental and iterative manner. The first evolutionary WBS is typically only done for the aspects of the project understood by the analyst. Later on, as the analyst understands more about the evolving development process, more details are added to the WBS. This encourages a more realistic view of both cost and schedule estimation. Third, because the structure of an evolutionary WBS is not tied to any specific project, evolutionary WBSs enable the comparison of the current project to earlier projects. This supports learning from past successes and failures.

In the case of the enhanced Unified Process, the workflows are the major points listed in the WBS. Next, each workflow is decomposed along the phases of the enhanced Unified Process. After that, each phase is decomposed along the tasks that are to be completed to create the deliverables associated with the phase. The tasks listed for each workflow depend upon the level of activity on the workflow during each phase (see Figure 1-11). For example, typical activities for the inception phase of the requirements workflow would be to interview stakeholders, develop a vision document, and develop use cases, whereas there are probably no tasks associated with the inception phase of the operations and support workflow. The template for the first two levels of an evolutionary WBS for the enhanced Unified Process would look like Figure 3-13. As each iteration through the development process is completed, additional tasks are added to the WBS to reflect the current understanding of the remaining tasks to be completed (i.e., the WBS evolves along with the evolving information system). As such, the workplan is developed in an incremental and iterative manner. [7] A sample evolutionary WBS for the planning of the inception phase of the enhanced Unified Process, based on Figures 1-11 and 3-13, is shown in Figure 3-14. Notice that the last two tasks for the project management workflow are "create workplan for first iteration of the elaboration phase" and "assess the inception phase;" that is, the last two things to do are to plan for the next iteration in the development of the evolving system and to assess the current iteration. As the project moves through the later phases, each workflow has tasks added. For example, the analysis workflow will have the creation of the functional, structural, and behavioral models created during the elaboration phase. This approach allows the workplan to evolve through the iterations and phases of the development process.

---

[7] A set of good sources that help explain this approach are Phillippe Krutchen, "Planning an Iterative Project," *The Rational Edge* (October 2002); and Eric Lopes Cordoza and D.J. de Villiers, "Project Planning Best Practices," *The Rational Edge* (August 2003).

I. Business Modeling	a. Inception	IX. Project Management
a. Inception	b. Elaboration	a. Inception
b. Elaboration	c. Construction	b. Elaboration
c. Construction	d. Transition	c. Construction
d. Transition	e. Production	d. Transition
e. Production	VI. Test	e. Production
II. Requirements	a. Inception	X. Environment
a. Inception	b. Elaboration	a. Inception
b. Elaboration	c. Construction	b. Elaboration
c. Construction	d. Transition	c. Construction
d. Transition	e. Production	d. Transition
e. Production	VII. Deployment	e. Production
III. Analysis	a. Inception	XI. Operations and Support
a. Inception	b. Elaboration	a. Inception
b. Elaboration	c. Construction	b. Elaboration
c. Construction	d. Transition	c. Construction
d. Transition	e. Production	d. Transition
e. Production	VIII. Configuration and Change Management	e. Production
IV. Design	a. Inception	XII. Infrastructure Management
a. Inception	b. Elaboration	a. Inception
b. Elaboration	c. Construction	b. Elaboration
c. Construction	d. Transition	c. Construction
d. Transition	e. Production	d. Transition
e. Production		e. Production
V. Implementation		

FIGURE 3-13
Evolutionary WBS Template for the Enhanced Unified Process

	Duration	Dependency
I. Business Modeling		
a. Inception		
1. Understand current business situation	0.50 days	
2. Uncover business process problems	0.25 days	
3. Identify potential projects	0.25 days	
b. Elaboration		
c. Construction		
d. Transition		
e. Production		
II. Requirements		
a. Inception		
1. Identify appropriate requirements analysis technique	0.25 days	
2. Identify appropriate requirements gathering techniques	0.25 days	
3. Identify functional and nonfunctional requirements		II.a.1, II.a.2
A. Perform JAD sessions	3 days	
B. Perform document analysis	5 days	II.a.3.A

FIGURE 3-14
Evolutionary WBS for a Single Iteration–based Inception Phase

	Duration	Dependency
C. Conduct interviews		II.a.3.A
1. Interview project sponsor	0.5 days	
2. Interview inventory system contact	0.5 days	
3. Interview special order system contact	0.5 days	
4. Interview ISP contact	0.5 days	
5. Interview CD Selection Web contact	0.5 days	
6. Interview other personnel	1 day	
D. Observe retail store processes	0.5 days	II.a.3.A
4. Analyze current systems	4 days	II.a.1, II.a.2
5. Create requirements definition		II.a.3, II.a.4
A. Determine requirements to track	1 day	
B. Compile requirements as they are elicited	5 days	II.a.5.A
C. Review requirements with sponsor	2 days	II.a.5.B
b. Elaboration		
c. Construction		
d. Transition		
e. Production		
III. Analysis		
a. Inception		
1. Identify business processes	3 days	
2. Identify use cases	3 days	III.a.1
b. Elaboration		
c. Construction		
d. Transition		
e. Production		
IV. Design		
a. Inception		
1. Identify potential classes	3 days	III.a
b. Elaboration		
c. Construction		
d. Transition		
e. Production		
V. Implementation		
a. Inception		
b. Elaboration		
c. Construction		
d. Transition		
e. Production		
VI. Test		
a. Inception		
b. Elaboration		
c. Construction		
d. Transition		
e. Production		

FIGURE 3-14
Continued

	Duration	Dependency
VII. Deployment		
a. Inception		
b. Elaboration		
c. Construction		
d. Transition		
e. Production		
VIII. Configuration and Change Management		
a. Inception		
1. Identify necessary access controls for developed artifacts	0.25 days	
2. Identify version control mechanisms for developed artifacts	0.25 days	
b. Elaboration		
c. Construction		
d. Transition		
e. Production		
IX. Project Management		
a. Inception		
1. Create workplan for the inception phase	1 day	
2. Create system request	1 day	
3. Perform feasibility analysis		IX.a.2
A. Perform technical feasibility analysis	1 day	
B. Perform economic feasibility analysis	2 days	
C. Perform organizational feasibility analysis	2 days	
4. Identify project size	0.50 days	IX.a.3
5. Identify staffing requirements	0.50 days	IX.a.4
6. Compute cost estimate	0.50 days	IX.a.5
7. Create workplan for first iteration of the elaboration phase	1 day	IX.a.1
8. Assess inception phase	1 day	I.a, II.a, III.a IV.a, V.a, VI.a VII.a, VIII.a, IX.a, X.a, XI.a XII.a
b. Elaboration		
c. Construction		
d. Transition		
e. Production		
X. Environment		
a. Inception		
1. Acquire and install CASE tool	0.25 days	
2. Acquire and install programming environment	0.25 days	
3. Acquire and install configuration and change management tools	0.25 days	
4. Acquire and install project management tools	0.25 days	
b. Elaboration		
c. Construction		

FIGURE 3-14
Continued

	Duration	Dependency
d. Transition		
e. Production		
XI. Operations and Support		
a. Inception		
b. Elaboration		
c. Construction		
d. Transition		
e. Production		
XII. Infrastructure Management		
a. Inception		
1. Identify appropriate standards and enterprise models	0.25 days	
2. Identify reuse opportunities, such as patterns, frameworks, and libraries	0.50 days	
3. Identify similar past projects	0.25 days	
b. Elaboration		
c. Construction		
d. Transition		
e. Production		

FIGURE 3-14
Continued

STAFFING THE PROJECT

Staffing the project includes determining how many people should be assigned to the project, matching people's skills with the needs of the project, motivating them to meet the project's objectives, and minimizing the conflict that will occur over time. The deliverables for this part of project management are a staffing plan, which describes the number and kinds of people who will work on the project, the overall reporting structure, and the project charter, which describes the project's objectives and rules.

Staffing Plan

The first step to staffing is determining the average number of staff needed for the project. To calculate this figure, divide the total person-months of effort by the optimal schedule. So to complete a forty-person-month project in ten months, a team should have an average of four full-time staff members, although this may change over time as different specialists enter and leave the team (e.g., business analysts, programmers, technical writers).

Many times, the temptation is to assign more staff to a project to shorten the project's length, but this is not a wise move. Adding staff resources does not translate into increased productivity; staff size and productivity share a disproportionate relationship, mainly because it is more difficult to coordinate a large number of staff members. The more a team grows, the more difficult it becomes to manage. Imagine how easy it is to work on a two-person project team: the team members share a single line of communication. But adding two people increases the number of communication lines to six, and greater increases lead to more dramatic gains in communication complexity. Figure 3-15 illustrates the impact of adding team members to a project team.

One way to reduce efficiency losses on teams is to understand the complexity that is created in numbers and to build in a *reporting structure* that tempers its effects. The rule of thumb is to keep team sizes of fewer than eight to ten people; therefore, if more people are needed,

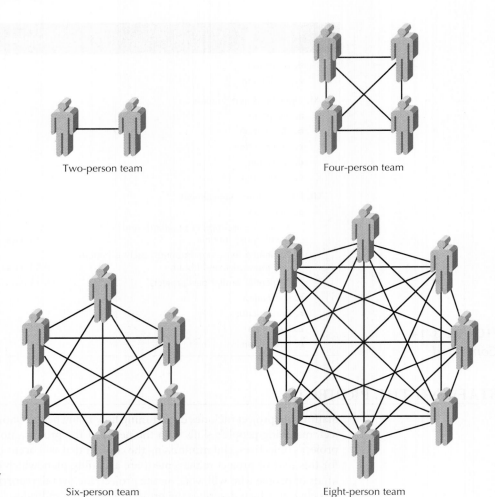

Two-person team

Four-person team

Six-person team

Eight-person team

FIGURE 3-15
Increasing Complexity
with Larger Teams

create subteams. In this way, the project manager can keep the communication effective within small teams, which, in turn, communicate to contact at a higher level in the project.

After the project manager understands how many people are needed for the project, he or she creates a *staffing plan* that lists the roles that are required for the project and the proposed reporting structure for the project. Typically, a project will have one project manager, who oversees the overall progress of the development effort, with the core of the team comprising the various types of analysts described in Chapter 1. A *functional lead* is usually assigned to manage a group of analysts, and a *technical lead* oversees the progress of a group of programmers and more technical staff members.

There are many structures for project teams; Figure 3-16 illustrates one possible configuration of a project team. After the roles are defined and the structure is in place, the project manager needs to think about which people can fill each role. Often, one person fills more than one role on a project team.

When you make assignments, remember that people have *technical skills* and *interpersonal skills*, and both are important on a project. Technical skills are useful when working with technical tasks (e.g., programming in Java) and in trying to understand the various roles that technology plays in the particular project (e.g., how a Web server should be configured on the basis of a projected number of hits from customers).

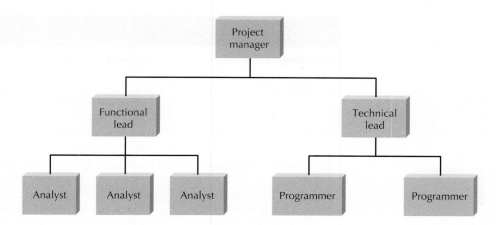

FIGURE 3-16
Possible Reporting
Structure

Interpersonal skills, on the other hand, include interpersonal and communication abilities that are used when dealing with business users, senior management executives, and other members of the project team. They are particularly critical when performing the requirements-gathering activities and when addressing organizational feasibility issues. Each project will require unique technical and interpersonal skills. For example, a Web-based project may require Internet experience or Java programming knowledge, whereas a highly controversial project may need analysts who are particularly adept at managing political or volatile situations.

Ideally, project roles are filled with people who have the right skills for the job. However, the people who fit the roles best may not be available; they may be working on other projects, or they may not exist in the company. Therefore, assigning project team members really is a combination of finding people with the appropriate skill sets and finding people who are available. When the skills of the available project team members do not match what is actually required by the project, the project manager has several options to improve the situation. First, people can be pulled off other projects, and resources can be shuffled around. This is the most disruptive approach from the organization's perspective. Another approach is to use outside help—such as a consultant or contractor—to train team members and start them off on the right foot. Training classes are usually available for both technical and interpersonal instruction if time is available. Mentoring may also be an option; a project team member can be sent to work on another similar project so that he or she can return with skills to apply to the current job.

YOUR TURN   3-3 Staffing Plan

Now it is time to staff the project that was described in Your Turn 3-1. On the basis of the effort required for the project (Your Turn 3-2), how many people will be needed on the project? Given this number, select classmates who will work with you on your project.

Questions

1. What roles will be needed to develop the project? List them and write short descriptions for each of these roles, almost as if you had to advertise the positions in a newspaper.

2. Which roles will each classmate perform? Will some people perform multiple roles?

3. What will the reporting structure be for the project?

Don'ts	Reasons
Assign unrealistic deadlines	Few people will work hard if they realize that a deadline is impossible to meet.
Ignore good efforts	People will work harder if they feel like their work is appreciated. Often, all it takes is public praise for a job well done.
Create a low-quality product	Few people can be proud of working on a project that is of low quality.
Give everyone on the project a raise	If everyone is given the same reward, then high-quality people will believe that mediocrity is rewarded—and they will resent it.
Make an important decision without the team's input	Buy-in is very important. If the project manager needs to make a decision that greatly affects the members of her team, she should involve them in the decision-making process.
Maintain poor working conditions	A project team needs a good working environment or motivation will go down the tubes. This includes lighting, desk space, technology, privacy from interruptions, and reference resources.

Source: Steve McConnell, *Adapted Rapid Development* (Redmond, WA: Microsoft Press, 1996).

FIGURE 3-17
Motivational Don'ts

Motivation

Assigning people to tasks isn't enough; project managers need to motivate the people to make the project a success. *Motivation* has been found to be the number one influence on people's performance,[8] but determining how to motivate the team can be quite difficult. You may think that good project managers motivate their staff by rewarding them with money and bonuses, but most project managers agree that this is the last thing that should be done. The more often managers reward team members with money, the more they expect it—and most times monetary motivation won't work.

Assuming that team members are paid a fair salary, technical employees on project teams are much more motivated by recognition, achievement, the work itself, responsibility, advancement, and the chance to learn new skills.[9] If a manager feels a need to give some kind of reward for motivational purposes, he or she might try a pizza or free dinner or even a kind letter or award. These rewards often have much more effective results than money. Figure 3-17 lists some other motivational don'ts to avoid to ensure that motivation on the project is as high as possible.

Handling Conflict

The third component of staffing is organizing the project to minimize conflict among group members. *Group cohesiveness* (the attraction that members feel to the group and to other members) contributes more to productivity than do project members' individual capabilities or experiences.[10] Clearly defining the roles on the project and holding team members

---

[8] Barry W. Boehm, *Software Engineering Economics* (Englewood Cliffs, NJ: Prentice Hall, 1981). One of the best books on managing project teams is that by Tom DeMarco and Timothy Lister, *Peopleware: Productive Projects and Teams* (New York: Dorset House, 1987).

[9] F. H. Hertzberg, "One More Time: How Do You Motivate Employees?" *Harvard Business Review* (January–February 1968).

[10] B. Lakhanpal, "Understanding the Factors Influencing the Performance of Software Development Groups: An Exploratory Group-Level Analysis," *Information and Software Technology* 35, no. 8 (1993): 468–473.

CONCEPTS IN ACTION

3-D RFID—Promising Technology?

Some animals are extremely valuable. For centuries, horse thieves have stolen horses, so now most horses have tattoos in their mouths. Likewise, purebred pets, such as dog show winners, are valuable animals. What if there were a better way to positively identify valuable animals?

Radio frequency identification (or RFID) has been used for many years in airplanes and on toll roads (like EZPass and FaneLane in the United States) as well as in libraries so that books and materials are not taken out of the library without being checked out. With RFID, a low-frequency radio transmitter, when bombarded with a radio wave, replies with a unique signal. Some animal owners have inserted RFID chips into their pets' shoulders so that they can be identified. The code is unique and cannot be changed. It would be possible to track a stolen race horse if the horse came into range of an RFID device. Likewise, a pet shop or a veterinarian could identify a valuable pet.

Questions

1. If you were working for a state consumer-protection agency, what requirements might you place on pet shops to ensure that animals for sale have not been stolen?
2. What technological requirements might be needed in the system proposal?
3. What ethical issues might be involved?
4. If your system project team did not have the correct technical background, what might you do?

accountable for their tasks is a good way to begin mitigating potential conflict on a project. Some project managers develop a *project charter*, which lists the project's norms and ground rules. For example, the charter may describe when the project team should be at work, when staff meetings will be held, how the group will communicate with each other, and the procedures for updating the workplan as tasks are completed. Figure 3-18 lists additional techniques that can be used at the start of a project to keep conflict to a minimum.

YOUR TURN

3-4 Project Charter

Get together with several of your classmates and pretend that you are all staffed on the project described in Your Turn 3-1. Discuss what would most motivate each of you to perform well on the project. List three potential sources of conflict that could surface as you work together.

Question

Develop a project charter that lists five rules that all team members will need to follow. How might these rules help avoid potential team conflict?

- Clearly define plans for the project.
- Make sure the team understands how the project is important to the organization.
- Develop detailed operating procedures and communicate these to the team members.
- Develop a project charter.
- Develop schedule commitments ahead of time.
- Forecast other priorities and their possible impact on project.

Source: H. J. Thamhain and D. L. Wilemon, "Conflict Management in Project Life Cycles," *Sloan Management Review* (Spring 1975).

FIGURE 3-18
Conflict-Avoidance Strategies

COORDINATING PROJECT ACTIVITIES

Like all project management responsibilities, the act of coordinating project activities continues throughout the entire project, until a system is delivered to the project sponsor and end users. This step includes putting efficient development practices in place and mitigating risk. These activities occur over the course of the entire SDLC, but it is at this point in the project when the project manager needs to put them in place. Ultimately, these activities ensure that the project stays on track and that the chance of failure is kept at a minimum.

Case Tools

Computer-aided software engineering (CASE) is a category of software that automates all or part of the development process. Some CASE software packages are used primarily during the analysis phase to create integrated diagrams of the system and to store information regarding the system components (often called *upper CASE*), whereas others are design-phase tools that create the diagrams and then generate code for database tables and system functionality (often called *lower CASE*). *Integrated CASE*, or I-CASE, contains functionality found in both upper CASE and lower CASE tools in that it supports tasks that happen throughout the SDLC. CASE comes in a wide assortment of flavors in terms of complexity and functionality, and there are many good programs available in the marketplace (e.g., Visible Analyst Workbench, Oracle Designer/2000, Rational Rose, Logic Works suite).

The benefits of using CASE are numerous. With CASE tools, tasks can be completed and altered much faster, development information is centralized, and information is illustrated through diagrams, which are typically easier to understand. Potentially, CASE can reduce maintenance costs, improve software quality, and enforce discipline, and some project teams even use CASE to assess the magnitude of changes to the project.

Of course, like anything else, CASE should not be considered a silver bullet for project development. The advanced CASE tools are complex applications that require significant training and experience to achieve real benefits. Often, CASE serves only as a glorified diagramming tool. Our experience has shown that CASE is a helpful way to support the communication and sharing of project diagrams and technical specifications—as long as it is used by trained developers who have applied CASE on past projects.

The central component of any CASE tool is the *CASE repository*, otherwise known as the information repository or data dictionary. The CASE repository stores the diagrams and other project information, such as screen and report designs, and it keeps track of how the diagrams fit together. For example, most CASE tools will warn you if you place a field on a screen design that doesn't exist in your data model. As the project evolves, project team members perform their tasks using CASE. As you read through the textbook, we will indicate when and how the CASE tool can be used so that you can see how CASE supports the project tasks.

YOUR TURN   **3-5 Computer-Aided Software Engineering Tool Analysis**

Select a CASE tool—either one that you will use for class, a program that you own, or a tool that you can examine over the Web. Create a list of the capabilities that are offered by the CASE tool.

Question

Would you classify the CASE as upper CASE, lower CASE, or I-CASE? Why?

Types of Standards	Examples
Documentation standards	The date and project name should appear as a header on all documentation.
	All margins should be set to 1 inch.
	All deliverables should be added to the project binder and recorded in its table of contents.
Coding standards	All modules of code should include a header that lists the programmer, last date of update, and a short description of the purpose of the code.
	Indentation should be used to indicate loops, if-then-else statements, and case statements.
	On average, every program should include one line of comments for every five lines of code.
Procedural standards	Record actual task progress in the work plan every Monday morning by 10 AM.
	Report to project update meeting on Fridays at 3:30 PM.
	All changes to a requirements document must be approved by the project manager.
Specification requirement standards	Name of program to be created
	Description of the program's purpose
	Special calculations that need to be computed
	Business rules that must be incorporated into the program
	Pseudocode
	Due date
User interface design standards	Labels will appear in boldface text, left-justified, and followed by a colon.
	The tab order of the screen will move from top left to bottom right.
	Accelerator keys will be provided for all updatable fields.

FIGURE 3-19

A Sampling of Project Standards

Standards

Members of a project team need to work together, and most project management software and CASE tools provide access privileges to everyone working on the system. When people work together, however, things can get pretty confusing. To make matters worse, people sometimes are reassigned in the middle of a project. It is important that their project knowledge does not leave with them and that their replacements can get up to speed quickly.

One way to make certain that everyone is performing tasks in the same way and following the same procedures is to create *standards* that the project team must follow. Standards can range from formal rules for naming files to forms that must be completed when goals are reached to programming guidelines. Figure 3-19 shows some examples of the types of standards that a project may create. When a team forms standards and then follows them, the project can be completed faster because task coordination becomes less complex.

Standards work best when they are created at the beginning of each major phase of the project and communicated clearly to the entire project team. As the team moves forward, new standards are added when necessary. Some standards (e.g., file naming conventions, status reporting) are applied to the entire SDLC, whereas others (e.g., programming guidelines) are appropriate only for certain tasks.

Documentation

A final technique that project teams put in place during the planning phase is good *documentation*, which includes detailed information about the tasks of the SDLC. Often, the documentation is stored in *project binder(s)* that contain all the deliverables and all the internal communication that takes place—the history of the project.

A poor project management practice is waiting until the last minute to create documentation; this typically leads to an undocumented system that no one understands. In fact, many problems that companies had updating their systems to handle the year 2000 crisis were the result of the lack of documentation. Good project teams learn to document a system's history as it evolves while the details are still fresh in their memory.

The first step to setting up documentation is to get some binders and include dividers with which to separate content according to the major phases of the project. An additional divider should contain internal communication, such as the minutes from status meetings, written standards, letters to and from the business users, and a dictionary of relevant business terms. Then, as the project moves forward, the deliverables from each task can be put into the project binder with descriptions so that someone outside of the project will be able to understand it. Also, a table of contents should be kept up to date with the content that is added. Documentation takes time up front, but it is a good investment that will pay off in the long run.

CONCEPTS　　3-E Poor Naming Standards

IN ACTION

I once started on a small project (four people) in which the original members of the project team had not set up any standards for naming electronic files. Two weeks into the project, I was asked to write a piece of code that would be referenced by other files that had already been written. When I finished my piece, I had to go back to the other files and make changes to reflect my new work. The only problem was that the lead programmer decided to name the files using his initials (e.g., GG1.prg, GG2.prg, GG3.prg)—and there were more than 200 files! I spent two days opening every one of those files because there was no way to tell what their contents were.

Needless to say, from then on, the team created a code for file names that provided basic information regarding the file's contents, and they kept a log that recorded the file name, its purpose, the date of the last update, and the programmer for every file on the project.

—Barbara Wixom

Question

Think about a program that you have written in the past. Would another programmer be easily able to make changes to it? Why or why not?

Managing Risk

One final facet of project management is *risk management*, the process of assessing and addressing the risks that are associated with developing a project. Many things can cause risks: weak personnel, scope creep, poor design, and overly optimistic estimates. The project team must be aware of potential risks so that problems can be avoided or controlled well ahead of time.

CONCEPTS 3-F The Real Names of the SDLC Phases

IN ACTION

*D*awn Adams, Senior Manager with Asymetrix Consulting, has renamed the SDLC phases:

1. Pudding (Planning)
2. Silly Putty (Analysis)
3. Concrete (Design)
4. Touch-this-and-you're-dead-sucker (Implementation)

Adams also uses icons, such as a skull and crossbones for the implementation phase. The funny labels lend a new depth of interest to a set of abstract concepts. But her names have had another benefit. "I had one participant who adopted the names wholeheartedly," she says,

"including my icons. He posted an icon on his office door for the duration of each of the phases, and he found it much easier to deal with requests for changes from the client, who could see the increasing difficulty of the changes right there on the door."

Source: Learning Technology Shorttakes 1, no. 2 (Wednesday, August 26, 1998).

Question

What would you do if your project sponsor demanded that an important change be made during the "touch-this-and-you're-dead-sucker" phase?

Typically, project teams create a risk assessment, or a document that tracks potential risks along with an evaluation of the likelihood of each risk and its potential impact on the project (Figure 3-20). A paragraph or two is also included to explain potential ways that the risk can be addressed. There are many options: the risk could be publicized, avoided, or even eliminated by dealing with its root cause. For example, imagine that a project team plans to use new technology but its members have identified a risk in the fact that its members do not have the right technical skills. They believe that tasks may take much longer to perform because of a high learning curve. One plan of attack could be to eliminate the root cause of the risk—the lack of technical experience by team members—by finding the time and resources needed to provide proper training to the team.

Most project managers keep abreast of potential risks, even prioritizing them according to their magnitude and importance. Over time, the list of risks will change as some items are removed and others surface. The best project managers, however, work hard to keep risks from having an impact on the schedule and costs associated with the project.

Risk Assessment

RISK 1:	The development of this system likely will be slowed considerably because project team members have not programmed in Java prior to this project.
Likelihood of risk:	High probability of risk.
Potential impact on the project:	This risk will probably increase the time to complete programming tasks by 50 percent.

Ways to address this risk:

It is very important that time and resources are allocated to up-front training in Java for the programmers who are used for this project. Adequate training will reduce the initial learning curve for Java when programming begins. Additionally, outside Java expertise should be brought in for at least some part of the early programming tasks. This person should be used to provide experiential knowledge to the project team so that JAVA-related issues (of which novice Java programmers would be unaware) are overcome.

RISK 2:   ...

FIGURE 3-20

Sample Risk Assessment

| PRACTICAL | Avoiding Classic Planning Mistakes |
| TIP | |

As Seattle University's David Umphress has pointed out, watching most organizations develop systems is like watching reruns of Gilligan's Island. At the beginning of each episode, someone comes up with a cockamamie scheme to get off the island that seems to work for a while, but something goes wrong and the castaways find themselves right back where they started—stuck on the island. Similarly, most companies start new projects with grand ideas that seem to work, only to make a classic mistake and deliver the project behind schedule, over budget, or both. Here we summarize four classic mistakes in the planning and project management aspects of the project and discuss how to avoid them:

1. *Overly optimistic schedule:* Wishful thinking can lead to an overly optimistic schedule that causes analysis and design to be cut short (missing key requirements) and puts intense pressure on the programmers, who produce poor code (full of bugs).

   Solution: Don't inflate time estimates; instead, explicitly schedule slack time at the end of each phase to account for the variability in estimates, using the margins of error from Figure 3-10.

2. *Failing to monitor the schedule:* If the team does not regularly report progress, no one knows if the project is on schedule.

Solution: Require team members to report progress (or the lack or progress) honestly every week. There is no penalty for reporting a lack of progress, but there are immediate sanctions for a misleading report.

3. *Failing to update the schedule:* When a part of the schedule falls behind (e.g., information gathering uses all the slack in item 1 plus 2 weeks), a project team often thinks it can make up the time later by working faster. It can't. This is an early warning that the entire schedule is too optimistic.

   Solution: Immediately revise the schedule and inform the project sponsor of the new end date or use timeboxing to reduce functionality or move it into future versions.

4. *Adding people to a late project:* When a project misses a schedule, the temptation is to add more people to speed it up. This makes the project take longer because it increases coordination problems and requires staff to take time to explain what has already been done.

   Solution: Revise the schedule, use timeboxing, throw away bug-filled code, and add people only to work on an isolated part of the project.

Source: Adapted from Steve McConnell, *Rapid Development* (Redmond, WA: Microsoft Press, 1996), pp. 29–50.

APPLYING THE CONCEPTS AT CD SELECTIONS

Alec Adams was very excited about managing the Internet Sales System project at CD Selections, but he realized that his project team would have very little time to deliver at least some parts of the system because the company wanted the application developed in time for the holiday season. Therefore, he decided that the project should follow an enhanced Unified Process–based approach (see Figure 1-11). In this way, he could be sure that some version of the product would be in the hands of the users within several months, even if the completed system would be delivered at a later date.

As project manager, Alec had to estimate the project's size, effort, and schedule—some of his least favorite jobs because of how tough they are to do at the very beginning of a project. But he knew that the users would expect at least general ranges for a product delivery date. He began by attempting to estimate the size of the project using function points and the function point estimation worksheet in Figure 3-3. For the Web part of the system to be used by customers, he could think of four main queries (searching by artist, by CD title, by song title, and by ad hoc criteria), three input screens (selecting a CD, entering information to put

System Components

Description	Total Number	Complexity			Total
		Low	Medium	High	
Inputs	6	0 * 3	4 * 4	2 * 6	28
Outputs	7	2 * 4	4 * 5	1 * 7	35
Queries	8	3 * 3	4 * 4	1 * 6	31
Files	4	0 * 7	4 * 10	0 * 15	40
Program Interfaces	3	0 * 5	2 * 7	1 * 10	24
Total Unadjusted Function Points (TUFP):					**158**

Assumed project adjusted processing complexity (APC) = 1.2

Total adjusted function points (TAFP): $1.2 * 158 = 189.6$

Estimated lines of code: $(.75 * 190 * 60) + (.25 * 190 * 43) = 10,600$ lines of code

COCOMO effort estimate: $(1.4 * 10.6) = 15$ person-months

Estimated schedule: $3.0 * 15^{1/3} = 7.5$ months

Final effort estimate after pad: $1.34 * 7.5 = 10$ months

Final personnel estimate: $15/10 = 1.5$

FIGURE 3-21
Size and Effort Estimate for the Internet Sales System

a CD on hold, and a special-order screen), four output screens (the home page with general information, information about CDs, information about the customer's special order, and the hold status), three files (CD information, inventory information, and customer orders), and two program interfaces (one to the company's special-order system and one that communicates hold information to the retail store systems). For the part of the system to be used by CD Selections staff (to maintain the marketing materials), he identified three additional inputs, three outputs, four queries, one file, and one program interface. Based on the complexity of each, he entered the numbers into the upper portion of the worksheet (see Figure 3-21). Based on the computation, there were 158 total unadjusted function points (TUFP).

Rather than attempt to assess the complexity of the system in detail, Alec chose to use a value of 1.20 for the adjusted processing complexity (APC). He reasoned that the system was of medium complexity and that the project team had little experience with Internet-based systems. This produced a total adjusted function points (TAFP) of about 190.

Converting function points into lines of code was challenging. The project would use a combination of Java (for most programs) and HTML for the Web screens. Alec decided to assume that about 75 percent of the function points would be Java and 25 percent would be HTML. Using the table in Figure 3-4, Alec estimated that there would be about 10,600 lines of code. Using the COCOMO formula, he found that this translated into about 15 person-months of effort. This in turn suggested a schedule time of about 7.5 months. Because the development team had very little experience in developing this type of system, Alec was not very sure of the estimate. After much deliberation, Alec decided to pad the estimate by 33 percent. As such, Alec estimated that the project would take about 10 months.

Once the estimation was underway, Alec began to create an evolutionary work breakdown structure and iterative workplan to identify the tasks that would be needed to complete the system. He started by reviewing the enhanced Unified Process phases and workflows (see Figure 1-11) and the evolutionary work breakdown structure template (see Figure 3-14). At that point in time, Alec did not know enough to create a complete workplan. As such, he included as much detail as he knew to be correct (see Figure 3-22).

	Duration	Dependency
I. Business Modeling		
a. Inception		
1. Understand current business situation		
2. Uncover business process problems		
3. Identify potential projects		
b. Elaboration		
c. Construction		
d. Transition		
e. Production		
II. Requirements		
a. Inception		
1. Identify appropriate requirements analysis technique		
2. Identify appropriate requirements gathering techniques		
3. Identify functional and nonfunctional requirements		II.a.1, II.a.2
4. Analyze current systems		II.a.1, II.a.2
5. Create requirements definition		II.a.3, II.a.4
A. Determine requirements to track		
B. Compile requirements as they are elicited		II.a.5.A
C. Review requirements with sponsor		II.a.5.B
b. Elaboration		
c. Construction		
d. Transition		
e. Production		
III. Analysis		
a. Inception		
1. Identify business processes		
2. Identify use cases		III.a.1
b. Elaboration		
c. Construction		
d. Transition		
e. Production		
IV. Design		
a. Inception		
1. Identify potential classes		III.a
b. Elaboration		
c. Construction		
d. Transition		
e. Production		
V. Implementation		
a. Inception		
b. Elaboration		
c. Construction		
d. Transition		
e. Production		
VI. Test		
a. Inception		
b. Elaboration		

FIGURE 3-22
Evolutionary Work Breakdown Structure for the Inception Phase for CD Selections.

	Duration	Dependency
c. Construction		
d. Transition		
e. Production		
VII. Deployment		
a. Inception		
b. Elaboration		
c. Construction		
d. Transition		
e. Production		
VIII. Configuration and change management		
a. Inception		
1. Identify necessary access controls for developed artifacts		
2. Identify version control mechanisms for developed artifacts		
b. Elaboration		
c. Construction		
d. Transition		
e. Production		
IX. Project management		
a. Inception		
1. Create workplan for the inception phase		
2. Create system request		
3. Perform feasibility analysis		IX.a.2
A. Perform technical feasibility analysis		
B. Perform economic feasibility analysis		
C. Perform organizational feasibility analysis		
4. Identify project size		IX.a.3
5. Identify staffing requirements		IX.a.4
6. Compute cost estimate		IX.a.5
7. Create workplan for first iteration of the elaboration phase		IX.a.1
8. Assess inception phase		I.a, II.a, III.a IV.a, V.a, VI.a VII.a, VIII.a, IX.a, X.a, XI.a XII.a
b. Elaboration		
c. Construction		
d. Transition		
e. Production		
X. Environment		
a. Inception		
1. Acquire and install CASE tool		
2. Acquire and install programming environment		
3. Acquire and install configuration and change management tools		
4. Acquire and install project management tools		
b. Elaboration		

(Continued)

	Duration	Dependency

   c. Construction
   d. Transition
   e. Production
 XI. Operations and Support
   a. Inception
   b. Elaboration
   c. Construction
   d. Transition
   e. Production
XII. Infrastructure Management
   a. Inception
     1. Identify appropriate standards and enterprise models
     2. Identify reuse opportunities, such as patterns, frameworks, and libraries
     3. Identify similar past projects
   b. Elaboration
   c. Construction
   d. Transition
   e. Production

FIGURE 3-22
Continued

For example, Alec felt confident about the estimation of time needed to create the requirements definition and to elicit the requirements. However, he would not know how long it will take to develop the functional, structural, or behavioral analysis models until after he had a better idea of the actual requirements. Until this determination can be made, any estimation as to the time required would be simply a guess. As time passes, Alec would expect to know much more about the development process and would add much more detail to the workplan. (Remember that the development process and the project management processes are iterative and incremental in nature.)

Staffing the Project

Alec next turned to the task of how to staff his project. On the basis of his earlier estimates, it appeared that about two people would be needed to deliver the system by the holidays (fifteen person–months over ten months of calendar time means two people, rounded up).

First, he created a list of the various roles that he needed to fill. He thought he would need a couple of analysts to work with the analysis and design of the system as well as an infrastructure analyst to manage the integration of the Internet sales system with CD Selections' existing technical environment. Alec also needed people who had good programmer skills and who could be responsible for ultimately implementing the system. Anne and Brian are two analysts with strong technical and interpersonal skills (although Anne is less balanced, having greater technical than interpersonal abilities), and Alec believed that they were available to bring onto this project. He wasn't certain if they had experience with the actual Web technology that would be used on the project, but he decided to rely on vendor training or an external consultant to build those skills when they were needed. The project was so small that Alec envisioned all the team members reporting to him because he would be serving as the project's manager.

Alec created a staffing plan that captured this information, and he included a special incentive structure in the plan (Figure 3–23). Meeting the holiday deadline was very

Role	Description	Assigned To
Project Manager	Oversees the project to ensure that it meets its objectives in time and within budget.	Alec
Infrastructure Analyst	Ensures the system conforms to infrastructure standards at CD Selections. Ensures that the CD Selections infrastructure can support the new system.	Anne
Systems Analyst	Designs the information system—with a focus on interfaces with the distribution system.	Anne
Systems Analyst	Designs the information system—with a focus on the data models and system performance.	Brian
Programmer	Codes system.	Anne

Reporting Structure: All project team members will report to Alec.

Special Incentives: If the deadline for the project is met, all team members who contributed to this goal will receive a free day off to be taken over the holiday season.

FIGURE 3-23

Staffing Plan for the Internet Sales System

important to the project's success, so he decided to offer a day off to the team members who contributed to meeting that date. He hoped that this incentive would motivate the team to work very hard. Alec also planned to budget money for pizza and sodas for times when the team worked long hours.

Before he left for the day, Alec drafted a project charter, to be fine-tuned after the team got together for its *kickoff meeting* (i.e., the first time the project team gets together). The charter listed several norms that Alec wanted to put in place from the start to eliminate any misunderstanding or problems that could come up otherwise (Figure 3–24).

Coordinating Project Activities

Alec wanted the Internet sales system project to be well coordinated, so he immediately put several practices in place to support his responsibilities. First, he acquired the CASE tool used at CD Selections and set up the product so that it could be used for the analysis tasks (e.g., drawing the functional, structural, and behavioral models). The team members would probably start creating diagrams and defining components of the system fairly early on. He pulled out some standards that he has used on all development projects and made a note to review them with his project team at the kickoff meeting for the system. He also had his assistant set up binders for the project deliverables that would start rolling in. Already, he was able to include the system request, feasibility analysis, initial workplan, staffing plan, project charter, standards list, and risk assessment.

Project objective: The Internet order system project team will create a working Web-based system to sell CDs to CD Selections' customers in time for the holiday season.

The Internet order system team members will:

1. Attend a staff meeting each Friday at 2 PM. to report on the status of assigned tasks.
2. Update the workplan with actual data each Friday by 5 PM.
3. Discuss all problems with Alec as soon as they are detected.
4. Agree to support each other when help is needed, especially for tasks that could hold back the progress of the project.
5. Post important changes to the project on the team bulletin board as they are made.

FIGURE 3-24

Project Charter for the Internet Order System

SUMMARY

Project Management

Project management is the second major component of planning of the SDLC, and it includes four steps: identifying the project size, creating and managing the workplan, staffing the project, and coordinating project activities. Project management is important in ensuring that a system is delivered on time, within budget, and with the desired functionality.

Identifying Project Size

The project manager estimates the amount of time and effort that will be needed to complete the project. First, the size is estimated by relying on past experiences or industry standards or by calculating the function points, a measure of program size based on the number and complexity of inputs, outputs, queries, files, and program interfaces. Next, the project manager calculates the effort for the project, which is a function of size and production rates. Algorithms such as the COCOMO model can be used to determine the effort value. Third, the optimal schedule for the project is estimated.

Creating And Managing the Workplan

Once a project manager has a general idea of the size and approximate schedule for the project, he or she creates a workplan, which is a dynamic schedule that records and keeps track of all the tasks that need to be accomplished over the course of the project. To create a workplan, the project manager first identifies the work breakdown structure, or the tasks that need to be accomplished, and then he or she determines how long the tasks will take. Important information about each task is entered into a workplan.

The workplan information can be presented graphically using Gantt and PERT charts. In the Gantt chart, horizontal bars are drawn to represent the duration of each task, and as people work on tasks, the appropriate bars are filled in proportionately to how much of the task is finished. PERT charts are the best way to communicate task dependencies because they lay out the tasks as a flowchart in the order in which they need to be completed. The longest path from the project inception to completion is referred to as the critical path.

Estimating what an IS development project will cost, how long it will take, and what the final system will actually do follows a hurricane model. The estimates become more accurate as the project progresses. One threat to the reliability of the estimates is scope creep, which occurs when new requirements are added to the project after the original project scope was defined and "frozen." If the final schedule will not result in delivery of the system in a timely fashion, timeboxing can be used. Timeboxing sets a fixed deadline for a project and delivers the system by that deadline no matter what, even if functionality must be reduced.

Evolutionary work breakdown structures and iterative workplans better fit the typical methodologies associated with object-oriented systems development. They allow the project manager to provide more realistic estimates for each iteration, or build, of a system. Furthermore, they allow the workplan to be decoupled from the architecture of the system, thus allowing projects to be comparable. By supporting comparability among projects, evolutionary WBSs enable organizational learning to take place.

Staffing the Project

Staffing involves determining how many people should be assigned to the project, assigning project roles to team members, developing a reporting structure for the team, and matching people's skills with the needs of the project. Staffing also includes motivating the team to meet the project's objectives and minimizing conflict among team members. Both motivation and cohesiveness have been found to greatly influence performance of team members in project situations. Team members are motivated most by such nonmonetary things as recognition, achievement, and the work itself. Conflict can be minimized by clearly defining the roles on a project and holding team members accountable for their tasks. Some managers create a project charter that lists the project's norms and ground rules.

Coordinating Project Activities

Coordinating project activities includes putting efficient development practices in place and mitigating risk, and these activities occur over the course of the entire SDLC. Three techniques are available to help coordinate activities on a project: CASE, standards, and documentation. CASE is a category of software that automates all or part of the development process; standards are formal rules or guidelines that project teams must follow during the project; and documentation includes detailed information about the tasks of the SDLC. Often, documentation is stored in project binders that contain all the deliverables and all the internal communication that takes place—the history of the project. A risk assessment is used to mitigate risk because it identifies potential risks and evaluates the likelihood of risk and its potential impact on the project.

KEY TERMS

Adjusted project complexity (APC)
Complexity
Computer-aided software
   engineering (CASE)
CASE repository
COCOMO model
Critical path method
Critical task
Documentation
Effort
Estimation
Evolutionary WBS
Function point
Function point approach
Functional lead
Gantt chart
Group cohesiveness

Hurricane model
Integrated CASE
Iterative workplan
Interpersonal skills
Kickoff meeting
Lower CASE
Methodology
Milestone
Motivation
Node
PERT Chart
Project binder
Project charter
Project management
Project management software
Project manager
Reporting structure

Risk assessment
Risk management
Scope creep
Staffing plan
Standards
Task dependency
Technical lead
Technical skills
Timeboxing
Total adjusted function
   points (TAFP)
Total unadjusted function
   points (TUFP)
Trade-offs
Upper CASE
Work breakdown structure (WBS)
Workplan

QUESTIONS

1. What are the trade-offs that project managers must manage?
2. Why do many projects end up having unreasonable deadlines? How should a project manager react to unreasonable demands?
3. What is a function point and how is it used?
4. Describe the three steps of the function point approach.

5. Name two ways to identify the tasks that need to be accomplished over the course of a project.
6. What are two basic ways to estimate the size of a project?
7. What is the formula for calculating the effort for a project?
8. Compare and contrast the Gantt chart with the PERT chart.

9. What is the difference between a methodology and a workplan? How are the two terms related?
10. What are the best ways to motivate a team? What are the worst ways?
11. What is scope creep, and how can it be managed?
12. What is timeboxing, and why is it used?
13. Describe the hurricane model.
14. What is an evolutionary WBS? How does it address the problems associated with a conventional WBS?
15. What are the problems associated with conventional WBSs?
16. List three techniques to reduce conflict.
17. Describe three technical skills and three interpersonal skills that would be very important to have on any project.
18. Describe the differences between a technical lead and a functional lead. How are they similar?
19. What is an iterative workplan?
20. What is the difference between upper CASE and lower CASE?
21. Create a list of potential risks that could affect the outcome of a project.
22. What belongs in the project binder? How is the project binder organized?
23. Describe three types of standards and provide examples of each.
24. Some companies hire consulting firms to develop the initial project plans and manage the project but use their own analysts and programmers to develop the system. Why do you think some companies do this?

EXERCISES

A. Consider the application that is used at your school to register for classes. Complete a function point worksheet to determine the size of such an application. You will need to make some assumptions about the application's interfaces and the various factors that affect its complexity.
B. Visit a project management Web site, such as the Project Management Institute (www.pmi.org). Most have links to project management software products, white papers, and research. Examine some of the links for project management to better understand a variety of Internet sites that contain information related to this chapter.
C. Pretend that the career services office at your university wants to develop a system that collects student résumés and makes them available to students and recruiters over the Web. Students should be able to input their résumé information into a standard résumé template. The information then is presented in a résumé format, and it also is placed in a database that can be queried using an online search form. You have been placed in charge of the project. Develop a plan for estimating the project. How long do you think it would take for you and three other students to complete the project? Provide support for the schedule that you propose.
D. Refer to the situation in exercise C. You have been told that recruiting season begins a month from today and that the new system must be used. How would you approach this situation? Describe what you can do as the project manager to make sure that your team does not burn out from unreasonable deadlines and commitments.
E. Suppose that you are in charge of the project that is described in exercise C and the project will be staffed by members of your class. Do your classmates have all the right skills to implement such a project? If not, how will you go about making sure that the proper skills are available to get the job done?
F. Consider the system described in exercise C. Create a workplan listing the tasks that will need to be completed to meet the project's objectives. Create a Gantt chart and a PERT chart in a project management tool (e.g., Microsoft Project) or using a spreadsheet package to graphically show the high level tasks of the project.
G. Select a specific project management topic such as CASE, project management software, or timeboxing and search for information on that topic using the Web. The URL listed in exercise A or any search engine (e.g., Yahoo!, AltaVista, Excite, InfoSeek) can provide a starting point for your efforts.
H. Read Your Turn 3-1. Create a risk assessment that lists the potential risks associated with performing the project, along with ways to address the risks.
I. Select two project management software packages and research them using the Web or trade magazines. Describe the features of the two packages. If you were a project manager, which one would you use to help support your job? Why?

J. Pretend that your instructor has asked you and two friends to create a Web page to describe the course to potential students and provide current class information (e.g., syllabus, assignments, readings) to current students. You have been assigned the role of leader, so you will need to coordinate your activities and those of your classmates until the project is completed. Describe how you would apply the project management techniques that you have learned in this chapter in this situation. Include descriptions of how you would create a workplan, staff the project, and coordinate all activities—yours and those of your classmates.

K. In 1997, Oxford Health Plans had a computer problem that caused the company to overestimate revenue and underestimate medical costs. Problems were caused by the migration of its claims processing system from the Pick operating system to a UNIX-based system that uses Oracle database software and hardware from Pyramid Technology. As a result, Oxford's stock price plummeted, and fixing the system became the number-one priority for the company. Suppose that you have been placed in charge of managing the repair of the claims processing system. Obviously, the project team will not be in good spirits. How will you motivate team members to meet the project's objectives?

L. Select two estimation software packages and research them using the Web or trade magazines. Describe the features of the two packages. If you were a project manager, which one would you use to help support your job? Why?

MINICASES

1. Emily Pemberton is an IS project manager facing a difficult situation. Emily works for the First Trust Bank, which has recently acquired the City National Bank. Prior to the acquisition, First Trust and City National were bitter rivals, fiercely competing for market share in the region. Following the acrimonious takeover, numerous staff were laid off in many banking areas, including IS. Key individuals were retained from both banks' IS areas, however, and were assigned to a new consolidated IS department. Emily has been made project manager for the first significant IS project since the takeover, and she faces the task of integrating staffers from both banks on her team. The project they are undertaking will be highly visible within the organization, and the time frame for the project is somewhat demanding. Emily believes that the team can meet the project goals successfully, but success will require that the team become cohesive quickly and that potential conflicts be avoided. What strategies do you suggest that Emily implement in order to help assure a successfully functioning project team?

2. HCL Technologies is a leading software outsourcing firm in India. It has been asked to bid on a small project by one of its US clients. The project will require HCL to write a custom production scheduling component to the large Enterprise Resource Planning (ERP) system that the client uses. Based on preliminary discussions with the client, HCL has developed this list of system elements:

Inputs: 1, low complexity; 2, medium complexity; 3, high complexity

Outputs: 4, medium complexity

Queries: 1, low complexity; 4, medium complexity; 4, high complexity

Files: 2, medium complexity

Program Interfaces: 3, medium complexity

Assume that an adjusted program complexity of 1.3 is appropriate for this project. Calculate the total adjusted function points for this project.

CHAPTER 4

REQUIREMENTS ANALYSIS

One of the first activities of an analyst is to determine the business requirements for a new system. This chapter begins by presenting the requirements definition, a document that lists the new system's capabilities. It then describes how to analyze requirements using business process automation, business process improvement, and business process reengineering techniques and how to gather requirements using interviews, JAD sessions, questionnaires, document analysis, and observation.

INTRODUCTION

SDLC is the process by which an organization moves from the current system (often called the *as-is system*) to the new system (often called the *to-be system*). The output of planning, discussed in Chapters 3 and 4, is the system request, which provides general ideas for the to-be system, defines the project's scope, and provides the initial workplan. The analysis phase takes the general ideas in the system request and refines them into a detailed requirements definition (this chapter), functional models (Chapter 5), structural models (Chapter 6), and behavioral models (Chapter 7) that together form the *system proposal*. The system proposal also includes revised project management deliverables, such as the feasibility analysis (Chapter 2) and the workplan (Chapter 3).

The system proposal is presented to the approval committee, who decides if the project is to continue. This usually happens at a system *walkthrough*, a meeting at which the concept for the new system is presented to the users, managers, and key decision makers. The goal of the walkthrough is to explain the system in moderate detail so that the users, managers, and key decision makers clearly understand it, can identify needed improvements, and are able to make a decision about whether the project should continue. If approved, the system proposal moves into the design phase, and its elements (requirements definition, functional, structural, and behavioral models) are used as inputs to the steps in design. This further refines them and defines in much more detail how the system will be built.

The line between analysis and design is very blurry. This is because the deliverables created during analysis are really the first step in the design of the new system. Many of the major design decisions for the new system are found in the analysis deliverables. In fact, a better name for analysis is really analysis and initial design, but because this is a rather long name and because most organizations simply call it analysis, we do too. Nonetheless, it is important to remember that the deliverables from analysis are really the first step in the design of the new system.

In many ways, the requirements determination step is the single most critical step of the entire SDLC because it is here that the major elements of the system first begin to emerge. During requirements determination, the system is easy to change because little work has been done yet. As the system moves through the other phases in the SDLC, it becomes harder and harder to return to requirements determination and to make major changes because of all of the rework that is involved. Several studies have shown that more than half of all system failures are due to problems with the requirements.[1] This is why the iterative approaches of many object-oriented methodologies are so effective—small batches of requirements can be identified and implemented in incremental stages, allowing the overall system to evolve over time. In this chapter, we focus on the requirements determination step of analysis. We begin by explaining what a requirement is and the overall process of requirements gathering and requirements analysis. We then present a set of techniques that can be used to analyze and gather requirements.

REQUIREMENTS DETERMINATION

The purpose of the *requirements determination* step is to turn the very high-level explanation of the business requirements stated in the system request into a more precise list of requirements that can be used as inputs to the rest of analysis (creating functional, structural, and behavioral models). This expansion of the requirements ultimately leads to the design phase.

Defining a Requirement

A *requirement* is simply a statement of what the system must do or what characteristic it must have. During analysis, requirements are written from the perspective of the businessperson, and they focus on the "what" of the system. Because they focus on the needs of the business user, they are usually called *business requirements* (and sometimes user

---

[1] For example, see *The Scope of Software Development Project Failures* (Dennis, MA: The Standish Group, 1995).

requirements). Later in design, business requirements evolve to become more technical, and they describe how the system will be implemented. Requirements in design are written from the developer's perspective, and they are usually called *system requirements*.

Before we continue, we want to stress that there is no black-and-white line dividing a business requirement and a system requirement—and some companies use the terms interchangeably. The important thing to remember is that a requirement is a statement of what the system must do, and requirements will change over time as the project moves from analysis to design to implementation. Requirements evolve from detailed statements of the business capabilities that a system should have to detailed statements of the technical way in which the capabilities will be implemented in the new system.

Requirements can be either functional or nonfunctional in nature. A *functional require-ment* relates directly to a process a system has to perform or information it needs to contain. For example, requirements that state that a system must have the ability to search for avail-able inventory or to report actual and budgeted expenses are functional requirements. Func-tional requirements flow directly into the next steps of analysis (functional, structural, and behavioral models) because they define the functions that the system must have.

Nonfunctional requirements refer to behavioral properties that the system must have, such as performance and usability. The ability to access the system using a Web browser is considered a nonfunctional requirement. Nonfunctional requirements may influence the rest of analysis (functional, structural, and behavioral models) but often do so only indirectly; nonfunctional requirements are used primarily in design when decisions are made about the user interface, the hardware and software, and the system's underlying physical architecture.

Figure 4-1 lists different kinds of nonfunctional requirements and examples of each kind. Notice that the nonfunctional requirements describe a variety of characteristics regarding the system: operational, performance, security, and cultural and political. For example, the project team needs to know if a system must be highly secure, requires sub-second response time, or has to reach a multicultural customer base.

These characteristics do not describe business processes or information, but they are very important in understanding what the final system should be like. Nonfunctional requirements primarily impact decisions that will be made during the design of a system. As such, we will return to this topic later in the book when we discuss design. The goal in this chapter is to identify any major issues.

Four topics that have recently influenced information system requirements are the Sarbanes-Oxley Act, COBIT compliance, ISO 9000 compliance, and Capability Maturity Model compliance. Depending on the system being considered, these four topics could impact the definition of a system's functional requirements, nonfunctional requirements, or both. The Sarbanes-Oxley Act, for example, mandates additional functional and non-functional requirements. These include additional security concerns (nonfunctional) and specific information requirements that management must now provide (functional). As such, when developing financial information systems, information system developers should be sure to include Sarbanes-Oxley expertise within the development team. In another exam-ple, a client could insist on COBIT compliance, ISO 9000 compliance, or that a specific Capability Maturity Model level had been reached for the firm to be considered as a possi-ble vendor to supply the system under consideration. Obviously, these types of require-ments add to the nonfunctional requirements. However, further discussion of these topics is beyond the scope of this book.[2]

---

[2] A concise discussion of the Sarbanes-Oxley Act is presented in G. P. Lander, *What is Sarbanes-Oxley?* (New York: McGraw-Hill, 2004). A good reference for Sarbanes-Oxley Act–based security requirements is D. C. Brewer, *Secu-rity Controls for Sarbanes-Oxley Section 404 IT Compliance: Authorization, Authentication, and Access* (Indianapolis, IN: Wiley, 2006). For detailed information on COBIT, see www.isaca.org; for ISO 9000, see www.iso.org; and for details on the Capability Maturity Model, see www.sei.cmu.edu/cmmi/.

Nonfunctional Requirement	Description	Examples
Operational	The physical and technical environments in which the system will operate	■ The system should be able to fit in a pocket or purse. ■ The system should be able to integrate with the existing inventory system. ■ The system should be able to work on any Web browser.
Performance	The speed, capacity, and reliability of the system	■ Any interaction between the user and the system should not exceed 2 seconds. ■ The system should receive updated inventory information every 15 minutes. ■ The system should be available for use 24 hours per day, 365 days per year.
Security	Who has authorized access to the system under what circumstances	■ Only direct managers can see personnel records of staff. ■ Customers can see their order history only during business hours.
Cultural and political	Cultural, political factors and legal requirements that affect the system	■ The system should be able to distinguish between United States and European currency. ■ Company policy says that we buy computers only from Dell. ■ Country managers are permitted to authorize customer user interfaces within their units. ■ The system shall comply with insurance industry standards.

Source: The Atlantic Systems Guild, http://www.systemsguild.com/GuildSite/Robs/Template.html

FIGURE 4-1   Nonfunctional Requirements

Another recent topic that influences requirements for some systems is the whole area of globalization. The idea of having a global information supply chain brings to bear a large number of additional nonfunctional requirements. For example, if the necessary operational environments do not exist for a mobile solution to be developed, it is important to adapt the solution to the local environment. Or, it may not be reasonable to expect to deploy a high-technology-based solution in an area that does not have the necessary power and communications infrastructure. In some cases, we may need to consider some parts of the global information supply chain to be supported with manual—rather than automated—information systems. Manual systems have an entirely different set of requirements that create different performance expectations and additional security concerns. Furthermore, cultural and political concerns are potentially paramount. A simple example that impacts the design of user interfaces is the proper use of color on forms (on a screen or paper). Different cultures interpret different colors differently. In other words, in a global, multicultural business environment, addressing cultural concerns goes well beyond simply having a multilingual user interface. As such, we must be able to adapt the global solution to the local realities. Friedman refers to these concerns as glocalization.[3] Otherwise, we will simply create another example of a failed information system development project.

[3] T. L. Friedman, *The World is Flat: A Brief History of the Twenty-First Century, Updated and Expanded Ed.* (New York: Farrar, Straus, and Giroux, 2006). For a criticism of Friedman's view, see R. Aronica and M. Ramdoo, *The World is FLAT? A Critical Analysis of Thomas L. Friedman's New York Times Bestseller* (Tampa, FL: Meghan-Kiffer Press, 2006).

YOUR TURN

4-1 Identifying Requirements

One of the most common mistakes by new analysts is to confuse functional and nonfunctional requirements. Pretend that you received the following list of requirements for a sales system.

Requirements for Proposed System
The system should

1. be accessible to Web users;
2. include the company standard logo and color scheme;
3. restrict access to profitability information;
4. include actual and budgeted cost information;
5. provide management reports;
6. include sales information that is updated at least daily;
7. have two-second maximum response time for pre-defined queries and ten-minute maximum response time for ad hoc queries;
8. include information from all company subsidiaries;
9. print subsidiary reports in the primary language of the subsidiary;
10. provide monthly rankings of salesperson performance.

Questions

1. Which requirements are functional business requirements? Provide two additional examples.
2. Which requirements are nonfunctional business requirements? What kind of nonfunctional requirements are they? Provide two additional examples.

Requirements Definition

The requirements definition report—usually just called the *requirements definition*—is a straightforward text report that simply lists the functional and nonfunctional requirements in an outline format. Figure 4-2 shows a sample requirements definition for a word processing program designed to compete against software such as Microsoft Word.

The requirements are numbered in a legal or outline format so that each requirement is clearly identified. The requirements are first grouped into functional and nonfunctional requirements; within each of those headings, they are further grouped by the type of nonfunctional requirement or by function.

CONCEPTS IN ACTION

4-A What Can Happen If You Ignore Nonfunctional Requirements

I once worked on a consulting project in which my manager created a requirements definition without listing nonfunctional requirements. The project was then estimated based on the requirements definition and sold to the client for $5,000. In my manager's mind, the system that we would build for the client would be a very simple stand-alone system running on current technology. It shouldn't take more than a week to analyze, design, and build.

Unfortunately, the clients had other ideas. They wanted the system to be used by many people in three different departments, and they wanted the ability for any number of people to work on the system concurrently. The technology they had in place was antiquated; nonetheless, they wanted the system to run effectively on the existing equipment. Because we didn't set the project scope properly by including our assumptions about nonfunctional requirements in the requirements definition, we basically had to do whatever they wanted.

The capabilities they wanted took weeks to design and program. The project ended up taking four months, and the final project cost was $250,000. Our company had to pick up the tab for everything except the agreed-upon $5,000. This was by far the most frustrating project situation I ever experienced.

Barbara Wixom

D. Nonfunctional Requirements

1. Operational Requirements
    1.1. The system will operate in Windows and Macintosh environments.
    1.2. The system will be able to read and write Word documents, RTF, and HTML.
    1.3. The system will be able to import Gif, Jpeg, and BMP graphics files.

2. Performance Requirements
    2.1. No special performance requirements are anticipated.

3. Security Requirements
    3.1. No special security requirements are anticipated.

4. Cultural and Political Requirements
    4.1. No special cultural and political requirements are anticipated.

C. Functional Requirements

1. Printing
    1.1. The user can select which pages to print.
    1.2. The user can view a preview of the pages before printing.
    1.3. The user can change the margins, paper size (e.g., letter, A4) and orientation
        on the page.

2. Spell Checking
    2.1. The user can check for spelling mistakes; the system can operate in one
        of two modes as selected by the users.
    2.1.1. Mode 1 (Manual): The user will activate the spell checker and it will move the
        user to the next misspelled word.
    2.1.2. Mode 2 (Automatic): As the user types, the spell checker will flag misspelled
        words so the user immediately see the misspelling.
    2.2. The user can add words to the dictionary.
    2.3. The user can mark words as not misspelled but not add them to the dictionary.

FIGURE 4-2
Sample Requirements
Definition

Sometimes, business requirements are prioritized on the requirements definition. They can be ranked as having high, medium, or low importance in the new system, or they can be labeled with the version of the system that will address the requirement (e.g., release 1, release 2, release 3). This practice is particularly important when using object-oriented methodologies because they deliver requirements in batches by developing incremental versions of the system.

The most obvious purpose of the requirements definition is to provide the information needed by the other deliverables in analysis, which include functional, structural, and behavioral models, and to support activities in the design phase. The most important purpose of the requirements definition, however, is to define the scope of the system. The document describes to the analysts exactly what the system needs to end up doing. When discrepancies arise, the document serves as the place to go for clarification.

Determining Requirements

Determining requirements for the requirements definition is both a business task and an information technology task. In the early days of computing, there was a presumption that the systems analysts, as experts with computer systems, were in the best position to define how a computer system should operate. Many systems failed because they did not adequately address the true business needs of the users. Gradually, the presumption changed so that the users, as the business experts, were seen as being the best position to define how a computer system should operate. However, many systems failed to deliver performance benefits because users simply automated an existing inefficient system, and they failed to incorporate new opportunities offered by technology.

A good analogy is building a house or an apartment. We have all lived in a house or apartment, and most of us have some understanding of what we would like to see in one. However, if we were asked to design one from scratch, it would be a challenge because we lack appropriate design skills and technical engineering skills. Likewise, an architect acting alone would probably miss some of our unique requirements.

Therefore, the most effective approach is to have both business people and analysts working together to determine business requirements. Sometimes, however, users don't know exactly what they want, and analysts need to help them discover their needs. Three kinds of techniques have become popular to help analysts do this: *business process automation (BPA)*, *business process improvement (BPI)*, and *business process reengineering (BPR)*. Analysts can use these tools when they need to guide the users in explaining what is wanted from a system.

The three kinds of techniques work similarly. They help users critically examine the current state of systems and processes (the as-is system), identify exactly what needs to change, and develop a concept for a new system (the to-be system). A different amount of change is associated with each technique; BPA creates a small amount of change, BPI creates a moderate amount of change, and BPR creates significant change that impacts much of the organization. All three are described in greater detail later in the chapter.

Although BPA, BPI, and BPR enable the analyst to help users create a vision for the new system, they are not sufficient for extracting information about the detailed business requirements that are needed to build it. Therefore, analysts use a portfolio of requirements-gathering techniques to acquire information from users. The analyst has many gathering techniques from which to choose: *interviews*, *questionnaires*, *observation*, *JAD*, *(joint application development)* and *document analysis*. The information gathered using these techniques is critically analyzed and used to craft the requirements definition report. The final section of this chapter describes each of the requirements-gathering techniques in greater depth.

Creating a Requirements Definition

Creating a requirements definition is an iterative and ongoing process whereby the analyst collects information with requirements-gathering techniques (e.g., interviews, document analysis), critically analyzes the information to identify appropriate business requirements for the system, and adds the requirements to the requirements definition report. The requirements definition is kept up to date so that the project team and business users can refer to it and get a clear understanding of the new system.

To create a requirements definition, the project team first determines the kinds of functional and nonfunctional requirements that they will collect about the system (of course, these may change over time). These become the main sections of the document. Next, the analysts use a variety of requirements-gathering techniques (e.g., interviews, observation) to collect information, and they list the business requirements that were identified from that information. Finally, the analysts work with the entire project team and the business users to verify, change, and complete the list and to help prioritize the importance of the requirements that were identified.

This process continues throughout analysis, and the requirements definition evolves over time as new requirements are identified and as the project moves into later phases of the SDLC. Beware: the evolution of the requirements definition must be carefully managed. The project team cannot keep adding to the requirements definition, or the system will keep growing and growing and never get finished. Instead, the project team carefully identifies requirements and evaluates which ones fit within the scope of the system. When a requirement reflects a real business need but is not within the scope of the current system or current release, it is either added on a list of future requirements or given a low priority. The management of requirements (and system scope) is one of the hardest parts of managing a project.

REQUIREMENTS ANALYSIS STRATEGIES

Before the project team can determine what requirements are appropriate for a given system, they need to have a clear vision of the kind of system that will be created and the level of change that it will bring to the organization. The basic process of *analysis* is divided into three steps: understanding the as-is system, identifying improvements, and developing requirements for the to-be system.

Sometimes the first step (i.e., understanding the as-is system) is skipped or done in a cursory manner. This happens when no current system exists, if the existing system and processes are irrelevant to the future system, or if the project team is using a RAD or agile development methodology in which the as-is system is not emphasized. Users of traditional design methods such as waterfall and parallel development (see Chapter 1) typically spend significant time understanding the as-is system and identifying improvements before moving to capture requirements for the to-be system. However, newer RAD, agile, and object-oriented methodologies, such as phased development, prototyping, throwaway prototyping, and extreme programming (see Chapters 1 and 2) focus almost exclusively on improvements and the to-be system requirements, and they spend little time investigating the current as-is system.

Three requirements analysis strategies—business process automation, business process improvement, and business process reengineering—help the analyst lead users through the three (or two) analysis steps so that the vision of the system can be developed. Requirements analysis strategies and requirements-gathering techniques go hand-in-hand. Analysts need to use requirements-gathering techniques to collect information; requirements analysis strategies drive the kind of information that is gathered and how it is ultimately analyzed. Although we now focus on the analysis strategies and then discuss requirements gathering at the end of the chapter, they happen concurrently and are complementary activities.

The choice of analysis technique to be used is based on the amount of change the system is meant to create in the organization. BPA is based on small change that improves process efficiency, BPI creates process improvements that lead to better effectiveness, and BPR revamps the way things work so that the organization is transformed on some level.

To move the users "from here to there," an analyst needs strong *critical thinking skills*. Critical thinking is the ability to recognize strengths and weaknesses and recast an idea in an improved form, and critical thinking skills are needed to really understand issues and develop new business processes. These skills are also needed to thoroughly examine the results of requirements gathering, to identify business requirements, and to translate those requirements into a concept for the new system.

Business Process Automation

BPA leaves the basic way in which the organization operates unchanged and uses computer technology to do some of the work. BPA can make the organization more efficient but has the least impact on the business. Planners in BPA projects spend a significant time understanding the current as-is system before moving on to improvements and to-be system requirements. Problem analysis and root cause analysis are two popular BPA techniques.

Problem Analysis   The most straightforward (and probably the most commonly used) requirements analysis technique is *problem analysis*. Problem analysis means asking the users and managers to identify problems with the as-is system and to describe how to solve them in the to-be system. Most users have a very good idea of the changes they would like to see, and most will be quite vocal about suggesting them. Most changes tend to solve problems rather than capitalize on opportunities, but the latter is possible as well. Improvements from problem analysis tend to be small and incremental (e.g., provide more space in which to type the customer's address; provide a new report that currently does not exist).

This type of improvement often is very effective at improving a system's efficiency or ease of use. However, it often provides only minor improvements in business value—the new system is better than the old, but it may be hard to identify significant monetary benefits from the new system.

Root Cause Analysis   The ideas produced by problem analysis tend to be solutions to problems. All solutions make assumptions about the nature of the problem, assumptions that may or may not be valid. In our experience, users (and most people in general) tend to quickly jump to solutions without fully considering the nature of the problem. Sometimes the solutions are appropriate, but many times they address a *symptom* of the problem, not the true problem or *root cause* itself.[4]

For example, suppose we notice that a lightbulb is burned out above our front door. We buy a new bulb, get out a ladder, and replace the bulb. A month later, we see that the

---

CONCEPTS   **4-B Success from Failure**

IN ACTION

*I*n the gold rush days of the late 1990s, getting on the Internet was a hot topic. There were many companies (many of which no longer exist) that created computers for the home Internet market (many with built-in dial-up connectivity and contracts for that connectivity. The AtHome company made such an Internet appliance. Taken out of the box, connected to a phone line, and provided with initial start-up, it "phoned home" (made the connection to an Internet service provider).

But, the days of the Internet appliance were short lived. Consumers wanted more than just Internet access—they wanted to be able to have and share files, photos, and materials with others. The basic AtHome Internet Appliance did not have any storage and was useful only for connecting to the Internet (by phone modem) and browsing the Internet.

The stock price dropped and sales dropped; even with prices that were comparable to giving away the device, consumers were not longer interested as 2001 came to an end.

When faced with such a situation, what does a company do? The company faced a real challenge—go out of business or reorganize. In this case AtHome, which had expertise in hardware and telecommunications, restructured into a security company. With September 11, 2001, bringing calls for better

security, AtHome scrambled to create a hardware device that would sit between the Internet connection and a business network. To keep their stock listed on the New York Stock Exchange, AtHome did a 15-to-1 reverse stock split and changed their name to indicate a new focus. Having been burned by consumers' whims, they set their goals on capturing major corporate business. It took two long years before their device started to be noticed—and just paying the employees almost ate up the available funds before sales of the new device started to kick in. The reorganized company is now recognized as a leader in the intrusion-prevention field.

A company that falls from consumer favor cannot always restructure itself to become successful in an alternative area. In this case, there was success from failure.

Questions

1. When should a company that has lost in the consumer marketplace re-create itself for the corporate market?

2. How might a systems analyst for the AtHome company learn to change with the times and adapt to the new environment?

---

[4] Two good books that discuss the difficulty in finding the root causes to problems are: E. M. Goldratt and J. Cox, *The Goal* (Croton-on-Hudson, NY: North River Press, 1986); and E. M. Goldratt, *The Haystack Syndrome* (Croton-on-Hudson, NY: North River Press, 1990).

same bulb is burned out, so we buy a new bulb, haul out the ladder, and replace it again. This happens several times. At this point, we have two choices. We can buy a large package of lightbulbs and a fancy lightbulb changer on a long pole so we don't need to haul the ladder out each time (thus saving a lot of trips to the store for new bulbs and a lot of effort in working with the ladder). Or, we can fix the light fixture that is causing the light to burn out in the first place. Buying the bulb changer is treating the symptom (the burned-out bulb), whereas fixing the fixture is treating the root cause.

In the business world, the challenge lies in identifying the root cause—few problems are as simple as the lightbulb problem. The solutions that users propose (or systems that analysts think of) may either address symptoms or root causes, but without a careful analysis, it is difficult to tell which one is addressed. And finding out that we just spent a million dollars on a new lightbulb changer is a horrible feeling!

Root cause analysis, therefore, focuses on problems, not solutions. The analyst starts by having the users generate a list of problems with the current system and then prioritize the problems in order of importance. Starting with the most important, the users and/or the analysts then generate all the possible root causes for the problems. Each possible root cause is investigated (starting with the most likely or easiest to check) until the true root causes are identified. If any possible root causes are identified for several problems, those should be investigated first, because there is a good chance they are the real root causes influencing the symptom problems.

In our lightbulb example, there are several possible root causes. A decision tree sometimes helps with the analysis. As Figure 4-3 shows, there are many possible root causes, so buying a new fixture may or may not address the true root cause. In fact, buying a lightbulb

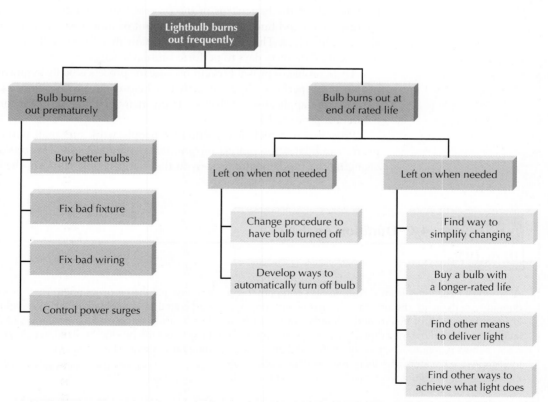

FIGURE 4-3   Root-Cause Analysis for the Example of the Burned-out Lightbulb

changer may actually address the root cause. The key point in root cause analysis is always to challenge the obvious.

Business Process Improvement

BPI makes moderate changes to the way in which the organization operates to take advantage of new opportunities offered by technology or to copy what competitors are doing. BPI can improve efficiency (i.e., doing things right) and improve effectiveness (i.e., doing the right things). Planners of BPI projects also spend time understanding the as-is system, but much less time than with BPA projects; their primary focus is on improving business processes, so time is spent on the as-is only to help with the improvement analyses and the to-be system requirements. *Duration analysis*, *activity-based costing*, and *informal benchmarking* are three popular BPI activities.

Duration Analysis   Duration analysis requires a detailed examination of the amount of time it takes to perform each process in the current as-is system. The analysts begin by determining the total amount of time it takes, on average, to perform a set of business processes for a typical input. They then time each of the individual steps (or subprocesses) in the business process. The time to complete the basic steps are then totaled and compared to the total for the overall process. A significant difference between the two—and in our experience the total time often can be 10 or even 100 times longer than the sum of the parts—indicates that this part of the process is badly in need of a major overhaul.

For example, suppose that the analysts are working on a home mortgage system and discover that on average, it takes thirty days for the bank to approve a mortgage. They then look at each of the basic steps in the process (e.g., data entry, credit check, title search, appraisal, etc.) and find that the total amount of time actually spent on each mortgage is about eight hours. This is a strong indication that the overall process is badly broken, because it takes thirty days to perform one day's work.

These problems probably occur because the process is badly fragmented. Many different people must perform different activities before the process finishes. In the mortgage example, the application probably sits on many people's desks for long periods of time before it is processed.

Processes in which many different people work on small parts of the inputs are prime candidates for *process integration* or *parallelization*. Process integration means changing the fundamental process so that fewer people work on the input, which often

---

CONCEPTS   **4-C Duration Analysis**

IN ACTION

A group of executives from a Fortune 500 company used duration analysis to discuss their procurement process. Using a huge wall of Velcro and a handful of placards, a facilitator mapped out the company's process for procuring a $50 software upgrade. Having quantified the time it took to complete each step, she then assigned costs based on the salaries of the employees involved. The fifteen-minute exercise left the group stunned. Their procurement process had gotten so convoluted that it took eighteen days, countless hours of paperwork, and nearly $22,000 in employee time to get the product ordered, received, and up and running on the requester's desktop.

Source: "For Good Measure" Debby Young, *CIO Magazine* (March 1, 1999).

requires changing the processes and retraining staff to perform a wider range of duties. Process parallelization means changing the process so that all the individual steps are performed at the same time. For example, in the mortgage application case, there is probably no reason that the credit check cannot be performed at the same time as the appraisal and title check.

Activity-Based Costing   Activity-based costing is a similar analysis, which examines the cost of each major process or step in a business process rather than the time taken.[5] The analysts identify the costs associated with each of the basic functional steps or processes, identify the most costly processes, and focus their improvement efforts on them.

Assigning costs is conceptually simple. Analysts simply examine the direct cost of labor and materials for each input. Materials costs are easily assigned in a manufacturing process, whereas labor costs are usually calculated based on the amount of time spent on the input and the hourly cost of the staff. However, as you may recall from a managerial accounting course, there are indirect costs such as rent, depreciation, and so on, that also can be included in activity costs.

Informal Benchmarking   Benchmarking refers to studying how other organizations perform a business process in order to learn how your organization can do something better. Benchmarking helps the organization by introducing ideas that employees may never have considered but have the potential to add value.

Informal benchmarking is fairly common for "customer-facing" business processes (i.e., processes that interact with the customer). With informal benchmarking, the managers and analysts think about other organizations or visit them as customers to watch how the business process is performed. In many cases, the business studied may be a known leader in the industry or simply a related firm. For example, suppose the team is developing a Web site for a car dealer. The project sponsor, key managers, and key team members would likely visit the Web sites of competitors as well as those of others in the car industry (e.g., manufacturers, accessories suppliers) and those in other industries that have won awards for their Web sites.

Business Process Reengineering

BPR means changing the fundamental way in which the organization operates, "obliterating" the current way of doing business and making major changes to take advantage of new ideas and new technology. Planners of BPR projects spend little time understanding the as-is, because their goal is to focus on new ideas and new ways of doing business. Outcome analysis, technology analysis, and activity elimination are three popular BPR activities.

Outcome Analysis   *Outcome analysis* focuses on understanding the fundamental outcomes that provide value to customers. Although these outcomes sound as though they should be obvious, they often aren't. For example, suppose you consider an insurance company. One of its customers has just had a car accident. What is the fundamental outcome

---

[5] Many books have been written on activity-based costing. Useful ones include K. B. Burk and D. W. Webster, *Activity-Based Costing* (Fairfax, VA: American Management Systems, 1994); and D. T. Hicks, *Activity-Based Costing: Making It Work for Small and Mid-sized Companies* (New York: Wiley, 1998). The two books by Eli Goldratt mentioned previously (*The Goal* and *The Haystack Syndrome*) also offer unique insights into costing.

from the *customer's* perspective? Traditionally, insurance companies have answered this question by assuming the customer wants to receive the insurance payment quickly. To the customer, however, the payment is only a *means* to the real outcome: a repaired car. The insurance company might benefit by extending its view of the business process past its traditional boundaries to include not paying for repairs, but performing the repairs or contracting with an authorized body shop to do them.

With this approach, system analysts encourage the managers and project sponsor to pretend they are customers and to think carefully about what the organization's products and services enable the customers to do—and what they could enable the customer to do.

Technology Analysis   Many major changes in business over the past decade have been enabled by new technologies. *Technology analysis* starts by having the analysts and managers develop a list of important and interesting technologies. Then the group systematically identifies how each and every technology could be applied to the business process and identifies how the business would benefit.

For example, one useful technology is the Internet. Saturn, the car manufacturer, took this idea and developed an extranet application for its suppliers. Rather than ordering parts for its cars, Saturn makes its production schedule available electronically to its suppliers, who ship the parts Saturn needs so that they arrive at the plant just in time. This saves Saturn significant costs because it eliminates the need for people to monitor the production schedule and issue purchase orders.

Activity Elimination   *Activity elimination* is exactly what it sounds like. The analysts and managers work together to identify how the organization could eliminate each and every activity in the business process, how the function could operate without it, and what effects are likely to occur. Initially, managers are reluctant to conclude that processes can be eliminated, but this is a "force-fit" exercise in that they must eliminate each activity. In some cases the results are silly; nonetheless, participants must address each and every activity in the business process.

For example, in the home mortgage approval process discussed earlier, the managers and analysts would start by eliminating the first activity, entering the data into the mortgage company's computer. This leads to two obvious possibilities: (1) eliminate the use of a computer system or (2) make someone else do the data entry (e.g., the customer over the Web). They would then eliminate the next activity, the credit check. Silly, right? After all, making sure the applicant has good credit is critical in issuing a loan. Not really. The real answer depends upon how many times the credit check identifies bad applications. If all or almost all applicants have good credit and are seldom turned down by a credit check, then the cost of the credit check may not be worth the cost of the few bad loans it prevents. Eliminating it may actually result in lower costs, even with the cost of bad loans.

Selecting Appropriate Strategies

Each technique discussed in this chapter has its own strengths and weaknesses (see Figure 4-4). No one technique is inherently better than the others, and in practice most projects use a combination of techniques.

Potential Business Value   *Potential business value* varies with the analysis strategy. Although BPA has the potential to improve the business, most of the benefits from BPA are

YOUR TURN

4-2 IBM Credit

*I*BM Credit was a wholly owned subsidiary of IBM responsible for financing mainframe computers sold by IBM. Although some customers bought mainframes outright or obtained financing from other sources, financing computers provided significant additional profit.

When an IBM sales representative made a sale, he or she would immediately call IBM Credit to obtain a financing quote. The call was received by a credit officer, who would record the information on a request form. The form would then be sent to the credit department to check the customer's credit status. This information would be recorded on the form, which was then sent to the business practices department, who would write a contract (sometimes reflecting changes requested by the customer). The form and the contract would then go to the pricing department, which used the credit information to establish an interest rate and recorded it on the form. The form and contract were then sent to the clerical group, where an administrator would prepare a cover letter quoting the interest rate and send the letter and contract via Federal Express to the customer.

The problem at IBM Credit was a major one. Getting a financing quote took anywhere for four to eight days (six days on average), giving the customer time to rethink the order or find financing elsewhere. While the quote was being prepared, sales representatives would often call to find out where the quote was in the process so they could tell the customer when to expect it. However, no one at IBM Credit could answer the question because the paper forms could be in any department, and it was impossible to locate one without physically walking through the departments and going through the piles of forms on everyone's desk.

IBM Credit examined the process and changed it so that each credit request was logged into a computer system and each department could record an application's status as they completed it and sent it to the next department. In this way, sales representatives could call the credit office and quickly learn the status of each application. IBM used some sophisticated management science queuing theory analysis to balance workloads and staff across the different departments so none would be overloaded. They also introduced performance standards for each department (e.g., the pricing decision had to be completed within one day after that department received an application).

However, process times got worse, even though each department was achieving almost 100 percent compliance on its performance goals. After some investigation, managers found that when people got busy, they conveniently found errors that forced them to return credit requests to the previous department for correction, thereby removing it from their time measurements.

Questions

1. What techniques can you use to identify improvements?
2. Choose one technique and apply it to this situation. What improvements did you identify?

Source: M. Hammer and J. Champy, *Reengineering the Corporation* (1993). New York, NY: Harper Business.

tactical and small in nature. Because BPA does not seek to change the business processes, it can only improve their efficiency. BPI usually offers moderate potential benefits, depending upon the scope of the project, because it seeks to change the business in some way. It can increase both efficiency and effectiveness. BPR creates large *potential* benefits because it seeks to radically improve the nature of the business.

FIGURE 4-4
Characteristics of Analysis Strategies

	Business Process Automation	Business Process Improvement	Business Process Reengineering
Potential business value	Low–moderate	Moderate	High
Project cost	Low	Low–moderate	High
Breadth of analysis	Narrow	Narrow–moderate	Very broad
Risk	Low–moderate	Low–moderate	Very high

Project Cost   *Project cost* is always important. In general, BPA has the lowest cost because it has the narrowest focus and seeks to make the fewest number of changes. BPI can be moderately expensive, depending upon the scope of the project. BPR is usually expensive, both because of the amount of time required of senior managers and the amount of redesign to business processes.

Breadth of Analysis   *Breadth of analysis* refers to the scope of analysis, or whether analysis includes business processes within a single business function, processes that cross the organization, or processes that interact with those in customer or supplier organizations. BPR takes a broad perspective, often spanning several major business processes, even across multiple organizations. BPI has a much narrower scope that usually includes one or several business functions. BPA typically examines a single process.

Risk   One final issue is *risk* of failure, which is the likelihood of failure due to poor design, unmet needs, or too much change for the organization to handle. BPA and BPI have low to moderate risk because the to-be system is fairly well defined and understood, and its potential impact on the business can be assessed before it is implemented. BPR projects, on the other hand, are less predictable. BPR is extremely risky and is not something to be undertaken unless the organization and its senior leadership are committed to making significant changes. Mike Hammer, the father of BPR, estimates that 70 percent of BPR projects fail.

YOUR TURN

4-3 Analysis Strategy

Suppose you are the analyst charged with developing a new Web site for a local car dealer who wants to be very innovative and try new things. What analysis strategies would you recommend? Why?

CONCEPTS IN ACTION

4-D Implementing a Satellite Data Network

A major retail store recently spent $24 million on a large private satellite communication system. The system provides state-of-the-art voice, data, and video transmission between stores and regional headquarters. When an item is sold, the scanner software updates the inventory system in real time. As a result, store transactions are passed on to regional and national headquarters instantly, which keeps inventory records up to date. One of their major competitors has an older system, where transactions are uploaded at the end of a business day. The first company feels such instant communication and feedback allows them to react more quickly to changes in the market and gives them a competitive advantage. For example, if an early winter snowstorm causes stores across the upper Midwest to start selling high-end (and high-profit) snowblowers, the nearest warehouse can quite quickly prepare next-day shipments to maintain a good inventory balance, whereas the competitor may not move quite as quickly and thus will lose out on such quick inventory turnover.

Questions

1. Do you think a $24 million investment in a private satellite communication system could be justified by a cost–benefit analysis? Could this be done with a standard communication line (with encryption)?
2. How might the competitor in this example attempt to close the information gap?

REQUIREMENTS-GATHERING TECHNIQUES

An analyst is very much like a detective (and business users are sometimes like elusive suspects). He or she knows that there is a problem to be solved and therefore must look for clues that uncover the solution. Unfortunately, the clues are not always obvious (and often missed), so the analyst needs to notice details, talk with witnesses, and follow leads just as Sherlock Holmes would have done. The best analysts will thoroughly gather requirements using a variety of techniques and make sure that the current business processes and the needs for the new system are well understood before moving into design. Analysts don't want to discover later that they have key requirements wrong—such surprises late in the SDLC can cause all kinds of problems.

The requirements-gathering process is used for building political support for the project and establishing trust and rapport between the project team building the system and the users who ultimately will choose to use or not use the system. Involving someone in the process implies that the project teams view that person as an important resource and value his or her opinions. All the key stakeholders (the people who can affect the system or who will be affected by the system) must be included in the requirements-gathering process. The stakeholders might include managers, employees, staff members, and even some customers and suppliers. If a key person is not involved, that individual may feel slighted, which can cause problems during implementation (e.g., How could they have developed the system without my input?).

The second challenge of requirements gathering is choosing the way(s) in which information is collected. There are many techniques for gathering requirements that vary from asking people questions to watching them work. In this chapter, we focus on the five most commonly used techniques: interviews, JAD sessions (a special type of group meeting), questionnaires, document analysis, and observation. Each technique has its own strengths and weaknesses, many of which are complementary, so most projects use a combination of techniques, probably most often interviews, JAD sessions, and document analysis.[6]

Interviews

An interview is the most commonly used requirements-gathering technique. After all, it is natural— if you need to know something, you usually ask someone. In general, interviews are conducted one-on-one (one interviewer and one interviewee), but sometimes, due to time constraints, several people are interviewed at the same time. There are five basic steps to the interview process: selecting interviewees, designing interview questions, preparing for the interview, conducting the interview, and postinterview follow-up.[7]

Selecting Interviewees   The first step in interviewing is to create an *interview schedule* listing all the people who will be interviewed, when, and for what purpose (see Figure 4-5). The schedule can be an informal list that is used to help set up meeting times or a formal list that is incorporated into the workplan. The people who appear on the interview schedule are selected based on the analyst's information needs. The project sponsor, key business users, and other members of the project team can help the analyst determine who in the organization can best provide important information about requirements. These people are listed on the interview schedule in the order in which they should be interviewed.

---

[6] Some excellent books that address the importance of gathering requirements and various techniques include Alan M. Davis, *Software Requirements: Objects, Functions, & States, Revision* (Englewood Cliffs, NJ: Prentice Hall, 1993); Gerald Kotonya and Ian Sommerville, *Requirements Engineering* (Chichester, England: Wiley, 1998); and Dean Leffingwell and Don Widrig, *Managing Software Requirements: A Unified Approach* (Reading, MA: Addison-Wesley, 2000).

[7] A good book on interviewing is that by Brian James, *The Systems Analysis Interview* (Manchester, England: NCC Blackwell, 1989).

Name	Position	Purpose of Interview	Meeting
Andria McClellan	Director, Accounting	Strategic vision for new accounting system	Mon., March 1 8:00–10:00 AM
Jennifer Draper	Manager, Accounts Receivable	Current problems with accounts receivable process; future goals	Mon., March 1 2:00–3:15 PM
Mark Goodin	Manager, Accounts Payable	Current problems with accounts payable process; future goals	Mon., March 1 4:00–5:15 PM
Anne Asher	Supervisor, Data Entry	Accounts receivable and payable processes	Wed., March 3 10:00–11:00 AM
Fernando Merce	Data Entry Clerk	Accounts receivable and payable processes	Wed., March 3 1:00–3:00 PM

FIGURE 4-5
Sample Interview Schedule

People at different levels of the organization will have different perspectives on the system, so it is important to include both managers who manage the processes and staff who actually perform the processes to gain both high-level and low-level perspectives on an issue. Also, the kinds of interview subjects needed may change over time. For example, at the start of the project, the analyst has a limited understanding of the as-is business process. It is common to begin by interviewing one or two senior managers to get a strategic view and then to move to midlevel managers, who can provide broad, overarching information about the business process and the expected role of the system being developed. Once the analyst has a good understanding of the "big picture," lower-level managers and staff members can fill in the exact details of how the process works. Like most other things about systems analysis, this is an iterative process— starting with senior managers, moving to midlevel managers, then staff members, back to midlevel managers, and so on, depending upon what information is needed along the way.

It is quite common for the list of interviewees to grow, often by 50 to 75 percent. As people are interviewed, more information that is needed and additional people who can provide the information will probably be identified.

CONCEPTS IN ACTION

4-E Selecting the Wrong People

In 1990, I led a consulting team for a major development project for the U.S. Army. The goal was to replace eight existing systems used on virtually every Army base across the United States. The as-is process and data models for these systems had been built, and our job was to identify improvement opportunities and develop to-be process models for each of the eight systems.

For the first system, we selected a group of midlevel managers (captains and majors) recommended by their commanders as being the experts in the system under construction. These individuals were the first- and second-line managers of the business function. The indi-

viduals were expert at managing the process but did not know the exact details of how the process worked. The resulting to-be process model was very general and nonspecific.

Alan Dennis

Question

Suppose you were in charge of the project. What interview schedule for the remaining seven projects would you use?

Designing Interview Questions   There are three types of interview questions: closed-ended questions, open-ended questions, and probing questions. *Closed-ended questions* are those that require a specific answer. They are similar to multiple-choice or arithmetic questions on an exam (see Figure 4-6). Closed-ended questions are used when an analyst is looking for specific, precise information (e.g., how many credit card requests are received per day). In general, precise questions are best. For example, rather than asking, Do you handle a lot of requests? it is better to ask, How many requests do you process per day?

Closed-ended questions enable analysts to control the interview and obtain the information they need. However, these types of questions don't uncover *why* the answer is the way it is, nor do they uncover information that the interviewer does not think to ask ahead of time.

Open-ended questions are those that leave room for elaboration on the part of the interviewee. They are similar in many ways to essay questions that you might find on an exam (see Figure 4-6 for examples). Open-ended questions are designed to gather rich information and give the interviewee more control over the information that is revealed during the interview. Sometimes the information that the interviewee chooses to discuss uncovers information that is just as important as the answer (e.g., if the interviewee talks only about other departments when asked for problems, it may suggest that he or she is reluctant to admit his or her own problems).

The third type of question is the *probing question*. Probing questions follow up on what has just been discussed in order to learn more, and they often are used when the interviewer is unclear about an interviewee's answer. They encourage the interviewee to expand on or to confirm information from a previous response, and they signal that the interviewer is listening and interested in the topic under discussion. Many beginning analysts are reluctant to use probing questions because they are afraid that the interviewee might be offended at being challenged or because they believe it shows that they didn't understand what the interviewee said. When done politely, probing questions can be a powerful tool in requirements gathering.

In general, an interviewer should not ask questions about information that is readily available from other sources. For example, rather than asking what information is used to perform to a task, it is simpler to show the interviewee a form or report (see the section on document analysis) and ask what information on it is used. This helps focus the interviewee on the task and saves time, because the interviewee does not need to describe the information detail—he or she just needs to point it out on the form or report.

Types of Questions	Examples
Closed-ended questions	• How many telephone orders are received per day? • How do customers place orders? • What information is missing from the monthly sales report?
Open-ended questions	• What do you think about the current system? • What are some of the problems you face on a daily basis? • What are some of the improvements you would like to see in a new system?
Probing questions	• Why? • Can you give me an example? • Can you explain that in a bit more detail?

FIGURE 4-6
Three Types of
Questions

No type of question is better than another, and a combination of questions is usually used during an interview. At the initial stage of an IS development project, the as-is process can be unclear, so the interview process begins with *unstructured interviews,* interviews that seek broad and roughly defined information. In this case, the interviewer has a general sense of the information needed but has few close-ended questions to ask. These are the most challenging interviews to conduct because they require the interviewer to ask open-ended questions and probe for important information "on the fly."

As the project progresses, the analyst comes to understand the business process much better and needs very specific information about how business processes are performed (e.g., exactly how a customer credit card is approved). At this time, the analyst conducts *structured interviews,* in which specific sets of questions are developed prior to the interviews. There usually are more close-ended questions in a structured interview then in the unstructured approach.

No matter what kind of interview is being conducted, interview questions must be organized into a logical sequence so that the interview flows well. For example, when trying to gather information about the current business process, it can be useful to move in logical order through the process or from the most important issues to the least important.

There are two fundamental approaches to organizing the interview questions: top-down or bottom-up (see Figure 4-7). With the *top-down interview,* the interviewer starts with broad, general issues and gradually works towards more specific ones. With the *bottom-up interview,* the interviewer starts with very specific questions and moves to broad questions. In practice, analysts mix the two approaches, starting with broad general issues, moving to specific questions, and then returning to general issues.

The top-down approach is an appropriate strategy for most interviews (it is certainly the most common approach). The top-down approach enables the interviewee to become accustomed to the topic before he or she needs to provide specifics. It also enables the interviewer to understand the issues before moving to the details because the interviewer may not have sufficient information at the start of the interview to ask very specific questions. Perhaps most importantly, the top-down approach enables the interviewee to raise a set of big-picture issues before becoming enmeshed in details, so the interviewer is less likely to miss important issues.

One case in which the bottom-up strategy may be preferred is when the analyst already has gathered a lot of information about issues and just needs to fill in some holes with details. Or, bottom-up interviewing may be appropriate if lower-level staff members feel threatened

FIGURE 4-7
Top-Down and
Bottom-up
Questioning Strategies

or unable to answer high-level questions. For example, How can we improve customer service? may be too broad a question for a customer service clerk, whereas a specific question is readily answerable (e.g., How can we speed up customer returns?). In any event, all interviews should begin with noncontroversial questions and then gradually move into more contentious issues after the interviewer has developed some rapport with the interviewee.

Preparing for the Interview   It is important to prepare for the interview in the same way that you would prepare to give a presentation. The interviewer should have a general interview plan listing the questions to be asked in the appropriate order; should anticipate possible answers and provide follow-up with them; and should identify segues between related topics. The interviewer should confirm the areas in which the interviewee has knowledge in order not to ask questions that he or she cannot answer. Review the topic areas, the questions, and the interview plan, and clearly decide which have the greatest priority in case time runs short.

In general, structured interviews with closed-ended questions take more time to prepare than unstructured interviews. So, some beginning analysts prefer unstructured interviews, thinking that they can "wing it." This is very dangerous and often counterproductive, because any information not gathered in the first interview will require follow-up efforts, and most users do not like to be interviewed repeatedly about the same issues.

The interviewer should be sure to prepare the interviewee as well. When the interview is scheduled, the interviewee should be told the reason for the interview and the areas that will be discussed far enough in advance so that he or she has time to think about the issues and organize his or her thoughts. This is particularly important when the interviewer is an outsider to the organization and for lower-level employees, who often are not asked for their opinions and who may be uncertain about why they are being interviewed.

Conducting the Interview   In starting the interview, the first goal is to build rapport with the interviewee, so that he or she trusts the interviewer and is willing to tell the whole truth, not just give the answers that he or she thinks are wanted. The interviewer should appear to be professional and an unbiased, independent seeker of information. The interview should start with an explanation of why the interviewer is there and why he or she has chosen to interview the person; then the interviewer should move into the planned interview questions.

It is critical to carefully record all the information that the interviewee provides. In our experience, the best approach is to take careful notes—write down *everything* the interviewee says, even if it does not appear immediately relevant. The interviewer shouldn't be afraid to ask the person to slow down or to pause while writing, because this is a clear indication that the interviewee's information is important. One potentially controversial issue is whether or not to tape-record an interview. Recording ensures that the interviewer does not miss important points, but it can be intimidating for the interviewee. Most organizations have policies or generally accepted practices about the recording of interviews, so they should be determined before an interview. If the interviewer is worried about missing information and cannot tape the interview, then he or she can bring along a second person to take detailed notes.

As the interview progresses, it is important to understand the issues that are discussed. If the interviewer does not understand something, he or she should be sure to ask. The interviewer should not be afraid to ask dumb questions, because the only thing worse than appearing dumb is to be dumb by not understanding something. If the interviewer doesn't understand something during the interview, he or she certainly won't understand it afterwards. Jargon should be recognized and defined; any jargon not understood should be clarified. One good strategy to increase understanding during an interview is to periodically summarize the key points that the interviewee is communicating. This avoids misunderstandings and also demonstrates that the interviewer is listening.

PRACTICAL
TIP

5-1 Developing Interpersonal Skills

Interpersonal skills are those skills that enable you to develop rapport with others, and they are very important for interviewing. They help you to communicate with others effectively. Some people develop good interpersonal skills at an early age; they simply seem to know how to communicate and interact with others. Other people are less lucky and need to work hard to develop their skills.

Interpersonal skills, like most skills, can be learned. Here are some tips:

- **Don't worry, be happy.** Happy people radiate confidence and project their feelings on others. Try interviewing someone while smiling and then interviewing someone else while frowning and see what happens.
- **Pay attention.** Pay attention to what the other person is saying (which is harder than you might think). See how many times you catch yourself with your mind on something other than the conversation at hand.
- **Summarize key points.** At the end of each major theme or idea that someone explains, repeat the

key points back to the speaker (e.g., Let me make sure I understand. The key issues are. . . ."). This demonstrates that you consider the information important—and also forces you to pay attention (you can't repeat what you didn't hear).

- **Be succinct.** When you speak, be succinct. The goal in interviewing (and in much of life) is to learn, not to impress. The more you speak, the less time you give to others.
- **Be honest.** Answer all questions truthfully, and if you don't know the answer, say so.
- **Watch body language (yours and theirs).** The way a person sits or stands conveys much information. In general, a person who is interested in what you are saying sits or leans forward, makes eye contact, and often touches his or her face. A person leaning away from you or with an arm over the back of a chair is disinterested. Crossed arms indicate defensiveness or uncertainty, while steepling (sitting with hands raised in front of the body with fingertips touching) indicates a feeling of superiority.

Finally, facts should be separated from opinion. The interviewee may say, for example, We process too many credit card requests. This is an opinion, and it is useful to follow this up with a probing question requesting support for the statement (e.g., Oh, how many do you process in a day?). It is helpful to check the facts because any differences between the facts and the interviewee's opinions can point out key areas for improvement. Suppose the interviewee complains about a high or increasing number of errors, but the logs show that errors have been decreasing. This suggests that errors are viewed as a very important problem that should be addressed by the new system, even if they are declining.

As the interview draws to a close, the interviewee should have time to ask questions or provide information that he or she thinks is important but was not part of the interview plan. In most cases, the interviewee will have no additional concerns or information, but in some cases this will lead to unanticipated, but important, information. Likewise, it can be useful to ask the interviewee if there are other people who should be interviewed. The interview should end on time (if necessary, some topics can be omitted or another interview can be scheduled).

As a last step in the interview, the interviewer should briefly explain what will happen next (see the next section). The interviewer shouldn't prematurely promise certain features in the new system or a specific delivery date, but he or she should reassure the interviewee that his or her time was well spent and very helpful to the project.

4-4 Interview Practice

*I*nterviewing is not as simple as it first appears. Select two people from class to go to the front of the room to demonstrate an interview. (This also can be done in groups.) Have one person be the interviewer and the other be the interviewee. The interviewer should conduct a five-minute interview regarding the school's course registration system. Gather information about the existing system and how the system can be improved. If there is time, repeat with another pair.

Questions

1. What was the body language of the interview pair like?
2. What kind of interview was conducted?
3. What kinds of questions were asked?
4. What was done well? How could the interview be improved?

Postinterview Follow-up   After the interview is over, the analyst needs to prepare an *interview report* that describes the information from the interview (Figure 4-8). The report contains *interview notes,* information that was collected over the course of the interview and is summarized in a useful format. In general, the interview report should be written within forty-eight hours of the interview, because the longer the interviewer waits, the more likely he or she is to forget information.

Often, the interview report is sent to the interviewee with a request to read it and inform the analyst of clarifications or updates. The interviewee needs to be convinced that the interviewer genuinely want his or her corrections to the report. Usually there are few changes, but the need for any significant changes suggests that a second interview will be required. Never distribute someone's information without prior approval.

Interview Notes Approved by: Linda Estey

Person Interviewed: Linda Estey,
     Director, Human Resources

Interviewer: Barbara Wixom

Purpose of Interview:

- Understand reports produced for Human Resources by the current system
- Determine information requirements for future system

Summary of Interview:

- Sample reports of all current HR reports are attached to this report. The information that is not used and missing information are noted on the reports.
- Two biggest problems with the current system are:
  1. The data are too old (the HR Department needs information within two days of month end; currently information is provided to them after a three-week delay)
  2. The data are of poor quality (often reports must be reconciled with departmental HR database)
- The most common data errors found in the current system include incorrect job level information and missing salary information.

Open Items:

- Get current employee roster report from Mary Skudrna (extension 4355).
- Verify calculations used to determine vacation time with Mary Skudrna.
- Schedule interview with Jim Wack (extension 2337) regarding the reasons for data quality problems.

Detailed Notes: See attached transcript.

FIGURE 4-8
Interview Report

Joint Application Development (JAD)

JAD is an information-gathering technique that allows the project team, users, and management to work together to identify requirements for the system. IBM developed the JAD technique in the late 1970s, and it is often the most useful method for collecting information from users.[8] Capers Jones claims that JAD can reduce scope creep by 50 percent, and it avoids the requirements for a system being too specific or too vague, both of which cause trouble during later stages of the SDLC.[9] JAD is a structured process in which ten to twenty users meet together under the direction of a *facilitator* skilled in JAD techniques. The facilitator is a person who sets the meeting agenda and guides the discussion but does not join in the discussion as a participant. He or she does not provide ideas or opinions on the topics under discussion in order to remain neutral during the session. The facilitator must be an expert in both group process techniques and systems analysis and design techniques. One or two *scribes* assist the facilitator by recording notes, making copies, and so on. Often the scribes will use computers and CASE tools to record information as the JAD session proceeds.

The JAD group meets for several hours, several days, or several weeks until all the issues have been discussed and the needed information is collected. Most JAD sessions take place in a specially prepared meeting room, away from the participants' offices so that they are not interrupted. The meeting room is usually arranged in a U-shape so that all participants can easily see each other (see Figure 4-9). At the front of the room (the open part of the U), are whiteboard, flip chart and/or overhead projector for use by the facilitator leading the discussion.

One problem with JAD is that it suffers from the traditional problems associated with groups; sometimes people are reluctant to challenge the opinions of others (particularly their boss), a few people often dominate the discussion, and not everyone participates. In a fifteen-member group, for example, if everyone participates equally, then each person can talk for only four minutes each hour and must listen for the remaining fifty-six minutes—not a very efficient way to collect information.

A new form of JAD called *electronic JAD*, or *e-JAD*, attempts to overcome these problems by using groupware. In an e-JAD meeting room, each participant uses special software on a networked computer to send anonymous ideas and opinions to everyone else. In this way, all participants can contribute at the same time without fear of reprisal from people with differing opinions. Initial research suggests that e-JAD can reduce the time required to run JAD sessions by 50 to 80 percent.[10]

Selecting Participants   Selecting JAD participants is done in the same basic way as selecting interview participants. Participants are selected based on the information they can contribute, to provide a broad mix of organizational levels, and to build political support for the new system. The need for all JAD participants to be away from their office at the same time can be a major problem. The office may need to be closed or operate with a skeleton staff until the JAD sessions are complete.

Ideally, the participants who are released from regular duties to attend the JAD sessions should be the very best people in that business unit. However, without strong management support, JAD sessions can fail because those selected to attend the JAD session are people who are less likely to be missed (i.e., the least competent people).

---

[8] More information on JAD can be found in J. Wood and D. Silver, *Joint Application Development* (New York: Wiley, 1989); and Alan Cline, "Joint Application Development for Requirements Collection and Management," http://www.carolla.com/wp-jad.htm.

[9] See Kevin Strehlo, "Catching up with the Jones and 'Requirement' Creep," *InfoWorld* (July 29, 1996); and Kevin Strehlo, "The Makings of a Happy Customer: Specifying Project X," *InfoWorld* (November 11, 1996).

[10] For more information on e-JAD, see A. R. Dennis, G. S. Hayes, and R. M. Daniels, "Business Process Modeling with Groupware," *Journal of Management Information Systems* 15, no. 4 (1999): 115–142.

FIGURE 4-9   JAD Meeting Room

The facilitator should be someone who is an expert in JAD or e-JAD techniques and, ideally, someone who has experience with the business under discussion. In many cases, the JAD facilitator is a consultant external to the organization because the organization may not have a recurring need for JAD or e-JAD expertise. Developing and maintaining this expertise in-house can be expensive.

Designing a JAD Session   JAD sessions can run from as little as half a day to several weeks, depending upon the size and scope of the project. In our experience, most JAD sessions tend to last five to ten days, spread over a three-week period. Most e-JAD sessions tend to last one to four days in a one-week period. JAD and e-JAD sessions usually go beyond the collection of information and move into analysis. For example, the users and the analysts collectively can create analysis deliverables, such as the functional models, structural models, or the requirements definition.

As with interviewing, success depends upon a careful plan. JAD sessions usually are designed and structured using the same principles as interviews. Most JAD sessions are designed to collect specific information from users, and this requires the development of a set of questions prior to the meeting. One difference between JAD and interviewing is that all JAD sessions are structured—they *must* be carefully planned. In general, closed-ended questions are seldom used because they do not spark the open and frank discussion that is typical of JAD. In our experience, it is better to proceed top-down in JAD sessions when gathering information. Typically thirty minutes is allocated to each separate agenda item, and frequent breaks are scheduled throughout the day because participants tire easily.

Preparing for a JAD Session   As with interviewing, it is important to prepare the analysts and participants for a JAD session. Because the sessions can go beyond the depth of a typical interview and are usually conducted off-site, participants can be more concerned about how to prepare. It is important that the participants understand what is expected of them. If the goal of the JAD session, for example, is to develop an understanding of the current system, then participants can bring procedure manuals and documents with them. If the goal is to identify improvements for a system, then they can think about how they would improve the system prior to the JAD session.

Conducting a JAD Session   Most JAD sessions try to follow a formal agenda, and most have formal *ground rules* that define appropriate behavior. Common ground rules include following the schedule, respecting others' opinions, accepting disagreement, and ensuring that only one person talks at once.

The role of a JAD facilitator can be challenging. Many participants come to a JAD session with strong feelings about the system to be discussed. Channeling these feelings so that the session moves forward in a positive direction and getting participants to recognize and accept—but not necessarily agree on—opinions and situations different from their own requires significant expertise in systems analysis and design, JAD, and interpersonal skills. Few systems analysts attempt to facilitate JAD sessions without being trained in JAD techniques, and most apprentice with a skilled JAD facilitator before they attempt to lead their first session.

The JAD facilitator performs three key functions. First, he or she ensures that the group sticks to the agenda. The only reason to digress from the agenda is when it becomes clear to the facilitator, project leader, and project sponsor that the JAD session has produced some new information that is unexpected and requires the JAD session (and perhaps the project) to move in a new direction. When participants attempt to divert the discussion away from the agenda, the facilitator must be firm but polite in leading discussion back to the agenda and getting the group back on track.

Second, the facilitator must help the group understand the technical terms and jargon that surround the system development process, and help the participants understand the specific analysis techniques used. Participants are experts in their area, or their part of the business, but they are not experts in systems analysis. The facilitator must, therefore, minimize the learning required and teach participants how to effectively provide the right information.

Third, the facilitator records the group's input on a public display area, which can be a whiteboard, flip chart, or computer display. He or she structures the information that the group provides and helps the group recognize key issues and important solutions. Under no circumstance should the facilitator insert his or her opinions into the discussion. The facilitator must remain neutral at all times and simply help the group through the process. The moment the facilitator offers an opinion on an issue, the group will see him or her not as a neutral party, but rather as someone who could be attempting to sway the group into some predetermined solution.

However, this does not mean that the facilitator should not try to help the group resolve issues. For example, if two items appear to be the same to the facilitator, the facilitator should not say, "I think these may be similar." Instead, the facilitator should ask, "Are these similar?" If the group decides they are, the facilitator can combine them and move on. However, if the group decides they are not similar (despite what the facilitator believes), the facilitator should accept the decision and move on. The group is *always* right, and the facilitator has no opinion.

PRACTICAL

TIP

Managing Problems in JAD Sessions

I have run more than a hundred JAD sessions and have learned several standard "facilitator tricks." Here are some common problems and some ways to deal with them.

- **Domination.** The facilitator should ensure that no one person dominates the group discussion. The only way to deal with someone who dominates is head on. During a break, approach the person, thank him or her for his or her insightful comments, and ask the person to help you make sure that others also participate.

- **Noncontributors.** Drawing out people who have participated very little is challenging because you want to bring them into the conversation so that they will contribute again. The best approach is to ask a direct factual question that you are certain they can answer. And it helps to ask the question in a long way to give them time to think. For example, Pat, I know you've worked shipping orders a long time. You've probably been in the shipping department longer than anyone else. Could you help us understand exactly what happens when an order is received in shipping?

- **Side discussions.** Sometimes participants engage in side conversations and fail to pay attention to the group. The easiest solution is simply to walk close to the people and continue to facilitate right in front of them. Few people will continue a side conversion when you are two feet from them and the entire group's attention is on you and them.

- **Agenda merry-go-round.** The merry-go-round occurs when a group member keeps returning to the same issue every few minutes and won't let go. One solution is to let the person have five minutes to ramble on about the issue while you carefully write down every point on a flip chart or computer file. This flip chart or file is then posted conspicuously on the wall. When the person brings up the issue again, you interrupt them, walk to the paper and ask them what to add. If they mention some-

thing already on the list, you quickly interrupt, point out that it is there, and ask what other information to add. Don't let them repeat the same point, but write any new information.

- **Violent agreement.** Some of the worst disagreements occur when participants really agree on the issues but don't realize that they agree because they are using different terms. An example is arguing whether a glass is half empty or half full; they agree on the facts but can't agree on the words. In this case, the facilitator has to translate the terms into different words and find common ground so the parties recognize that they really agree.

- **Unresolved conflict.** In some cases, participants don't agree and can't understand how to determine what alternatives are better. You can help by structuring the issue. Ask for criteria by which the group will identify a good alternative (e.g., Suppose this idea really did improve customer service. How would I recognize the improved customer service?). Then once you have a list of criteria, ask the group to assess the alternatives using them.

- **True conflict.** Sometimes, despite every attempt, participants just can't agree on an issue. The solution is to postpone the discussion and move on. Document the issue as an open issue and list it prominently on a flip chart. Have the group return to the issue hours later. Often the issue will have resolved itself by then and you haven't wasted time on it. If the issue cannot be resolved later, move it to the list of issues to be decided by the project sponsor or some other more senior member of management.

- **Humor.** Humor is one of the most powerful tools a facilitator has and thus must be used judiciously. The best JAD humor is always in context; never tell jokes but take the opportunity to find the humor in the situation.

Alan Dennis

4-5 JAD Practice

*O*rganize yourselves into groups of four to seven people, and pick one person in each group to be the JAD facilitator. Using a blackboard, whiteboard or flip chart, gather information about how the group performs some process (e.g., working on a class assignment, making a sandwich, paying bills, getting to class).

Questions

1. How did the JAD session go?
2. Based on your experience, what are pros and cons of using JAD in a real organization?

Post-JAD Follow-up   As with interviews, a JAD *postsession report* is prepared and circulated among session attendees. The postsession report is essentially the same as the interview report in Figure 4-8. Because the JAD sessions are longer and provide more information, it usually takes a week or two after the JAD session before the report is complete.

Questionnaires

A questionnaire is a set of written questions used to obtain information from individuals. Questionnaires are often used when there is a large number of people from whom information and opinions are needed. In our experience, questionnaires are a common technique with systems intended for use outside the organization (e.g., by customers or vendors) or for systems with business users spread across many geographic locations. Most people automatically think of paper when they think of questionnaires, but today more questionnaires are being distributed in electronic form, either via e-mail or on the Web. Electronic distribution can save a significant amount of money as compared to distributing paper questionnaires.

Selecting Participants   As with interviews and JAD sessions, the first step is to identify the individuals to whom the questionnaire will be sent. However, it is not usual to select every person who could provide useful information. The standard approach is to select a *sample*, or subset, of people who are representative of an entire group. Sampling guidelines are discussed in most statistics books, and most business schools include courses that cover the topic, so we do not discuss it here. The important point in selecting a sample, however, is to realize that not everyone who receives a questionnaire will actually complete it. On average, only 30 to 50 percent of paper and e-mail questionnaires are returned. Response rates for Web-based questionnaires tend to be significantly lower (often only 5 to 30 percent).

Designing a Questionnaire   Developing good questions is critical for questionnaires because the information on a questionnaire cannot be immediately clarified for a confused respondent. Questions on questionnaires must be very clearly written and leave little room for misunderstanding, so closed-ended questions tend to be most commonly used. Questions must clearly enable the analyst to separate facts from opinions. Opinion questions often ask the respondent the extent to which they agree or disagree (e.g., Are network problems common?), whereas factual questions seek more precise values (e.g., How often does a network problem occur: once an hour, once a day, once a week?). See Figure 4-10 for guidelines on questionnaire design.

Perhaps the most obvious issue—but one that is sometimes overlooked—is to have a clear understanding of how the information collected from the questionnaire will be analyzed and used. This issue must be addressed before the questionnaire is distributed, because it is too late afterward.

FIGURE 4-10
Good Questionnaire
Design

- Begin with nonthreatening and interesting questions.
- Group items into logically coherent sections.
- Do not put important items at the very end of the questionnaire.
- Do not crowd a page with too many items.
- Avoid abbreviations.
- Avoid biased or suggestive items or terms.
- Number questions to avoid confusion.
- Pretest the questionnaire to identify confusing questions.
- Provide anonymity to respondents.

Questions should be relatively consistent in style, so that the respondent does not have to read instructions for each question before answering it. It is generally good practice to group related questions together to make them simpler to answer. Some experts suggest that questionnaires should start with questions important to respondents, so that the questionnaire immediately grabs their interest and induces them to answer it. Perhaps the most important step is to have several colleagues review the questionnaire and then pretest it with a few people drawn from the groups to whom it will be sent. It is surprising how often seemingly simple questions can be misunderstood.

Administering the Questionnaire   The key issue in administering the questionnaire is getting participants to complete the questionnaire and send it back. Dozens of marketing research books have been written about ways to improve response rates. Commonly used techniques include clearly explaining why the questionnaire is being conducted and why the respondent has been selected; stating a date by which the questionnaire is to be returned; offering an inducement to complete the questionnaire (e.g., a free pen); and offering to supply a summary of the questionnaire responses. Systems analysts have additional techniques to improve response rates inside the organization, such as personally handing out the questionnaire and personally contacting those who have not returned them after a week or two, as well as requesting the respondents' supervisors to administer the questionnaires in a group meeting.

Questionnaire Follow-up   It is helpful to process the returned questionnaires and develop a questionnaire report soon after the questionnaire deadline. This ensures that the analysis process proceeds in a timely fashion and that respondents who requested copies of the results receive them promptly.

YOUR TURN

4-6 Questionnaire Practice

*O*rganize yourselves into small groups. Have each person develop a short questionnaire to collect information about how often group members perform some process (e.g., working on a class assignment, making a sandwich, paying bills, getting to class), how long it takes them, how they feel about the process, and opportunities for improving the process.

Once everyone has completed his or her questionnaire, ask each member to pass it to the right and then complete his or her neighbor's questionnaire. Pass the questionnaire back to the creator when it is completed.

Questions

1. How did the questionnaire you completed differ from the one you created?

2. What are the strengths of each questionnaire?

3. How would you analyze the survey results if you had received fifty responses?

4. What would you change about the questionnaire that you developed?

Document Analysis

Project teams often use document analysis to understand the as-is system. Under ideal circumstances, the project team that developed the existing system will have produced documentation, which was then updated by all subsequent projects. In this case, the project team can start by reviewing the documentation and examining the system itself.

Unfortunately, most systems are not well documented because project teams fail to document their projects along the way, and when the projects are over, there is no time to go back and document. Therefore, there may not be much technical documentation about the current systems available, or it may not contain updated information about recent system changes. However, there are many helpful documents that do exist in an organization: paper reports, memorandums, policy manuals, user-training manuals, organization charts, forms, and, of course, the user interface with the existing system.

But these documents tell only part of the story. They represent the *formal system* that the organization uses. Quite often, the real, or *informal, system* differs from the formal one, and these differences, particularly large ones, give strong indications of what needs to be changed. For example, forms or reports that are never used should probably be eliminated. Likewise, boxes or questions on forms that are never filled in (or are used for other purposes) should be rethought. See Figure 4-11 for an example of how a document can be interpreted.

The most powerful indication that the system needs to be changed is when users create their own forms or add additional information to existing ones. Such changes clearly demonstrate the need for improvements to existing systems. Thus, it is useful to review both blank and completed forms to identify these deviations. Likewise, when users access multiple reports to satisfy their information needs, it is a clear sign that new information or new information formats are needed.

Observation

Observation, the act of watching processes being performed, is a powerful tool for gathering information about the as-is system because it enables the analyst to see the reality of a situation, rather than listening to others describe it in interviews or JAD

CONCEPTS IN ACTION

4-F Publix Credit-Card Forms

At my neighborhood Publix grocery store, the cashiers always hand-write the total amount of the charge on every credit-card charge form, even though it is printed on the form. Why? Because the "back office" staff people who reconcile the cash in the cash drawers with the amount sold at the end of each shift find it hard to read the small print on the credit-card forms. Writing in large print makes it easier for them to add the values up. However, cashiers sometimes make mistakes and write the wrong amount on the forms, which causes problems.

Barbara Wixom

Questions

1. What does the credit-card charge form indicate about the existing system?
2. How can you make improvements with a new system?

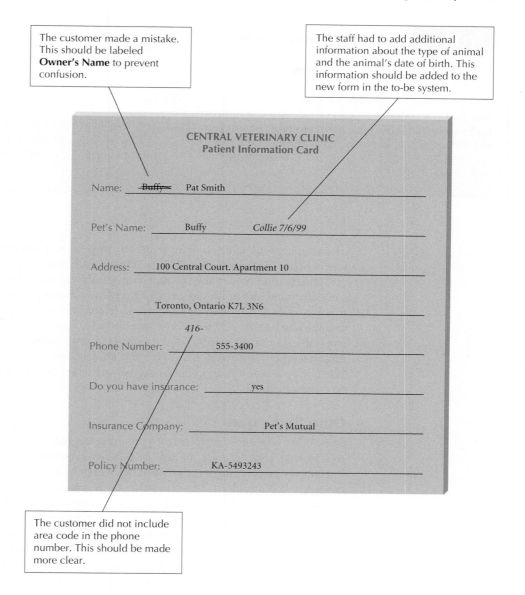

The customer made a mistake. This should be labeled **Owner's Name** to prevent confusion.

The staff had to add additional information about the type of animal and the animal's date of birth. This information should be added to the new form in the to-be system.

CENTRAL VETERINARY CLINIC
Patient Information Card

Name: ~~Buffy~~ Pat Smith

Pet's Name: Buffy *Collie 7/6/99*

Address: 100 Central Court. Apartment 10

Toronto, Ontario K7L 3N6

416-
Phone Number: 555-3400

Do you have insurance: yes

Insurance Company: Pet's Mutual

Policy Number: KA-5493243

The customer did not include area code in the phone number. This should be made more clear.

FIGURE 4-11
Performing a Document Analysis

sessions. Several research studies have shown that many managers really do not remember how they work and how they allocate their time. (Quick, how many hours did you spend last week on each of your courses?) Observation is a good way to check the validity of information gathered from indirect sources such as interviews and questionnaires.

In many ways, the analyst becomes an anthropologist as he or she walks through the organization and observes the business system as it functions. The goal is to keep a low profile, to not interrupt those working, and to not influence those being observed. Nonetheless, it is important to understand that what analysts observe may not be the normal day-to-day routine because people tend to be extremely careful in their behavior when they are being watched. Even though normal practice may be to break formal

YOUR	4-7 Observation Practice
TURN	

Visit the library at your college or university and observe how the book checkout process occurs. First watch several students checking books out, and then check one out yourself. Prepare a brief summary report of your observations.

   When you return to class, share your observations with others.

Questions

1. Why might the reports present different information?
2. How would the information be different had you used the interview or JAD technique?

organizational rules, the observer is unlikely to see this. (Remember how you drove the last time a police car followed you?) Thus, what you see may *not* be what you get.

   Observation is often used to supplement interview information. The location of a person's office and its furnishings give clues as to the person's power and influence in the organization and can be used to support or refute information given in an interview. For example, an analyst might become skeptical of someone who claims to use the existing computer system extensively if the computer is never turned on while the analyst visits. In most cases, observation will support the information that users provide in interviews. When it does not, it is an important signal that extra care must be taken in analyzing the business system.

Other Techniques

Some other very useful requirements-gathering techniques include throwaway prototyping, role-playing CRC cards with use case–based scenarios, and mind/concept mapping. Throwaway prototyping was described in Chapter 1. In essence, throwaway prototypes are created to better understand some aspect of the new system. In many cases, they are used to test out some technical aspect of a nonfunctional requirement, such as connecting a client workstation to a server. If you have never done this before, it will be a lot easier to develop a very small example system to test out the necessary design of the connection from the client workstation to the server instead of trying to do it the first time with the full blown system. As such, throwaway prototyping is very useful when designing the physical architecture of the system (see Chapter 12). Throwaway prototyping can also be very useful in designing user interfaces (see Chapter 11).

   Role-playing CRC cards with use case–based scenarios are very useful when creating functional (see Chapter 5), structural (see Chapter 6), and behavioral (see Chapter 7) models. We describe this approach in Chapter 6.

   Concept mapping is an educational psychology technique that has been used in schools, corporations, and health-care agencies to facilitate learning, understanding, and knowledge creation.[11] Concept maps represent meaningful relationships between concepts.

---

[11] For more information on concept mapping, see J. D. Novak and D. B. Gowin, *Learning How to Learn* (Cambridge, UK: Cambridge University Press, 1984); and J. D. Novak, *Learning, Creating, and Using Knowledge: Concept Maps™ as Facilitative Tools in Schools and Corporations* (Mahwah, NL: Lawrence Erlbaum Associates, Publishers, 1998). Also, a concept mapping tool is available from the Institute of Human and Machine Cognition at cmap.ihmc.us.

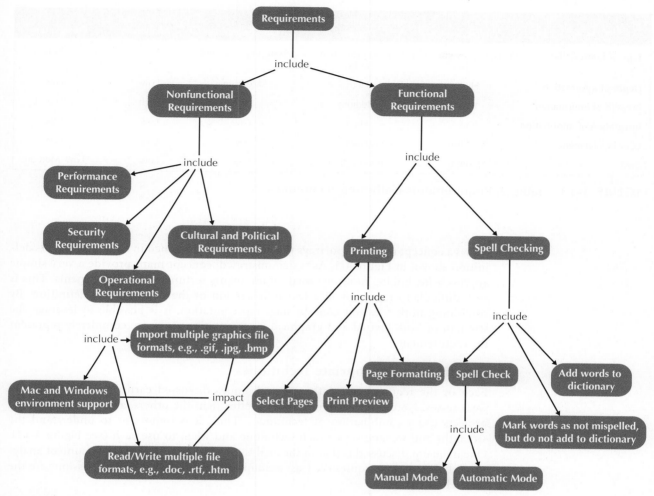

FIGURE 4-12 Sample Concept Map

They are useful for focusing individuals on the small number of key ideas on which they should concentrate. A concept map is essentially a node-and-arc representation, where the nodes represent the individual requirements and the arcs represent the relationships among the requirements. Each arc is labeled with a relationship name. Concept maps also have been recommended as a possible technique to support modeling requirements for object-oriented systems development and knowledge management systems.[12] The advantage of the concept mapping approach to representing requirements over the typical textual approach (see Figure 4-2) is that a concept map is not limited to a hierarchical representation. Concept maps allow the relationships among the functional and nonfunctional requirements to be explicitly represented. For example, Figure 4-12

---

[12] See B. Henderson-Sellers, A. Simons, and H. Younessi, *The OPEN Toolbox of Techniques* (Harlow, England: Addison-Wesley, 1998)

	Interviews	Joint Application Design	Questionnaires	Document Analysis	Observation
Type of information	As-is, improvements, to-be	As-is, improvements, to-be	As-is, improvements	As-is	As-is
Depth of information	High	High	Medium	Low	Low
Breadth of information	Low	Medium	High	High	Low
Integration of information	Low	High	Low	Low	Low
User involvement	Medium	High	Low	Low	Low
Cost	Medium	Low–Medium	Low	Low	Low–Medium

FIGURE 4-13   Table of Requirements-Gathering Techniques

shows a concept map that portrays the information contained in the requirements definition shown in Figure 4-1. As is demonstrated, concept maps provide a very simple approach for linking nonfunctional requirements to functional requirements. This is very difficult to represent in a text-only version of the requirements definition. By combining both text and concept map representation, it is possible to leverage the strength of both textual and graphical representations to more completely represent the requirements.

Selecting the Appropriate Techniques

Each of the requirements-gathering techniques discussed earlier has strengths and weaknesses. No one technique is always better than the others, and in practice most projects use a combination of techniques. Thus, it is important to understand the strengths and weaknesses of each technique and when to use each (see Figure 4-13). One issue not discussed is that of the analysts' experience. In general, document analysis and observation require the least amount of training, whereas JAD sessions are the most challenging.

Type of Information   The first characteristic is type of information. Some techniques are more suited for use at different stages of the analysis process, whether understanding the as-is system, identifying improvements, or developing the to-be system. Interviews and JAD are commonly used in all three stages. In contrast, document analysis and observation usually are most helpful for understanding the as-is, although occasionally they provide information about current problems that need to be improved. Questionnaires are often used to gather information about the as-is system as well as general information about improvements.

Depth of Information   The depth of information refers to how rich and detailed the information is that the technique usually produces and the extent to which the technique is useful for obtaining not only facts and opinions, but also an understanding of *why* those facts and opinions exist. Interviews and JAD sessions are very useful for providing a good depth of rich and detailed information and helping the analyst to understand the reasons behind them. At the other extreme, document analysis and observation are useful for obtaining facts, but little beyond that. Questionnaires can provide a medium depth of information, soliciting both facts and opinions with little understanding of why they exist.

Breadth of Information   Breadth of information refers to the range of information and information sources that can be easily collected using the chosen technique. Questionnaires and document analysis are both easily capable of soliciting a wide range of information from a large number of information sources. In contrast, interviews and observation require the analyst to visit each information source individually and, therefore, take more time. JAD sessions are in the middle because many information sources are brought together at the same time.

Integration of Information   One of the most challenging aspects of requirements gathering is the integration of information from different sources. Simply put, different people can provide conflicting information. Combining this information and attempting to resolve differences in opinions or facts is usually very time consuming because it means contacting each information source in turn, explaining the discrepancy, and attempting to refine the information. In many cases, the individual wrongly perceives that the analyst is challenging his or her information, when in fact it is another user in the organization who is doing so. This can make the user defensive and make it hard to resolve the differences.

All techniques suffer integration problems to some degree, but JAD sessions are designed to improve integration because all information is integrated when it is collected, not afterwards. If two users provide conflicting information, the conflict becomes immediately obvious, as does the source of the conflict. The immediate integration of information is the single most important benefit of JAD that distinguishes it from other techniques, and this is why most organizations use JAD for important projects.

User Involvement   User involvement refers to the amount of time and energy the intended users of the new system must devote to the analysis process. It is generally agreed that as users become more involved in the analysis process, the chance of success

CONCEPTS **4-G Campus Technology Updates**

IN ACTION

Colleges and universities need to stay current with technologies. Many campuses have gone with laptop programs, where students are expected to purchase or lease a particular model of laptop that will be preloaded with appropriate software and used for the students' collegiate careers. Likewise, the campuses need to update their infrastructure—such as increasing bandwidth (to handle more video, such as YouTube)—and to provide wireless communication.

The University of Northern Wisconsin is a campus that is trying to remain current with technology. Campus budgets are almost always tight. UNW offers programs from its Superior, Wisconsin, main campus as well as programs on two satellite campuses in Ashland and Rhinelander. Users on the two satellite campuses frequently

do not get the same level of service as students on the main campus. Internet access is generally slower and not all the software is the same. For example, students at the main campus have access to Blomberg systems for analysis of financial trading data. The campus opted to build an Internet portal for all students to get to the same software and systems, set up by student ID and student profiles and permissions.

Questions

1. What technologies would be needed to make your campus a premier technology-oriented school?
2. How might a college campus be like a business with multiple locations and software needs?

increases. However, user involvement can have a significant cost, and not all users are willing to contribute valuable time and energy. Questionnaires, document analysis, and observation place the least burden on users, whereas JAD sessions require the greatest effort.

Cost   Cost is always an important consideration. In general, questionnaires, document analysis, and observation are low-cost techniques (although observation can be quite time consuming). The low cost does not imply that they are more or less effective than the other techniques. Interviews and JAD sessions generally have moderate costs. In general, JAD sessions are much more expensive initially, because they require many users to be absent from their offices for significant periods of time, and they often involve highly paid consultants. However, JAD sessions significantly reduce the time spent in information integration and thus cost less in the long term.

Combining Techniques   In practice, requirements gathering combines a series of different techniques. Most analysts start by using interviews with senior manager(s) to gain an understanding of the project and the big-picture issues. From these interviews, it becomes clear whether large or small changes are anticipated. These interviews are often followed with analysis of documents and policies to gain some understanding of the as-is system. Usually interviews come next to gather the rest of the information needed for the as-is picture.

In our experience, identifying improvements is most commonly done using JAD sessions because the JAD session enables the users and key stakeholders to work together through an analysis technique and come to a shared understanding of the possibilities for the to-be system. Occasionally, these JAD sessions are followed by questionnaires sent to a much wider set of users or potential users to see whether the opinions of those who participated in the JAD sessions are widely shared.

Developing the concept for the to-be system is often done through interviews with senior managers, followed by JAD sessions with users of all levels to make sure the key needs of the new system are well understood.

THE SYSTEM PROPOSAL

A system proposal brings together into a single comprehensive document the material created during planning and analysis. The system proposal typically includes an executive summary, the system request, the workplan, the feasibility analysis, the requirements definition, and the evolving models that describe the new system.[13] The executive summary provides all critical information in a very concise form. It can be thought of as a summary of the complete proposal. Its purpose is to allow a busy executive to quickly read through it and determine which parts of the proposal that they need to go through more thoroughly. Figure 4-14 provides a template for a system proposal and references to where the other sections of the proposal are described.

[13] Depending on the client, much more detailed specifications may be required; for example Department of Defense, NASA, IEEE/ANSI, and the Naval Research Laboratory all have very specific formats that must be followed. For more information on these more detailed specifications see A. M Davis, *Software Requirements, Revision* (Upper Saddle River, NJ: Prentice Hall, 1993); G. Kotonya and I. Sommerville, *Requirements Engineering* (Chichester, England: Wiley, 1998); and R. H. Thayer and M. Dorfman (eds.), *Software Requirements Engineering*, 2nd ed. (Los Alamitos, CA: IEEE Computer Society Press, 1997).

1. Table of Contents

2. Executive Summary

A summary of all the essential information in the proposal so a busy executive can read it quickly and decide what parts of the plan to read in more depth.

3. System Request

The revised system request form (see Chapter 2).

4. Workplan

The original workplan, revised after having completed the analysis phase (see Chapter 3).

5. Feasibility Analysis

A revised feasibility analysis, using the information from the analysis phase (see Chapter 2).

6. Requirements Definition

A list of the functional and nonfunctional business requirements for the system (this chapter).

7. Functional Model

An activity diagram, a set of use case descriptions, and a use case diagram that illustrate the basic processes or external functionality that the system needs to support (see Chapter 5).

8. Structural Models

A set of CRC cards, class diagram and object diagrams that describe the structural aspects of the to-be system (see Chapter 6). This may also include structural models of the current as-is system that will be replaced.

9. Behavioral Models

A set of sequence diagrams, communication diagrams, behavioral state machines, and a CRUD matrix that describe the internal behavior of the to-be system (see Chapter 7). This may include behavioral models of the as-is system that will be replaced.

Appendices

These contain additional material relevant to the proposal, often used to support the recommended system. This might include results of a questionnaire survey or interviews, industry reports and statistics, and so on.

FIGURE 4-14
System Proposal
Template

APPLYING THE CONCEPTS AT CD SELECTIONS

Once the CD Selections steering committee approved the system proposal and feasibility analysis, the project team began performing analysis activities. These included gathering requirements using a variety of techniques, and analyzing the requirements that were gathered. An Internet marketing and sales consultant, Chris Campbell, was hired to advise Alec, Margaret, and the project team during the inception phase. Some highlights of the project team's activities are presented next.

Requirements Analysis Strategies

Margaret suggested that the project team conduct several JAD sessions with store managers, marketing analysts, and Web-savvy members of the IT staff. Together, the groups could work through some BPI strategies and brainstorm how improvements could be made to the current order process using a new Web-based system.

Alec facilitated three JAD sessions that were conducted over the course of a week. Alec's past facilitation experience helped the eight-person meetings run smoothly and stay on track. First, Alec used technology analysis and suggested several important Web technologies that could be used for the system. The JAD session generated ideas about how CD Selections could apply each of the technologies to the Internet order project.

Alec had the group categorize the ideas into three sets: definite ideas that would have a good probability of providing business value; possible ideas that might add business value; and unlikely ideas.

Next, Alec applied informal benchmarking by introducing the Web sites of several leading retailers and pointing out the features that they offered online. He selected some sites based on their success with Internet sales and others based on their similarity to the vision for CD Selections' new system. The group discussed the features that were common across most retailers versus unique functionality, and they created a list of suggested business requirements for the project team.

Requirements-Gathering Techniques

Alec believed that it would be important to understand the order processes and systems that already existed in the organization because they would have to be closely integrated with the Web order system. Three requirements-gathering techniques proved to be helpful in understanding the current systems and processes—document analysis, interviews, and observation.

First, the project team collected existing reports (e.g., order forms, screenshots of the online order screens) and system documentation (functional, structural, and behavioral models) that shed light on the as-is system. They were able to gather a good amount of information about the brick-and-mortar order processes and systems in this way. When questions arose, they conducted short interviews with the person who provided the documentation for clarification.

Next, Alec interviewed the senior analysts for the order and inventory systems to get a better understanding of how those systems worked. He asked if they had any ideas for the new system as well as for any integration issues that would need to be addressed. Alex also interviewed a contact from the ISP and the IT person who supported CD Selections' current Web site; both provided information about the existing communications infrastructure at CD Selections and its Web capabilities. Finally, Alex spent a half day visiting two of the retail stores and observing exactly how the order and hold processes worked in the brick-and-mortar facilities.

Requirements Definition

Throughout all these activities, the project team collected information and tried to identify the business requirements for the system from the information. As the project progressed, requirements were added to the requirements definition and grouped by requirement type. When questions arose, the team worked with Margaret, Chris, and Alec to confirm that requirements were within the scope. The requirements that fell outside of the scope of the current system were typed into a separate document that would be saved for future use.

After gathering and documenting the requirements, the requirements definition was distributed to Margaret, two marketing employees who would work with the system on the business side, and several retail store managers. This group then met for a two-day JAD session to clarify, finalize, and prioritize business requirements.[14]

The project team spent time creating functional, structural and behavioral models (Chapters 5, 6, and 7) that depicted the objects in the future system. Members of marketing and IT departments reviewed the documents during interviews with the project team. Figure 4-15 shows a portion of the final requirements definition, and Figure 4-16 represents the requirements in the form of a concept map.

---

[14] This JAD session was not originally planned. As such, the workplan (see Figure 3-22) should be modified.

Nonfunctional Requirements

1. Operational Requirements

   1.1 The Internet sales system will draw information from the main CD information database, which contains basic information about CDs (e.g., title, artist, ID number, price, quantity in inventory). The Internet sales system will not write information to the main CD information database.

   1.2 The Internet sales system will store orders for new CDs in the special order system and will rely on the special order system to complete the special orders generated.

   1.3 A new module for the in-store system will be written to manage the "holds" generated by the Internet sales system. The requirements for this new module will be documented as part of the Internet sales system because they are necessary for the Internet sales system to function.

2. Performance Requirements

   No special performance requirements are anticipated.

3. Security Requirements

   No special security requirements are anticipated.

4. Cultural and Political Requirements.

   No special cultural and political requirements are anticipated.

Functional Requirements

1. Maintain CD Information

   1.1 The Internet sales system will need a database of basic information about the CDs that it can sell over the Internet, similar to the CD database at each of the retail stores (e.g., title, artist, ID number, price, quantity in inventory).

   1.2 Every day, the Internet sales system will receive an update from the distribution system that will be used to update this CD database. Some new CDs will be added, some will be deleted, and others will be revised (e.g., a new price).

   1.3 The electronic marketing (EM) manager (a position that will need to be created) will also have the ability to update information (e.g., prices for sales).

2. Maintain CD Marketing Information

   2.1 The Internet sales system provides an additional opportunity to market CDs to current and new customers. The system will provide a database of marketing materials about selected CDs that will help Web users learn more about them (e.g., music reviews, links to Web sites, artist information, and sample sound clips). When information about a CD that has additional marketing information is displayed, a link will be provided to the additional information.

   2.2 Marketing materials will be supplied primarily by vendors and record labels so that we can better promote their CDs. The EM manager of the marketing department will determine what marketing materials will be placed in the system and will be responsible for adding, changing, and deleting the materials.

3. Place CD Orders

   3.1 Customers will access the Internet sales system to look for CDs of interest. Some customers will search for specific CDs or CDs by specific artists, whereas other customers will want to browse for interesting CDs in certain categories (e.g., rock, jazz, classical).

   3.2 When the customer has found all the CDs he or she wants, the customer will "check out" by providing personal information (e.g., name, e-mail, address, credit card), and information regarding the order (e.g., the CDs to purchase, and the quantity for each item).

   3.3 The system will verify the customer's credit card information with an online credit card clearance center and either accept the order or reject it.

   3.4 Customers will also be able check to see if their preferred stores have the CDs in stock. They will use zip code to find stores close to their location. If the CD is available at a preferred store, a customer can immediately place a hold on the CD in stock and then come into the store and pick it up.

   3.5 If the CD is not available in the customer's preferred store, the customer can request that the CD be special ordered to that store for later pickup. The customer will be notified by e-mail when the requested CD arrives at the requested store; the CD will be placed on hold (which will again expire after 7 days). This process will work similarly to the current special order systems already available in the regular stores.

   3.6 Alternatively, the customer can mail order the CD (see requirement 4).

4. Fill Mail Orders

   4.1 When a CD is mail-ordered, the Internet sales system will send the mail order to the mail order distribution system.

   4.2 The mail-order distribution system will handle the actual sending of CDs to customers; it will notify the Internet sales system and e-mail the customer.

   4.3 Weekly reports can be run by the EM manager to check the order status.

FIGURE 4-15   CD Selections Requirements Definition

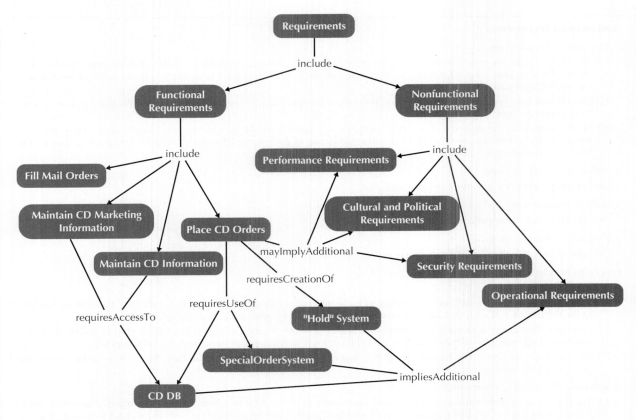

FIGURE 4-16   Concept Map Requirements Model

System Proposal

Alec reviewed the requirements definition and the other deliverables that the project team created during the inception phase. Given Margaret's desire to have the system operating before next year's Christmas season, Alec decided to timebox the project, and he determined what functionality could be included in the system by that schedule deadline (see Chapter 3). He suggested that the project team develop the system in three versions rather than attempting to develop a complete system that provided all the features initially. The first version, to be operational well before the holidays, would implement a basic system that would have the standard order features of other Internet retailers. The second version, planned for late spring or early summer, would have several features unique to CD Selections. The third version would add more advanced features, such as the ability to listen to a sample of music over the Internet, to find similar CDs, and to write reviews.

Alec revised the workplan accordingly, and he worked with Margaret and the folks in marketing to review the feasibility analysis and update it where appropriate. Using the system proposal template in Figure 4-14, Alec combined all of the deliverables from the project and submitted a system proposal to the steering committee for approval. Figure 4-17 shows the outline of the CD Selections system proposal. Margaret and Alec met with the committee and presented the highlights of what was learned during the

1. **Table of Contents**

2. **Executive Summary**
   To be completed once everything else is done.

3. **System Request**
   See Figure 2-13.

4. **Workplan**
   See Figure 3-22.

5. **Feasibility Analysis**
   See Figures 2-14 and 2-15.

6. **Requirements Definition**
   See Figure 4-15.

7. **Functional Model**
   To be completed in the future (see Chapter 5).

8. **Structural Models**
   To be completed in the future (see Chapter 6).

9. **Behavioral Model**
   To be completed in the future (see Chapter 7).

Appendices
   A. Size and Effort Estimate
      See Figure 3-21.

   B. Staffing Plan
      See Figure 3-23.

   C. Project Charter
      See Figure 3-24.

FIGURE 4-17
Outline of the CD
Selections System
Proposal

inception phase and the final concept of the new system. Based on the proposal and presentation, the steering committee decided that they would continue to fund the Internet sales system.

SUMMARY

Requirements Determination

Requirements determination is the part of analysis whereby the project team turns the very high-level explanation of the business requirements stated in the system request into a more precise list of requirements. A requirement is simply a statement of what the system must do or what characteristic it needs to have. Business requirements describe the what of the systems, and system requirements describe how the system will be implemented. A functional-requirement relates directly to a process the system has to perform or information it needs to contain. Nonfunctional requirements refer to behavioral properties that the system must have, such as performance and usability. All the functional and nonfunctional business requirements that fit within the scope of the system are written in the requirements definition, which is used to create other analysis deliverables and leads to the initial design for the new system.

Requirements Analysis Strategies

The basic process of analysis is divided into three steps: understanding the as-is system, identifying improvements, and developing requirements for the to-be system. Three requirements analysis strategies—BPA, BPI, and BPR—help the analyst lead users through the three (or two) analysis steps so that the vision of the system can be developed. BPA means leaving the basic way in which the organization operates unchanged and using computer technology to do some of the work. Problem analysis and root-cause analysis are two popular BPA techniques. BPI means making moderate changes to the way in which the organization operates to take advantage of new opportunities offered by technology or to copy what competitors are doing. Duration analysis, activity-based costing, and information benchmarking are three popular BPI activities. BPR means changing the fundamental way in which the organization operates. Outcome analysis, technology analysis, and activity elimination are three popular BPR activities.

Requirements-Gathering Techniques

Five techniques can be used to gather the business requirements for the proposed system: interviews, joint application development, questionnaires, document analysis, and observation. Interviews involve meeting one or more people and asking them questions. There are five basic steps in the interview process: selecting interviewees, designing interview questions, preparing for the interview, conducting the interview, and postinterview follow-up. JAD allows the project team, users, and management to work together to identify requirements for the system. Electronic JAD attempts to overcome common problems associated with groups by using groupware. A questionnaire is a set of written questions for obtaining information from individuals. Questionnaires are often used when there is a large number of people from whom information and opinions are needed. Document analysis entails reviewing the documentation and examining the system itself. It can provide insights into the formal and informal system. Observation, the act of watching processes being performed, is a powerful tool for gathering information about the as-is system because it enables the analyst to see the reality of a situation firsthand.

The System Proposal

The system proposal documents the results of the planning and analysis activities in a single, comprehensive document. The actual format of the system proposal depends somewhat on the client. For example, the federal government has very specific requirements that a system proposal must meet, whereas a small locally owned bike shop would be willing to use a much simpler format.

KEY TERMS

Activity elimination	Bottom-up interview	Business requirements
Activity-based costing	Breadth of analysis	Closed-ended question
Analysis	Business process automation (BPA)	Critical thinking skills
As-is system	Business process improvement (BPI)	Document analysis
Benchmarking	Business process reengineering (BPR)	Duration analysis

Electronic JAD (e-JAD)
Facilitator
Formal system
Functional requirements
Ground rules
Informal benchmarking
Informal system
Interpersonal skills
Interview
Interview notes
Interview report
Interview schedule
JAD (joint application development)
Nonfunctional requirements

Observation
Open-ended question
Outcome analysis
Parallelization
Process Integration
Postsession report
Potential business value
Probing question
Problem analysis
Project cost
Questionnaire
Requirement
Requirements definition
Requirements determination

Risk
Root cause
Root-cause analysis
Sample
Scribe
Structured interview
System proposal
System requirements
Technology analysis
To-be system
Top-down interview
Unstructured interview
Walkthrough

QUESTIONS

1. What are the three basic steps of the analysis process? Which step is sometimes skipped or done in a cursory fashion? Why?
2. What are the important factors in selecting an appropriate analysis strategy?
3. What are the key deliverables that are created during the analysis phase? What is the final deliverable from the analysis phase, and what does it contain?
4. What is the purpose of the requirements definition?
5. Compare and contrast the business goals of BPA, BPI, and BPR.
6. What is the difference between an as-is system and a to-be system?
7. Assuming time and money were not important concerns, would BPR projects benefit from additional time spent understanding the as-is system? Why or why not?
8. Compare and contrast problem analysis and root-cause analysis. Under what conditions would you use problem analysis? Under what conditions would you use root-cause analysis?
9. Compare and contrast duration analysis and activity-based costing.
10. Describe the five major steps in conducting interviews.
11. How can you differentiate between facts and opinions? Why can both be useful?

12. What is document analysis?
13. Explain the differences between unstructured interviews and structured interviews. When would you use each approach?
14. What are typical response rates for questionnaires and how can you improve them?
15. Explain the difference between a closed-ended question, an open-ended question, and a probing question. When would you use each?
16. What are the key aspects of using observation in the information gathering process?
17. Describe the five major steps in conducting JAD sessions.
18. How are participants selected for interviews and JAD sessions?
19. What are the three primary things that a facilitator does in conducting the JAD session?
20. How does a JAD facilitator differ from a scribe?
21. How does designing questions for questionnaires differ from designing questions for interviews or JAD sessions?
22. What is e-JAD and why might a company be interested in using it?
23. Explain factors that can be used to select information gathering techniques.

EXERCISES

A. Suppose your university is having a dramatic increase in enrollment and is having difficulty finding enough seats in courses for students. Perform a technology analysis to identify new ways to help students complete their studies and graduate.

B. Review the Amazon.com Web site. Develop the requirements definition for the site. Create a list of functional business requirements that the system meets. What different kinds of non-functional business requirements does the system meet? Provide examples for each kind.

C. Suppose you are going to build a new system that automates or improves the interview process for the career services department of your school. Develop a requirements definition for the new system. Include both functional and nonfunctional system requirements. Pretend you will release the system in three different versions. Prioritize the requirements accordingly.

D. Suppose you are the analyst charged with developing a new system to help senior managers make better strategic decisions. What requirements-gathering techniques will you use? Describe in detail how you would apply the techniques.

E. Suppose you are the analyst charged with developing a new system for the university bookstore so students can order books online and have them delivered to their dorms or off-campus housing. What requirements-gathering techniques will you use? Describe in detail how you would apply the techniques.

F. Describe in very general terms the as-is business process for registering for classes at your university. What BPA technique would you use to identify improvements? With whom would you use the BPA technique? What requirements-gathering technique would help you apply the BPA technique? List some examples of improvements that you would expect to find.

G. Describe in very general terms the as-is business process for registering for classes at your university. What BPI technique would you use to identify improvements?

With whom would you use the BPI technique? What requirements-gathering technique would help you apply the BPI technique? List some examples of improvements that you would expect to find.

H. Describe in very general terms the as-is business process for registering for classes at your university. What BPR technique would you use to identify improvements? With whom would you use the BPR technique? What requirements gathering technique would help you apply the BPR technique? List some examples of improvements that you would expect to find.

I. Find a questionnaire on the Web that has been created to capture customer information. Describe the purpose of the survey, the way questions are worded, and how the questions have been organized. How can it be improved? How will the responses be analyzed?

J. Develop a questionnaire that will help gather information regarding processes at a popular restaurant, or the college cafeteria (e.g., ordering, customer service). Give the questionnaire to ten to fifteen students, analyze the responses, and write a brief report that describes the results.

K. Contact the career services department at your university and find all the pertinent documents designed to help students find permanent and/or part-time jobs. Analyze the documents and write a brief report.

L. Find a partner and interview each other about what tasks each did in the last job you held (full-time, part-time, past, or current). If you haven't worked before, then assume your job is being a student. Before you do this, develop a brief interview plan. After your partner interviews you, identify the type of interview, interview approach, and types of questions used.

M. Find a group of students and run a 60-minute JAD session on improving alumni relations at your university. Develop a brief JAD plan, select two techniques that will help identify improvements, and then develop an agenda. Conduct the session using the agenda, and write your postsession report.

MINICASES

1. Brian Callahan, IS project manager, is just about ready to depart for an urgent meeting called by Joe Campbell, manager of manufacturing operations. A major BPI project sponsored by Joe recently cleared the approval hurdle, and Brian helped bring the project through project initiation. Now that the approval committee has given the go-ahead, Brian has been working on the project's analysis plan.

One evening, while playing golf with a friend who works in the manufacturing operations department,

Brian learned that Joe wants to push the project's time frame up from Brian's original estimate of 13 months. Brian's friend overheard Joe say, "I can't see why that IS project team needs to spend all that time analyzing things. They've got two weeks scheduled just to look at the existing system! That seems like a real waste. I want that team to get going on building my system."

Because Brian has a little inside knowledge about Joe's agenda for this meeting, he has been considering how to handle Joe. What do you suggest Brian tell Joe?

2. The Victoria Public Service Council is a small organization with a membership of 1000 public sector workers in the Australian state of Victoria. Most members work in small towns across the state. The Council was formed in 1922 with the goal of providing education and training to small town government employees (e.g., city managers, police chiefs, fire fighters). The Council provides online materials and also organizes a conference each year, bringing together members from all over the state. Members' dues are billed annually. The Council's bookkeeping is handled by the elected treasurer, Jack Archibald. Jack runs unopposed each year at the election, because no one wants to take over the tedious and time-consuming job of tracking memberships for the small $4000 annual stipend. Several Council members feel that the Council needs to improve its systems before Jack decides to step down from his position.

The system that is currently used to track the billing and receipt of funds was developed many years ago and is not user-friendly. The company that wrote the system is no longer in business, Questions from members concerning their statements cannot be easily answered. Usually Jack just jots down the enquiry and calls back the member with the answer. Sometimes he has to calculate manually because the system was not programmed to handle certain types of queries. For example, the membership report does not alphabetize members by city; only the names of cities are listed in alphabetical order.

What requirements analysis strategy or strategies would you recommend for this situation? Explain your answer.

3. Barry has recently been assigned to a project team that will be developing a new retail store management system for a chain of submarine sandwich shops. Barry has several years of experience in programming, but he has not done much analysis in his career. He was a little nervous about the new work he would be doing, but was confident he could handle any assignment he was given.

One of Barry's first assignments was to visit one of the submarine sandwich shops and prepare an observation report on how the store operates. Barry planned to arrive at the store around noon, but he chose a store in an area of town he was unfamiliar with, and due to traffic delays and difficulty in finding the store, he did not arrive until 1:30. The store manager was not expecting him and refused to let a stranger behind the counter until Barry had her contact the project sponsor (the director of store management) at company headquarters to verify who he was and what his purpose was.

After finally securing permission to observe, Barry stationed himself prominently in the work area behind the counter so that he could see everything. The staff had to maneuver around him as they went about their tasks, but there were only minor occasional collisions. Barry noticed that the store staff seemed to be going about their work very slowly and deliberately, but he supposed that was because the store wasn't very busy. At first, Barry questioned each worker about what he or she was doing, but the store manager eventually asked him not to interrupt their work so much—he was interfering with their service to the customers.

By 3:30, Barry was a little bored. He decided to leave, figuring he could get back to the office and prepare his report before 5:00 that day. He was sure his team leader would be pleased with his quick completion of his assignment. As he drove, he reflected, There really won't be much to say in this report. All they do is take the order, make the sandwich, collect the payment, and hand over the order. It's really simple! Barry's confidence in his analytical skills soared as he anticipated his team leader's praise.

Back at the store, the store manager shook her head, commenting to her staff, "He comes here at the slowest time of day on the slowest day of the week. He never even looked at all the work I was doing in the back room while he was here—summarizing yesterday's sales, checking inventory on hand, making up resupply orders for the weekend . . . plus he never even considered our store-opening and -closing procedures. I hate to think that the new store management system is going to be built by someone like that. I'd better contact Chuck (the director of store management) and let him know what went on here today." Evaluate Barry's conduct of the observation assignment.

4. Anne has been given the task of conducting a survey of sales clerks who will be using a new order-entry system being developed for a household products catalog

company. The goal of the survey is to identify the clerks' opinions on the strengths and weaknesses of the current system. There are about 50 clerks who work in three different cities, so a survey seemed like an ideal way of gathering the needed information from the clerks.

Anne developed the questionnaire carefully and pretested it on several sales supervisors who were available at corporate headquarters. After revising it based on their suggestions, she sent a paper version of the questionnaire to each clerk, asking that it be returned within one week. After one week, she had only three completed questionnaires returned. After another week, Anne received just two more completed questionnaires. Feeling somewhat desperate, Anne then sent out an e-mail version of the questionnaire, again to all the clerks, asking them to respond to the questionnaire by e-mail as soon as possible. She received two e-mail questionnaires and three messages from clerks who had completed the paper version expressing annoyance at being bothered with the same questionnaire a second time. At this point, Anne has just a 14 percent response rate, which she is sure will not please her team leader. What suggestions do you have that could have improved Anne's response rate to the questionnaire?

PART TWO

ANALYSIS

Analysis modeling answers the questions of *who* will use the system, *what* the system will do, and *where* and *when* it will be used. During this phase, the project team will learn about the system. The team then produces the functional model (activity diagrams, use-case descriptions and diagram), structural model (CRC cards and class and object diagrams), and behavioral models (sequence diagrams, communication diagrams, behavioral state machines, and a CRUD matrix).

CHAPTER 5 □
USE CASE
DIAGRAMS

Functional
Models

CHAPTER 6 ■
CLASS
DIAGRAMS

Structural
Models

CHAPTER 7 □
INTERACTION
DIAGRAMS

Behavioral
Models

CHAPTER 5

USE CASE DIAGRAMS

Functional models describe business processes and the interaction of an information system with its environment. In object-oriented systems development, two types of models are used to describe the functionality of an information system: activity diagrams and use cases. Activity diagrams support the logical modeling of business processes and workflows. Use cases are used to describe the basic functions of the information system. Both activity diagrams and use cases can be used to describe the current as-is system and the to-be system being developed. This chapter describes functional modeling as a means to document and understand requirements and to understand the function or external behavior of the system. This chapter also introduces use case points as a basis for project size estimation.

OBJECTIVES

- Understand the rules and style guidelines for activity diagrams.
- Understand the rules and style guidelines for use cases and use-case diagrams.
- Understand the process used to create use cases and use-case diagrams
- Be able to create functional models using activity diagrams, use cases, and use-case diagrams.

CHAPTER OUTLINE

Introduction
Business Process Modeling with
  Activity Diagrams
  Elements of an Activity Diagram
  Guidelines for Creating Activity
    Diagrams
Use-Case Descriptions
  Types of Use Cases
  Elements of a Use-Case Description
  Guidelines for Creating Use-Case
    Descriptions
Use-Case Diagrams
  Actors
  Association
  Use Case
  System Boundary
Creating Use-Case Descriptions
  and Use-Case Diagrams

Identifying the Major Use Cases
Expanding the Major Use Cases
Confirming the Major Use Cases
Creating a Use-Case Diagram
Refining Project Size and Effort
  Estimation Using Use-Case Points
Applying the Concepts
  at CD Selections
  Business Process Modeling with
    Activity Diagrams
  Identifing the Major Use Cases
  Expanding the Major Use Cases
  Confirming the Major Use Cases
  Creating the Use-Case Diagram
  Refining Project Size and Effort
    Estimation Using Use-Case Points
Summary

INTRODUCTION

The previous chapter discussed the more popular requirements-gathering techniques, such as interviewing, JAD, and observation. Using these techniques, the analyst determined the requirements and created a requirements definition. The requirements definition defined what the system is to do. In this chapter, we discuss how the information that is gathered using these techniques is organized and presented in the form of activity diagrams and use cases. Because UML has been accepted as the standard notation by the OMG, almost all object-oriented development projects today utilize activity diagrams and use cases to document and organize the requirements that are obtained during the analysis phase.[1]

An *activity diagram* can be used for any type of process-modeling activity.[2] In this chapter, we describe their use in the context of business process modeling. *Process models* depict how a business system operates. They illustrate the processes or activities that are performed and how objects (data) move among them. A process model can be used to document a current system (i.e., as-is system) or a new system being developed (i.e., to-be system), whether computerized or not. There are many different process-modeling techniques in use today:[3]

A *use case* is a formal way of representing the way in which a business system interacts with its environment. It illustrates the activities performed by the users of the system. As such, use-case modeling is often thought of as an external or functional view of a business process in that it shows how the users view the process, rather than the internal mechanisms by which the process and supporting systems operate. Like activity diagrams, use cases can document the current system (i.e., as-is system) or the new system being developed (i.e., to-be system).

Activity diagrams and use cases are *logical models*—models that describe the business domain's activities without suggesting how they are conducted. Logical models are sometimes referred to as problem domain models. Reading an activity diagram or use case, in principle, should not indicate if an activity is computerized or manual; if a piece of information is collected by paper form or via the Web; or if information is placed in a filing cabinet or a large database. These physical details are defined during design when the logical models are refined into *physical models*. These models provide information that is needed to ultimately build the system. By focusing on logical activities first, analysts can focus on how the business should run without being distracted with implementation details.

As a first step, the project team gathers requirements from the users (see Chapter 4). Next, using the gathered requirements, the project team models the overall business process using activity diagrams. Next the team uses the identified activities to pinpoint the use cases that occur in the business. They then prepare *use-case descriptions* and *use-case diagrams*

---

[1] Other similar techniques that are commonly used in non-UML projects are task modeling—see Ian Graham, *Migrating to Object Technology* (Reading, MA: Addison-Wesley, 1995); and Ian Graham, Brian Henderson-Sellers, and Houman Younessi, *The OPEN Process Specification*, (Reading, MA: Addison-Wesley, 1997)—and scenario-based design—see John M. Carroll, *Scenario-Based Design: Envisioning Work and Technology in System Development* (New York: Wiley, 1995).

[2] We actually used an activity diagram to describe a simple process in Chapter 1 (see Figure 1-1).

[3] Another commonly used process-modeling technique is IDEF0. IDEF0 is used extensively throughout the U.S. federal government. For more information about IDEF0, see FIPS 183: Integration Definition for Function Modeling (IDEF0), Federal Information Processing Standards Publications (Washington, DC: U.S. Department of Commerce, 1993). Also, from an object-oriented perspective, a good book that utilizes the UML to address business process modeling is Hans-Erik Eriksson and Magnus Penker, *Business Modeling with UML* (New York: Wiley, 2000).

for each use case. Use cases are the discrete activities that the users perform, such as selling CDs, ordering CDs, and accepting returned CDs from customers. Users work closely with the project team to create the business process models in the form of activity diagrams and use-case descriptions that contain all the information needed to start modeling the system. Once the activity diagrams and use-case descriptions are prepared, the systems analysts transform them into a use-case diagram, which portrays the external behavioral or functional view of the business process. Next, the analyst typically creates class diagrams (see Chapter 6) to create a structural model of the business problem domain.

In this chapter, we first describe business process modeling and activity diagrams. Second we describe use cases, their elements, and a set of guidelines for creating use cases. Third, we describe use-case diagrams and how to create them. Fourth, using use cases as a basis, we revisit the topic of project size estimation.

BUSINESS PROCESS MODELING WITH ACTIVITY DIAGRAMS

Business process models describe the different activities that, when combined together, support a business process. Business processes typically cut across functional departments (e.g., the creation of a new product will involve many different activities that will combine the efforts of many employees in many departments). Furthermore, from an object-oriented perspective, they cut across multiple objects. As such, many of the earlier object-oriented systems development approaches tended to ignore business process modeling. Instead, they focused only on modeling processes via use cases (see later in this chapter) and behavioral models (see Chapter 7). However, today we realize that modeling business processes themselves is a very constructive activity that can be used to make sense out of the gathered requirements (see Chapter 4). The one potential problem of building business process models, from an object-oriented systems development perspective, is that they tend to reinforce a functional decomposition mindset. However, as long as they are used properly, they are a very powerful tool for communicating the analyst's current understanding of the requirements with the user. In this section of the chapter, we introduce the use of UML 2.0's activity diagrams as a means to build business process models.

Activity diagrams are used to model the behavior in a business process independent of objects. In many ways, activity diagrams can be viewed as sophisticated data flow diagrams that are used in conjunction with structured analysis.[4] However, unlike data flow diagrams, activity diagrams include notation that addresses the modeling of parallel, concurrent activities and complex decision processes. As such, activity diagrams can be used to model everything from a high-level business workflow that involves many different use cases, to the details of an individual use case, all the way down to the specific details of an individual method. In a nutshell, activity diagrams can be used to model any type of process.[5] In this chapter, we restrict our coverage of activity diagrams to the modeling of high-level business processes.

[4] For a good introduction to data flow diagrams and structured approaches to systems analysis and design, see Alan Dennis, Barbara Haley Wixom, and Roberta M. Roth, *Systems Analysis & Design,* 3rd ed. (New York: Wiley, 2006).

[5] Technically speaking, activity diagrams combine process modeling ideas from many different techniques including event models, statecharts, and Petri nets. However, UML 2.0's activity diagram has more in common with Petri nets than the other process modeling techniques. For a good description of using Petri nets to model business workflows, see Wil van der Aalst and Kees van Hee, *Workflow Management: Models, Methods, and Systems* (Cambridge, MA: MIT Press, 2002).

An action: ■ Is a simple, nondecomposable piece of behavior. ■ Is labeled by its name.	Action
An activity: ■ Is used to represent a set of actions. ■ Is labeled by its name.	Activity
An object node: ■ Is used to represent an object that is connected to a set of object flows. ■ Is labeled by its class name.	Class Name
A control flow: ■ Shows the sequence of execution.	→
An object flow: ■ Shows the flow of an object from one activity (or action) to another activity (or action).	⇢
An initial node: ■ Portrays the beginning of a set of actions or activities.	●
A final-activity node: ■ Is used to stop all control flows and object flows in an activity (or action).	◉
A final-flow node: ■ Is used to stop a specific control flow or object flow.	⊗
A decision node: ■ Is used to represent a test condition to ensure that the control flow or object flow only goes down one path. ■ Is labeled with the decision criteria to continue down the specific path.	◇ [Decision Criteria] [Decision Criteria]
A merge node: ■ Is used to bring back together different decision paths that were created using a decision node.	◇

FIGURE 5-1   Syntax for an Activity Diagram

Elements of An Activity Diagram

Activity diagrams portray the primary activities and the relationships among the activities in a process. Figure 5-1 shows the syntax of an activity diagram. Figure 5-2 presents a simple activity diagram that represents part of an appointment system for a doctor's office.[6]

---

[6] Due to the actual complexity of the syntax of activity diagrams, we follow a minimalist philosophy in our coverage [see John M. Carrol, *The Nurnberg Funnel: Designing Minimalist Instruction for Practical Computer Skill* (Cambridge, MA: MIT Press, 1990)]. However, the material contained in this section is based on the *Unified Modeling Language: Superstructure Version 2.0, ptc/03-08-02* (www.uml.org). Additional useful references include Michael Jesse Chonoles and James A. Schardt, *UML 2 for Dummies* (Indianapolis, IN: Wiley, 2003); Hans-Erik Eriksson, Magnus Penker, Brian Lyons, and David Fado, *UML 2 Toolkit* (Indianapolis, IN: Wiley, 2004); and Kendall Scott, *Fast Track UML 2.0* (Berkeley, CA: Apress, 2004). For a complete description of all diagrams, see www.uml.org.

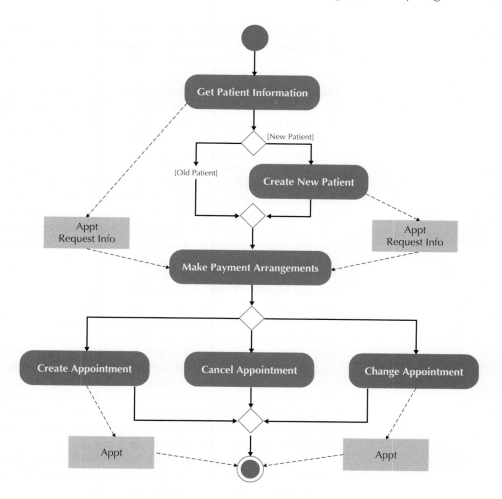

FIGURE 5-2
Activity Diagram for
Appointment System

Actions and Activities   *Actions* and *activities* are performed for some specific business reason. Actions and activities can represent manual or computerized behavior. They are depicted in an activity diagram as a rounded rectangle (See Figure 5-1). Furthermore, they should have a name that begins with a verb and ends with a noun (e.g., Make Appointment or Make Payment Arrangements). Names should be short, yet contain enough information so that the reader can easily understand exactly what they do. The only difference between an action and an activity is that an activity can be decomposed further into a set of activities and/or actions, whereas an action represents a simple nondecomposable piece of the overall behavior being modeled. Typically, only activities are used for business process or workflow modeling. Furthermore, in most cases, each activity is associated with a use case. The activity diagram in Figure 5-2 shows six separate but related activities for a typical appointment system used in a doctor's office: Get Patient Information, Create New Patient, Make Payment Arrangements, Create Appointment, Cancel Appointment, and Change Appointment.

Object Nodes   Activities and actions typically modify or transform objects. *Object nodes* model these objects in an activity diagram. Object nodes are portrayed in an activity diagram as rectangles (See Figure 5-1). The name of the class of the object is written inside the rectangle. Essentially, object nodes represent the flow of information from one activity to another activity. The simple appointment system portrayed in Figure 5-2 shows object nodes flowing from the Create Appointment and Change Appointment activities.

Control Flows and Object Flows   There are two different types of flows in activity diagrams: control and object (See Figure 5-1). *Control flows* model the paths of execution through a business process. A control flow is portrayed as a solid line with an arrowhead on it showing the direction of flow. Control flows can be attached only to actions or activities. Figure 5-2 portrays a set of control flows through a doctor's office's appointment system. *Object flows* model the flow of objects through a business process. Because activities and actions modify or transform objects, object flows are necessary to show the actual objects that flow into and out of the actions or activities.[7] An object flow is depicted as a dashed line with an arrowhead on it showing the direction of flow. An individual object flow must be attached to an action or activity on one end and an object node on the other end. Figure 5-2 portrays a set of control and object flows through the appointment system of a doctor's office.

Control Nodes   There are seven different types of *control nodes* in an activity diagram: initial, final-activity, final-flow, decision, merge, fork, and join (see Figure 5-1). An *initial node* portrays the beginning of a set of actions or activities.[8] An initial node is shown as a small, filled-in circle. A *final-activity node* is used to stop the process being modeled. Any time a final-activity node is reached, all actions and activities are ended immediately, regardless of whether they are completed. A final-activity node is represented as a circle surrounding a small, filled-in circle, making it resemble a bull's eye. A new control node added to the activity diagram in UML 2.0 is the final-flow node. A *final-flow* node is similar to a final-activity node, except that it stops a specific path of execution through the business process but allows the other concurrent or parallel paths to continue. A final-flow node is shown as a small circle with an X in it.

The decision and merge nodes support modeling the decision structure of a business process. The *decision node* is used to represent the actual test condition that determines which of the paths exiting the decision node is to be traversed. In this case, each exiting path must be labeled with a guard condition. A *guard condition* represents the value of the test for that particular path to be executed. For example, in Figure 5-2, the decision node immediately below the Get Patient Information activity has two mutually exclusive paths that could be executed: one for old, or previous, patients, the other for new patients. The *merge node* is used to bring back together multiple mutually exclusive paths that have been split based on an earlier decision (e.g., the old- and new-patient paths in Figure 5-2 are brought back together). However, sometimes, for clarity purposes, it may be better not to use a merge node. For example, in Figure 5-3, which of the two activity diagrams, both representing an overview level of an order process, is easier to understand, the one on the left or the one on the right? The one on the left contains a merge node for the More Items on Order question, but the one on the right does not. In a sense, the decision node is playing double duty in the diagram on the right. It also serves as a merge node. Technically speaking, we should not omit the merge node, however, sometimes being technically correct according to the UML's diagramming rules may actually cause the diagram to become confusing. From a business process modeling perspective, a good deal of common sense can go a long way.

---

[7] These are identical to data flows in data flow diagrams.

[8] For those familiar with IBM flowcharts, this is similar to the start node.

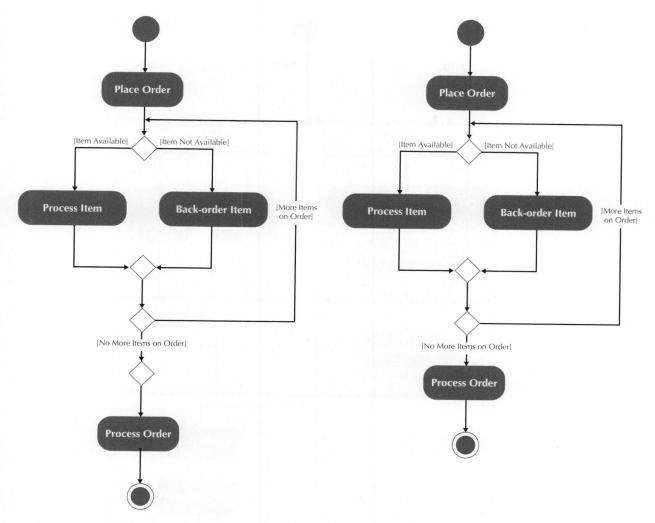

FIGURE 5-3   Two Very Similar Activity Diagrams

The fork and join nodes allow parallel and concurrent processes to be modeled (see Figure 5-1). The *fork node* is used to split the behavior of the business process into multiple parallel or concurrent flows. Unlike the decision node, the paths are not mutually exclusive (i.e., both paths are executed concurrently). For example, in Figure 5-4, the fork node is used to show that two concurrent, parallel processes are to be executed. In this case, each process is executed by two separate processors (parents). The purpose of the join node is similar to that of the merge node. The *join node* simply brings back together the separate parallel or concurrent flows in the business process into a single flow.

Swimlanes   As already described, activity diagrams can model a business process independent of any object implementation. However, there are times when it helps to break up an activity diagram in such a way that it can be used to assign responsibility to objects or individuals who would actually perform the activity. This is especially useful when modeling a business workflow and is accomplished through the use of *swimlanes*. In Figure 5-4, the

FIGURE 5-4
Activity Diagram for Making a School Box Lunch

swimlanes are used to break up among two parents the making of a school lunch comprising a peanut butter and jelly sandwich, a drink, and dessert. In this case, we use vertical swimlanes. We could also draw the activity diagram using more of a left-to-right orientation instead of a top-down orientation. In that case, the swimlanes would be drawn horizontally.

In an actual business workflow, there would be activities that should be associated with roles of individuals involved in the business workflow (e.g., employees or customers) and the activities to be accomplished by the information system being created. This association

YOUR **5-1 Activity Diagrams**

TURN

*L*ook at the activity diagram for the appointment system in Figure 5-2. Think of one more activity that a user might ask for when gathering requirements for this system (e.g., maintaining patient insurance information).

Questions

1. How would you depict this on the existing diagram?

2. After adding the activity to the diagram, what is your recommendation?

3. Would you keep the new activity within the scope of this system? Why or why not?

of activities with external roles, internal roles, and the system is very useful when creating the use-case descriptions and diagrams described later in this chapter.

Guidelines for Creating Activity Diagrams

Scott Ambler suggests the following guidelines when creating activity diagrams:[9]

1. Because an activity diagram can be used to model any kind of process, you should set the context or scope of the activity being modeled. Once you have determined the scope, you should give the diagram an appropriate title.

2. You must identify the activities, control flows, and object flows that occur between the activities.

3. You should identify any decisions that are part of the process being modeled.

4. You should attempt to identify any prospects for parallelism in the process.

5. You should draw the activity diagram.

When drawing an activity diagram, the diagram should be limited to a single initial node that starts the process being modeled. This node should be placed at the top or top left of the diagram, depending on the complexity of the diagram. Furthermore, for most business processes, there should only be a single final-activity node. This node should be placed at the bottom or bottom right of the diagram (see Figures 5-2, 5-3, and 5-4). Because most high-level business processes are sequential, not parallel, the use of a final-flow node should be limited.

When modeling high-level business processes or workflows, only the more important decisions should be included in the activity diagrams. In those cases, the guard conditions associated with the outflows of the decision nodes should be mutually exclusive. Furthermore, the outflows and guard conditions should form a complete set (i.e., all potential values of the decision are associated with one of the flows).

As in decision modeling, forks and joins should be included only to represent the more important parallel activities in the process. For example, an alternative version of Figure 5-4 may not include the forks and joins associated with the Get Jelly, Get Bread, Get Peanut Butter, Get Drink, and Get Dessert activities. This would greatly simplify the diagram.[10]

---

[9] The guidelines presented here are based on work done by Scott Ambler. For more details, see Scott W. Ambler, *The Object Primer: The Application Developer's Guide to Object Orientation and the UML,* 2nd ed. (Cambridge, UK: Cambridge University Press/SIGS Books, 2001); and Scott W. Ambler, *The Elements of UML Style* (Cambridge, UK: Cambridge University Press, 2003).

[10] In fact, the only reason we depicted the diagram in Figure 5-3 with the multiple fork and join nodes was to demonstrate that it could be done.

When laying out the activity diagram, line crossings should be minimized to enhance the readability of the diagram. The activities on the diagram should also be laid out in a left-to-right and/or top-to-bottom order, based on the order in which the activities are executed. For example, in Figure 5-4, the Create Sandwich activity takes place before the Create Lunch activity.

Swimlanes should be used only to simplify the understanding of an activity diagram. Furthermore, the swimlanes should enhance the readability of a diagram. For example, when using a horizontal orientation for swimlanes, the top swimlane should represent the most important object or individual involved with the process. The order of the remaining swimlanes should be based on minimizing the number of flows crossing the different swimlanes. Also, when there are object flows among the activities associated with the different individuals (swimlanes) executing the activities of the process, it is useful to show the actual object flowing from one individual to another individual by including an object node between the two individuals (i.e., between the two swimlanes). This, of course, impacts how the swimlanes should be placed on the diagram.

Finally, any activity that does not have any outflows or any inflows should be challenged. Activities with no outflows are referred to as *black-hole activities*. If the activity is truly an end point in the diagram, the activity should have a control flow from it to a final-activity or final-flow node. An activity that does not have any inflow is known as a *miracle activity*. In this case, the activity is either missing an inflow from the initial node of the diagram or from another activity.

USE-CASE DESCRIPTIONS

Use cases are simple descriptions of a system's functions from the bird's-eye view of the users.[11] Use-case diagrams are functional diagrams in that they portray the basic functions of the system—that is, what the users can do and how the system should respond to the user's actions.[12] Creating use-case diagrams is a two-step process: First, the users work with the project team to write text-based use-case descriptions; second, the project team translates the use-case descriptions into formal use-case diagrams. Both the use-case descriptions and the use-case diagrams are based on the identified requirements and the activity diagram description of the business process. Use-case descriptions contain all the information needed to produce use-case diagrams. Although it is possible to skip the use-case description step and move directly to creating use-case diagrams and the other diagrams that follow, users often have difficulty describing their business processes using only use-case diagrams. Through the creation of use-case descriptions, users can describe the required details of each individual use case. As to which should come first—use-case descriptions or a use-case diagram—technically speaking, it really does not matter. Both should be done to fully describe the requirements that an information system must meet. In this text, we present the creation of a use-case description first.[13]

---

[11] For a more detailed description of use-case modeling, see Alistair Cockburn, *Writing Effective Use Cases* (Reading, MA: Addison-Wesley, 2001).

[12] Nonfunctional requirements, such as reliability requirements and performance requirements, are often documented outside of the use case through more traditional requirements documents. See Gerald Kotonya and Ian Sommerville, *Requirements Engineering* (Chichester, England: Wiley, 1998); Benjamin L. Kovitz, *Practical Software Requirements: A Manual of Content & Style* (Greenwich, CT: Manning, 1999); Dean Leffingwell and Don Widrig, *Managing Software Requirements: A Unified Approach* (Reading, MA: Addison-Wesley, 2000); and Richard H. Thayer, M. Dorfman, and Sidney C. Bailin (eds.), *Software Requirements Engineering*, 2nd ed. (Los Alamitos, CA: IEEE Computer Society, 1997).

[13] Practically speaking, the analyst and users will iterate between the descriptions and diagram. Do not worry about which comes first. Whichever seems to work with the user community of the current project is the way that it should be done.

Use cases are the primary drivers for all the UML diagramming techniques. A use case communicates at a high level what the system needs to do, and all the UML diagramming techniques build on this by presenting the use-case functionality in a different way for a different purpose. Use cases are the building blocks by which the system is designed and built.

Use cases capture the typical interaction of the system with the system's users (end users and other systems). These interactions represent the external, or functional, view of the system from the perspective of the user. Each use case describes one and only one function in which users interact with the system,[14] although a use case may contain several paths that a user can take while interacting with the system (e.g., when searching for a book in a Web bookstore, the user might search by subject, by author, or by title). Each possible execution path through the use case is referred to as a *scenario*. Another way to look at a scenario is as if a scenario is an instantiation of a specific use case. Scenarios are used extensively in behavioral modeling (see Chapter 7). Finally, by identifying all scenarios and trying to execute them through role playing CRC cards (see Chapter 6), you will be testing the clarity and completeness of your evolving understanding of the system being developed.[15]

When creating use cases, the project team must work closely with the users to gather the requirements needed for the use cases. This is often done through interviews, JAD sessions, and observation. Gathering the requirements needed for a use case is a relatively simple process, but it takes considerable practice. A good place to look for potential use cases is the activity diagram representation of the business process. In many cases, the activities identified in the activity diagram will become the use cases in the business process being modeled. The key thing to remember is that each use case is associated with one and only one *role* that users have in the system. For example, a receptionist in a doctor's office may play multiple roles—he or she can make appointments, answer the telephone, file medical records, welcome patients, and so on. Furthermore, it is possible that multiple users will play the same role. As such, use cases should be associated with the roles played by the users and not with the users themselves.

Types of Use Cases

There are many different types of use cases. We suggest two separate dimensions on which to classify a use case based on the purpose of the use case and the amount of information that the use case contains: (1) overview versus detail; and (2) essential versus real.

An *overview use case* is used to enable the analyst and user to agree on a high-level overview of the requirements. Typically, they are created very early in the process of understanding the system requirements, and they document only basic information about the use case, such as its name, ID number, primary actor, type, and a brief description.

---

[14] This is one key difference between traditional structured analysis and design approaches and object-oriented approaches. If you have experience using traditional structured approaches (or have taken a course on them), then this is an important change for you. If you have no experience with structured approaches, then you can skip this footnote.

The traditional structured approach is to start with one overall view of the system and to model processes via functional decomposition—the gradual decomposition of the overall view of the system into the smaller and smaller parts that make up the whole system. On the surface, this is similar to business process modeling using activity diagrams. However, functional decomposition is *not* used with object-oriented approaches. Instead, each of the use cases documents one individual piece of the system; there is no overall use case that documents the entire system in the same way that a level 0 data flow diagram attempts to document the entire system. By removing this overall view, object-oriented approaches make it easier to decouple the system's objects so they can be designed, developed, and reused independently of the other parts of the system. Although this lack of an overall view may prove unsettling initially, it is very liberating over the long term.

[15] For presentation purposes, we defer discussion of role-playing to Chapter 6, walkthroughs to Chapter 8, and testing to Chapter 13.

Once the user and the analyst agree upon a high-level overview of the requirements, the overview use cases can be converted to detail use cases. A *detail use case* typically documents, as far as possible, all the information needed for the use case.

An *essential use case* is one that describes only the minimum essential issues necessary to understand the required functionality. A *real use case* goes further and describes a specific set of steps. For example, an essential use case in a dentist office might say that the receptionist should attempt to match the patient's desired appointment times with the available times, whereas a real use case might say that the receptionist should look up the available dates on the calendar using MS Exchange to determine if the requested appointment times were available. The primary difference is that essential use cases are implementation independent, whereas real use cases are detailed descriptions of how to use the system once it is implemented. As such, real use cases tend to be used only in detailed design, implementation, and testing.

Elements of a Use-Case Description

A use-case description contains all the information needed to build the diagrams that follow, but it expresses the information in a less formal way that is usually simpler for users to understand. Figure 5-5 shows a sample use-case description.[16] A use-case description has three basic parts: overview information, relationships, and the flow of events.

Overview Information   The overview information identifies the use case and provides basic background information about the use case. The *use-case name* should be a verb–noun phrase (e.g., Make Appointment). The *use-case ID number* provides a unique way to find every use case and also enables the team to trace design decisions back to a specific requirement. As already stated, the *use-case type* is either overview or detail and essential or real. The *primary actor* is usually the trigger of the use case—the person or thing that starts the execution of the use case. The primary purpose of the use case is to meet the goal of the primary actor. The *brief description* is typically a single sentence that describes the essence of the use case.

The *importance level* can be used to prioritize the use cases. As described in Chapter 1, object-oriented development tends to follow a RAD-phased development approach, in which some parts of the system are developed first and other parts are developed only in later versions. The importance level enables the users to explicitly prioritize which business functions are most important and need to be part of the first version of the system and which are less important and can wait until later versions if necessary.

The importance level can use a fuzzy scale, such as high, medium, and low (e.g., in Figure 5-5 we have assigned an importance level of high to the Make Appointment use case). It can also be done more formally using a weighted average of a set of criteria. For example, Larman[17] suggests rating each use case over the following criteria using a scale from zero to five:

- The use case represents an important business process.
- The use case supports revenue generation or cost reduction.
- Technology needed to support the use case is new or risky and therefore will require considerable research.

[16] Currently, there is no standard set of elements for a use case. The elements described in this section are based on recommendations contained in Alistair Cockburn, *Writing Effective Use Cases* (Reading, MA: Addison-Wesley, 2001); Craig Larman, *Applying UML and Patterns: An Introduction to Object-Oriented Analysis and Design and the Unified Process,* 2nd ed. (Upper Saddle River, NJ: Prentice Hall, 2002); Brian Henderson-Sellers and Bhuvan Unhelkar, *OPEN modeling with UML* (Reading, MA: Addison-Wesley, 2000); Graham, *Migrating to Object Technology*; and Alan Dennis and Barbara Haley Wixom, *Systems Analysis and Design,* 2nd ed. (New York: Wiley, 2003).
[17] Larman, *Applying UML and Patterns: An Introduction to Object-Oriented Analysis and Design.*

Use-Case Name: Make appointment		ID: _2_	Importance Level: _High_
Primary Actor: Patient		Use Case Type: Detail, essential	

Stakeholders and Interests:
Patient - wants to make, change, or cancel an appointment
Doctor - wants to ensure patients nee ds are met in a timely manner

Brief Description: This use case describes how we make an appointment as well as changing or canceling an appointment.

Trigger: Patient calls and asks for a new appointment or asks to cancel or change an existing appointment.

Type: External

Relationships:
    Association: Patient
    Include: Make Payment Arrangements
    Extend: Create New Patient
    Generalization:

Normal Flow of Events:
1. The Patient contacts the office regarding an appointment.
2. The Patient provides the Receptionist with his or her name and address.
3. The Receptionist validates that the Patient exists in the Patient database.
4. The Receptionist executes the Make Payment Arrangements use case.
5. The Receptionist asks Patient if he or she would like to make a new appointment, cancel an existing appointment, or change an existing appointment.
    If the patient wants to make a new appointment,
        the S-1: new appointment subflow is performed.
    If the patient wants to cancel an existing appointment,
        the S-2: cancel appointment subflow is performed.
    If the patient wants to change an existing appointment,
        the S-3: change appointment subflow is performed.
6. The Receptionist provides the results of the transaction to the Patient.

Subflows:
    S-1: New Appointment
        1. The Receptionist asks the Patient for possible appointment times.
        2. The Receptionist matches the Patient's desired appointment times with available dates and times and schedules the new appointment.
    S-2: Cancel Appointment
        1. The Receptionist asks the Patient for the old appointment time.
        2. The Receptionist finds the current appointment in the appointment file and cancels it.
    S-3: Change Appointment
        1. The Receptionist performs the S-2: cancel appointment subflow.
        2. The Receptionist performs the S-1: new appointment subflow.

Alternate/Exceptional Flows:
    3a: The Receptionist executes the Create New Patient use case.
    S-1, 2a1: The Receptionist proposes some alternative appointment times based on what is available in the appointment schedule.
    S-1, 2a2: The Patient chooses one of the proposed times or decides not to make an appointment.

TEMPLATE
can be found at
www.wiley.com
/college/dennis

FIGURE 5-5 Sample Use-Case Description

- Functionality described in the use case is complex, risky, and/or time critical. Depending on a use case's complexity, it may be useful to consider splitting its implementation over several different versions.
- The use case could increase understanding of the evolving design relative to the effort expended.

A use case may have multiple *stakeholders* that have an interest in the use case. As such, each use case lists each of the stakeholders with each one's interest in the use case (e.g., Patient and Doctor). The stakeholders' list always includes the primary actor (e.g., Patient).

Each use case typically has a *trigger* —the event that causes the use case to begin (e.g., Patient calls and asks for a new appointment or asks to cancel or change an existing appointment). A trigger can be an *external trigger*, such as a customer placing an order or the fire alarm ringing, or it can be a *temporal trigger*, such as a book being overdue at the library or the need to pay the rent.

Relationships   Use-case relationships explain how the use case is related to other use cases and users. There are four basic types of *relationships:* association, extend, include, and generalization. An *association relationship* documents the communication that takes place between the use case and the actors that use the use case. An actor is the UML representation for the role that a user plays in the use case. For example, in Figure 5-5, the Make Appointment use case is associated with the actor Patient. In this case, a patient makes an appointment. All actors involved in the use case are documented with the association relationship.

An *extend relationship* represents the extension of the functionality of the use case to incorporate optional behavior. In Figure 5-5, the Make Appointment use case conditionally uses the Create New Patient use case. This use case is executed only if the patient does not exist in the patient database. As such, it is not part of the normal flow of events and should be modeled with an extend relationship and an alternate/exceptional flow.

An *include relationship* represents the mandatory inclusion of another use case. The include relationship enables *functional decomposition*—the breaking up of a complex use case into several simpler ones. For example, in Figure 5-5, the Make Payment Arrangements use case was considered to be complex and complete enough to be factored out as a separate use case that could be executed by the Make Appointment use case. The include relationship also enables parts of use cases to be reused by creating them as separate use cases.

The *generalization relationship* allows use cases to support *inheritance*. For example, the use case in Figure 5-5 could be changed such that a new patient could be associated with a specialized use case called Make New Patient Appointment and an old patient could be associated with a Make Old Patient Appointment. The common, or generalized, behavior that both the Make New Patient Appointment and Make Old Patient Appointment use cases contain would be placed in the generalized Make Appointment use case—for example, the include relationship to the Make Payment Arrangements use case. In other words, the Make New Patient Appointment and Make Old Patient Appointment use cases would inherit the Make Payment Arrangements use case from the Make Appointment use case. The specialized behavior would be placed in the appropriate specialized use case. For example, the extend relationship to the Create New Patient use case would be placed with the specialized Make New Patient Appointment use case as an included use case.

Flow of Events   Finally, the individual steps within the business process are described. There are three different categories of steps, or *flows of events*, that can be documented: normal flow of events, subflows, and alternate, or exceptional, flows:

1. The *normal flow of events* includes only those steps that normally are executed in a use case. The steps are listed in the order in which they are performed. In Figure 5-5, the patient and the receptionist have a conversation regarding the patient's name, address, and action to be performed.
2. In some cases, the normal flow of events should be decomposed into a set of *subflows* to keep the normal flow of events as simple as possible. In Figure 5-5, we have identified three subflows: New Appointment, Cancel Appointment, and

Change Appointment. Each of the steps of the subflows is listed. These subflows are based on the control flow logic in the activity diagram representation of the business process (see Figure 5-2). Alternatively, we could replace a subflow with a separate use case that could be incorporated via the include relationships (see the earlier discussion). However, this should be done only if the newly created use case makes sense by itself. For example, in Figure 5-5, does it make sense to factor out a Create New Appointment, Cancel Appointment, and/or Change Appointment use case? If it does, then the specific subflow(s) should be replaced with a call to the related use case, and the use case should be added to the include relationship list.

3. *Alternate or exceptional flows* are ones that do happen but are not considered to be the norm. These must be documented. For example, in Figure 5-5, we have identified four alternate or exceptional flows. The first one simply addresses the need to create a new patient before the new patient can make an appointment. Normally, the patient would already exist in the patient database. Like the subflows, the primary purpose of separating out alternate or exceptional flows is to keep the normal flow of events as simple as possible. Again, as with the subflows, replace the alternate or exceptional flows with separate use cases that could be integrated via the extend relationship (see the earlier discussion).

When should events be factored out from the nomal flow of events into subflows or subflows and/or alternate or exceptional flows be factored out into separate use cases? Or, when should things simply be left alone? The primary criteria should be based on the level of complexity that the use case entails. The more difficult it is to understand the use case, the more likely events should be factored out into subflows, or subflows and/or alternate or exceptional flows should be factored out into separate use cases that are called by the current use case. This, of course, will create more use cases. As such, the use case diagram (see the following discussion) will become more cluttered. Practically speaking, we must decide which makes more sense. This will vary greatly, depending on the problem and the client. Remember, we are trying to represent, in as complete and concise manner as possible, our understanding of the business processes that we are investigating so that the client can validate the requirements that we are modeling. Therefore, there really is no single right answer. It really depends on the analyst, the client, and the problem.

Optional Characteristics   There are other characteristics of use cases that can be documented. These include the level of complexity of the use case, the estimated amount of time it takes to execute the use case, the system with which the use case is associated, specific data flows between the primary actor and the use case, any specific attribute, constraint, or operation associated with the use case, any preconditions that must be satisfied for the use case to execute, or any guarantees that can be made based on the execution of the use case. As we noted at the beginning of this section, there is no standard set of characteristics of a use case that must be captured. In this book, we suggest that the information contained in Figure 5-5 is the minimal amount to be captured.

Guidelines for Creating Use-Case Descriptions

The essence of a use case is the flow of events. Writing the flow of events in a manner that is useful for later stages of development generally comes with experience. Figure 5-6 provides a set of guidelines that have proven to be useful.[18]

---

[18] These guidelines are based on Cockburn, *Writing Effective Use Cases,* and Graham, *Migrating to Object Technology.*

FIGURE 5-6

Guidelines for Writing
Effective Use-Case
Descriptions

1. Write each set in the form of subject-verb-direct object (and sometimes preposition-indirect object).
2. Make sure it is clear who the initiator of the step is.
3. Write the steps from the perspective of the independent observer.
4. Write each step at about the same level of abstraction.
5. Ensure the use case has a sensible set of steps.
6. Apply the KISS principle liberally.
7. Write repeating instructions after the set of steps to be repeated.

First, write each individual step in the form subject–verb–direct object and, optionally, preposition–indirect object. This form has become known as *SVDPI* sentences. This form of sentence has proven to be useful in identifying classes and operations (see Chapter 6). For example, in Figure 5-5, the first step in the normal flow of events, the Patient contacts the office regarding an appointment, suggests the possibility of three classes of objects: Patient, Office, and Appointment. This approach simplifies the process of identifying the classes in the structural model (described in Chapter 6). SVDPI sentences cannot be used for all steps, but they should be used whenever possible.

Second, make clear who or what is the initiator of the action and who is the receiver of the action in each step. Normally, the initiator should be the subject of the sentence and the receiver should be the direct object of the sentence. For example, in Figure 5-5, the second step, Patient provides the Receptionist with his or her name and address, clearly portrays the Patient as the initiator and the Receptionist as the receiver.

Third, write the step from the perspective of an independent observer. To accomplish this, each step may have to be written first from the perspective of both the initiator and the receiver. Based on the two points of view, the bird's eye view version can then be written. For example, in Figure 5-5, the Patient provides the Receptionist with his or her name and address, neither the patient's nor the receptionist's perspective is represented.

Fourth, write each step at the same level of abstraction. Each step should make about the same amount of progress toward completing the use case as each of the other steps. On high-level use cases, the amount of progress could be very substantial, whereas in a low-level use case, each step could represent only incremental progress. For example, in Figure 5-5, each step represents about the same amount of effort to complete.

Fifth, ensure that the use case contains a sensible set of actions. Each use case should represent a transaction. Therefore, each use case should comprise four parts:

1. The primary actor initiates the execution of the use case by sending a request (and possibly data) to the system.
2. The system ensures that the request (and data) is valid.
3. The system processes the request (and data) and possibly changes its own internal state.
4. The system sends the primary actor the result of the processing.

For example, in Figure 5-5, (1) the patient requests an appointment (steps 1 and 2), (2) the receptionist determines if the patient exists in the database (step 3), (3) the receptionist sets up the appointment transaction (steps 4 and 5), and (4) the receptionist gives the time and date of the appointment to the patient (step 6).

The sixth guideline is the KISS principle. If the use case becomes too complex and/or too long, the use case should be decomposed into a set of use cases. Furthermore, if the normal flow of events of the use case becomes too complex, subflows should be used. For example, in Figure 5-5, the fifth step in the normal flow of events was sufficiently complex to decompose it into three separate subflows. However, as stated previously, care must be taken to avoid the possibility of decomposing too much. Most decomposition should be done with classes (see Chapter 6).

The seventh guideline deals with repeating steps. Normally, in a programming language such as Visual Basic or C, we put loop definition and controls at the beginning of the loop. However, because the steps are written in simple English, it is normally better to simply write Repeat steps A through E until some condition is met after step E. It makes the use case more readable to people unfamiliar with programming.

USE-CASE DIAGRAMS

In the previous section, we learned what a use case is and discussed a set of guidelines for writing effective use-case descriptions. In this section, we learn how the use case is the building block for the use-case diagram, which summarizes, in one picture, all the use cases for the part of the system being modeled. An analyst can use the use-case diagram to better understand the functionality of the system at a very high level. Typically, the use-case diagram is drawn early on in the SDLC when gathering and defining requirements for the system because it provides a simple, straightforward way of communicating to the users exactly what the system will do. In this manner, the use-case diagram can encourage the users to provide additional requirements that the written use case may not uncover.

A use-case diagram illustrates in a very simple way the main functions of the system and the different kinds of users that will interact with it. Figure 5-7 describes the basic syntax rules for a use-case diagram. Figure 5-8 presents a use-case diagram for a doctor's office appointment system. We can see from the diagram that patients, doctors, and management personnel will use the appointment system to make appointments, record availability, and produce schedule information, respectively.

Actors

The labeled stick figures on the diagram represent actors (see Figure 5-7). An *actor* is not a specific user, but instead is a role that a user can play while interacting with the system. An actor can also represent another system in which the current system interacts. In this case, the actor optionally can be represented by a rectangle containing <<actor>> and the name of the system. Basically, actors represent the principal elements in the environment in which the system operates. Actors can provide input to the system, receive output from the

An actor: ■ Is a person or system that derives benefit from and is external to the subject. ■ Is depicted as either a stick figure (default) or if a nonhuman actor is involved, as a rectangle with <<actor>> in it (alternative). ■ Is labeled with its role. ■ Can be associated with other actors using a specialization/superclass association, denoted by an arrow with a hollow arrowhead. ■ Is placed outside the subject boundary.	Actor/Role <<actor>> Actor/Role
A use case: ■ Represents a major piece of system functionality. ■ Can extend another use case. ■ Can include another use case. ■ Is placed inside the system boundary. ■ Is labeled with a descriptive verb–noun phrase.	Use Case
A subject boundary: ■ Includes the name of the subject inside or on top. ■ Represents the scope of the subject, e.g., a system or an individual business process.	Subject
An association relationship: ■ Links an actor with the use case(s) with which it interacts.	* ———— *
An include relationship: ■ Represents the inclusion of the functionality of one use case within another. ■ Has an arrow drawn from the base use case to the used use case.	<<include>> ←---------
An extend relationship: ■ Represents the extension of the use case to include optional behavior. ■ Has an arrow drawn from the extension use case to the base use case.	<<extend>> ---------→
A generalization relationship: ■ Represents a specialized use case to a more generalized one. ■ Has an arrow drawn from the specialized use case to the base use case.	⇐————

FIGURE 5-7   Syntax for Use-Case Diagram

system, or both. The diagram in Figure 5-8 shows that three actors will interact with the appointment system (a patient, a doctor, and management).

Sometimes an actor plays a specialized role of a more general type of actor. For example, there may be times when a new patient interacts with the system in a way that is somewhat different than a general patient. In this case, a *specialized actor* (i.e., new patient) can be placed on the model, shown using a line with a hollow triangle at the end of the more general actor (i.e., patient). The specialized actor will inherit the behavior of the more general actor and extend it in some way (see Figure 5-9). Can you think of some ways in which a new patient may behave differently than an existing patient?

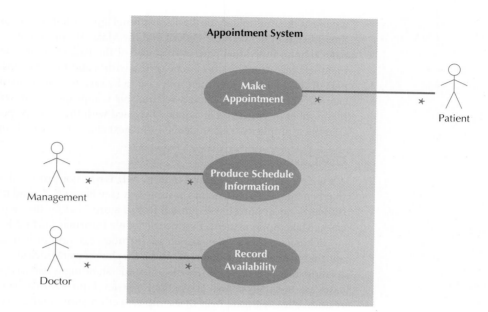

FIGURE 5-8
Use-Case Diagram
for the Appointment
System

Association

Use cases are connected to actors through association relationships; these relationships show with which use cases the actors interact (see Figure 5-7). A line drawn from an actor to a use case depicts an association. The association typically represents two-way communication between the use case and the actor. If the communication is only one way, then a solid arrowhead can be used to designate the direction of the flow of information. For example, in Figure 5-8 the Patient actor communicates with the Make Appointment use case. Because there are no arrowheads on the association, the communication is two-way. Finally, it is possible to

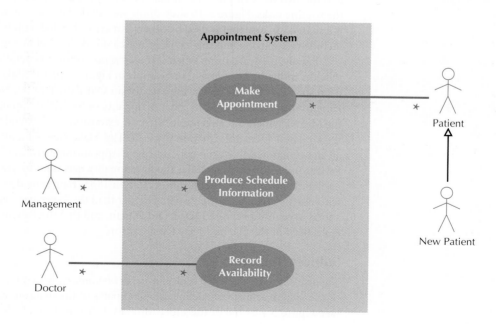

FIGURE 5-9
Use-Case System with
Specialized Actor

represent the multiplicity of the association. Figure 5-8 shows an asterisk (*) at either end of the association between the Patient and the Make Appointment use case. This simply designates that an individual patient (instance of the Patient actor) executes the Make Appointment use case as many times as he or she wishes and that it is possible for the appointment part of the Make Appointment use case to be executed by many different patients. In most cases, this type of many-to-many relationship is appropriate. However, it is possible to restrict the number of patients that can be associated with the Make Appointment use case. We discuss the multiplicity issue in detail in the next chapter in regards to class diagrams.

Use Case

A use case, depicted by an oval in the UML, is a major process that the system will perform that benefits an actor or actors in some way (see Figure 5-7), and it is labeled using a descriptive verb–noun phrase. We can tell from Figure 5-8 that the system has three primary use cases: Make Appointment, Produce Schedule Information, and Record Availability.

There are times when a use case includes, extends, or generalizes the functionality of another use case in the diagram. These are shown using include, extend, and generalization relationships. To increase the ease of understanding a use-case diagram, higher-level use cases are normally drawn above the lower-level ones. It may be easier to understand these relationships with the help of examples. Let's assume that every time a patient makes an appointment, the patient is asked to verify payment arrangements. However, it is occasionally necessary to actually make new payment arrangements. Therefore, we may want to have a use case called Make Payment Arrangements that *extends* the Make Appointment use case to include this additional functionality. In Figure 5-10, an arrow labeled with extend was drawn between the Make Payment Arrangements use case and the Make Appointment use case to denote this special use-case relationship. Furthermore, the Make Payment Arrangements use case was drawn lower than the Make Appointment use case.

Similarly, there are times when a single use case contains common functions that are used by other use cases. For example, suppose there is a use case called Manage Schedule that performs some routine tasks needed to maintain the doctor's office appointment schedule, and the two use cases Record Availability and Produce Schedule Information both perform the routine tasks. Figure 5-10 shows how we can design the system so that Manage Schedule is a shared use case that is used by others. An arrow labeled with include is used to denote the include relationship and the included use case is drawn below the use cases that contain it.

Finally, there are times when it makes sense to use a generalization relationship to simplify the individual use cases. For example in Figure 5-10, the Make Appointment use case has been specialized to include a use case for an Old Patient and a New Patient. The Make Old Patient Appt use case inherits the functionality of the Make Appointment use case (including the Make Payment Arrangements use-case extension) and extends its own functionality with the Update Patient Information use case. The Make New Patient Appt use case also inherits all the functionality of the generic Make Appointment use case and calls the Create New Patient use case, which includes the functionality necessary to insert the new patient into the patient database. The generalization relationship is represented as an unlabeled hollow arrow with the more general use case being higher than the lower-use cases. Also notice that we have added a second specialized actor, Old Patient, and that the Patient actor is now simply a generalization of the Old and New Patient actors.

Subject Boundary

The use cases are enclosed within a *subject boundary*, which is a box that defines the scope of the system and clearly delineates what parts of the diagram are external or internal to it (see Figure 5-7). One of the more difficult decisions to make is where to draw the subject

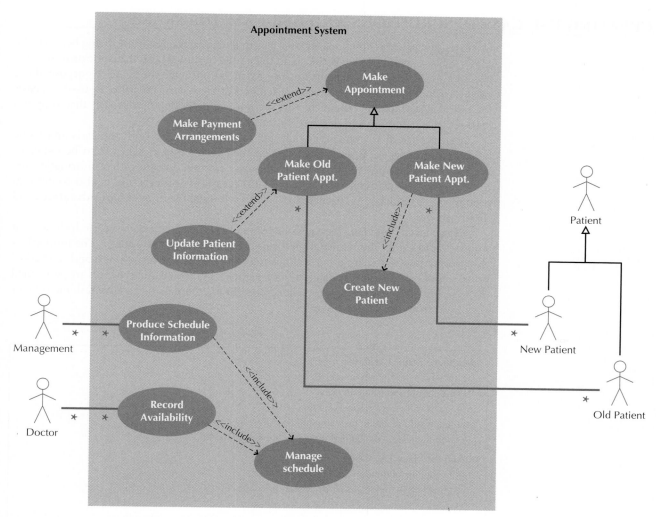

FIGURE 5-10   Extend and Include Relationships

boundary. A subject boundary can be used to separate a software system from its environment, a subsystem from other subsystems within the software system, or an individual process in a software system. They also can be used to separate an information system, including both software and internal actors, from its environment. As such, care should be taken to decide carefully what the scope of the information system is to be.

The name of the subject can appear either inside or on top of the box. The subject boundary is drawn based on the scope of the system. In the appointment system, the Management and Doctor actors may not be drawn. Remember that actors are outside of the scope of the system. In Figure 5-8, we listed a receptionist as being part of the use case. However, in this case, we assumed that the receptionist is an internal actor who is part of the Make Appointment use case. As such, the receptionist is not drawn on the diagram. [19]

[19] In other non-UML approaches to object-oriented systems development, it is possible to represent external actors along with internal actors. In this example, the receptionist would be considered an internal actor (see Graham, *Migrating to Object Technology*, and Graham, Henderson-Sellers, and Younessi, *The OPEN Process Specification*).

CREATING USE-CASE DESCRIPTIONS AND USE-CASE DIAGRAMS

Use cases describe the functionality of a system and are a model of the dialog between the actors and the system. As such, they tend to be used to model both the contexts of the system and the detailed requirements for the system. Even though the primary purpose of use cases is to document the functional requirements of the system, they also are used as a basis for testing the evolving system. In this section, we describe an approach that supports requirements gathering and documentation with use cases.

It is important to remember that use cases are used for both the as-is and to-be behavioral models. As-is use cases focus on the current system, whereas to-be use cases focus on the desired new system. The two most common ways to gather information for the use cases are through interviews and JAD sessions (observation is also sometimes used for as-is models). As discussed in Chapter 4, these techniques have advantages and disadvantages.

Regardless of whether interviews or JAD sessions are used, recent research shows that some ways to gather the information for use cases are better than others. The most effective process involves using thirteen steps to create use-case descriptions[20] and four additional steps to create use-case diagrams (see Figure 5-11). These thirteen steps are performed in order, but of course the analyst often cycles among them in an iterative fashion as he or she moves from use case to use case.

Identify the Major Use Cases

1. Review the activity diagram.
2. Find the subject's boundaries.
3. Identify the primary actors and their goals.
4. Identify and write the overviews of the major use cases for the above.
5. Carefully review the current use cases. Revise as needed.

Expand the Major Use Cases

6. Choose one of the use cases to expand.
7. Start filling in the details of the chosen use case.
8. Write the normal flow of events of the use case.
9. If the normal flow of events is too complex or long, decompose into subflows.
10. List the possible alternate or exceptional flows.
11. For each alternate or exceptional flow, list how the actor and/or system should react.

Confirm the Major Use Cases

12. Carefully review the current set of use cases. Revise as needed.
13. Start at the top again.

Create the Use-Case Diagram

1. Draw the subject boundary.
2. Place the use cases on the diagram.
3. Place the actors on the diagram.
4. Draw the associations.

FIGURE 5-11

Steps for Writing Effective Use-Case Descriptions and Use-Case Diagrams

---

[20] The approach in this section is based on the work of Cockburn, *Writing Effective Use Cases*; Graham, *Migrating to Object Technology*; George Marakas and Joyce Elam, "Semantic Structuring in Analyst Acquisition and Representation of Facts in Requirements Analysis," *Information Systems Research* 9, no. 1 (1998): 37–63; and Alan Dennis, Glenda Hayes, and Robert Daniels, "Business Process Modeling with Group Support Systems," *Journal of Management Information Systems* 15, no. 4 (1999): 115–142.

Identifying the Major Use Cases

The first step is to review the activity diagram. This helps the analyst to get a complete overview of the underlying business process being modeled.

The second step is to identify the subject's boundaries. This helps the analyst to identify the scope of the system. However, as we work through the SDLC, the boundary of the system will change.

The third step is to identify the primary actors and their goals. The primary actors involved with the system will come from a list of stakeholders and users. However, recall that an actor is a role that a stakeholder or user plays, not a specific user (e.g., doctor, not Dr. Jones). The goals represent the functionality that the system must provide the actor for the system to be a success. Identifying what the tasks are that each actor must perform can facilitate this. For example, does the actor need to create, read, update, or delete (CRUD)[21] any information currently in the system, are there any external changes of which an actor must inform the system, or is there any information that the system should give the actor? Steps 2 and 3 are intertwined. As actors are identified and their goals are uncovered, the boundary of the system will change.

The fourth step is to identify and write the major use cases, with basic information about each, rather than jumping into one use case and describing it completely (i.e., only an overview use case). Recall from the previous description of the elements of a use case that overview use cases contain only the use case name, ID number, type, primary actor, importance level, and brief description. Creating only overview use cases at this time prevents the users and analysts from forgetting key use cases and helps the users explain the overall set of business processes for which they are responsible. It also helps them understand how to describe the use cases and reduces the chance of overlap between use cases. It is important at this point to understand and define acronyms and jargon so that the project team and others from outside the user group can clearly understand the use cases. Again, the activity diagram is a very useful beginning point for this step.

The fifth step is to carefully review the current set of use cases. It may be necessary to split some of them into multiple use cases or merge some of them into a single use case. Also, based on the current set, a new use case may be identified. You should remember that identifying use cases is an iterative process, with users often changing their minds about what a use case is and what it includes. It is very easy to get trapped in the details at this point, so you need to remember that the goal at this step is to identify the *major* use cases and that you are creating only overview use cases. For example, in the doctors' office example in Figure 5-10, we defined one use case as Make Appointment. This use case included the cases for both new patients and existing patients, as well as for when a patient changes or cancels an appointment. We could have defined each of these activities (makes an appointment, changes an appointment, or cancels an appointment) as separate use cases, but this would have created a huge set of small use cases.

The trick is to select the right size so that you end up with three to nine use cases in each system. If the project team discovers many more than eight use cases, this suggests that the use cases are too small or that the system boundary is too big. If more than nine use cases exist, the use cases should be grouped together into *packages* (i.e., logical groups of use cases) to make the diagrams easier to read and keep the models at a reasonable level of complexity. It is simple at that point to sort the use cases and group together these small use cases into larger use cases that include several small ones or to change the system boundaries.[22]

---

[21] We describe the use of CRUD matrices in Chapter 7.

[22] For those familiar with structured analysis and design, packages serve a similar purpose as the leveling and balancing processes used in data flow diagramming. Packages are described in Chapter 8.

Expanding the Major Use Cases

The sixth step is to choose one of the major use cases to expand. Using the importance level of the use case can help do this. For example, in Figure 5-5, the Make Appointment use case has an importance level of high. As such, it should be one of the earlier use cases to be expanded. The criteria suggested by Larman[23] can also be used to set the prioritization of the use cases, as noted earlier.

The seventh step is to start filling in the details on the use-case template. For example, list all of the identified stakeholders and their interests in the use case, the level of importance of the use case, a brief description of the use case, the trigger information for the use case, and the relationships in which the use case participates.

The eighth step is to fill in the steps of the normal flow of events required to describe each use case. The steps focus on what the business process does to complete the use case, as opposed to what actions the users or other external entities do. In general, the steps should be listed in the order in which they are performed, from first to last. Remember to write the steps in an SVDPI form whenever possible. In writing the use case, remember the seven guidelines described earlier. The goal at this point is to describe how the chosen use case operates. One of the best ways to begin to understand how an actor works through a use case is to visualize performing the steps in the use case—that is, role-play. The techniques of visualizing how to interact with the system and of thinking about how other systems work (informal benchmarking) are important techniques that help analysts and users understand how systems work and how to write a use case. Both techniques (visualization and informal benchmarking) are common in practice. It is important to remember that at this point in the development of a use case, we are interested only in the typical successful execution of the use case. If we try to think of all of the possible combinations of activities that could go on, we will never get anything written down. At this point, we are looking only for the three to seven *major* steps. As such, focus only on performing the typical process that the use case represents.

The ninth step is to ensure that the steps listed in the normal flow of events are not too complex or too long. Each step should be about the same size as the others. For example, if we were writing steps for preparing a meal, steps such as take fork out of drawer and put fork on table are much smaller than prepare cake using mix. If we end up with more than seven steps or steps that vary greatly in size, we should go back and review each step carefully and possibly rewrite the steps.

One good approach to produce the steps for a use case is to have the users visualize themselves actually performing the use case and to have them write down the steps as if they were writing a recipe for a cookbook. In most cases the users will be able to quickly define what they do in the as-is model. Defining the steps for to-be use cases may take a bit more coaching. In our experience, the descriptions of the steps change greatly as users work through a use case. Our advice is to use a blackboard or whiteboard (or paper with pencil) that can be easily erased to develop the list of steps, and then write the list on the use-case form. It should be written on the use-case form only after the set of steps is fairly well defined.

The tenth step focuses on identifying the alternate or exceptional flows. Alternate or exceptional flows are those flows of success that represent optional or exceptional behavior. They tend to occur infrequently or as a result of a normal flow failing. They should be labeled so that there is no doubt as to which normal flow of events it is related. For example in Figure 5-5, alternate/exceptional flow 3a executes when step 3 fails (i.e., the Patient does not exist in the Patient database).

---

[23] Larman, *Applying UML and Patterns: An Introduction to Object-Oriented Analysis and Design.*

The eleventh step is simply to write the description of any alternate or exceptional flows. In other words, if an alternate or exceptional flow is to be executed, describe the response that the actor or system should produce. Like the normal flows and subflows, alternate or exceptional flows should be written in the SVDPI form whenever possible.

Confirming the Major Use Cases[24]

The twelfth step is to carefully review the current set of use cases and to confirm that the use case is correct as written, which means reviewing the use case with the users to make sure each step is correct. The review should look for opportunities to simplify a use case by decomposing it into a set of smaller use cases, merging it with others, looking for common aspects in both the semantics and syntax of the use cases, and identifying new use cases. This is also the time to look into adding the include, extend, or generalization relationships between use cases. The most powerful way to confirm a use case is to ask the user to role-play, or execute the process using the written steps in the use case. The analyst will hand the user pieces of paper labeled as the major inputs to the use case and will have the user follow the written steps like a recipe to make sure that those steps really can produce the outputs defined for the use case using its inputs.

The thirteenth and final step is to *iterate* over the entire set of steps again. Users will often change their minds about what is a use case and what it includes. It is very easy to get trapped in the details at this point, so remember that the goal is to just address the major use cases. Therefore, the analyst should continue to iterate over these steps until he or she and the users believe that a sufficient number of use cases has been documented to begin identifying candidate classes for the structural model (see Chapter 6). As candidate classes are identified, it is likely that additional use cases will be uncovered. Remember, object-oriented systems development is both iterative and incremental. As such, do not worry about identifying each and every possible use case at this point in the development of the system.

Creating a Use-Case Diagram

Basically, drawing the use-case diagram is very straightforward once use cases have been detailed. The actual use-case diagram encourages the use of information hiding. The only parts drawn on the use-case diagram are the system boundary, the use-cases themselves, the actors, and the various associations between these components. The major strength of the use-case diagram is that it provides the user with an overview of the detailed use cases. However, remember that anytime a use case changes, it could impact the use case diagram.

There are four major steps in drawing a use-case diagram. First, the use-case diagram starts with the subject boundary. This forms the border of the subject, separating use cases (i.e., the subject's functionality) from actors (i.e., the roles of the external users).

Second, the use cases are drawn on the diagram. These are taken directly from the detailed use-case descriptions. Special use-case associations (include, extend, or generalization) are also added to the model at this point. Be careful in laying out the diagram. There is no formal order to the use cases, so they can be placed in whatever fashion needed to make the diagram easy to read and to minimize the number of lines that cross. It often is necessary to redraw the diagram several times with use cases in different places to make the diagram easy to read. Also, for understandability purposes, there should be no more than three to nine use cases on the model so the diagram is as simple as possible. These includes those use cases that have been factored out and now are associated with another use case through the include, extend, or generalization relationships.

---

[24] This process is related to role-playing, which is discussed in Chapter 6, walkthroughs discussed in Chapter 8, and testing discussed in Chapter 13.

Third, the actors are placed on the diagram. The actors are taken directly from the primary actor element on the detailed use-case description. Like use-case placement, to minimize the number of lines that will cross on the diagram, the actors should be placed near the use cases with which they are associated.

The fourth and last step is to draw lines connecting the actors to the use cases with which they interact. No order is implied by the diagram, and the items added along the way do not have to be placed in a particular order; therefore, it may help to rearrange the symbols a bit to minimize the number of lines that cross, making the diagram less confusing.

REFINING PROJECT SIZE AND EFFORT ESTIMATION USING USE-CASE POINTS[25]

In Chapter 3, we used function points and the COCOMO method to estimate the size and effort of a systems development project, respectively. However, neither of these approaches were developed or validated with object-oriented systems in mind. As such, even though they are very popular approaches to estimation, their application to object-oriented systems is questionable. Because we now know about use cases, we introduce a size and effort *estimation* technique that was developed around their use. This technique, known as use-case points, was originally developed by Gustav Karner of Objectory AB.[26] *Use-case points* are based on the same ideas from which function points were developed. However, they have been refined to take into consideration the unique features of use cases and object orientation.

To estimate size and effort using use-case points, at least the set of essential use cases and the use-case diagram must have been created. Otherwise, the information required will not be available. Once the use cases and use case diagram have been created, the actors and use cases must be classified as simple, average, or complex.

Simple actors are separate systems with which the current system must communicate through a well-defined *application program interface (API)*. *Average actors* are separate systems that interact with the current system using standard communication protocols, such as TCP/IP, FTP, or HTTP, or an external database that can be accessed using standard SQL. *Complex actors* are typically end users communicating with the system using some type of GUI. Once all the actors have been classified, the appropriate numbers should be entered into the unadjusted actor weighting table contained in the use-case point–estimation worksheet (See Figure 5-12). For example, reviewing the Make Appointment use case (see Figure 5-5) and the use-case diagram for the Appointment

[25] The material in this section is based on descriptions of use-case points contained in Raul R. Reed, Jr., *Developing Applications with Java and UML* (Reading, MA: Addison-Wesley, 2002); Geri Schneider and Jason P. Winters, *Applying Use Cases: A Practical Guide* (Reading, MA: Addison-Wesley, 1998); and Kirsten Ribu, "Estimating Object-Oriented Software Projects with Use Cases" (master's thesis, University of Oslo, 2001).

[26] Objectory AB was acquired by Rational in 1995. Rational has now been acquired by IBM.

Unadjusted Actor Weighting Table:

Actor Type	Description	Weighting Factor	Number	Result
Simple	External System with well-defined API	1		
Average	External System using a protocol-based interface, e.g., HTTP, TCT/IP, or a database	2		
Complex	Human	3		
		Unadjusted Actor Weight Total (UAW)		

Unadjusted Use Case Weighting Table:

Use-Case Type	Description	Weighting Factor	Number	Result
Simple	1–3 transactions	5		
Average	4–7 transactions	10		
Complex	>7 transactions	15		
		Unadjusted Use-Case Weight Total (UUCW)		

Unadjusted Use Case Points (UUCP) = UAW + UUCW

Technical Complexity Factors:

Factor Number	Description	Weight	Assigned Value (0–5)	Weighted Value	Notes
T1	Distributed system	2.0			
T2	Response time or throughput performance objectives	1.0			
T3	End-user online efficiency	1.0			
T4	Complex internal processing	1.0			
T5	Reusability of code	1.0			
T6	Easy to install	0.5			
T7	Ease of use	0.5			
T8	Portability	2.0			
T9	Ease of change	1.0			
T10	Concurrency	1.0			
T11	Special security objectives included	1.0			
T12	Direct access for third parties	1.0			
T13	Special user training required	1.0			
		Technical Factor Value (TFactor)			

Technical Complexity Factor (TCF) = 0.6 + (0.01 * TFactor)

Environmental Factors:

Factor Number	Description	Weight	Assigned Value (0 – 5)	Weighted Value	Notes
E1	Familiarity with system development process being used	1.5			
E2	Application experience	0.5			
E3	Object-oriented experience	1.0			
E4	Lead analyst capability	0.5			
E5	Motivation	1.0			
E6	Requirements stability	2.0			
E7	Part time staff	−1.0			
E8	Difficulty of programming language	−1.0			
		Environmental Factor Value (EFactor)			

Environmental Factor (EF) = 1.4 + (−0.03 * EFactor)
Adjusted Use Case Points (UCP) = UUCP * TCF * ECF
Effort in Person Hours = UCP * PHM

TEMPLATE
can be found at
www.wiley.com
/college/dennis

FIGURE 5-12   Use-Case Point–Estimation Worksheet

system (see Figure 5-10) we see that we have four human actors interacting with the system, giving an *unadjusted actor weight total (UAW)* of 12. This is computed by summing the individual results that were computed by multiplying the weighting fact by the number of actors of each type. In this case there were zero simple actors, zero average actors, and four complex actors.

Depending on the number of unique transactions that the use case must address, a use case is classified as a *simple use case* (1–3), an *average use case* (4–7), or a *complex use case* (>7). In the original formulation of use-case point estimation, Karner suggested that included and extended use cases should be ignored. However, today the recommendation is to use all use cases, regardless of their type in the estimation calculation. For example, the Make Appointment use case described in Figure 5-5 must deal with activities including making a new appointment, canceling an existing appointment, changing an existing appointment, creating a new patient, and making payment arrangements. In this case, because the Make Appointment use case has to deal with five separate transactions, it is classified as an average use case (see the unadjusted use-case weighting table in Figure 5-12). Based on the Appointment System use-case diagram (see Figure 5-10), we have three simple use cases (Produce Schedule, Make Old Patient Appointment, and Record Availability), four average use cases (Make Appointment, Make New Patient Appointment, Manage Schedule, and Make Payment Arrangements), and one complex use case (Update Patient Information). By multiplying by the appropriate weights and summing the results, we get an *unadjusted use-case weight total (UUCW)* value of 70.

The value of the *unadjusted use-case points (UUCP)* is simply the sum of the unadjusted actor weight total (12) and the unadjusted use-case weight total (70). As such, UUCP is equal to 82.

In the spirit of function point analysis, use-case point–based estimation also has a set of factors that are used to adjust the use-case point value. In this case, there are two sets of factors: *technical complexity factors (TCF)* and *environmental factors (EF)*. There are thirteen separate technical factors and eight separate environmental factors. The purpose of these factors is to allow the project as a whole to be evaluated for complexity and experience levels of the staff, respectively. Obviously, these types of factors can impact the effort required by a team to develop a system. Each of these factors is assigned a value between 0 and 5, 0 representing that the factor is irrelevant to the system under consideration and 5 representing that the factor is essential for the system to be successful. The assigned values are then multiplied by their respective weights. These weighted values are then summed up to create a *technical factor value (TFactor)* and an *environmental factor value (EFactor)*.

The technical factors include the following:

- Whether the system is going to be a distributed system
- The importance of response time
- The efficiency level of the end user using the system
- The complexity of the internal processing of the system
- The importance of code reuse
- How easy the installation process has to be
- The importance of the ease of using the system
- How important it is for the system to be able to be ported to another platform
- Whether system maintenance is important
- Whether the system is going to have to handle parallel and concurrent processing
- The level of special security required

- The level of system access by third parties
- Whether special end user training is to be required (see Figure 5-12)

The environmental factors include the following:

- The level of experience the development staff has with the development process being used
- The application being developed
- The level of object-oriented experience
- The level of capability of the lead analyst
- The level of motivation of the development team to deliver the system
- The stability of the requirements
- Whether part-time staff have to be included as part of the development team
- The difficulty of the programming language being used to implement the system (see Figure 5-12)

Continuing the appointment example, the values for the technical factors were T1 (0), T2 (5), T3 (3), T4 (1), T5 (1), T6 (2), T7 (4), T8 (0), T9 (2), T10 (0), T11 (0), T12 (0), and T13 (0). Summing the weighted values in Figure 5-12 gives a TFactor value of 15. Plugging this value into the technical complexity factor (TCF) equation of the use-case point worksheet gives a value of .75 for the TCF of the appointment system.

The values for the environmental factors were E1 (4), E2 (4), E3 (4), E4 (5), E5 (5), E6 (5), E7 (0), and E8 (4), giving an EFactor of 25.5. Like the TFactor, Efactor is simply the sum of the weighted values. Using the environmental factor (EF) equation of the use-case point worksheet produces a value of .635 for the EF of the appointment system. Plugging the TCF and EF values, along with the UUCP value computed earlier, into the adjusted use-case points equation of the worksheet yields a value of 33.3375 *adjusted use- case points (UCP)*.

Now that we know the estimated size of the system by means of the value of the adjusted use-case points, we are ready to estimate the *effort* required to build the system. In Karner's original work, he suggested simply multiplying the number of use-case points by 20 to estimate the number of person-hours required to build the system. However, based on additional experiences using use-case points, a decision rule to determine the value of the *person-hours multiplier (PHM)* has been created that suggests using either 20 or 28, based on the values assigned to the individual environmental factors. The decision rule is:

> If the sum of (number of Efactors E1 through E6 assigned value < 3) and
>   (number of Efactors E7 and E8 assigned value > 3)
> $\leqslant 2$
>   PHM = 20
> Else If the sum of (number of Efactors E1 through E6 assigned value < 3) and
>   (number of Efactors E7 and E8 assigned value > 3)
>   = 3 or 4
>   PHM = 28
> Else
>
>   Rethink project; it has too high of a risk for failure

Based on these rules, because none of Efactors E1 through E6 have a value less than 3 and only Efactors E8 has a value greater than 3, the sum of the number EFactors is 1. Thus, the appointment system should use a PHM of 20. This gives an estimated number of person-hours of 666.75 hours (20 * 33.3375). The filled-out worksheet is given in Figure 5-13.

Unadjusted Actor Weighting Table:

Actor Type	Description	Weighting Factor	Number	Result
Simple	External system with well-defined API	1	0	0
Average	External system using a protocol-based interface, e.g., HTTP, TCT/IP, or a database	2	0	0
Complex	Human	3	4	12
		Unadjusted Actor Weight Total (UAW)		**12**

Unadjusted Use-Case Weighting Table:

Use Case Type	Description	Weighting Factor	Number	Result
Simple	1–3 transactions	5	3	15
Average	4–7 transactions	10	4	40
Complex	>7 transactions	15	1	15
		Unadjusted Use Case Weight Total (UUCW)		**70**

Unadjusted use-case points (UUCP) = UAW + UUCW   82 = 12 + 70

Technical Complexity Factors:

Factor Number	Description	Weight	Assigned Value (0–5)	Weighted Value	Notes
T1	Distributed system	2	0	0	
T2	Response time or throughput performance objectives	1	5	5	
T3	End-user online efficiency	1	3	3	
T4	Complex internal processing	1	1	1	
T5	Reusability of code	1	1	1	
T6	Easy to install	0.5	2	1	
T7	Ease of use	0.5	4	2	
T8	Portability	2	0	0	
T9	Ease of change	1	2	2	
T10	Concurrency	1	0	0	
T11	Special security objectives included	1	0	0	
T12	Direct access for third parties	1	0	0	
T13	Special user training required	1	0	0	
			Technical Factor Value (TFactor)	**15**	

Technical complexity factor (TCF) = 0.6 + (0.01 ∗ TFactor)   0.75 = 0.6 + (0.01 ∗ 15)

Environmental Factors:

Factor Number	Description	Weight	Assigned Value (0 – 5)	Weighted Value	Notes
E1	Familiarity with system development process being used	1.5	4	6	
E2	Application experience	0.5	4	2	
E3	Object-oriented experience	1	4	4	
E4	Lead analyst capability	0.5	5	2.5	
E5	Motivation	1	5	5	
E6	Requirements stability	2	5	10	
E7	Part-time staff	−1	0	0	
E8	Difficulty of programming language	−1	4	−4.0	
			Environmental Factor Value (EFactor)	**25.5**	

Environmental factor (EF) = 1.4 + (−0.03 ∗ EFactor)   0.635 = 1.4 + (−0.03 ∗ 25.5)
Adjusted use case points (UCP) = UUCP ∗ TCF ∗ ECF   33.3375 = 70 ∗ 0.75 ∗ 0.635
Effort in person-hours = UCP ∗ PHM   666.75 = 20 ∗ 33.3375

FIGURE 5-13　Use-Case Point Estimation for the Appointment System

The primary advantages of using use-case points over traditional estimation techniques are that they are based in object-oriented systems development and use cases rather than traditional approaches to systems development. However, the risk of using use-case points is that their use does not have as much history in comparison to the traditional approaches described in Chapter 3. As such, the suggested classifications used for simple, average, and complex actors and use cases, the weighting factors for simple, average, and complex actors and use cases, and the weightings associated with the computation of the technical complexity and environmental factors may need to be modified in the future. However, at this point the suggested values seem to be the best available.

YOUR TURN   5-4 Estimating Using Use-Case Points

Consider the use case you created in Your Turn 5-2. Using the worksheet in Figure 5-12, estimate the project size and effort for your use case.

YOUR TURN   5-5 Campus Housing

Create a set of use cases for the following high-level processes in a housing system run by the campus housing service. The campus housing service helps students find apartments. Apartment owners fill in information forms about the rental units they have available (e.g., location, number of bedrooms, monthly rent), which are then entered into a database. Students can search through this database via the Web to find apartments that meet their needs (e.g., a two-bedroom apartment for $400 or less per month within a half mile of campus). They then contact the apartment owners directly to see the apartment and possibly rent it. Apartment owners call the service to delete their listing when they have rented their apartment(s).

YOUR TURN   5-6 Drawing a Use-Case Diagram

In Your Turn 5-5, you identified use cases for a campus housing service that helps students find apartments. Based on those use cases, create a use-case diagram.

YOUR TURN   5-7 Estimating Using Use-Case Points

Consider the use cases you created in Your Turn 5-5. Using the worksheet in Figure 5-12, estimate the project size and effort for your use cases.

APPLYING THE CONCEPTS AT CD SELECTIONS

The basic functional and nonfunctional requirements for the CD Selections Internet Sales System were developed previously. Carefully review these requirements (see Figures 2-13, 2-14, 2-15, 3-22, and 4-15).

Business Process Modeling with Activity Diagrams

Based on the functional requirements identified for the Internet Sales System, Alec and his team decided that there were at least four high-level activities that needed to be addressed: Place CD Orders, Maintain CD Information, Maintain CD Marketing Information, and Fill Mail Orders. As a first step toward developing a functional model of the functional requirements, Alec decided to model the overall flow of execution of the business processes as an activity diagram. Upon close review of the Place CD Orders requirements, the team identified a decision and two additional activities: Place InStore Hold and Place Special Order. These activities were based on the requirements laid out in point 3.5 of Figure 4-15. Furthermore, the team noticed that there seemed to be three separate concurrent paths of execution implied in the requirements. The first path dealt with orders, the second addressed the maintenance of marketing information, and the third focused on maintaining the information about the CDs. Based on these new insights, Alec and his team drew the activity diagram for the Internet Sales system shown in Figure 5-14.

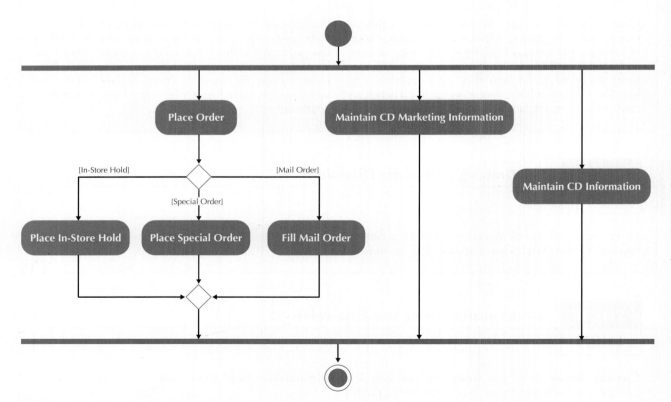

FIGURE 5-14   Activity Diagram for the CD Selections To-Be Internet Sales System

Identify the Major Use Cases

The first four steps in writing use cases deal with reviewing the activity diagram, finding the subject boundaries, listing the primary actors and their goals, and identifying and writing overviews of the major use cases based on these results. These use cases will form the basis of the use cases that are contained on the use-case diagram. Take a minute and go back and review the requirements identified in Figure 4-15 and the activity diagram we just completed (see Figure 5-14) to identify the three to nine major use cases before you continue reading.

To begin with, it seems that the subject boundary should be drawn in such a manner that anything that is not part of CD Selections' Internet Sales System, such as the vendors and customers, should be identified as primary actors. Therefore, these are considered outside of the scope of the system. The other potential actors identified could be the distribution system, electronic marketing (EM) manager, and the current CD Selections stores.

Upon closer review of Figure 4-15, it seems that the distribution system and the current CD Selections stores should be outside the scope of the Internet Sales System. As such, they also should be identified as primary actors. It also seems that the EM manager should be considered as part of the Internet Sales System and, therefore, should not be identified as a primary actor. Remember, primary actors are only those that can be viewed as being outside of the scope of the system. The decision on whether the EM manager, the current CD Selections stores, or the distribution system is inside or outside of the system is somewhat arbitrary. From a customer's perspective, the distribution system, and the current CD Selections stores could be seen as being inside of the overall system, and it could be argued that the EM manager is a primary user of the Internet Sales System.

At this point in the process, it is important to make a decision and to move on. During the process of writing use cases, there are ample opportunities to revisit this decision to determine whether the set of use cases identified are necessary and sufficient to describe the requirements of the Internet Sales System for CD Selections. As you can see, based on the preceding, finding the systems boundaries and listing the primary actors are heavily intertwined.

Based on the functional requirements and the activity diagram, Alec and his team identified four overview major use cases: Maintain CD Information, Maintain CD Marketing Information, Place Order, and Fill Order. You might have considered adding new CDs to the database, deleting old ones, and changing information about CDs as three separate use cases—which, in fact, they are. In addition to these three, we also need to have use cases for finding CD information and printing reports about CDs. However, our goal at this point is to develop a small set of essential use cases for the Internet Sales System. The same pattern should be evident for the marketing materials. We have the same processes for recording new marketing material for a CD, creating it, reading it, updating it, deleting it, and printing it. These activities are usually required any time that you have information to be stored in a database.

The project team at CD Selections identified these same four major use cases. At this point in the process, the project team began writing the overview use cases for these four. Remember that an overview use case only has five pieces of information: use case name, ID number, primary actor, type, and a brief description. At this point in time, we have already identified the primary actors and have associated them with the four use cases. Furthermore, because we are just beginning the development process, all four use cases will be overview and essential types. Because the ID numbers are simply used for identification purposes (i.e., they act as a key in a database), their values can be assigned in a sequential manner. This leaves us with only two pieces of information for each use case to write. The use-case name should be an action verb–noun phrase combination (e.g., Make Appointment).

Use Case Name:	Maintain CD information		ID:	1	Importance Level:
Primary Actor:	Distribution system		Use Case Type:	Overview, essential	
Stakeholders and Interests:					
Brief Description:	This adds, deletes, and modifies the basic information about the CDs we have available for sale (e.g., album name, artist(s), price, quantity on hand, etc.).				
Trigger:					
Relationships					
Normal Flow					
Subflows:					
Alternate/exc					

Use Case Name:	Maintain marketing information		ID:	2	Importance Level:
Primary Actor:	Vendor		Use Case Type:	Overview, essential	
Stakeholders and Interests:					
Brief Description:	This adds, delete, and modifies the additional marketing material available for some CDs (e.g., reviews).				
Trigger:					
Relationships					
Normal Flow					
Subflows:					
Alternate/exc					

Use Case Name:	Place order		ID:	3	Importance Level:
Primary Actor:	Customer		Use Case Type:	Overview, essential	
Stakeholders and interests:					
Brief Description:	This use case describes how customers can search the Web site and place orders.				
Trigger:					
Relationships					
Normal Flow					
Subflows:					
Alternate/exc					

Use Case Name:	Fill order		ID:	4	Importance Level:
Primary Actor:	Distribution system		Use Case Type:	Overview, essential	
Stakeholders and Interests:					
Brief Description:	This describes how orders move from the Internet Sales System into the distribution system and how status information will be updated from the distribution system.				
Trigger:					
Type:					
Relationships:					
Association:					
Include:					
Extend:					
Generalization:					
Normal Flow of Events:					
Subflows:					

FIGURE 5-15   Overview of Major Use Cases for CD Selections

In the CD Selections Internet Sales situation, Maintain CD Information, Maintain CD Marketing Information, Place Order, and Fill Order seem to capture the essence of each of the use cases. Finally, a brief description is written to describe the purpose of the use case or the goal of the primary actor using the use case. This description can range from a sentence to a short essay. The idea is to capture the primary issues in the use case and make them explicit. (See Figure 5-15).

The fifth step is to carefully review the current set of use cases. Take a moment to review the use cases and make sure you understand them. Based on the descriptions of all four use cases, there seems to be an obvious missing use case: Maintain Order. Because the project team did not include language in the brief description of the Place Order use case, it seems that the project team believes that maintaining an order is sufficiently different from placing an order that it should have its own use case to describe it. Furthermore, it does not seem reasonable that a customer would see placing an order and maintaining an order as the same use case. This type of interactive and incremental process goes on until the project team feels that they have identified the major use cases for the Internet Sales System.

Expanding the Major Use Cases

The sixth step is to choose one of the major use cases to expand. To help the project team to make this choice, they identified the trigger and the stakeholders and their interests for each of the major use cases.

The first use case, Maintain CD Information, was triggered by the Distribution System distributing new information for use in the CD database. Besides the Distribution System, another stakeholder was identified: EM Manager. The second use case, Maintain CD Marketing Information, has a similar structure. The receipt of marketing materials from vendors triggers it. Again, it seemed to the project team that the EM Manager was an interested stakeholder. The third use case, Place Order, is more interesting. The trigger is the action of a Customer. Again, the EM Manager is an interested stakeholder. This use case has many more inputs. The fourth use case, Fill Order, is based on the decision logic identified in the activity diagram. However, upon further reflection, the team decided to replace this use case with three separate use cases, one for each approach to filling an order: Place Special Order, Place In-Store Hold, and Fill Mail Order. The first two are controlled by actions of the Customer. However, the Fill Mail Order use case has a temporal trigger: every hour the Internet Sales System downloads orders into the Distribution System. The final use case, Maintain Order, is triggered by the action of a Customer. However, on close review, it seems that it also has much in common with the other maintenance-related use cases: Maintain CD Information and Maintain CD Marketing Information. And like all of the other use cases, the EM Manager has an interest.

Based on their review of the major use cases, the project team decided that the Place Order use case was the most interesting. The next step, step 7, is to start filling in the details of the chosen use case. At this point in time, the project team had the information to fill in the stakeholders and interests, the trigger, and the association relationship. (*Note:* Remember that the association relationship is simply the communication link between the primary actor and the use case.) In this use case, the association relationship is with the Customer.

The project team then needed to gather and organize the information needed to define the Place Order use case in more detail. Specifically, they needed to begin writing the normal flow of events; that is, they should perform the eighth step for writing effective use cases (see Figure 5-11). This was done based on the results of the earlier analyses described in Chapter 4 as well as through a series of JAD meetings with the project sponsor and the key marketing department managers and staff who would ultimately operate the system.

The goal at this point is to describe how the chosen use-case operates: Place Order. Alec and the team decided to visualize placing a CD order over the Web and to think about how other electronic commerce Web sites work—that is, to role-play. As they role-played the Place Order use case, they realized that after the customer connected to the Web site, he or she probably began searching—perhaps for a specific CD, perhaps for a category of music—but in any event, they entered some information for their search. The Web site then should present a list of CDs matching their request along with some basic information about the CDs (e.g., artist, title, and price). If one of the CDs is of interest to them, they

Use Case Name:	Place Order		ID:	3	importance level:	High

Primary Actor:	Customer	Use case type:	Detail, Essential

Stakeholders and Interests:	Customer - wants to search Web site to purchase CD EM manager - wants to maximize customer satisfaction

Brief Description: This use case describes how customers can search the Web site and place orders.

Trigger: Customer visits Web site and places order

Type: External

Relationships:
  Association: Customer
  Include: Maintain order
  Extend:
  Generalization:

Normal Flow of Events:
  1. The Customer submits a search request to the system.
  2. The System provides the Customer a list of recommended CDs.
  3. The Customer chooses one of the CDs to find out additional information.
  4. The System provides the Customer with basic information and reviews on the CD.
  5. The Customer adds the CD to his or her shopping cart.
  6. The Customer decides how to "fill" the order.
  7. The Customer iterates over 3 through 5 until done shopping.
  8. The Customer executes the Maintain Order use case.
  9. The Customer logs in to check out.
  10. The System validates the Customer's credit card information.
  11. The System sends an order confirmation to the Customer.
  12. The Customer leaves the Web site.

Subflows:

Alternate/exceptional Flows:

FIGURE 5-16   Places Order Use Case after Step 8

might seek more information about it, such as the list of songs, liner notes, reviews, etc. Once they found a CD they liked, they would add it to their order and perhaps continue looking for more CDs. Once they were done—perhaps immediately—they would check out by presenting their order with information on the CDs they want and giving additional information such as mailing address, credit card, etc.

When the team wrote the use case's normal flow of events, they paid close attention to the seven guidelines described earlier. Alec realized that the first step was to present the customer with the home page or a form to fill in to search for an album. Even though this was technically correct, this type of step was very small compared to the other steps that followed.[27] It was analogous to making the first step, hand the user a piece of paper. At this point, the team was looking only for the three to seven *major* steps. Based on the role-playing and the application of the earlier principles (see Figure 5-6), the team successfully identified a set of steps (see Figure 5-16).

[27] Because it is so small, it violates the fourth principle (see Figure 5-6).

The first major step performed by the system is to respond to the customer's search inquiry, which might include a search for a specific album name or albums by a specific artist. Or, it might be the customer wanting to see all the classical or alternative CDs in stock. Or, it might be a request to see the list of special deals or CDs on sale. In any event, the system finds all the CDs matching the request and shows a list of CDs in response. The user will view this response and perhaps will decide to seek more information about one or more CDs. He or she will click on it, and the system will provide additional information. Perhaps the user will also want to see any extra marketing material that is available as well. The user will then select one or more CDs for purchase, decide how to take delivery of each CD, and perhaps continue with a new search. These steps correspond to events 1–7 in Figure 5-16.

The user may later make changes to the CDs selected, either by dropping some or changing the number ordered. This seems to be similar to an already identified separate use case: Maintain Order. As such, we can simply reuse that use case to handle those types of details. At some point the user checks out by verifying the CDs he or she has selected for purchase and providing information about himself or herself (e.g., name, mailing address, credit card). The system calculates the total payment and verifies the credit-card information with the credit-card clearance center. At this point in the transaction, the system sends an order confirmation to the customer, and the customer typically leaves the Web site. Figure 5-16 shows the use case at this point. Note that the normal flow of events has been added to the form, but nothing else has changed.

The ninth step for writing uses cases (see Figure 5-11) is to try and simplify the current normal flow of events. Currently, the team has twelve events, which is a little high. Therefore, the team decided to see if there were any steps that they could merge, delete, rearrange, and or leave out. Based on this review, they decided to create a separate use case that could handle the checkout process (events 9, 10, and 11): Checkout. They also realized that events 5 and 6 could be viewed as a part of the Maintain Order use case previously identified. As such they deleted events 5 and 6 and moved event 8 into their place. At this point in time, the Place Order use case had eight events. Given the purpose of this use case, it seemed to be a reasonable number of events.

The next two steps in writing a use case deals with alternate or exceptional flows. (*Note:* Remember the normal flow of events captures only the typical set of events that end in a successful transaction.) With the Place Order use case, the development team defined success as a new order being placed. However, the team identified two sets of events that were exceptions to the normal flow. First, event 3 assumed that the list of recommended CDs was acceptable to the customer. However, as one of the team members pointed out, that is an unrealistic assumption. As such, two exceptional flows have been identified and written (3a-1 and 3a-2 in Figure 5-17) to handle this specific situation. Second, a customer may want to abort the entire order instead of going through the checkout process. In this case, exceptional flow 7a was created for this case.[28]

Confirming the Major Use Cases

Once all the use cases had been defined, the final step in the JAD session was to confirm that they were accurate. The project team had the users role-play the use cases again. A few minor problems were discovered and were easily fixed. However, one major problem was

---

[28] Another approach would be to force the customer through the Maintain Order or Checkout use cases. However, the marketing representatives on the project team were concerned with customer frustration. As such, the project team included it in the Place Order use case.

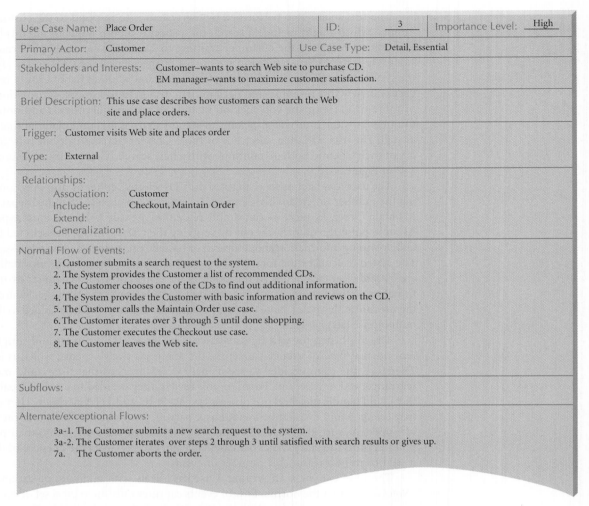

FIGURE 5-17   Places Order Use Case after Step 11

discovered: how is a new customer created? This was easily fixed by creating a new use case: Create New Customer. The new use case was added as an extension to the Checkout use case. Figure 5-18 shows the revised use cases. At this point in time, the use-case development process can actually start all over again, or we can move on to drawing the use-case diagram.

YOUR TURN   5-8 CD Selections Internet Sales System

Complete the detailed use-case descriptions for the remaining overview use cases in Figure 5-18. Remember that it is possible to uncover additional functionality when you iterate through the use cases. As such, be sure to review all of them once you have finished. Furthermore, you may also need to modify the activity diagram contained in Figure 5-14 once you have completed the detailed use case descriptions.

Use Case Name: Maintain CD Information		ID: ___1___	Importance Level:
Primary Actor: Distribution System		Use Case Type: Detail, Essential	
Stakeholders and Interests:			

Brief Description: This adds, deletes, and modifies the basic information about the CDs we have available for sale (e.g., album name, artist(s), price, quantity on hand, etc.).

Trigger: Downloads from the Distribution System

Type: External

Relationships:
    Association: Distribution System
    Include:

Use Case Name: Maintain Marketing Information		ID: ___2___	Importance Level: _High_
Primary Actor: Vendor		Use Case Type: Detail, Essential	

Stakeholders and Interests: Vendor–wants to ensure marketing information is as current as possible.
                           EM Manager–wants marketing information to be correct to maximize sales.

Brief Description: This adds, deletes, and modifies the marketing material available for some CDs (e.g., reviews).

Trigger: Materials arrive from vendors, distributors, wholesalers, record companies, and articles from trade magazines

Type: External

Relationships:
    Association: Vendor
    Include:

Use Case Name: Place Order		ID: ___3___	Importance Level: _High_
Primary Actor: Customer		Use Case Type: Detail, Essential	

Stakeholders and Interests: Customer–wants to search Web site to purchase CD.
                           EM Manager–wants to maximize Customer satisfaction.

Brief Description: This use case describes how customers can search the Web site and place orders.

Trigger: Customer visits Web site and places order.

Type: External

Relationships:
    Association: Customer
    Include: Checkout, Maintain Order
    Extend:
    Generalization :

FIGURE 5-18   Revised Major Use Cases for CD Selections

Use Case Name:	Maintain Order		ID:	5	Importance Level:	High
Primary Actor: Customer			Use Case Type: Detail, Essential			

Stakeholders and Interests: Customer–wants to modify order.
EM Manager–wants to ensure high customer satisfaction.

Brief Description: This use case describes how a Customer can cancel or modify an open or existing order.

Trigger: Customer visits Web site and requests to modify a current order.

Type: External

Relationships:
    Association:
    Include:

Use Case Name:	Checkout		ID:	6	Importance Level:	High
Primary Actor:	Customer		Use Case Type:	Detail, Essential		

Stakeholders and Interests: Customer–wants to finalize the order.
Credit Card Center–wants to provide effective and efficient service to CD Selections.
EM Manager–wants to maximize order closings.

Brief Description: This use case describes how the customer completes an order including credit card authorization.

Trigger: Customer signals the system they want to finalize their order.

Type: External

Relationships:
    Association: Credit Card Center
    Include: Maintain Order

Use Case Name:	Create New Customer		ID:	7	Importance Level:	High
Primary Actor:	Customer		Use Case Type:	Detail, Essential		

Stakeholders and Interests: Customer–wants to be able to purchase CDs from CD Selections.
EM Manager–wants to increase CD Selections Customer base.

Brief Description: This use case describes how a customer is added to the Customer database.

Trigger: An unknown Customer attempts to checkout.

Type: External

Relationships:
    Association:
    Include:
    Extend: Customer
    Generalization :

FIGURE 5-18   Revised Major Use Cases for CD Selections *(Continued)*

Use Case Name:	Place Special Order		ID:	8	Importance Level:	High

Primary Actor: Customer		Use Case Type: Detail, Essential

Stakeholders and Interests: Customer–wants to be able to place a special order of CDs for in store pick up.
                                EM manager–wants to increase sales associated with the Internet Sales system.
                                Bricks and Mortar Store Manager–wants to increase sales associated with the store.

Brief Description: This use case describes how a Customer places a special order using the Internet Sales system.

Trigger: Customer selects CD on order for a special order at bricks and mortar store.

Type: External

Relationships:
    Association: Bricks and mortar store
    Include:

Use Case Name:	Place InStore Hold		ID:	9	Importance Level:	High

Primary Actor: Customer		Use Case Type: Detail, Essential

Stakeholders and Interests: Customer–wants to be able to place an in store hold a CD for In Store pick up.
                                EM manager–wants to increase sales associated with the Internet Sales system.
                                Bricks and Mortar Store Manager–wants to increase sales associated with the store.

Brief Description: This use case describes how a Customer places an in store hold using the Internet Sales system.

Trigger: Customer selects CD on order for an in store hold to be picked up at bricks and mortar store.

Type: External

Relationships:
    Association: Bricks and mortar store
    Include:

Use Case Name:	Fill Mail Order		ID:	10	Importance Level:	High

Primary Actor: Customer		Use Case Type: Detail, Essential

Stakeholders and Interests: Mail Order Distribution System–wants to complete order processing in a timely manner.
                                Customer–wants to receive order in a timely manner.
                                EM Manager–wants to maximize order throughput.

Brief Description: This describes how mail orders move from the Internet Sales system into the distribution system and
                   how status information will be updated from the distribution system.

Trigger: Every hour the Distribution System will initiate a trading of information with the Internet Sales system.

Type: Temporal

Relationships:
    Association: Distribution System
    Include:
    Extend: Maintain Order
    Generalization :

FIGURE 5-18   Revised Major Use Cases for CD Selections (Continued)

Creating the Use-Case Diagram

Because use-case diagrams provide only a pictorial overview of the detailed use cases, the team found that creating the use-case diagram from the detailed use-case descriptions was very easy. (*Note:* Remember that a use-case diagram shows only the subject boundary, the use cases themselves, the actors, and the various associations between these components.) The team followed the four major steps for drawing a use-case diagram given in Figure 5-11.

First, they placed a box on the use-case diagram to represent the Internet Sales System and placed the system's name at the top of the box. Second, they added the use cases to the diagram. Third, they placed the actors near the use cases with which they were associated. Fourth, they drew in the association connecting the actors with their respective use cases. Figure 5-19 portrays the use case diagram that the team created.

Refining Project Size and Effort Estimation using Use-Case Points

Now that the functional modeling aspect of the system development effort has been completed, Alec and his team decided that they needed to refine the estimation of the size and effort to build the system. Based on the completed use cases (see Figure 5-18) and the use-case diagram (see Figure 5-19) and using the use-case point template (see Figure 5-13), Alec could now use the use cases as the basis for the estimation instead of using the function point estimate that he had done earlier (see Figure 3-21).

First, Alec had to classify each actor and use case as being simple, average, or complex. In the case of the actors, Bricks and Mortar store and Distribution System had a well-defined API. As such they were classified as simple actors. The Credit-Card center was considered to be average, whereas the vendor and customer actors were classified as being complex. This gave an unadjusted actor weight total value of 10. (See Figure 5-20.)

Second, Alec classified each use case based on the number of unique transactions that each had to handle. In this case, there were three simple use cases (Place InStore Hold, Place Special Order, and Fill Mail Order), one average use case (Create New Customer), and five complex use cases (Place Order, Checkout, Maintain Order, Maintain CD Information, and Maintain CD Marketing Information). Based on these, the unadjusted use-case weight total was computed to be 100 (see Figure 5-20).

Third, Alec computed a value of 110 for the unadjusted use-case points. Fourth, he rated each of the technical complexity factors, rated each of the environmental factors, and computed the values for TCF and EF (see Figure 5-20). Fifth, using the unadjusted use-case points and the TCF and EF values, Alec calculated a value of 134.992 for adjusted use-case points. Sixth, based on the decision rule to determine whether to use 20 or 28 as the value of the person-hours multiplier, Alec realized that he should use 28. Consequently, Alec estimated the remaining effort for the project to be 3,779.776 person-hours. Converting the person-hours to person-months (3,779.776/160) and using the estimated schedule equation that he had

YOUR TURN   5-9 Identifying Generalization Associations in Use Cases and Specialized Actors

The use-case diagram for the Internet Sales Systems does not include any generalization relationships. See if you can come up with one example for a use case and another example for an actor that may be helpful for CD Selections to add to the use-case diagram shown in Figure 5-19. Describe how the development effort may benefit from including your examples.

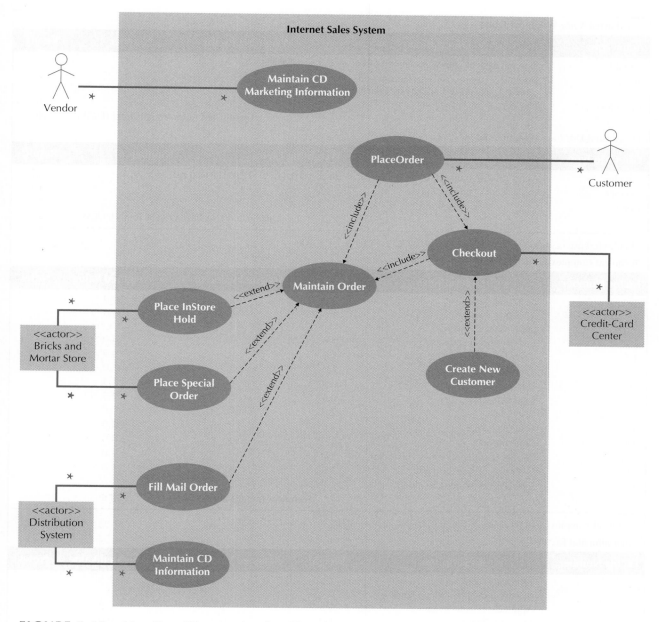

FIGURE 5-19   Use-Case Diagram for the CD Selections To-Be Internet Sales System

used in the earlier estimate, 3.0 * person-months$^{1/3}$ (see Figure 3-21), Alec estimated that the schedule would still be under the 10 months (8.6 months) that he had originally estimated. However, if he had not padded the estimate, he realized that he would have had to make significant changes to the current schedule. This could potentially have put the project into jeopardy of not being completed on time. Even though he currently considers himself fairly smart for padding the estimate, he realizes that his pad of 2.5 months has now been reduced to 1.4 months and he has not even delivered the analysis models yet. As such, Alec realizes that to prevent any schedule slippage, he must carefully manage the team.

Unadjusted Actor Weighting Table:

Actor Type	Description	Weighting Factor	Number	Result
Simple	External System with well-defined API	1	2	2
Average	External System using a protocol-based interface, e.g., HTTP, TCT/IP, or a database	2	1	2
Complex	Human	3	2	6
		Unadjusted Actor Weight Total (UAW)		**10**

Unadjusted Use Case Weighting Table:

Use Case Type	Description	Weighting Factor	Number	Result
Simple	1–3 transactions	5	3	15
Average	4–7 transactions	10	1	10
Complex	>7 transactions	15	5	75
		Unadjusted Use Case Weight Total (UUCW)		**100**

Unadjusted use case points (UUCP) = UAW + UUCW   110 = 10 + 100

Technical Complexity Factors:

Factor Number	Description	Weight	Assigned Value (0–5)	Weighted Value	Notes
T1	Distributed system	2.0	5	10.0	
T2	Response time or throughput performance objectives	1.0	5	5.0	
T3	End-user online efficiency	1.0	5	5.0	
T4	Complex internal processing	1.0	4	4.0	
T5	Reusability of code	1.0	3	3.0	
T6	Easy to install	0.5	3	1.5	
T7	Ease of use	0.5	5	2.5	
T8	Portability	2.0	4	8.0	
T9	Ease of change	1.0	3	3.0	
T10	Concurrency	1.0	3	3.0	
T11	Special security objectives included	1.0	5	5.0	
T12	Direct access for third parties	1.0	5	5.0	
T13	Special User training required	1.0	3	3.0	
		Technical Factor Value (TFactor)		**58.0**	

Technical complexity factor (TCF) = 0.6 + (0.01 × TFactor)   1.18 = 0.6 + (0.01 × 58)

Environmental Factors:

Factor Number	Description	Weight	Assigned Value (0 – 5)	Weighted Value	Notes
E1	Familiarity with system development process being used	1.5	1	1.5	
E2	Application experience	0.5	2	1.0	
E3	Object-oriented experience	1.0	0	0.0	
E4	Lead analyst capability	0.5	3	1.5	
E5	Motivation	1.0	4	4.0	
E6	Requirements stability	2.0	4	8.0	
E7	Part time staff	–1.0	0	0.0	
E8	Difficulty of programming language	–1.0	4	–4.0	
		Environmental Factor Value (EFactor)		**12.0**	

Environmental factor (EF) = 1.4 + (–0.03 ∗ EFactor)   1.04 = 1.4 + (–.03 × 12)
Adjusted use case points (UCP) = UUCP ∗ TCF ∗ ECF   134.992 = 110 × 1.18 × 1.04
Person hours multiplier (PHM) PHM = 28
Person hours = UPC × PHM   3,779.776 = 134.992 × 28

FIGURE 5-20   Use-Case Points Estimation for the Internet Sales Systems

SUMMARY

Business Process Modeling with Activity Diagrams

Even from an object-oriented systems development point of view, business process modeling has been shown to be valuable. The one major hazard of business process modeling is that it focuses the systems development effort process in a functional decomposition direction. However, if used carefully, it can enhance the development of object-oriented systems. UML 2.0 supports process modeling using an activity diagram. Activity diagrams comprise activities or actions, objects, control flows, object flows, and a set of seven different control nodes (initial, final-activity, final-flow, decision, merge, fork, and join). Furthermore, swimlanes can be used to enhance the readability of the diagrams. An activity diagram is very useful for helping the analyst to identify the relevant use cases for the information system being developed.

Use-Case Descriptions

Use cases are the primary method of documenting requirements for object-oriented systems. They represent a functional view of the system being developed. There are overview and detail use cases. Overview use cases comprise the use-case name, ID number, primary actor, type, and a brief description. Detailed use cases extend the overview use case with the identification and description of the stakeholders and their interest, the trigger and its type, the relationships in which the use case participates (association, extend, generalization, and include), the normal flow of events, the subflows, and any alternate or exception flows to the normal flow of events. There are seven guidelines (see Figure 5-6) and thirteen steps (see Figure 5-11) for writing effective use cases. The thirteen steps can be summed up into three aggregate groups of steps: identify the major use cases (steps 1–5), expand the major use cases (steps 6–11), and confirm the major use cases (steps 12–13).

Use-Case Diagrams

Use-case diagrams are simply a graphical overview of the set of use cases contained in the system. They illustrate the main functionality of a system and the actors that interact with the system. The diagram includes actors, which are people or things that derive value from the system, and use cases that represent the functionality of the system. The actors and use cases are separated by a system boundary and connected by lines representing associations. At times, actors are specialized versions of more general actors. Similarly, use cases can extend or use other use cases. Creating use-case diagrams is a four-step process (see Figure 5-8) whereby the analyst draws the system boundary, adds the use cases to the diagram, identifies the actors, and finally adds appropriate associations to connect use cases and actors together. This process is simple because the written description of the use cases is created first.

Refining Project Size and Effort Estimation with Use-Case Points

Use-case points are a relatively new size- and effort-estimation technique that is based on ideas similar to function point analysis. Use-case points are founded on the two primary constructs associated with use-case analysis: actors and use cases. Like function points, use-case points have a set of factors used to modify their raw value: technical complexity factors and environmental factors. Technical complexity factors address the complexity of the project under consideration, whereas the environmental factors deal with the level of experience of the development staff. Based on the number of use-case points, the estimated effort required can be computed.

KEY TERMS

Action
Activity
Activity diagram
Actor
Adjusted use-case points (UCP)
Alternate flows
Application program interface (API)
Association relationship
Average actors
Average use case
Black-hole activities
Brief description
Complex actors
Complex use case
Control flow
Control node
Decision node
Detail use case
Effort
Environmental factor (EF)
Environmental factor value (EFactor)
Essential use case
Estimation
Exceptional flows
Extend relationship
External trigger

Final-activity node
Final-flow node
Flow of events
Fork node
Functional decomposition
Generalization relationship
Guard condition
Importance level
Include relationship
Inheritance
Initial node
Iterate
Join node
Logical model
Merge node
Miracle activity
Normal flow of events
Object flow
Object node
Overview use cases
Packages
Person-hours multiplier (PHM)
Physical model
Primary actor
Process models
Real use case

Relationships
Role
Scenario
Simple actors
Simple use case
Specialized actor
Stakeholders
Subflows
Subject boundary
SVDPI
Swim lanes
Technical complexity factor (TCF)
Technical factor value (TFactor)
Temporal trigger
Trigger
Unadjusted actor weight total (UAW)
Unadjusted use-case points (UUCP)
Unadjusted use-case weight
    total (UUCW)
Use case
Use-case description
Use-case diagram
Use-case ID number
Use-case name
Use-case points
Use-case type

QUESTIONS

1. What is the purpose of an activity diagram?
2. What is the difference between an activity and an action?
3. Why is business process modeling important?
4. What are the different types of control nodes?
5. What is an object node?
6. What is the difference between a control flow and an object flow?
7. What is the purpose of a fork node?
8. What is CRUD? Why is this useful?
9. How is use case diagramming related to functional modeling?
10. Every association must be connected to at least one _____ and one _____. Why?
11. Explain the following terms. Use layperson's language, as though you were describing them to a user: (a) actor;

(b) use case; (c) system boundary; (d) relationship.
12. Which of the following could be an actor found on a use-case diagram? Why?

Ms. Mary Smith
Supplier
Customer
Internet customer
Mr. John Seals
Data entry clerk
Database administrator

13. What are some guidelines for designing a set of use cases? Give two examples of the extend associations on a use-case diagram. Give two examples for the include associations.
14. How does an essential use case differ from a real use case?

15. How does a detail use case differ from an overview use case?
16. What process do we use to estimate systems development based on use cases?
17. What is a use-case point? For what is it used?
18. What are the major elements of a detail use case?
19. What are the major elements of an overview use case?
20. How do you create use cases?
21. What are some heuristics for creating a use case diagram?
22. Why is iteration important in creating use cases?
23. What is the viewpoint of a use case, and why is it important?
24. Why do we strive to have about three to nine major use cases in a business process?

EXERCISES

A. Create an activity diagram and a set of detail use-case descriptions for the process of buying glasses from the viewpoint of the patient, but do not bother to identify the flow of events within each use case. The first step is to see an eye doctor who will give you a prescription. Once you have a prescription, you go to a optical dispensary, where you select your frames and place the order for your glasses. Once the glasses have been made, you return to the store for a fitting and pay for the glasses.
B. Draw a use-case diagram for the process of buying glasses in exercise A.
C. Investigate the Object Management Group. Write a brief memo describing what it is, its purpose, and its influence on UML and the object approach to systems development. (*Hint:* A good resource is www.omg.org.)
D. Investigate the Web site for Rational Software (www.306 .ibm.com/software/rational/) and its repository of information about UML. Write a paragraph news brief on the current state of UML (e.g., the current version and when it will be released; future improvements; etc.).
E. Create an activity diagram and a set of detail use-case descriptions for the following dentist's office system, but do not bother to identify the flow of events within each use case. Whenever new patients are seen for the first time, they complete a patient information form that asks their name, address, phone number, and brief medical history, which are stored in the patient information file. When a patient calls to schedule a new appointment or change an existing appointment, the receptionist checks the appointment file for an available time. Once a good time is found for the patient, the appointment is scheduled. If the patient is a new patient, an incomplete entry is made in the patient file; the full information will be collected when they arrive for their appointment. Because appointments are often made far in advance, the receptionist usually

mails a reminder postcard to each patient two weeks before their appointment.
F. Draw a use-case diagram for the dentist's office system in exercise E.
G. Complete the detail use-case descriptions for the dentist's office system in exercise E by identifying the normal flow of events, subflows, and alternate/exceptional flows within the use cases.
H. Create an activity diagram and a set of detail use-case descriptions for the following system. A Real Estate Inc. (AREI) sells houses. People who want to sell their houses sign a contract with AREI, and provide information on their house. This information is kept in a database by AREI, and a subset of this information is sent to the citywide multiple listing service used by all real estate agents. AREI works with two types of potential buyers. Some buyers have an interest in one specific house. In this case, AREI prints information from its database, which the real estate agent uses to help show the house to the buyer (a process beyond the scope of the system to be modeled). Other buyers seek AREI's advice in finding a house that meets their needs. In this case, the buyer completes a buyer information form that is entered into a buyer data base, and AREI real estate agents use its information to search AREI's data base and the multiple listing service for houses that meet their needs. The results of these searches are printed and used to help the real estate agent show houses to the buyer.
I. Draw a use-case diagram for the real estate system in exercise H.
J. Create an activity diagram and a set of detail use case descriptions for an online university registration system. The system should enable the staff of each academic department to examine the courses offered by their department, add and remove courses, and change the information about them (e.g., the maximum number

of students permitted). It should permit students to examine currently available courses, add and drop courses to and from their schedules, and examine the courses for which they are enrolled. Department staff should be able to print a variety of reports about the courses and the students enrolled in them. The system should ensure that no student takes too many courses and that students who have any unpaid fees are not permitted to register (assume that a fees data store is maintained by the university's financial office, which the registration system accesses but does not change).

K. Draw a use-case diagram for the online university registration system in exercise J.

L. Create an activity diagram and a set of detail use-case descriptions for the following system. A Video Store (AVS) runs a series of fairly standard video stores. Before a video can be put on the shelf, it must be cataloged and entered into the video database. Every customer must have a valid AVS customer card in order to rent a video. Customers rent videos for three days at a time. Every time a customer rents a video, the system must ensure that he or she does not have any overdue videos. If so, the overdue videos must be returned and an overdue fee paid before customer can rent more videos. Likewise, if the customer has returned overdue videos but has not paid the overdue fee, the fee must be paid before new videos can be rented. Every morning, the store manager prints a report that lists overdue videos. If a video is two or more days overdue, the manager calls the customer to remind him or her to return the video. If a video is returned in damaged condition, the manager removes it from the video database and may sometimes charge the customer.

M. Draw a use-case diagram for the video system in exercise L.

N. Create an activity diagram and a set of detail use-case descriptions for a health club membership system. When members join the health club, they pay a fee for a certain length of time. Most memberships are for one year, but memberships as short as two months are available. Throughout the year, the health club offers a variety of discounts on their regular membership prices (e.g., two memberships for the price of one for Valentine's day). It is common for members to pay different amounts for the same length of membership. The club wants to mail out reminder letters to members asking them to renew their memberships one month before their memberships expire. Some members have become angry when asked to renew at a much higher rate than their original membership con-

tract, so the club wants to track the prices paid so that the manager can override the regular prices with special prices when members are asked to renew. The system must track these new prices so that renewals can be processed accurately. One of the problems in the health club industry is the high turnover rate of members. Although some members remain active for many years, about half of the members do not renew their memberships. This is a major problem, because the health club spends a lot in advertising to attract each new member. The manager wants the system to track each time a member comes into the club. The system will then identify the heavy users and generate a report so the manager can ask them to renew their memberships early, perhaps offering them a reduced rate for early renewal. Likewise, the system should identify members who have not visited the club in more than a month, so the manager can call them and attempt to reinterest them in the club.

O. Draw a use-case diagram for the system in exercise N.

P. Create an activity diagram and a set of detail use-case descriptions for the following system. Picnics R Us (PRU) is a small catering firm with five employees. During a typical summer weekend, PRU caters fifteen picnics with twenty to fifty people each. The business has grown rapidly over the past year and the owner wants to install a new computer system for managing the ordering and buying process. PUR has a set of ten standard menus. When potential customers call, the receptionist describes the menus to them. If the customer decides to book a picnic, the receptionist records the customer information (e.g., name, address, phone number, etc.) and the information about the picnic (e.g., place, date, time, which one of the standard menus, total price) on a contract. The customer is then faxed a copy of the contract and must sign and return it along with a deposit (often a credit card or by check) before the picnic is officially booked. The remaining money is collected when the picnic is delivered. Sometimes, the customer wants something special (e.g., birthday cake). In this case, the receptionist takes the information and gives it to the owner, who determines the cost; the receptionist then calls the customer back with the price information. Sometimes the customer accepts the price, other times, the customer requests some changes that have to go back to the owner for a new cost estimate. Each week, the owner looks through the picnics scheduled for that weekend and orders the supplies (e.g., plates) and food (e.g., bread, chicken) needed to make them. The owner would like to use the system for marketing as well. It should be able to track how customers learned about PUR and identify repeat customers, so that PUR can mail special offers to them. The owner also wants to track the picnics for which PUR sent a contract, but the customer never signed the contract and actually booked a picnic.

Q. Draw a use-case diagram for the system in exercise P.

R. Create an activity diagram and a set of detail use case descriptions for a university library borrowing system (do not worry about catalogue searching, etc.). The system will record the books owned by the library and will record who has borrowed which books. Before someone can borrow a book, he or she must show a valid ID card, which is checked against the student database maintained by the registrar's office (for student borrowers), the faculty/staff database maintained by the personnel office (for faculty/staff borrowers), or against the library's own guest database (for individuals issued a guest card by the library) to ensure that it is still valid. The system must also check to ensure that the borrower does not have any overdue books or unpaid fines before he or she can borrow another book. Every Monday, the library prints and mails postcards to those people with overdue books. If a book is overdue by more than two weeks, a fine will be imposed and a librarian will telephone the borrower to remind him or her to return the book(s). Sometimes books are lost or are returned in damaged condition. The manager must then remove them from the database and will sometimes impose a fine on the borrower.

S. Draw a use-case diagram for the system in exercise R.

T. Create an activity diagram and a set of detail use-case descriptions for the following system. Of-the-Month Club (OTMC) is an innovative young firm that sells memberships to people who have an interest in certain products. People pay membership fees for one year and each month receive a product by mail. For example, OTMC has a coffee-of-the-month club that sends members one pound of special coffee each month. OTMC currently has six memberships (coffee, wine, beer, cigars, flowers, and computer games), each of which costs a different amount. Customers usually belong to just one, but some belong to two or more. When people join OTMC, the telephone operator records the name, mailing address, phone number, e-mail address, credit-card information, start date, and membership service(s) (e.g., coffee). Some customers request a double or triple membership (e.g., two pounds of coffee, three cases of beer). The computer game membership operates a bit differently from the others. In this case, the member must also select the type of game (action, arcade, fantasy/science fiction, educational, etc.) and age level. OTMC is planning to greatly expand the number of memberships it offers (e.g., video games, movies, toys, cheese, fruit, and vegetables), so the system needs to accommodate this future expansion. OTMC is also planning to offer three-month and six-month memberships.

U. Draw a use-case diagram for the system in exercise T.

V. Consider the application that is used at your school to register for classes. Complete a use-case point worksheet to estimate the effort to build such an application. You will need to make some assumptions about the application's interfaces and the various factors that affect its complexity.

W. For exercises E, H, J, L, N, P, R, and T, complete a use-case point worksheet to estimate the effort to build such an application.

MINICASES

1. Kunsan Hanvit Inc. is a small custom manufacturing firm located in Kunsan South Korea. When Chung-Hee Ko, the owner, first brought computers into the business office, the firm was very small and simple. He was able to use an inexpensive PC-based accounting system to handle the basic information-processing needs of the firm. As time went on, the firm grew and the work has become much more complex. The firm's business contracts are as complex as the custom products it manufactures. The simple accounting software is no longer able to keep track of many of the company's sophisticated contracts with its customers.

Mr. Ko has a staff of four in the business office who are familiar with the intricacies of the company's record-keeping requirements. He recently discussed with his staff his plan to hire an information system (IS) consultancy to evaluate Kunsan Hanvit's IS needs and propose an upgrading of its computer system. The staff are excited about the prospect of a new system, because the current system has caused much aggravation. However, they are wary of the consultants who will be conducting the project.

Assume that you are a systems analyst on the consulting team assigned to the Kunsan Hanvit Inc. project. At your first meeting with the staff, you want to be sure that they understand the work that your team will be performing and how they will participate in that work.

a. Explain, in clear, nontechnical terms, the goals of the analysis of the project.

b. Explain, in clear, nontechnical terms, how use cases and a use-case diagram will be used by the project team. Explain what these models are, what they represent in the system, and how they will be used by the team.

2. Professional and Scientific Staff Management (PSSM) is a unique type of temporary staffing agency. Many organizations today hire highly skilled, technical employees on a short-term, temporary basis to assist with special projects or to provide a needed technical skill. PSSM negotiates contracts with its client companies in which it agrees to provide temporary staff in specific job categories for a specified cost. For example, PSSM has a contract with an oil and gas exploration company in which it agrees to supply geologists with at least a master's degree for $5,000 per week. PSSM has contracts with a wide range of companies and can place almost any type of professional or scientific staff members, from computer programmers to geologists to astrophysicists.

When a PSSM client company determines that it will need a temporary professional or scientific employee, it issues a staffing request against the contract it had previously negotiated with PSSM. When a staffing request is received by PSSM's contract manager, the contract number referenced on the staffing request is entered into the contract database. Using information from the database, the contract manager reviews the terms and conditions of the contract and determines whether the staffing request is valid. The staffing request is valid if the contract has not expired, the type of professional or

scientific employee requested is listed on the original contract, and the requested fee falls within the negotiated fee range. If the staffing request is not valid, the contract manager sends the staffing request back to the client with a letter stating why the staffing request cannot be filled, and a copy of the letter is filed. If the staffing request is valid, the contract manager enters the staffing request into the staffing request database as an outstanding staffing request. The staffing request is then sent to the PSSM placement department.

In the placement department, the type of staff member, experience, and qualifications requested on the staffing request are checked against the database of available professional and scientific staff. If a qualified individual is found, he or she is marked "reserved" in the staff database. If a qualified individual cannot be found in the database, or is not immediately available, the placement department creates a memo that explains the inability to meet the staffing request and attaches it to the staffing request. All staffing requests are then sent to the arrangements department.

In the arrangements department the prospective temporary employee is contacted and asked to agree to the placement. After the placement details have been worked out and agreed to, the staff member is marked "placed" in the staff database. A copy of the staffing request and a bill for the placement fee is sent to the client. Finally, the staffing request, the "unable-to-fill" memo (if any), and a copy of the placement fee bill is sent to the contract manager. If the staffing request was filled, the contract manager closes the open staffing request in the staffing request database. If the staffing request could not be filled, the client is notified. The staffing request, placement fee bill, and unable-to-fill memo are then filed in the contract office.

a. Create an activity diagram for the business process described here.

b. Develop a use case for each major activity identified in the activity diagram.

c. Create a use-case diagram for the system described here.

CHAPTER 6

CLASS DIAGRAMS

A structural, or conceptual, model describes the structure of the data that supports the business processes in an organization. During analysis, the structural model presents the logical organization of data without indicating how the data are stored, created, or manipulated so that analysts can focus on the business, without being distracted by technical details. Later during design, the structural model is updated to reflect exactly how the data will be stored in databases and files. This chapter describes *class-responsibility-collaboration (CRC)* cards, *class diagrams,* and *object diagrams,* which are used to create the structural model.

OBJECTIVES

- Understand the rules and style guidelines for creating CRC cards, class diagrams, and object diagrams.
- Understand the processes used to create CRC cards, class diagrams, and object diagrams.
- Be able to create CRC cards, class diagrams, and object diagrams.
- Understand the relationship between the structural and use case models.

CHAPTER OUTLINE

INTRODUCTION

During analysis, analysts create functional models to represent how the business system will behave. At the same time, analysts need to understand the information that is used and created by the business system (e.g., customer information, order information). In this chapter, we discuss how the data underlying the behavior modeled in the use cases are organized and presented.

A *structural model* is a formal way of representing the objects that are used and created by a business system. It illustrates people, places, or things about which information is captured and how they are related to each other. The structural model is drawn using an iterative process in which the model becomes more detailed and less conceptual over time. In analysis, analysts draw a *conceptual model,* which shows the logical organization of the objects without indicating how the objects are stored, created, or manipulated. Because this model is free from any implementation or technical details, the analysts can focus more easily on matching the model to the real business requirements of the system.

In design, analysts evolve the conceptual structural model into a design model that reflects how the objects will be organized in databases and files. At this point, the model is checked for redundancy and the analysts investigate ways to make the objects easy to retrieve. The specifics of the design model are discussed in detail in the design chapters.

In this chapter, we focus on creating a conceptual structural model of the objects using CRC cards and class diagrams. Using these techniques, it is possible to show all the objects of a business system. We first describe structural models and their elements. Then, we describe *CRC cards,* class diagrams, and object diagrams. Finally, we describe how to create structural models using CRC cards and class diagrams and how the structural model relates to the use-case descriptions and the use-case diagram that we learned about in Chapter 5.

STRUCTURAL MODELS

Every time a systems analyst encounters a new problem to solve, the analyst must learn the underlying problem domain. The goal of the analyst is to discover the key data contained in the problem domain and to build a structural model of the objects. Object-oriented modeling allows the analyst to reduce the semantic gap between the underlying problem domain and the evolving structural model. However, the real world and the world of software are very different. The real world tends to be messy, whereas the world of software must be neat and logical. As such, an exact mapping between the structural model and the problem domain may not be possible. In fact, it may not even be desirable.

One of the primary purposes of the structural model is to create a vocabulary that can be used by the analyst and the users. Structural models represent the things, ideas, or concepts—that is, the objects—contained in the domain of the problem. They also allow the representation of the relationships between the things, ideas, or concepts. By creating a structural model of the problem domain, the analyst creates the vocabulary necessary for the analyst and users to communicate effectively.

One important thing to remember is that at this stage of development, the structural model does not represent software components or classes in an object-oriented programming language, even though the structural model does contain analysis classes, attributes, operations, and relationships among the analysis classes. The refinement of these initial classes into programming-level objects comes later. Nonetheless, the structural model at this point should represent the responsibilities of each class and the collaborations between the classes. Typically, structural models are depicted using CRC cards, class diagrams, and, in some cases, object diagrams. However, before describing CRC cards, class diagrams, and object diagrams, we describe the basic elements of structural models: classes, attributes, operations, and relationships.[1]

---

[1] For a more complete description of the fundamental characteristics of object-oriented systems, see the appendix.

Classes, Attributes, and Operations

A *class* is a general template that we use to create specific instances, or *objects,* in the problem domain. All objects of a given class are identical in structure and behavior but contain different data in their attributes. There are two different general kinds of classes of interest during analysis: concrete and abstract. Normally, when an analyst describes the application domain classes, they are referring to concrete classes; that is, *concrete classes* are used to create objects. *Abstract classes* do not actually exist in the real world; they are simply useful abstractions. For example, from an employee class and a customer class, we may identify a generalization of the two classes and name the abstract class person. We may not actually instantiate the person class in the system itself, instead creating and using only employees and customers.[2] We describe generalization in more detail with the following relationships.

A second classification of classes is the type of real-world thing that a class represents. There are domain classes, user interface classes, data structure classes, file structure classes, operating environment classes, document classes, and various types of multimedia classes. At this point in the development of our evolving system, we are interested only in domain classes. Later in design and implementation, the other types of classes become more relevant.

An *attribute* of an analysis class represents a piece of information that is relevant to the description of the class within the application domain of the problem being investigated. An attribute contains information the analyst or user feels the system should store. For example, a possible relevant attribute of an employee class is employee name, whereas one that may not be as relevant is hair color. Both describe something about an employee, but hair color is probably not all that useful for most business applications. Only those attributes that are important to the task should be included in the class. Finally, only attributes that are primitive or atomic types (i.e., integers, strings, doubles, date, time, boolean, etc.) should be added. Most complex or compound attributes are really placeholders for relationships between classes. Therefore, they should be modeled as relationships, not attributes (see the following).

The behavior of an analysis class is defined in an *operation,* or service. In later phases, the operations are converted to *methods.* However, because methods are more related to implementation, at this point in the development we use the term operation to describe the actions to which the *instances* of the class are capable of responding. Like attributes, only problem domain–specific operations that are relevant to the problem being investigated should be considered. For example, it is normally required that classes provide means of creating instances, deleting instances, accessing individual attribute values, setting individual attribute values, accessing individual relationship values, and removing individual relationship values. However, at this point in the development of the evolving system, the analyst should avoid cluttering up the definition of the class with these basic types of operations and focus only on relevant problem domain–specific operations.

Relationships

There are many different types of relationships that can be defined, but all can be classified into three basic categories of data abstraction mechanisms: generalization relationships, aggregation relationships and association relationships. These data abstraction mechanisms allow the analyst to focus on the important dimensions while ignoring nonessential dimensions. Like attributes, the analyst must be careful to include only relevant relationships.

---

[2] Because abstract classes are essentially not necessary and are not instantiated, there have been arguments made that it would be better not to include any of them in the description of the evolving system at this stage of development (see J. Evermann and Y. Wand, "Towards Ontologically Based Semantics for UML Constructs," in H. S. Junii, S. Jajodia, and A. Solvberg (eds.) *ER 2001, Lecture Notes in Computer Science 2224* (Berlin: Springer-Verlag, 2001): 354–367. However, because abstract classes have been included traditionally at this stage of development, we also include them.

Generalization Relationships   The generalization abstraction enables the analyst to create classes that inherit attributes and operations of other classes. The analyst creates a *superclass* that contains the basic attributes and operations that will be used in several *subclasses*. The subclasses inherit the attributes and operations of their superclass and can also contain attributes and operations that are unique just to them. For example, a customer class and an employee class can be generalized into a person class by extracting the attributes and operations they have in common and placing them into the new superclass, person. In this way, the analyst can reduce the redundancy in the class definitions so that the common elements are defined once and then reused in the subclasses. Generalization is represented with the *a-kind-of* relationship, so that we say that an employee is a-kind-of person.

The analyst also can use the flip side of generalization, specialization, to uncover additional classes by allowing new subclasses to be created from an existing class. For example, an employee class can be specialized into a secretary class and an engineer class. Furthermore, generalization relationships between classes can be combined to form generalization hierarchies. Based on the previous examples, a secretary class and an engineer class can be subclasses of an employee class, which in turn could be a subclass of a person class. This would be read as a secretary and an engineer are a-kind-of employee and a customer and an employee are a-kind-of person.

The generalization data abstraction is a very powerful mechanism that encourages the analyst to focus on the properties that make each class unique by allowing the similarities to be factored into superclasses. However, to ensure that the semantics of the subclasses are maintained, the analyst should apply the principle of *substitutability.* By this we mean that the subclass should be capable of substituting for the superclass anywhere that uses the superclass (e.g., anywhere we use the employee superclass, we could also logically use its secretary subclass). By focusing on the a-kind-of interpretation of the generalization relationship, the principle of substitutability is applied.

Aggregation Relationships   There have been many different types of aggregation or composition relationships proposed in data modeling, knowledge representation, and linguistics. For example, a-part-of (logically or physically), a-member-of (as in set membership), contained-in, related-to, and associated-with. However, generally speaking, all aggregation relationships relate *parts* to *wholes* or parts to *assemblies.* For our purposes, we use the *a-part-of* or *has-parts* semantic relationship to represent the aggregation abstraction. For example, a door is a-part-of a car, an employee is a-part-of a department, or a department is a-part-of an organization. Like the generalization relationship, aggregation relationships can be combined into aggregation hierarchies. For example, a piston is a-part-of an engine, and an engine is a-part-of a car.

Aggregation relationships are bidirectional in nature. The flip side of aggregation is *decomposition.* The analyst can use decomposition to uncover parts of a class that should be modeled separately. For example, if a door and engine are a-part-of a car, then a car has-parts door and engine. The analyst can bounce around between the various parts to uncover new parts. For example, the analyst can ask, What other parts are there to a car? or To which other assemblies can a door belong?

Association Relationships   There are other types of relationships that do not fit neatly into a generalization (a-kind-of) or aggregation (a-part-of) framework. Technically speaking, these relationships are usually a weaker form of the aggregation relationship. For example, a patient schedules an appointment. It could be argued that a patient is a-part-of an appointment. However, there is a clear semantic difference between this type of relationship and one that models the relationship between doors and cars or even workers and unions. As such, they are simply considered to be *associations* between instances of classes.

CRC CARDS

CRC cards are used to document the responsibilities and collaborations of a class. We use an extended form of the CRC card to capture all relevant information associated with a class.[3] We describe the elements of our CRC cards later, after we explain responsibilities and collaborations.

Responsibilities and Collaborations

Responsibilities of a class can be broken into two separate types: knowing and doing. *Knowing responsibilities* are those things that an instance of a class must be capable of knowing. An instance of a class typically knows the values of its attributes and its relationships. *Doing responsibilities* are those things that an instance of a class must be capable of doing. In this case, an instance of a class can execute its operations or it can request a second instance, which it knows about, to execute one of its operations on behalf of the first instance.

The structural model describes the objects necessary to support the business processes modeled by the use cases. Most use cases involve a set of several classes, not just one class. These classes form a *collaboration*. Collaborations allow the analyst to think in terms of clients, servers, and contracts.[4] A *client* object is an instance of a class that sends a request to an instance of another class for an operation to be executed. A *server* object is the instance that receives the request from the client object. A *contract* formalizes the interactions between the client and server objects. For example, a patient makes an appointment with a doctor. This sets up an obligation for both the patient and doctor to appear at the appointed time. Otherwise, consequences, such as billing the patient for the appointment regardless of whether he or she appears, can be dealt out. Also, the contract should spell out what the benefits of the contract will be, such as a treatment being prescribed for whatever ails the patient and a payment to the doctor for the services provided. Chapter 9 provides a more detailed explanation of contracts and examples of their use.

An analyst can use the idea of class responsibilities and client–server–contract collaborations to help identify the classes, along with the attributes, operations, and relationships, involved with a use case. One of the easiest ways in which to use CRC cards in developing a structural model is through anthropomorphism—pretending that the classes have human characteristics. This is done by having analyst and/or the user role-play and pretend that they are an instance of the class being considered. They then can either ask questions of themselves or be asked questions by other members of the development team. For example:

> Who or what are you?
> What do you know?
> What can you do?

The answers to the questions are then used to add detail to the evolving CRC cards.

---

[3] Our CRC cards are based on the work of D. Bellin and S. S. Simone, *The CRC Card Book* (Reading, MA: Addison-Wesley, 1997); I. Graham, *Migrating to Object Technology* (Wokingham, England: Addison-Wesley, 1995); and B. Henderson-Sellers and B. Unhelkar, *OPEN modeling with UML* (Harlow, England: Addison-Wesley, 2000).

[4] For more information, see K. Beck and W. Cunningham, "A Laboratory for Teaching Object-Oriented Thinking," *Proceedings of OOPSLA, SIGPLAN Notices*, 24, no. 10 (1989): 1–6; B. Henderson-Sellers and B. Unhelkar, *OPEN Modeling with UML* (Harlow, England: Addison-Wesley, 2000); C. Larman, *Applying UML and Patterns: An Introduction to Object-Oriented Analysis and Design* (Englewood Cliffs, NJ: Prentice Hall, 1998); B. Meyer, *Object-Oriented Software Construction* (Englewood Cliffs, NJ: Prentice Hall, 1994); and R. Wirfs-Brock, B. Wilkerson, and L. Wiener, *Designing Object-Oriented Software* (Englewood Cliffs, NJ, Prentice Hall, 1990).

Elements of a CRC Card

The set of CRC cards contains all the information necessary to build a logical structural model of the problem under investigation. Figure 6-1 shows a sample CRC card. Each CRC card captures and describes the essential elements of a class. The front of the card allows the recording of the class's name, ID, type, description, list of associated use cases, responsibilities, and collaborators. The name of a class should be a noun (but not a proper noun, such as the name of a specific person or thing). Just like the use cases, in later stages of development, it is important to be able to trace back design decisions to specific requirements. In conjunction with the list of associated use cases, the ID number for each class can be used to accomplish this. The description is simply a brief statement that can be used as a textual definition for the class. The responsibilities of the class tend to be the operations that the class must contain (i.e., the doing responsibilities).

The back of a CRC card contains the attributes and relationships of the class. The attributes of the class represent the knowing responsibilities that each instance of the class

Front:

Class Name: Patient	ID: 3	Type: Concrete, Domain
Description: An individual that needs to receive or has received medical attention		Associated Use Cases: 2

Responsibilities	Collaborators
Make appointment	Appointment
Calculate last visit	
Change status	
Provide medical history	Medical history

Back:

Attributes:
Amount (double)
Insurance carrier (text)

Relationships:
Generalization (a-kind-of): Person
Aggregation (has-parts): Medical History
Other Associations: Appointment

FIGURE 6-1
Sample CRC Card

will have to meet. Typically, the data type of each attribute is listed with the name of the attribute (e.g., the amount attribute is double and the insurance carrier is text). There are three types of relationships typically captured at this point in time, generalization, aggregation, and other associations. In Figure 6-1, we see that a Patient is a-kind-of Person and that a Patient is associated with Appointments.

Again, CRC cards are used to document the essential properties of a class. However, once the cards are filled out, the analyst can use the cards and anthropomorphism in role-playing to uncover missing properties by executing the different scenarios associated with the use cases (see Chapter 5). This can be used as a basis to test the clarity and completeness of the evolving representation of the system.[5]

CLASS DIAGRAMS

A *class diagram* is a *static model* that shows the classes and the relationships among classes that remain constant in the system over time. The class diagram depicts classes, which include both behaviors and states, with the relationships between the classes. The following sections first present the elements of the class diagram, followed by the way in which a class diagram is drawn.

Elements of a Class Diagram

Figure 6-2 shows a class diagram that was created to reflect the classes and relationships needed for the set of use cases that describe the appointment system in Chapter 5.

Class   The main building block of a class diagram is the *class*, which stores and manages information in the system (see Figure 6-3). During analysis, classes refer to the people, places, events, and things about which the system will capture information. Later, during design and implementation, classes can refer to implementation-specific artifacts such as windows, forms, and other objects used to build the system. Each class is drawn using three part rectangles, with the class's name at top, attributes in the middle, and operations at the bottom. We can see that Person, Medical Staff, Doctor, Patient, Medical History, Appointment, Bill, Diagnosis, Health Problem, and Symptom are classes in Figure 6-2. The attributes of a class and their values define the state of each object created from the class, and the behavior is represented by the operations.

Attributes are properties of the class about which we want to capture information (see Figure 6-3). Notice that the Person class in Figure 6-2 contains the attributes lastname, firstname, address, phone, and birthdate. At times, you may want to store *derived attributes*, which are attributes that can be calculated or derived; these special attributes are denoted by placing a slash (/) before the attribute's name. Notice how the person class contains a derived attribute called /age, which can be derived by subtracting the patient's birth date from the current date. It is also possible to show the *visibility* of the attribute on the diagram. Visibility relates to the level of information hiding to be enforced for the attribute. The visibility of an attribute can either be public (+), protected (#), or private (−). A *public* attribute is one that is not hidden from any other object. As such, other objects can modify its value. A *protected* attribute is one that is hidden from all other classes except its immediate subclasses. A *private* attribute is one that is hidden from all other classes. The default visibility for an attribute is normally private.

---

[5] For presentation purposes, we defer discussing walkthroughs and inspections to Chapter 8 and testing to Chapter 13.

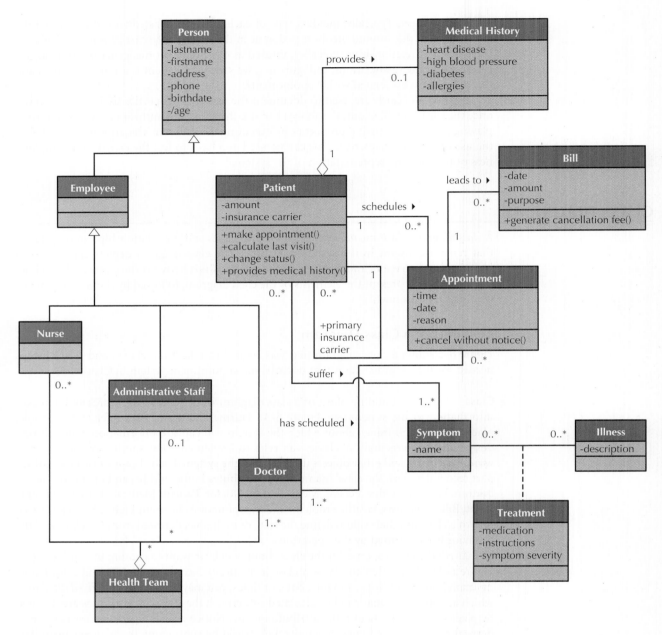

FIGURE 6-2   Sample Class Diagram

Operations are actions or functions that a class can perform (see Figure 6-3). The functions that are available to all classes (e.g., create a new instance, return a value for a particular attribute, set a value for a particular attribute, delete an instance) are not explicitly shown within the class rectangle. Instead, only those operations unique to the class are included, such as the cancel without notice operation in the Appointment class and the generate cancellation fee operation in the Bill class in Figure 6-2. Notice that both the operations are followed by parentheses, which contain the parameter(s) needed by the operation. If an

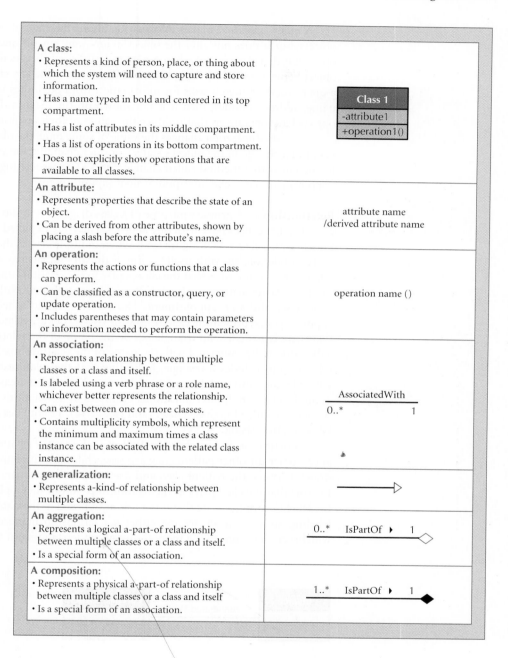

A class: • Represents a kind of person, place, or thing about which the system will need to capture and store information. • Has a name typed in bold and centered in its top compartment. • Has a list of attributes in its middle compartment. • Has a list of operations in its bottom compartment. • Does not explicitly show operations that are available to all classes.	**Class 1** -attribute1 +operation1()
An attribute: • Represents properties that describe the state of an object. • Can be derived from other attributes, shown by placing a slash before the attribute's name.	attribute name /derived attribute name
An operation: • Represents the actions or functions that a class can perform. • Can be classified as a constructor, query, or update operation. • Includes parentheses that may contain parameters or information needed to perform the operation.	operation name ()
An association: • Represents a relationship between multiple classes or a class and itself. • Is labeled using a verb phrase or a role name, whichever better represents the relationship. • Can exist between one or more classes. • Contains multiplicity symbols, which represent the minimum and maximum times a class instance can be associated with the related class instance.	AssociatedWith 0..* 1
A generalization: • Represents a-kind-of relationship between multiple classes.	———————▷
An aggregation: • Represents a logical a-part-of relationship between multiple classes or a class and itself. • Is a special form of an association.	0..* IsPartOf ▸ 1 ◇
A composition: • Represents a physical a-part-of relationship between multiple classes or a class and itself • Is a special form of an association.	1..* IsPartOf ▸ 1 ◆

FIGURE 6-3
Class Diagram Syntax

operation has no parameters, the parentheses are still shown but are empty. As with attributes, the visibility of an operation can be designated as public, protected, or private. The default visibility for an operation is normally public.

There are three kinds of operations that a class can contain: constructor, query, and update. A *constructor operation* creates a new instance of a class. For example, the patient class may have a method called insert (), which creates a new patient instance as patients are entered into the system. As we just mentioned, if an operation implements one of the basic functions (e.g., create a new instance), it is not explicitly shown on the class diagram, so typically we do not see constructor methods explicitly on the class diagram.

A *query operation* makes information about the state of an object available to other objects, but it does not alter the object in any way. For instance, the calculate last visit () operation that determines when a patient last visited the doctor's office will result in the object being accessed by the system, but it will not make any change to its information. If a query method merely asks for information from attributes in the class (e.g., a patient's name, address, phone), then it is not shown on the diagram because we assume that all objects have operations that produce the values of their attributes.

An *update operation* changes the value of some or all the object's attributes, which may result in a change in the object's state. Consider changing the status of a patient from new to current with a method called change status () or associating a patient with a particular appointment with schedule appointment (appointment).

Relationships   A primary purpose of a class diagram is to show the relationships, or associations, that classes have with one another. These are depicted on the diagram by drawing lines between classes (see Figure 6-3). When multiple classes share a relationship (or a class shares a relationship with itself), a line is drawn and labeled with either the name of the relationship or the roles that the classes play in the relationship. For example, in Figure 6-2 the two classes patient and appointment are associated with one another whenever a patient schedules an appointment. Thus, a line labeled schedules connects patient and appointment, representing exactly how the two classes are related to each other. Also, notice that there is a small solid triangle beside the name of the relationship. The triangle allows a direction to be associated with the name of the relationship. In Figure 6-2, the schedules relationship includes a triangle, indicating that the relationship is to be read as patient schedules appointment. Inclusion of the triangle simply increases the readability of the diagram. In Figure 6-4, three additional examples of associations are portrayed: an Invoice is AssociatedWith a Purchase Order (and vice versa), a Pilot Flies an Aircraft, and a Spare Tire IsLocatedIn a Trunk.

Sometimes a class is related to itself, as in the case of a patient being the primary insurance carrier for other patients (e.g., spouse, children). In Figure 6-2, notice that a line was drawn between the patient class and itself and called primary insurance carrier to depict the role that the class plays in the relationship. Notice that a plus (+) sign is placed before the label to communicate that it is a role as opposed to the name of the relationship. When labeling an association, we use either a relationship name or a role name (not both), whichever communicates a more thorough understanding of the model.

FIGURE 6-4
Sample Association

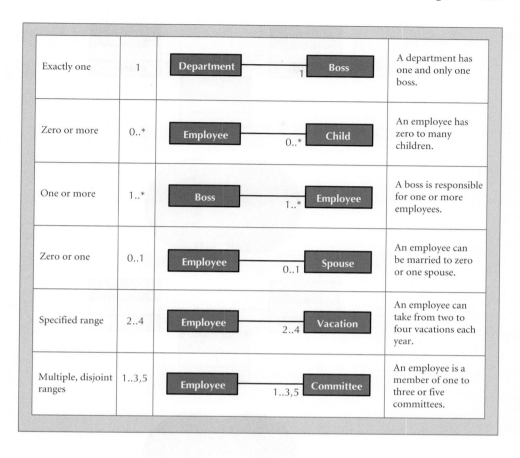

FIGURE 6-5
Multiplicity

Relationships also have *multiplicity*, which documents how an instance of an object can be associated with other instances. Numbers are placed on the association path to denote the minimum and maximum instances that can be related through the association in the format minimum number .. maximum number (see Figure 6-5). The numbers specify the relationship from the class at the far end of the relationship line to the end with the number. For example, in Figure 6-2, there is a 0..* on the appointment end of the patient schedules appointment relationship. This means that a patient can be associated with zero through many different appointments. At the patient end of this same relationship there is a 1, meaning that an appointment must be associated with one and only one (1) patient. In Figure 6-4, we see that an instance of the Invoice class must be AssociatedWith one instance of the Purchase Order class and that an instance of the Purchase Order class may be AssociatedWith zero or more instances of the Invoice class, an instance of the Pilot class Flies zero or more instances of the Aircraft class and that an instance of the Aircraft class may be flown by zero or more instances of the Pilot class; finally, we see that an instance the Spare Tire class IsLocatedIn zero or one instance of the Trunk class, whereas an instance of the Trunk class can contain zero or one instance of the Spare Tire class.

There are times when a relationship itself has associated properties, especially when its classes share a many-to-many relationship. In these cases, a class called an *association class* is formed, which has its own attributes and operations. It is shown as a rectangle attached by a dashed line to the association path, and the rectangle's name matches the label of the association. Think about the case of capturing information about illnesses and symptoms. An illness (e.g., the flu) can be associated with many symptoms (e.g., sore throat, fever),

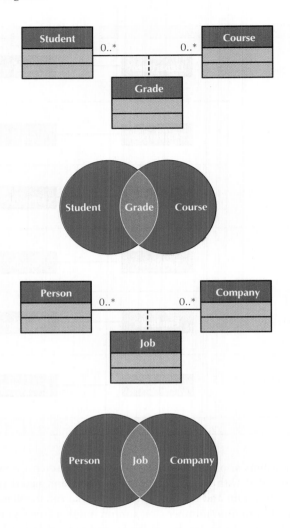

FIGURE 6-6
Sample Association
Classes

and a symptom (e.g., sore throat) can be associated with many illnesses (e.g., the flu, strep throat, the common cold). Figure 6-2 shows how an association class can capture information about remedies that change depending on the various combinations. For example, a sore throat caused by strep throat will require antibiotics; whereas, treatment for a sore throat from the flu or a cold could be throat lozenges or hot tea. Another way to decide when to use an association class is when attributes that belong to the intersection of the two classes involved in the association must be captured. We can visually think about an association class as a Venn diagram. For example, in Figure 6-6, the Grade idea is really an intersection of the Student and Course classes, because a grade exists only at the intersection of these two ideas. Another example shown in Figure 6-6 is that a job may be viewed as the intersection between a Person and a Company. Most often, classes are related through a normal association; however, there are two special cases of an association that you will see appear quite often: generalization and aggregation.

Generalization and Aggregation Associations   A *generalization association* shows that one class (subclass) inherits from another class (superclass), meaning that the properties and operations of the superclass are also valid for objects of the subclass. The generalization

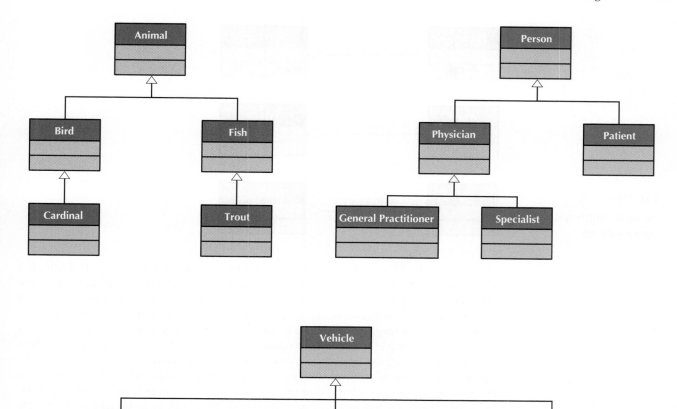

FIGURE 6-7   Sample Generalizations

path is shown with a solid line from the subclass to the superclass and a hollow arrow pointing at the superclass (see Figure 6-3). For example, Figure 6-2 communicates that doctors, nurses, and administrative personnel are all kinds of employees and those employees and patients are kinds of persons. Remember that the generalization relationship occurs when you need to use words like *is a kind of* to describe the relationship. Some additional examples of generalization are given in Figure 6-7. For example, Cardinal is a kind of Bird, which is a kind of Animal; a General Practitioner is a kind of Physician, which is a kind of Person; and a Truck is a kind of Land (Vehicle), which is a kind of Vehicle.

An *aggregation association* is used when classes actually comprise other classes. For example, think about a doctor's office that has decided to create health-care teams that include doctors, nurses, and administrative personnel. As patients enter the office, they are

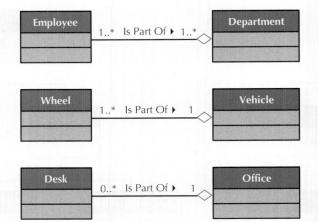

FIGURE 6-8
Sample Aggregation
Associations

assigned to a health-care team, which cares for their needs during their visits. Figure 6-2 shows how this relationship is denoted on a class diagram. A diamond is placed nearest the class representing the aggregation (health-care team), and lines are drawn from the arrow to connect the classes that serve as its parts (doctors, nurses, and administrative personnel). Typically, you can identify these kinds of associations when you need to use words like is a part of or is made up of to describe the relationship. However, from a *UML* perspective, there are two types of aggregation associations: aggregation and composition (see Figure 6-3). Aggregation is used to portray logical a part of relationships and is depicted on a UML class diagram by a hollow or white diamond. For example in Figure 6-8, three logical aggregations are shown. Logical implies that it is possible for a part be associated with multiple wholes or that is relatively simple for the part to be removed from the whole. For example, an instance of the Employee class IsPartOf an instance of at least one instance of the Department class, an instance of the Wheel class IsPartOf an instance of the Vehicle class, and an instance of the Desk class IsPartOf an instance of the Office class. Obviously, it is relatively easy to remove a wheel from a vehicle or move a desk from an office. Composition is used to portray a physical part of relationships and is shown by a black diamond. Physical implies that the part can be associated with only a single whole. For example in Figure 6-9, three physical compositions are illustrated: an instance of a door can be a part of only a single instance of a car, an instance of a room can be a part of an instance only of

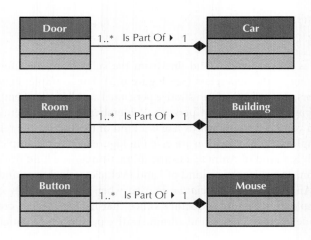

FIGURE 6-9
Sample Composition
Associations

a single building; and an instance of a button can be a part of only a single mouse. However, in many cases, the distinction that you can achieve by including aggregation (white diamonds) and composition (black diamonds) in a class diagram may not be worth the price of adding additional graphical notation for the client to learn. As such, many UML experts view the inclusion of aggregation and composition notation to the UML class diagram as simply "syntactic sugar" and, as such, not necessary because the same information can always be portrayed by simply using the association syntax.

Simplifying Class Diagrams

When a class diagram is fully populated with all the classes and relationships for a real-world system, the class diagram can become very difficult to interpret (i.e., be very complex). When this occurs, it is sometimes necessary to simplify the diagram. One way to simplify the class diagram is to show only concrete classes.[6] However, depending on the number of associations that are connected to abstract classes—and thus inherited down to the concrete classes—this particular suggestion could make the diagram more difficult to comprehend.

A second way to simplify the class diagram is through the use of a *view* mechanism. Views were developed originally with relational database management systems and are simply a subset of the information contained in the database. In this case, the view would be a useful subset of the class diagram, such as a use case view that shows only the classes and relationships relevant to a particular use case. A second view could be to show only a particular type of relationship: aggregation, association, or generalization. A third type of view is to restrict the information shown with each class, for example, show only the name of the class, the name and attributes, or the name and operations. These view mechanisms can be combined to further simplify the diagram.

A third approach to simplifying a class diagram is through the use of *packages* (i.e., logical groups of classes). To make the diagrams easier to read and keep the models at a reasonable level of complexity, the classes can be grouped together into packages. Packages are general constructs that can be applied to any of the elements in UML models. In Chapter 5, we introduced them to simplify use-case diagrams. In the case of class diagrams, it is simple to sort the classes into groups based on the relationships that they share.[7]

Object Diagrams

Although class diagrams are necessary to document the structure of the classes, there are times when a second type of *static structure diagram*, called an object diagram, can be useful. An object diagram is essentially an instantiation of all or part of a class diagram. Instantiation means to create an instance of the class with a set of appropriate attribute values.

Object diagrams can be very useful when trying to uncover details of a class. Generally speaking, it is easier to think in terms of concrete objects (instances) rather than abstractions of objects (classes). For example in Figure 6-10, a portion of the class diagram in Figure 6-2 has been copied and instantiated. The top part of the figure simply is a copy of a small view of the overall class diagram. The lower portion is the object diagram that instantiates that subset of classes. By reviewing the actual instances involved, John Doe, Appt1, and Dr. Smith, we may discover additional relevant attributes, relationships, and/or

[6] See Footnote 2.

[7] For those familiar with structured analysis and design, packages serve a similar purpose as the leveling and balancing processes used in data flow diagramming. Packages and package diagrams are described in more detail in Chapter 8.

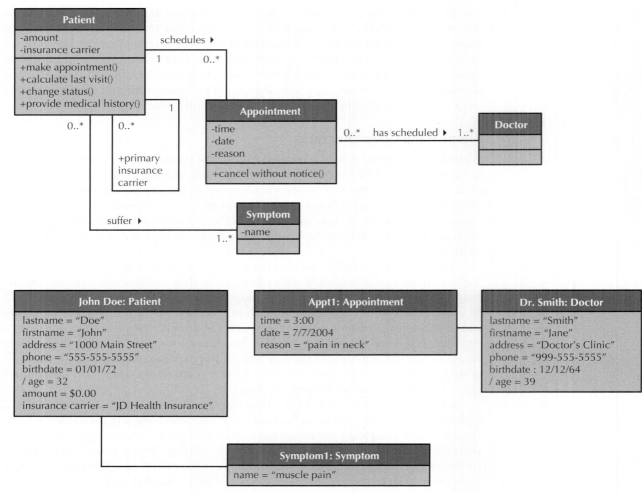

FIGURE 6-10   Sample Object Diagram

operations or possibly misplaced attributes, relationships, and/or operations. For example, an appointment has a reason attribute. Upon closer examination, the reason attribute may have been better modeled as an association with the Symptom class. Currently, the Symptom class is associated with the Patient class. After reviewing the object diagram, this seems to be in error. As such, we should modify the class diagram to reflect this new understanding of the problem.

CREATING CRC CARDS AND CLASS DIAGRAMS

Creating a structural model is an iterative process whereby the analyst makes a rough cut of the model and then refines it over time. Structural models can become quite complex—in fact, there are systems that have class diagrams containing hundreds of classes. In this section, we go through one iteration of creating a structural model, but we would expect that the model would change dramatically as we communicate with users and fine-tune our understanding of the system. However, before we do this, we first present a set of

- A common or improper noun implies a class of objects.
- A proper noun or direct reference implies an instance of a class.
- A collective noun implies a class of objects made up of groups of instances of another class.
- An adjective implies an attribute of an object.
- A doing verb implies an operation.
- A being verb implies a classification relationship between an object and its class.
- A having verb implies an aggregation or association relationship.
- A transitive verb implies an operation.
- An intransitive verb implies an exception.
- A predicate or descriptive verb phrase implies an operation.
- An adverb implies an attribute of a relationship or an operation.

Source: These guidelines are based on Russell J. Abbott, "Program Design by Informal English Descriptions," *Communications of the ACM* 26, no. 11 (1983): 882–894; Peter P-S Chen, "English Sentence Structure and Entity-Relationship Diagrams," *Information Sciences: An International Journal* 29, no. 2–3 (1983): 127–149; and Graham, *Migrating to Object Technology.*

FIGURE 6-11
Textual Analysis
Guidelines

approaches to identify and refine a set of candidate objects and provide a set of steps based on these approaches that can be used to create the initial rough cut structural model.

Object Identification

Many different approaches have been suggested to aid the analyst in identifying a set of candidate objects for the structural model. The three most common approaches are textual analysis, common object lists, and patterns. Most analysts use a combination of all three techniques to make sure that no important objects and object attributes, operations, and relationships have been overlooked.

Textual Analysis   *Textual analysis* is an analysis of the text in the use-case descriptions. The analyst starts by reviewing the use-case descriptions and the use-case diagrams. The text in the descriptions is examined to identify potential objects, attributes, operations, and relationships. The nouns in the use case suggest possible classes, whereas the verbs suggest possible operations. Figure 6-11 presents a summary of guidelines we have found useful. The textual analysis of use-case descriptions has been criticized as being too simple, but because its primary purpose is to create an initial rough-cut structural model, its simplicity is a major advantage.

Common Object List   As its name implies, a *common object list* is simply a list of objects common to the business domain of the system. In addition to looking at the specific use cases, the analyst also thinks about the business, separately from the use cases. Several categories of objects have been found to help the analyst in the creation of the list, such as physical or tangible things, incidents, roles, and interactions.[8] Analysts should first look for physical, or *tangible, things* in the business domain. These could include books, desks, chairs, and office equipment. Normally, these types of objects are the easiest to identify. *Incidents* are events that occur in the business domain, such as meetings, flights, performances, or accidents. Reviewing the use cases can readily identify the *roles* that the people play in the problem, such as doctor, nurse, patient, or receptionist. Typically, an *interaction* is a transaction that takes

[8] For example, see C. Larman, *Applying UML and Patterns: An Introduction to Object-Oriented Analysis and Design* (Englewood Cliffs, NJ: Prentice Hall, 1998); and S. Shlaer and S. J. Mellor, *Object-Oriented Systems Analysis: Modeling the World in Data* (Englewood Cliffs, NJ: Yourdon Press, 1988).

FIGURE 6-12 Sales Order Contract Pattern

place in the business domain, such as a sales transaction. Other types of objects that can be identified include places, containers, organizations, business records, catalogs, and policies. In rare cases, processes themselves may need information stored about them. In these cases processes may need an object, in addition to a use case, to represent them.

Patterns   The idea of using patterns is a relatively new area in object-oriented systems development.[9] There have been many definitions of exactly what a pattern is. From our perspective, a *pattern* is simply a useful group of collaborating classes that provide a solution to a commonly occurring problem. Because patterns provide a solution to commonly occurring problems, they are reusable.

An architect, Christopher Alexander, has inspired much of the work associated with using patterns in object-oriented systems development. According to Alexander and his colleagues,[10] it is possible to make very sophisticated buildings by stringing together commonly found patterns, rather than creating entirely new concepts and designs. In a very similar manner, it is possible to put together commonly found object-oriented patterns to form elegant object-oriented information systems. For example, many business transactions involve the same type of objects and interactions. Virtually all transactions would require a transaction class, a transaction line item class, an item class, a location class, and a participant class. By simply reusing these existing patterns of classes, we can more quickly and more completely define the system than if we start with a blank piece of paper.

Many different types of patterns have been proposed, ranging from high-level business-oriented patterns to more low-level design patterns. For example, Figure 6-12 depicts a very useful pattern: the Sales Order Contract pattern[11]. Figure 6-13 lists some common business domains for which patterns have been developed and their source. If we are developing a business information system in one of these business domains, then the patterns developed for that domain may be a very useful starting point in identifying needed classes and their attributes, operations, and relationships.

---

[9] There have been many books devoted to this topic; for example, see Peter Coad, David North, and Mark Mayfield, *Object Models: Strategies, Patterns, & Applications,* 2nd ed. (Englewood Cliffs, NJ: Prentice Hall, 1997); Hans-Erik Eriksson and Magnus Penker, *Business Modeling with UML: Business Patterns at Work* (New York: Wiley, 2000); Martin Fowler, *Analysis Patterns: Reusable Object Models* (Reading, MA: Addison-Wesley, 1997); Erich Gamma, Richard Helm, Ralph Johnson, and John Vlissides, *Design Patterns: Elements of Reusable Object-Oriented Software* (Reading, MA: Addison-Wesley, 1995); David C. Hay, *Data Model Patterns: Conventions of Thought* (New York: Dorset House, 1996).

[10] Christopher Alexander, Sara Ishikawa, Murray Silverstein, Max Jacobson, Ingrid Fiksdahl-King, and Shlomo Angel, *A Pattern Language* (New York: Oxford University Press, 1977).

[11] This pattern is based on the sales order contract pattern described in David C. Hay, *Data Model Patterns: Conventions of Thought* (New York: Dorset House, 1996).

Business Domains	Sources of Patterns
Accounting	3, 4
Actor-Role	2
Assembly-Part	1
Container-Content	1
Contract	2, 4
Document	2, 4
Employment	2, 4
Financial Derivative Contracts	3
Geographic Location	2, 4
Group-Member	1
Interaction	1
Material Requirements Planning	4
Organization and Party	2, 3
Plan	1, 3
Process Manufacturing	4
Trading	3
Transactions	1, 4

1. Peter Coad, David North, and Mark Mayfield, *Object Models: Strategies, Patterns, and Applications,* 2nd ed. (Englewood Cliffs, NJ: Prentice Hall, 1997).

2. Hans-Erik Eriksson and Magnus Penker, *Business Modeling with UML: Business Patterns at Work,* (New York: Wiley, 2000).

3. Martin Fowler, *Analysis Patterns: Reusable Object Models* (Reading, MA: Addison-Wesley, 1997).

4. David C. Hay, *Data Model Patterns: Conventions of Thought* (New York, NY, Dorset House, 1996).

FIGURE 6-13
Useful Patterns

Building CRC Cards and Class Diagrams

It is important to remember that CRC cards and class diagrams can be used to describe both the as-is and to-be structural models of the evolving system, but they are most often used for the to-be model. There are many different ways to identify a set of candidate objects and to create CRC cards and class diagrams. Today most object identification begins with the use cases identified for the problem (See Chapter 5). In this section, we describe a seven-step process used to create the structural model of the problem (see Figure 6-14).

1. Create CRC cards by performing textual analysis on the use cases.

2. Brainstorm additional candidate classes, attributes, operations, and relationships by using the common object list approach.

3. Role-play each use case using the CRC cards.

4. Create the class diagram based on CRC cards.

5. Review the structural model for missing and/or unnecessary classes, attributes, operations, and relationships.

6. Incorporate useful patterns.

7. Review the structural model.

FIGURE 6-14
Steps for Object
Identification and
Structural Modeling

Step 1: Create CRC Cards We could begin creating the structural model with a class diagram instead of CRC cards. However, due to the ease of role-playing with CRC cards, we prefer to create the CRC cards first and then transfer the information from the CRC cards into a class diagram later. As a result, the first step of our recommended process is to create CRC cards. This is done by performing textual analysis on the use case descriptions. If you recall, the normal flow of events, subflows, and alternate/exceptional flows written as part of the use case were written in a special form called **S**ubject–**V**erb–**D**irect object–**P**reposition– **I**ndirect object (*SVDPI*). By writing use-case events in this form, it is easier to use the guidelines for use case analysis in Figure 6-11 to identify the objects. Reviewing the primary actors, stakeholders and interests, and brief descriptions of each use case allows additional candidate objects to be identified. Furthermore, it is useful to go back and review the original requirements to look for information that was not included in the text of the use cases. Record all the uncovered information for each candidate object on a CRC Card.

Step 2: Examine Common Object Lists The second step is to brainstorm additional candidate objects, attributes, operations, and relationships using the common object list approach. What are the tangible things associated with the problem? What are the roles played by the people in the problem domain? What incidents and interactions take place in the problem domain? As you can readily see, by beginning with the use cases, many of these questions already have partial answers. For example, the primary actors and stakeholders are the roles that are played by the people in the problem domain. However, it is possible to uncover additional roles not thought of previously. This obviously would cause the use cases and the use-case model to be modified and possibly expanded. As in the previous step, be sure to record all the uncovered information onto the CRC cards. This includes any modifications uncovered for any previously identified candidate objects and any information regarding any new candidate objects identified.

Step 3: Role-Play the CRC Cards The third step is to role-play each use-case scenario using the CRC cards.[12] Each CRC card should be assigned to an individual, who will perform the operations for the class on the CRC card. As the performers play out their roles, the system will tend to break down. When this occurs, additional objects, attributes, operations, or relationships will be identified. Again, like the previous steps, any time any new information is discovered, new CRC cards are created or modifications to existing CRC cards are made.

Step 4: Create the Class Diagram The fourth step is to create the class diagram based on the CRC cards. This is equivalent to creating the use-case diagram from the use case descriptions. Information contained on the CRC cards is simply transferred to the class diagrams. The responsibilities are transferred as operations, the attributes are drawn as attributes, and the relationships are drawn as generalization, aggregation, or association relationships. However, the class diagram also requires that the visibility of the attributes and operations be known. As a general rule of thumb, attributes are private and operations are public. Therefore, unless the analyst has a good reason to change the default visibility of these properties, then the defaults should be accepted. Finally, the analyst should examine the model for additional opportunities to use aggregation or generalization relationships.

---

[12] For more information on role playing scenarios using CRC cards, see D. Bellin and S. Suchman Simone, *The CRC Card Book* (Reading, MA: Addison-Wesley, 1997); G. Kotonya and I. Sommerville, *Requirements Engineering* (Chichester, England: Wiley, 1998), and D. Leffingwell and D. Widrig, *Managing Software Requirements: A Unified Approach* (Reading, MA: Addison-Wesley, 2000).

These types of relationships can simplify the individual class descriptions. As in the previous steps any and all changes must be recorded on the CRC cards.

Step 5: Review the Class Diagram   The fifth step is to review the structural model for missing and/or unnecessary classes, attributes, operations, and relationships. Up until this step, the focus of the process has been on adding information to the evolving model. At this point in time, the focus begins to switch from simply adding information to also challenging the reasons for including the information contained in the model.

Step 6: Incorporate Patterns   The sixth step is to incorporate useful patterns into the evolving structural model. A useful pattern is one that would allow the analyst to more fully describe the underlying domain of the problem being investigated. Looking at the collection of patterns available (Figure 6-13) and comparing the classes contained in the patterns with those in the evolving class diagram does this. After identifying the useful patterns, the analyst incorporates the identified patterns into the class diagram and modifies the affected CRC cards. This includes adding and removing classes, attributes, operations, and/or relationships.

Step 7: Review the Model   The seventh and final step is to validate the structural model, including both the CRC cards and the class diagram. This is best accomplished during a formal review meeting using a walkthrough approach, in which an analyst presents the model to a team of developers and users. The analyst walks through the model, explaining each part of the model and all the reasoning that went into the decision to include each of the classes in the structural model. This explanation includes justifications for the attributes, operations, and relationships associated with the classes. Finally, each class should be linked back to at least one use case; otherwise, the purpose of including the class in the structural model will not be understood. It is often best if the review team also includes people outside the development team that produced the model because these individuals can bring a fresh perspective to the model and uncover missing objects.[13]

YOUR	6-1 Appointment System
TURN	

Using Figure 6-1 as a template, complete the CRC cards   for the remaining identified classes in Figure 6-2.

YOUR	6-2 Campus Housing
TURN	

In Your Turn 5-2, you created a set of the use cases from the campus housing service that helps students find apartments. Using the same use cases, create a structural model (CRC cards and class diagram). See if you can identify at least one potential derived attribute, aggregation relationship, generalization relationship, and association relationship for the model.

---

[13] We describe walkthroughs, verification, and validation in Chapter 8 and testing in Chapter 13.

6-A Health and Insurance Medical Provider—Implementing an EIM System

A large direct health and insurance medical provider needed an Enterprise Information Management (EIM) system to enable enterprisewide information management and support the effective use of data for critical cross-functional decision making. In addition, the client needed to resolve issues related to data redundancy, inconsistency and unnecessary expenditure. The client faced information challenges such as these: The company data resided in multiple sources and was developed for department-specific use, with limited enterprise access. In addition, departments created varied data definitions and data were being managed by multiple departments within the company.

Source: http://www.deloitte.com/dtt/case_study/0,1005,sid% 253D26562%02526cid%253D132760,00.html

Questions

1. Should the company assess their current information management?

2. How would a structural model aid the firm in understanding their current information management? What solution would you propose?

APPLYING THE CONCEPTS AT CD SELECTIONS

The previous chapter described the CD Selections Internet Sales System (see Figures 5-14 through 5-20). In this section, we demonstrate the process of structural modeling using one of the use cases identified: Place Order (see Figure 5-17). Even though we are using just one of the use cases for our structural model, you should remember that to create a complete structural model, all use cases should be used.

Step 1: Create CRC Cards

The first step Alec and the team took was to create the set of CRC cards by performing textual analysis on the use cases. To begin with, Alec chose the Place Order use case (see Figure 5-17). He and his team then used the textual analysis rules (see Figure 6-11) to identify the candidate classes, attributes, operations, and relationships. Using these rules on the Normal Flow of Events, they identified Customer, Search Request, CD, List, and Review as candidate classes. They uncovered three different types of search requests: Title Search, Artist Search, and Category Search. By applying the textual analysis rules to the Brief Description, an additional candidate class was discovered: Order. By reviewing the verbs contained in this use case, they saw that a Customer places an Order and that a Customer makes a Search Request.

To be as thorough as possible, Alec and his team also reviewed the original requirements used to create the use case. The original requirements are contained in Figure 4-15. After reviewing this information, they identified a set of attributes for the Customer (name, address, e-mail, and credit card) and Order (CDs to purchase and quantity) classes and uncovered additional candidate classes: CD Categories and Credit-Card Clearance Center. Furthermore, they realized that the Category Search class used the CD Categories class. Finally, they also identified three subclasses of CD Categories: Rock, Jazz, and Classical. Alec's goal, at this point in time, was to be as complete as possible. As such, he realized that they might have identified many candidate classes, attributes, operations, and relationships that might not be included in the final structural model.

Step 2: Examine Common Object Lists

The second step for Alec and his team was to brainstorm additional candidate classes, attributes, operations, and relationships. He asked the team members to take a minute and think about what information they would like to keep about CDs. The information that they thought of was a set of attributes—for example, title, artist, quantity on hand, price, and category.

He then asked them to take another minute and think about the information that they should store about orders and an order's responsibilities. The responsibilities they identified were a set of operations, including calculate tax, calculate extension price, calculate shipping, and calculate total. Currently, the attributes (CDs to purchase and quantity) of Order implied that a customer should be allowed to order multiple copies of the same CD and allow different CDs to be ordered on the same order. However, the current structural model did not allow this. As such, they created a new class that was associated with both the Order class and the CD class: Order Line Item. This new class only had one attribute, quantity, but it had two relationships: one with Order and the other with CD.

When they reviewed the Customer class, they decided that the name and address attributes needed to be expanded; name should become last name, first name, and middle initial, and address should become street address, city, state, country, and zip code. The updated Customer class and Order class CRC cards are shown in Figures 6-15 and 6-16, respectively.

Front:		
Class Name: Customer	**ID:** 1	**Type:** Concrete, Domain
Description: An individual that may or has purchased merchandise from the CD Selections Internet sales system		**Associated Use Cases:** 3

Responsibilities	Collaborators
_____	_____
_____	_____
_____	_____
_____	_____
_____	_____
_____	_____
_____	_____

Back:

Attributes:

First name	State
Middle initial	Country
Last name	Zip code
Street address	E-mail
City	Credit card

Relationships:

Generalization (a-kind-of): _____

Aggregation (has-parts): _____

Other Associations: Order; Search Request

FIGURE 6-15
Customer Class
CRC Card

Front:

Class Name: Order		ID: 2	Type: Concrete, Domain
Description: An order that has been placed by a customer which includes the individual items purchased by the customer			Associated Use Cases: 3

Responsibilities	Collaborators
Calculate tax	
Calculate shipping	
Calculate total	

Back:

Attributes:
Tax
Shipping
Total

Relationships:
Generalization (a-kind-of):

Aggregation (has-parts):

Other Associations: Order Item; Customer

FIGURE 6-16
Order Class CRC Card

Step 3: Role-Play the CRC Cards

The third step was to role-play the classes recorded on the CRC cards. The purpose of this step was to validate the current state of the evolving structural model. Alec handed out the CRC cards to different members of his team. Using the CRC cards, they began executing the different use cases (see Figure 5-18), one at a time, to see if the current structural model could support each use case or whether the use case caused the system to crash. Anytime

YOUR TURN   6-3 CD Selections Internet Sales System

Complete the CRC cards for the remaining identified classes.

the system crashed, there was something missing: a class, an attribute, a relationship, or an operation. They would then add the missing information to the structural model and try executing the use case again.

First, Alec and the team decided that the customer had requested the system to perform a search for all the CDs associated with a specific artist. Based on the current CRC cards, the team felt that the system would produce an accurate list of CDs. They then tried to ask the system for a set of reviews of the CDs. At this point in the exercise, the system crashed. The CRC cards did not have the Review class associated with the CD class. Therefore, there was no way to retrieve the requested information. This observation raised another question: was there other marketing information that should be made available to the customer—for example, artist information and sample clips?

Next, the team realized that vendor information should be a separate class that was associated with a CD rather than an additional attribute of a CD. This was because vendors had additional information and operations themselves. If the team had modeled the vendor information as an attribute of CD, then the additional information and operations would have been lost. They continued role-playing each of the use cases until they were comfortable with the structural model's ability to support each and every one.

Step 4: Create the Class Diagram

The fourth step was to create the class diagram from the CRC cards. Figure 6-17 shows the current state of the evolving structural model as depicted in a class diagram based on the Places Order use case.

Step 5: Review the Class Diagram

The fifth step was to review the structural model for missing and/or unnecessary classes, attributes, operations, and relationships. At this point, the team challenged all components of the model (class, attribute, relationship, or operation) that did not seem to be adding anything useful to the model. If a component could not be justified sufficiently, then they removed it from the structural model. By carefully reviewing the current state of the structural model, they were able to challenge more than a third of the classes contained in the class diagram (see Figure 6-17). It seemed that the CD categories and their subclasses were not really necessary. There were no attributes or operations for these classes. As such, the idea of CD categories was modeled as an attribute of a CD. The category attribute for the CD class was previously uncovered during the brainstorming step. Also, upon further review of the Search Request class and its subclasses, it was decided that the subclasses were really nothing more than a set of operations of the Search Request class. This was an example of process decomposition creeping into the modeling process. From an object-oriented perspective, we must always be careful to not allow this to occur. However, during the previous steps in the modeling process, Alec wanted to include as much information as possible in the model. He felt that it was more beneficial to remove this type of information after it had crept into the model than to take a chance on not capturing the information required to solve the problem.

Step 6: Incorporate Patterns

The sixth step was to incorporate any useful pattern into the structural model. One pattern that could be useful is the Contract pattern listed in Figure 6-13. By reviewing this pattern (see Figure 6-12), Alec and his team uncovered two subclasses of the Customer class: Individual and Organizational. Furthermore, Peter Coad has identified twelve transaction patterns that also could be useful for Alec and his team to investigate further.[14]

---

[14] See Peter Coad, David North, and Mark Mayfield, *Object Models: Strategies, Patterns, and Applications,* 2nd ed. (Englewood Cliffs, NJ: Prentice Hall, 1997).

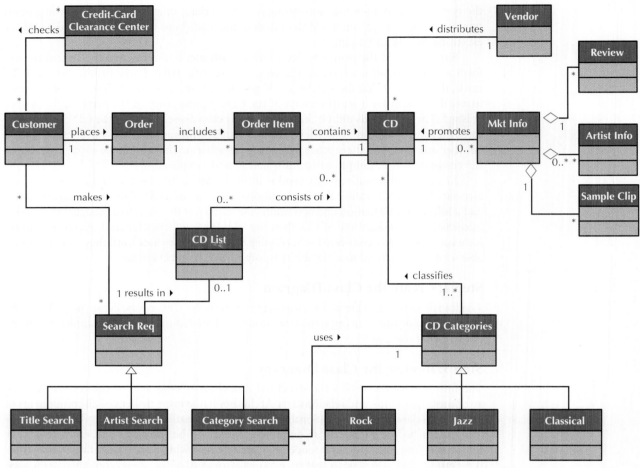

FIGURE 6-17 Preliminary CD Selections Internet Sales System Class Diagrams (Places Order Use-Case View)

Step 7: Review the Model

The seventh and final step is to carefully review the structural model. Figure 6-18 shows the Places Order use-case view of the structural model as portrayed in a class diagram developed by Alec and his team. This version of the class diagram incorporates all the modifications described previously.

YOUR TURN   6-4 CD Selections Internet Sales System

*I*n Your Turn 5-8, you completed the detailed use cases for the CD Selections Internet Sales System. Using these detailed use cases, complete the structural model (CRC cards and class diagram) for the remaining use cases in Figure 5-18.

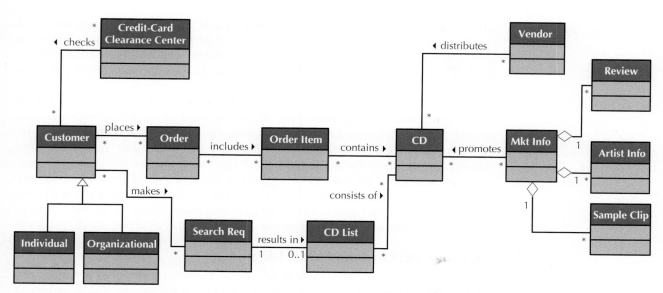

FIGURE 6-18   CD Selections Internet Sales System Class Diagram (Places Order Use-Case View)

SUMMARY

Structural Models

Structural models describe the underlying data structure of an object-oriented system. Whereas use-case models provide an external functional view of the evolving system (i.e., what the system does), structural models provide an internal static view of the evolving system (i.e., how the objects are organized in the system). At this point in the development of the system, the structural model represents only a logical model of the underlying problem domain. One of the primary purposes of the structural model is to create a vocabulary that allows the users and developers to communicate effectively about the problem under investigation. Structural models comprise classes, attributes, operations, and relationships. There are three basic types of relationships that normally are depicted on structural models: aggregation, generalization, and association. Structural models typically are represented by CRC cards, class diagrams, and, in some cases, object diagrams.

CRC Cards

CRC cards model the classes, their responsibilities, and their collaborations. There are two different types of responsibilities: knowing and doing. Knowing responsibilities are associated mostly with attributes of instances of the class, whereas doing responsibilities are associated mostly with operations of instances of the class. Collaborations support the concepts of clients, servers, and contracts between objects. CRC cards capture all the essential elements of the instances of a class. The front of the card allows the recording of the class's name, ID, type, description, list of associated use cases, responsibilities, and collaborators, whereas the back of the card contains the attributes and relationships.

Class Diagrams

A class diagram is a graphical description of the information contained on the CRC cards. It shows the classes and the relationships between the classes. The class diagram portrays additional information, not included on the CRC cards: the visibility of the attributes and operations and the multiplicity of the relationships. Also, there are times that a relationship itself contains information. In this case, an associated class is created. There are special arcs for each of the relationships (aggregation, generalization, and association) contained in the diagram.

In real-world systems that can have over 100 classes, the class diagram can become overly complicated. To allow for the simplification of the diagram, a view mechanism can be used. A view restricts the amount of information portrayed on the diagram. Some useful views are: hiding all information about the class except for its name and relationships; showing only the classes that are associated with a particular use case; and limiting the relationships included to only one specific type (aggregation, generalization, and association).

When attempting to uncover additional information about the details of a class, it is useful to portray specific instances of a class instead of the classes themselves. An object diagram is used to depict a set of objects that represent an instantiation of all or part of a class diagram.

Creating CRC Cards and Class Diagrams

Creating a structural model is an iterative process. The process described is a seven-step process. The steps include textual analysis of use cases, brainstorming additional objects using common object lists, role-playing each use case using the CRC cards, creating class diagrams, and incorporating useful patterns.

KEY TERMS

A-kind-of
A-part-of
Abstract class
Aggregation association
Assemblies
Association
Association class
Attribute
Class
Class diagram
Client
Collaboration
Common object list
Conceptual model
Concrete class
Constructor operation
Contract
Class-Responsibility-
    Collaboration (CRC)
CRC cards

Decomposition
Derived attribute
Doing responsibility
Generalization
    association
Has-parts
Incidents
Instance
Interactions
Knowing responsibility
Method
Multiplicity
Object
Object diagram
Operation
Package
Parts
Pattern
Private
Protected

Public
Query operation
Responsibility
Roles
Server
Static model
Static structure diagram
Structural model
Subclass
Substitutability
Superclass
SVDPI
Tangible things
Textual analysis
UML
Update operation
View
Visibility
Wholes

QUESTIONS

1. Give three examples of derived attributes that may exist on a class diagram. How would they be denoted on the class diagram?

2. Give two examples of aggregation, generalization, and association relationships. How is each type of association depicted on a class diagram?

3. Contrast the following sets of terms:
   object; class; instance
   property; method; attribute
   superclass; subclass
   concrete class; abstract class
4. Describe to a businessperson the multiplicity of a relationship between two classes.
5. What is an association class?
6. For what is an association class used in a class diagram? Give an example of an association class that may be found in a class diagram that captures students and the courses that they have taken.
7. What are the different types of visibility? How would they be denoted on a class diagram?
8. How do you designate the reading direction of a relationship on a class diagram?
9. Why are assumptions important to a structural model?
10. Draw the relationships that are described by the following business rules. Include the multiplicities for each relationship.

    A patient must be assigned to only one doctor and a doctor can have one or many patients.

    An employee has one phone extension, and a unique phone extension is assigned to an employee.

    A movie theater shows at least one movie, and a movie can be shown at up to four other movie theaters around town.

    A movie either has one star, two costars, or more than ten people starring together. A star must be in at least one movie.

11. Identify the following operations as constructor, query, or update. Which operations would not need to be shown in the class rectangle?

    Calculate employee raise (raise percent)
    Calculate sick days ()
    Increment number of employee vacation days ()
    Locate employee name ()
    Place request for vacation (vacation day)
    Find employee address ()
    Insert employee ()
    Change employee address ()
    Insert spouse ()

EXERCISES

A. Draw a class diagram for the following classes. Consider that the entities represent a system for a patient billing system. Include only the attributes that would be appropriate for this context.

   Patient (age, name, hobbies, blood type, occupation, insurance carrier, address, phone)
   Insurance carrier (name, number of patients on plan, address, contact name, phone)
   Doctor (specialty, provider identification number, golf handicap, age, phone, name)

B. Create an object diagram based on the class diagram you drew for exercise A.

C. Draw the following relationships:
   1. A patient must be assigned to only one doctor and a doctor can have many patients.
   2. An employee has one phone extension, and a unique phone extension is assigned to an employee.
   3. A movie theater shows many different movies, and the same movie can be shown at different movie theaters around town.

D. Draw class diagrams for the following classes:
   Movie (title, producer, length, director, genre)
   Ticket (price, adult or child, showtime, movie)
   Patron (name, adult or child, age)

E. Create an object diagram based on the class diagram you drew for exercise D.

F. Draw a class diagram for each of the following situations:
   1. Whenever new patients are seen for the first time, they complete a patient information form that asks their name, address, phone number and insurance carrier, which are stored in the patient information file. Patients can be signed up with only one carrier, but they must be signed up to be seen by the doctor. Each time a patient visits the doctor, an insurance claim is sent to the carrier for payment. The claim must contain information about the visit, such as the date, purpose, and cost. It would be possible for a patient to submit two claims on the same day.
   2. The state of Georgia is interested in designing a system that will track its researchers. Information of interest includes: researcher name, title, position; university name, location, enrollment; and research interests. Researchers are associated with one institution, and each researcher has several research interests.
   3. A department store has a bridal registry. This registry keeps information about the customer (usually the bride), the products that the store carries, and

the products each customer registers for. Customers typically register for a large number of products and many customers register for the same products.

4. Jim Smith's dealership sells Fords, Hondas, and Toyotas. The dealership keeps information about each car manufacturer with whom they deal so that they can get in touch with them easily. The dealership also keeps information about the models of cars that they carry from each manufacturer. They keep information such as list price, the price the dealership paid to obtain the model, and the model name and series (e.g., Honda Civic LX). They also keep information about all sales that they have made (for instance, they will record the buyer's name, the car they bought, and the amount they paid for the car). In order to contact the buyers in the future, contact information is also kept (e.g., address, phone number).

G. Create object diagrams based on the class diagrams you drew for exercise F.

H. Examine the class diagrams that you created for exercise F. How would the models change (if at all) based on these new assumptions?

1. Two patients have the same first and last names.
2. Researchers can be associated with more than one institution.

3. The store would like to keep track of purchase items.
4. Many buyers have purchased multiple cars from Jim over time because he is such a good dealer.

I. Visit a Web site that allows customers to order a product over the Web (e.g., Amazon.com). Create a structural model (CRC cards and class diagram) that the site must need to support its business process. Include classes to show what they need information about. Be sure to include the attributes and operations to represent the type of information they use and create. Finally, draw relationships, making assumptions about how the classes are related.

J. Create a structural model (CRC cards and class diagram) for Exercise A in Chapter 5.

K. Create a structural model for exercise E in Chapter 5.

L. Create a structural model for exercise H in Chapter 5.

M. Create a structural model for exercise J in Chapter 5.

N. Create a structural model for exercise L in Chapter 5.

O. Create a structural model for exercise N in Chapter 5.

P. Create a structural model for exercise P in Chapter 5.

Q. Create a structural model for exercise R in Chapter 5.

R. Create a structural model for exercise T in Chapter 5.

MINICASES

1. Holiday Travel Vehicles sells new recreational vehicles and travel trailers. When new vehicles arrive at Holiday Travel Vehicles, a new vehicle record is created. Included in the new vehicle record are a vehicle serial number, name, model, year, manufacturer, and base cost.

When a customer arrives at Holiday Travel Vehicles, he or she works with a salesperson to negotiate a vehicle purchase. When a purchase has been agreed upon, a sales invoice is completed by the salesperson. The invoice summarizes the purchase, including full customer information, information on the trade-in vehicle (if any), the trade-in allowance, and information on the purchased vehicle. If the customer requests dealer-installed options, they are listed on the invoice as well. The invoice also summarizes the final negotiated price, plus any applicable taxes and license fees. The transaction concludes with a customer signature on the sales invoice.

a. Identify the classes described in the preceding scenario (you should find six). Create CRC cards for each class.

Customers are assigned a customer ID when they make their first purchase from Holiday Travel Vehicles. Name, address, and phone number are recorded for the customer. The trade-in vehicle is described by a serial number, make, model, and year. Dealer-installed options are described by an option code, description, and price.

b. Develop a list of attributes for each class. Place the attributes onto the CRC cards.

Each invoice lists just one customer. A person does not become a customer until he or she purchases a vehicle. Over time, a customer may purchase a number of vehicles from Holiday Travel Vehicles.

Every invoice must be filled out by only one salesperson. A new salesperson may not have sold any vehicles, but experienced salespeople have probably sold many vehicles.

Each invoice only lists one new vehicle. If a new vehicle in inventory has not been sold, there will be no invoice for it. Once the vehicle sells, there will be just one invoice for it.

A customer may decide to have no options added to the vehicle or may choose to add many options. An option may be listed on no invoices, or it may be listed on many invoices.

A customer may trade in no more than one vehicle on a purchase of a new vehicle. The trade-in vehicle may be sold to another customer, who later trades it in on another Holiday Travel vehicle.

c. Based on the preceding business rules in force at Holiday Travel Vehicles and CRC cards, draw a class diagram and document the relationships with the appropriate multiplicities. Remember to update the CRC cards.

2. You are the project manager of an IS consulting firm that has been working with Hock Hai Tan, the general manager of Singapore Importers Pte Ltd. Singapore Importers acts as an intermediary between Western firms and low-cost suppliers in China. It has extensive knowledge of Chinese industry and can offer low-cost manufacturing services to small and medium-size Western firms that lack knowledge of Chinese companies, but want to do business in China. Singapore Importers operates 10 offices across China to identify potential suppliers and monitor the performance of the companies they deal with.

Your team has been hard at work on a project that will eventually link all the Singapore Importers' offices into one unified, networked system. The team has developed a use-case diagram of the current system.

This model has been carefully checked. Last week, the team invited several system users to role-play the various use cases, which were refined to the users' satisfaction. Right now, you feel confident that the as-is system has been adequately represented in the use-case diagram.

Mr. Tan sat in on the role-playing of the use cases and was very pleased by the thorough job your team had done in developing the model. He made it clear to you that he was anxious to see your team begin work on the use cases for the to-be system. He was skeptical that your team had to spend time modeling the current system in the first place but grudgingly admitted that your team seemed to understand the business after having gone through that phase.

The methodology you are following, however, specifies that the team should now turn its attention to developing the structural models for the as-is system. When you mentioned this to Mr. Tan he seemed confused and a little irritated. "You are going to spend even more time looking at the current system? I thought you were done with that! Why is this necessary? I want to see some progress on the way things will work in the future!"

What is your response to Mr. Tan? Why do we perform structural modeling? Is there any benefit to developing a structural model of the current system at all? How do the use cases and use-case diagram help us develop the structural model?

CHAPTER 7

INTERACTION DIAGRAMS

Behavioral models describe the internal dynamic aspects of an information system that supports the business processes in an organization. During analysis, behavioral models describe what the internal logic of the processes is without specifying how the processes are to be implemented. Later, in the design and implementation phases, the detailed design of the operations contained in the object is fully specified. In this chapter, we describe three UML 2.0 diagrams that are used in behavioral modeling: *sequence diagrams, communication diagrams,* and *behavioral state machines.*

OBJECTIVES

■ Understand the rules and style guidelines for sequence and communication diagrams and behavioral state machines.
■ Understand the processes used to create sequence and communication diagrams and behavioral state machines.
■ Be able to create sequence and communication diagrams and behavioral state machines.
■ Understand the relationship between the behavioral models and the structural and functional models.

CHAPTER OUTLINE

INTRODUCTION

The previous two chapters discussed functional models and structural models. Systems analysts utilize functional models to describe the external behavioral view of an information system, whereas they use structural models to depict the static view of an information system. In this chapter, we discuss how analysts use *behavioral models* to represent the internal behavior or dynamic view of an information system.

There are two types of behavioral models. First, there are behavioral models used to represent the underlying details of a business process portrayed by a use-case model. In

UML, interaction diagrams (sequence and communication) are used for this type of behavioral model. Second, there is a behavioral model that is used to represent the changes that occur in the underlying data. UML uses behavioral state machines for this.

During analysis, analysts use behavioral models to capture a basic understanding of the dynamic aspects of the underlying business process. Traditionally, behavioral models have been used primarily during design, where analysts refine the behavioral models to include implementation details (see Chapter 8). For now, our focus is on *what* the dynamic view of the evolving system is and not on *how* the dynamic aspect of the system will be implemented.

In this chapter, we concentrate on creating behavioral models of the underlying business process. Using the interaction diagrams (sequence and communication diagrams) and behavioral state machines, it is possible to give a complete view of the dynamic aspects of the evolving business information system. We first describe behavioral models and their components. We then describe each of the diagrams, how they are created, and how they are related to the functional and structural models described in Chapters 5 and 6.

BEHAVIORAL MODELS

When an analyst is attempting to understand the underlying application domain of a problem, he or she must consider both structural and behavioral aspects of the problem. Unlike other approaches to the development of information systems, object-oriented approaches attempt to view the underlying application domain in a holistic manner. By viewing the problem domain as a set of *use cases* that are supported by a set of collaborating objects, object-oriented approaches allow an analyst to minimize the semantic gap between the real-world set of objects and the evolving object-oriented model of the problem domain. However, as we pointed out in the previous chapter, the real world tends to be messy, which makes perfectly modeling the application domain practically impossible in software. This is because software must be neat and logical to work.

One of the primary purposes of behavioral models is to show how the underlying objects in a problem domain will work together to form a *collaboration* to support each of the *use cases*. Whereas structural models represent the objects and the relationships between them, behavioral models depict the internal view of the business process that a use case describes. The process can be shown by the interaction that takes place between the objects that collaborate to support a use case through the use of interaction (sequence and communication) diagrams. It is also possible to show the effect that the set of use cases that make up the system has on the objects in the system through the use of behavioral state machines.

Creating behavioral models is an iterative process that iterates not only over the individual behavioral models [e.g., interaction (sequence and communication) diagrams and behavioral state machines] but also over the functional (see Chapter 5) and structural models (see Chapter 6). As the behavioral models are created, it is not unusual to make changes to the functional and structural models. In the next three sections, we describe interaction diagrams and behavioral state machines and when to use each.

INTERACTION DIAGRAMS

One of the primary differences between class diagrams and interaction diagrams, besides the obvious difference that one describes structure and the other behavior, is that the modeling focus on a class diagram is at the class level, whereas the interaction diagrams focus on the object level. In this section we review objects, operations, and messages and we cover the two different diagrams (sequence and communication) that can be used to model the interactions that take place between the objects in an information system.

Objects, Operations, and Messages

An object is an instantiation of a *class*, that is, an actual person, place, event, or thing about which we want to capture information. If we were building an appointment system for a doctor's office, classes might include doctor, patient, and appointment. The specific patients, such as Jim Maloney, Mary Wilson, and Theresa Marks, are considered objects— that is, *instances* of the patient class.

Each object has *attributes* that describe information about the object, such as a patient's name, birth date, address, and phone number. Each object also has *behaviors*. At this point in the development of the evolving system, the behaviors are described by *operations*. An operation is nothing more than an action that an object can perform. For example, an appointment object can probably schedule a new appointment, delete an appointment, and locate the next available appointment. Later on during the development of the evolving system, the behaviors will be implemented as methods.

Each object also can send and receive messages. *Messages* are information sent to objects to tell an object to execute one of its behaviors. Essentially, a message is a function or procedure call from one object to another object. For example, if a patient is new to the doctor's office, the system will send an insert message to the application. The patient object will receive the instruction (the message) and do what it needs to do to go about inserting the new patient into the system (the behavior).

Sequence Diagrams

Sequence diagrams are one of two types of interaction diagrams. They illustrate the objects that participate in a use case and the messages that pass between them over time for *one* use case. A sequence diagram is a *dynamic model* that shows the explicit sequence of messages that are passed between objects in a defined interaction. Because sequence diagrams emphasize the time-based ordering of the activity that takes place among a set of objects, they are very helpful for understanding real-time specifications and complex use cases.

The sequence diagram can be a *generic sequence diagram* that shows all possible scenarios[1] for a use case, but usually each analyst develops a set of *instance sequence diagrams,* each of which depicts a single *scenario* within the use case. If you are interested in understanding the flow of control of a scenario by time, you should use a sequence diagram to depict this information. The diagrams are used throughout both the analysis and design phases. However, the design diagrams are very implementation specific, often including database objects or specific user interface components as the classes.

Elements of a Sequence Diagram   Figure 7-1 shows an instance sequence diagram that depicts the objects and messages for the Make Appointment use case, which describes the process by which a patient creates a new appointment or cancels or reschedules an appointment for the doctor's office appointment system. In this specific instance, the Make Appointment process is portrayed.

---

[1] Remember that a scenario is a single executable path through a use case.

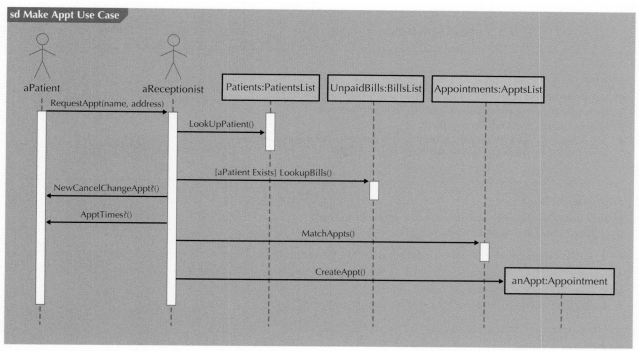

FIGURE 7-1   Example Sequence Diagram

Actors and *objects* that participate in the sequence are placed across the top of the diagram using actor symbols from the use-case diagram and object symbols from the object diagram (see Figure 7-2). Notice that the *actors* and objects in Figure 7-1 are aPatient, aReceptionist, Patients, UnpaidBills, Appointments, and anAppt.[2] They are not placed in any particular order, although it is nice to organize them in some logical way, such as the order in which they participate in the sequence. For each of the objects, the name of the class of which they are an instance is given after the object's name (e.g., Patients: PatientsList means that Patients is an instance of the PatientsList class that contains individual patient objects).

A dotted line runs vertically below each actor and object to denote the *lifeline* of the actors/objects over time (see Figure 7-1).[3] Sometimes an object creates a *temporary object;* in this case an X is placed at the end of the lifeline at the point the object is destroyed (not shown). For example, think about a shopping cart object for a Web commerce application. The shopping cart is used for temporarily capturing line items for an order, but once the order is confirmed, the shopping cart is no longer needed. In this case, an X would be located at the point at which the shopping cart object is destroyed. When objects continue to exist in the system after they are used in the sequence diagram, then the lifeline continues to the bottom of the diagram (this is the case with all of the objects in Figure 7-1).

[2] In some versions of the sequence diagram, object symbols are used as surrogates for the actors. However, for purposes of clarity, we recommend using actor symbols for actors instead.

[3] Technically speaking, in UML 2.0 the lifeline actually refers to both the object (actor) and the dashed line drawn vertically underneath the object (actor). However, we prefer to use the older terminology because it is more descriptive of what is actually being represented.

Term and Definition	Symbol
An actor: ■ Is a person or system that derives benefit from and is external to the system. ■ Participates in a sequence by sending and/or receiving messages. ■ Is placed across the top of the diagram. ■ Is depicted either as a stick figure (default) or, if a nonhuman actor is involved, as a rectangle with <<actor>> in it (alternative).	anActor <<actor>> Actor/Role
An object: ■ Participates in a sequence by sending and/or receiving messages. ■ Is placed across the top of the diagram.	anObject : aClass
A lifeline: ■ Denotes the life of an object during a sequence. ■ Contains an X at the point at which the class no longer interacts.	
An execution occurrence: ■ Is a long narrow rectangle placed atop a lifeline. ■ Denotes when an object is sending or receiving messages.	
A message: ■ Conveys information from one object to another one. ■ A operation call is labeled with the message being sent and a solid arrow, whereas a return is labeled with the value being returned and shown as a dashed arrow.	aMessage() Return Value
A guard condition: ■ Represents a test that must be met for the message to be sent.	[aGuardCondition]: aMessage()
For object destruction: ■ An X is placed at the end of an object's lifeline to show that it is going out of existence.	X
A frame: ■ Indicates the context of the sequence diagram.	Context

FIGURE 7-2   Sequence Diagram Syntax

A thin rectangular box, called the *execution occurrence*, is overlaid onto the lifeline to show when the classes are sending and receiving messages (see Figure 7-2). A message is a communication between objects that conveys information with the expectation that activity will ensue. There are many different types of messages that can be portrayed on a sequence diagram. However, in the case of using sequence diagrams to model use cases, two types of messages are typically used: operation call and return. *Operation call messages* passed between classes are shown using solid lines connecting two objects with an arrow on the line showing which way the message is being passed. Argument values for the message are placed in parentheses next to the message's name. The order of messages goes from the top to the bottom of the page, so messages located higher on the diagram represent messages that occur earlier on in the sequence, versus the lower messages that occur later. A *return message* is depicted as a dashed line with an arrow on the end of the line portraying the direction of the return. The information being returned is used to label the arrow. However, because adding return messages tends to clutter the diagram, unless the return messages add a lot of information to the diagram, they can be omitted. For example, in Figure 7-1, no return messages are depicted.[4] In Figure 7-1, 'LookUpPatient()' is a message sent from the actor aReceptionist to the object Patients, which is a container for the current patients to determine whether the aPatient actor is a current patient.

At times a message is sent only if a *condition* is met. In those cases, the condition is placed between a set of brackets, [ ] {e.g., [aPatient Exists] LookupBills()}. The condition is placed in front of the message name. However, when using a sequence diagram to model a specific scenario, conditions are typically not shown on any single sequence diagram. Instead, conditions are implied only through the existence of different sequence diagrams. There are times that a message is repeated. This is designated with an asterisk (*) in front of the message name (e.g., * Request CD). An object can send a message to itself. This is known as *self-delegation.*

Sometimes, an object will create another object. This is shown by the message being sent directly to an object instead of its lifeline. In Figure 7-1, the actor aReceptionist creates an object anAppt.

Figure 7-3 portrays two additional examples of instance-specific sequence diagrams. The first one is related to the Make Lunch use case that was described in the activity diagram portrayed in Figure 5-4. The second one is related to the Place Order use case associated with the activity diagram in Figure 5-3. In both examples, the diagrams simply represent a single scenario. Notice, in the Make Lunch sequence diagram there is a message being sent from an actor to itself [CreateSandwich()]. Depending on the complexity of the scenario being modeled, this particular message could have been eliminated. Obviously, both the process of making a lunch and placing an order can be quite a bit more complex. However, from a learning point of view, you should be able to see how the sequence diagrams and the activity diagrams relate to one another.

Building a Sequence Diagram   In this section we describe a six-step process used to create a sequence diagram[5] (see Figure 7-4). The first step in the process is to determine the context of the sequence diagram. The context of the diagram can be a system, a use case, or a scenario of a use case. The context of the diagram is depicted as a labeled *frame* around the diagram (see Figures 7-1, 7-2, and 7-3). Most commonly, it is one use-case scenario. Figure 7-1 portrays the instance-specific sequence diagram for the scenario from the Make Appointment use case given in Figure 5-5 for making a new appointment for an existing patient. For each possible scenario for the Make Appointment use case, a separate instance-specific sequence diagram would be created. On the surface, this seems to be a lot of potentially redundant and useless work. However, at this

---

[4] However, some CASE tools require the return messages to be displayed. Obviously, when using these tools, you will have to include the return messages on the diagram.

[5] The approach described in this section are adapted from Grady Booch, James Rumbaugh, Ivar Jacobson, *The Unified Modeling Language User Guide* (Reading, MA: Addison-Wesley, 1999).

FIGURE 7-3
Additional Sample
Instance-Specific
Sequence Diagrams

point in the representation of a system, we are still trying to completely understand the problem. As such, this process of creating instance-specific sequence diagrams for each scenario instead of creating a single generic sequence diagram for the entire use case will enable the developers to attain a more complete understanding of the problem being addressed. Furthermore, each instance-specific sequence diagram is fairly simple to interpret, whereas a generic sequence diagram can be very complex. As such, the testing of a specific use case is accomplished in a much easier manner by validating and verifying the completeness of the set of instance-specific sequence diagrams instead of trying to work through a single complex generic sequence diagram.[6]

---

[6] Additional discussion of testing is included in Chapter 13. We also discuss walkthroughs and inspections in Chapter 8.

1. Set the context.
2. Identify which objects will participate.
3. Set the lifeline for each object.
4. Lay out the messages from the top to the bottom of the diagram based on the order in which they are sent.
5. Add the execution occurrence to each object's lifeline.
6. Validate the sequence diagram.

FIGURE 7-4

Steps for Building Sequence Diagrams

The second step is to identify the objects that participate in the sequence being modeled—that is, the objects that interact with each other during the use-case scenario. The objects are identified during the development of the structural model (see Chapter 6). These are the classes on which the objects of the sequence diagram for this scenario will be based. One very useful approach to identifying all of the scenarios associated with a use case is to role-play the CRC cards (see Chapter 6). This can help you identify potentially missing operations that are necessary to support the business process, which the use case is representing, in a complete manner. Also, during role playing, it is likely that new classes, and hence new objects, will be uncovered.[7] Don't worry too much about identifying all the objects perfectly; remember that the behavioral modeling process is iterative. Usually, the sequence diagrams are revised multiple times during the behavioral modeling processes.

The third step is to set the lifeline for each object. To do this, you need to draw a vertical dotted line below each class to represent the class's existence during the sequence. An X should be placed below the object at the point on the lifeline where the object goes out of existence.

The fourth step is to add the messages to the diagram. This is done by drawing arrows to represent the messages being passed from object to object, with the arrow pointing in the message's transmission direction. The arrows should be placed in order from the first message (at the top) to the last (at the bottom) to show time sequence. Any parameters passed along with the messages should be placed in parentheses next to the message's name. If a message is expected to be returned as a response to a message, then the return message is not explicitly shown on the diagram.

The fifth step is to place the execution occurrence on each object's lifeline by drawing a narrow rectangle box over top of the lifelines to represent when the classes are sending and receiving messages.

The sixth and final step is to validate the sequence diagram. The purpose of this step is to guarantee that the sequence diagram completely represents the underlying process. This is done by guaranteeing that the diagram depicts all the steps in the process.[8]

YOUR TURN

7-1 Drawing a Sequence Diagram

*I*n Your Turn 5-2, you were asked to create a set of use cases and a use-case diagram for the campus housing service that helps students find apartments. In Your Turn 6-2, you were asked to create a structural model (CRC cards and class diagram) for those use-cases. Select one of the use cases from the use-case diagram and create a set of instance-specific sequence diagrams that represents the interaction among classes in the different scenarios of the use case.

[7] This obviously will cause you to go back and modify the structural model (see Chapter 6).
[8] We describe validation in more detail in Chapter 8.

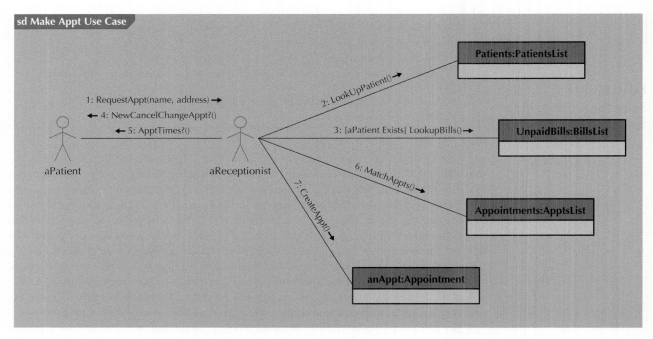

FIGURE 7-5   Sample Communication Diagram

Communication Diagrams

Communication diagrams, like sequence diagrams, essentially provide a view of the dynamic aspects of an object-oriented system. As such, they can show how the members of a set of objects collaborate to implement a use case or a use-case scenario. Furthermore, they can be used to model all the interactions among a set of collaborating objects, i.e., a collaboration (see CRC cards in Chapter 6). In this case, a communication diagram can portray how dependent the different objects are on one another.[9] A communication diagram is essentially an object diagram that shows message-passing relationships instead of aggregation or generalization associations. Communication diagrams are very useful to show process patterns (i.e., patterns of activity that occur over a set of collaborating classes).

Communication diagrams are equivalent to sequence diagrams, but they emphasize the flow of messages through a set of objects, whereas the sequence diagrams focus on the time ordering of the messages being passed. Therefore, to understand the flow of control over a set of collaborating objects or to understand which objects collaborate to support business processes, a communication diagram can be used. For time ordering the messages, a sequence diagram should be used. In some cases, both can be used to more fully understand the dynamic activity of the system.

Elements of a Communication Diagram   Figure 7-5 shows a communication diagram for the Make Appointment use case. Like the sequence diagram in Figure 7-1, the Create Appointment process is portrayed.

---

[9] We return to this idea of dependency in Chapters 8 and 9.

Term and Definition	Symbol
An actor: ■ Is a person or system that derives benefit from and is external to the system. ■ Participates in a collaboration by sending and/or receiving messages. ■ Is depicted either as a stick figure (default) or, if a nonhuman actor is involved, as a rectangle with <<actor>> in it (alternative).	anActor <<actor>> Actor/Role
An object: ■ Participates in a collaboration by sending and/or receiving messages. ■ Is placed across the top of the diagram.	anObject:aClass
An association: ■ Shows an association between actors and/or objects. ■ Is used to send messages.	
A message: ■ Conveys information from one object to another one. ■ Has direction shown using an arrowhead. ■ Has sequence shown by a sequence number.	SeqNumber: aMessage →
A guard condition: ■ Represents a test that must be met for the message to be sent.	SeqNumber: [aGuardCondition]: aMessage →
A frame: ■ Indicates the context of the communication diagram.	Context

FIGURE 7-6   Communication Diagram Syntax

Actors and objects that collaborate to execute the use case are placed on the communication diagram in a manner to emphasize the message passing that takes place between them. Notice that the actors and objects in Figure 7-5 are the same ones in Figure 7-1: aPatient, aReceptionist, Patients, UnpaidBills, Appointments, and anAppt.[10] Again, as with the sequence diagram, for each of the objects, the name of the class of which they are an instance is given after the object's name (e.g., Patients: PatientsList). (The communication diagram syntax is given in Figure 7-6.) Unlike the sequence diagram, the communication diagram does not have a means to explicitly show an object being deleted or created. It is assumed that when a delete, destroy, or remove message is sent to an object, it will go out of existence, and a create or new message will cause a new object to come into existence.

[10] In some versions of the communication diagram, object symbols are used as surrogates for the actors. However, again we recommend using actor symbols for actors instead.

FIGURE 7-7

Additional Sample
Communication
Diagrams

Another difference between the two interaction diagrams is that the communication diagram never shows returns from message sends, whereas the sequence diagram can optionally show them.

An *association* is shown between actors and objects with an undirected line. For example, there is an association shown between the aPatient and aReceptionist actors. Messages are shown as labels on the associations. Included with the labels are lines with arrows showing the direction of the message being sent. For example, in Figure 7-5, the aPatient actor sends the RequestAppt() message to the aReceptionist actor, and the aReceptionist actor, sends the NewCancelChangeAppt?() and the ApptTimes?() messages to the aPatient actor. The sequence of the message sends is designated with a sequence number. In Figure 7-5, the RequestAppt() message is the first message sent, whereas the NewCancelChangeAppt?() and the ApptTimes?() messages are the fourth and fifth message sent, respectively.

Like the sequence diagram, the communication diagram can represent conditional messages. For example, in Figure 7-5, the LookupBills() message is sent only if the [aPatient exists] condition is met. If a message is repeatedly sent, an asterisk is placed after the sequence number. Finally, an association that loops onto an object shows self-delegation. The message is shown as the label of the association.

When a communication diagram is fully populated with all the objects, it can become very complex and difficult to understand. When this occurs, it is necessary to simplify the diagram. One approach to simplifying a communication diagram, like use-case diagrams (see Chapter 5) and class diagrams (see Chapter 6), is through the use of *packages* (i.e., logical groups of classes). In the case of communication diagrams, its objects are grouped together based on the messages sent to and received from the other objects.[11]

Figure 7-7 provides two additional examples of communication diagrams. These diagrams are equivalent to the sequence diagrams contained in Figure 7-3. However, when comparing the communication diagrams to the sequence diagrams in these figures, you see that there is quite a bit of information lost. For example, the CreateSandwich() message is nowhere to be found. However, the primary purpose of the communication diagram is to show how the different actors and classes interact, and this is exactly the information that is included.

---

[11] For those familiar with structured analysis and design, packages serve a similar purpose as the leveling and balancing processes used in data flow diagramming. Packages and package diagrams are described in Chapter 8.

FIGURE 7-8

Steps for Building
Communication
Diagrams

1. Set the context.
2. Identify which objects (actors) and the associations between the objects participate in the collaboration.
3. Lay out the communication diagram.
4. Add messages.
5. Validate the communication diagram.

Building a Communication Diagram[12]   Remember that a communication diagram is basically an object diagram that shows message-passing relationships instead of aggregation or generalization associations. In this section, we describe a five-step process used to build a communication diagram (see Figure 7-8). The first step in the process is to determine the context of the communication diagram. Like a sequence diagram, the context of the diagram can be a system, a use case, or a scenario of a use case. The context of the diagram is depicted as a labeled frame around the diagram (see Figures 7-5, 7-6, and 7-7).

The second step is to identify the objects (actors) and the associations that link the objects (actors) that participate in the collaboration together. Remember, the objects that participate in the collaboration are instances of the classes identified during the development of the structural model (see Chapter 6). Like the sequence-diagramming process, it is likely that additional objects, and hence classes, will be discovered. Again, this is normal because the underlying development process is iterative and incremental. As such, in addition to the communication diagram being modified, the sequence diagrams and structural model will probably also have to be modified. Furthermore, additional functional requirements may also be uncovered, hence requiring the functional models to be modified as well (see Chapter 5).

The third step is to lay out the objects (actors) and their associations on the communication diagram by placing them together based on the associations that they have with the other objects in the collaboration. By focusing on the associations between the objects (actors) and minimizing the number of associations that cross over one another, we can increase the understandability of the diagram.

The fourth step is to add the messages to the associations between the objects. We do this by adding the name of the message(s) to the association link between the objects and an arrow showing the direction of the message being sent. Furthermore, each message has a sequence number associated with it to portray the time-based ordering of the message.[13]

The fifth and final step is to validate the communication diagram. The purpose of this step is to guarantee that the communication diagram faithfully portrays the underlying process(es). This is done by ensuring that all steps in the process are depicted on the diagram.[14]

YOUR
TURN

7-2 Drawing a Communication Diagram

In Your Turn 7-1 you were asked to create a set of instance specific sequence diagrams for a use case of the campus housing service. Draw a communication diagram for the same situation.

---

[12] The approach described in this section is adapted from Booch, Rumbaugh, and Jacobson, *The Unified Modeling Language User Guide*.

[13] However, remember the sequence diagram portrays the time-based ordering of the messages in a top-down manner. As such, if your focus is on the time-based ordering of the messages, we recommend that you also use the sequence diagram.

[14] We describe validation in Chapter 8.

BEHAVIORAL STATE MACHINES

Some of the classes in the *class diagrams* represent a set of objects that are quite dynamic in that they pass through a variety of states over the course of their existence. For example, a patient can change over time from being 'new' to 'current,' to 'former,' and so on, based on his or her status with the doctor's office. A behavioral state machine is a dynamic model that shows the different states through which a single object passes during its life in response to events, along with its responses and actions. Typically, behavioral state machines are not used for all objects, but just to further define complex objects to help simplify the design of algorithms for their methods. The behavioral state machine shows the different states of the object and what events cause the object to change from one state to another. In comparison to the interaction diagrams, behavioral state machines should be used to help understand the dynamic aspects of a single class and how its instances evolve over time[15] and not with seeing how a particular use case or use-case scenario is executed over a set of classes.

In this section, we describe states, events, transitions, actions, and activities and the use of the behavioral state machine to model the state changes through which complex objects pass. As in the creation of the interaction diagrams, when we create a behavioral state machine for an object, it is possible that we will uncover additional events that need to be included in your functional model (see Chapter 5), additional operations that need to be included in the structural model (see Chapter 6), so possibly our interaction diagrams may have to be modified again. Once more, because object-oriented development is iterative and incremental, this continuous modification of the evolving models (functional, structural, and behavioral) of the system is to be expected.

States, Events, Transitions, Actions, and Activities

The *state* of an object is defined by the value of its attributes and its relationships with other objects at a particular point in time. For example, a patient might have a state of new, current, or former. The attributes or properties of an object affect the state that it is in; however, not all attributes or attribute changes will make a difference. For example, think about a patient's address. Those attributes make very little difference as to changes in a patient's state. However, if states were based on a patient's geographic location (e.g., in-town patients were treated differently than out-of-town patients), changes to the patient's address would influence state changes.

An *event* is something that takes place at a certain point in time and changes a value or values that describe an object, which, in turn, changes the object's state. It can be a designated condition becoming true, the receipt of the call for a method by an object, or the passage of a designated period of time. The state of the object determines exactly what the response will be.

A *transition* is a relationship that represents the movement of an object from one state to another state. Some transitions have a guard condition. A *guard condition* is a Boolean expression that includes attribute values, which allows a transition to occur only if the condition is true. An object typically moves from one state to another based on the outcome of an action triggered by an event. An action is an atomic, nondecomposable process that cannot be interrupted. From a practical perspective, actions take zero time, and they are associated with a transition. In contrast, an *activity* is a nonatomic, decomposable process that can be interrupted. Activities take a long period of time to complete, they can be started and stopped by an action, and they are associated with states.

---

[15] Some authors refer to this as modeling an object's life cycle.

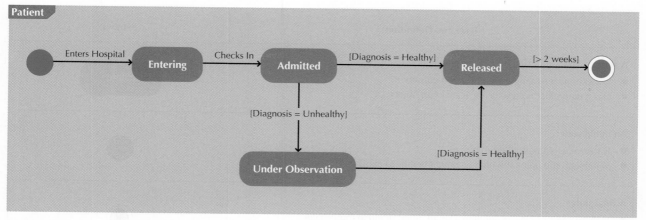

FIGURE 7-9   Sample Behavioral State Machine Diagram

Elements of a Behavioral State Machine

Figure 7-9 presents an example of a behavioral state machine representing the patient class in the context of a hospital environment. From this diagram we can tell that a patient enters a hospital and is admitted after checking in. If a doctor finds the patient to be healthy, he or she is released and is no longer considered a patient after two weeks elapses. If a patient is found to be unhealthy, he or she remains under observation until the diagnosis changes.

A state is a set of values that describes an object at a specific point in time, and it represents a point in an object's life in which it satisfies some condition, performs some action, or waits for something to happen (see Figure 7-10). In Figure 7-9 states include entering, admitted, released, and under observation. A state is depicted by a *state symbol,* which is a rectangle with rounded corners with a descriptive label that communicates a particular state. There are two exceptions. An *initial state* is shown using a small, filled-in circle, and an object's *final state* is shown as a circle surrounding a small, filled-in circle. These exceptions depict when an object begins and ceases to exist, respectively.

Arrows are used to connect the state symbols, representing the transitions between states. Each arrow is labeled with the appropriate event name and any parameters or conditions that may apply. For example, the two transitions from admitted to released and under observation contain guard conditions. As in the other behavioral diagrams, in many cases it is useful to explicitly show the context of the behavioral state machine using a frame.

Figure 7-11 depicts two additional behavioral state machines. The first one is for the lunch object that was associated with the Make Lunch use-case scenario of Figures 7-3 and 7-7. In this case, there is obviously additional information that has been captured about the lunch object. For example, the scenario of Figures 7-3 and 7-7 did not include information regarding the lunch being taken out of the box or being eaten. This implies additional use cases and/or use-case scenarios that would have to be included in a system that dealt with lunch processing. The second behavioral state machine deals with the life cycle of an order. The order object is associated with the submit order use-case scenario described in Figures 7-3 and 7-7. As in the lunch example, there is quite a bit of additional information contained in this behavioral state machine. As such, for an order-processing system, quite a few additional sequence and communication diagrams

Term and Definition	Symbol
A state: ■ Is shown as a rectangle with rounded corners. ■ Has a name that represents the state of an object.	aState
An initial state: ■ Is shown as a small, filled-in circle. ■ Represents the point at which an object begins to exist.	●
A final state: ■ Is shown as a circle surrounding a small, filled-in circle (bull's-eye). ■ Represents the completion of activity.	◉
An event: ■ Is a noteworthy occurrence that triggers a change in state. ■ Can be a designated condition becoming true, the receipt of an explicit signal from one object to another, or the passage of a designated period of time. ■ Is used to label a transition.	anEvent
A transition: ■ Indicates that an object in the first state will enter the second state. ■ Is triggered by the occurrence of the event labeling the transition. ■ Is shown as a solid arrow from one state to another, labeled by the event name.	⟶
A frame: ■ Indicates the context of the behavioral state machine.	Context

FIGURE 7-10   Behavioral State Machine Diagram Syntax

would be necessary to completely represent all the processing associated with an order object. Obviously, because behavioral state machines can uncover additional processing requirements, they can be very useful in filling out the complete description of an evolving system.

Sometimes, states and subclasses can be confused. For example, in Figure 7-12, are the classes Freshman, Sophomore, Junior, and Senior subclasses of the class Undergraduate or are they simply states that an instance of the Undergraduate class goes through during its lifetime? In this case, the latter is the better answer. When trying to identify all potential classes when the structural model is created (see Chapter 6), you may actually identify states of the relevant superclass instead of subclasses. This is another example of how tightly intertwined the functional, structural, and behavioral models can be. From a modeling perspective, even though we had to remove the Freshman, Sophomore, Junior, and Senior subclasses from the structural model, it was better to capture that information as part of the structural model and remove it when we were creating the behavioral models than to omit it and take the chance of missing a crucial piece of information about the problem domain. Remember,

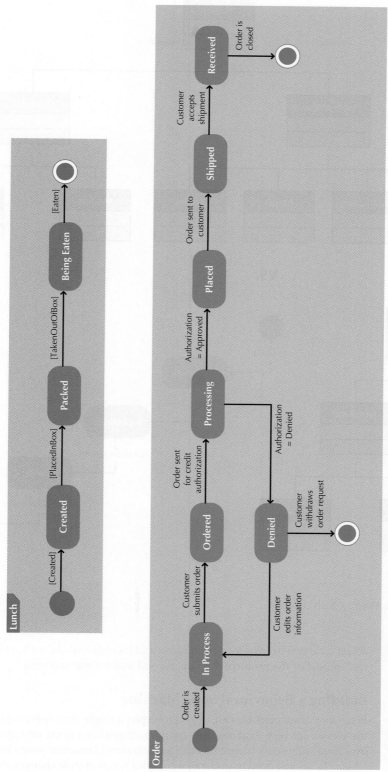

FIGURE 7-11   Additional Sample: Behavioral State Machine Diagrams

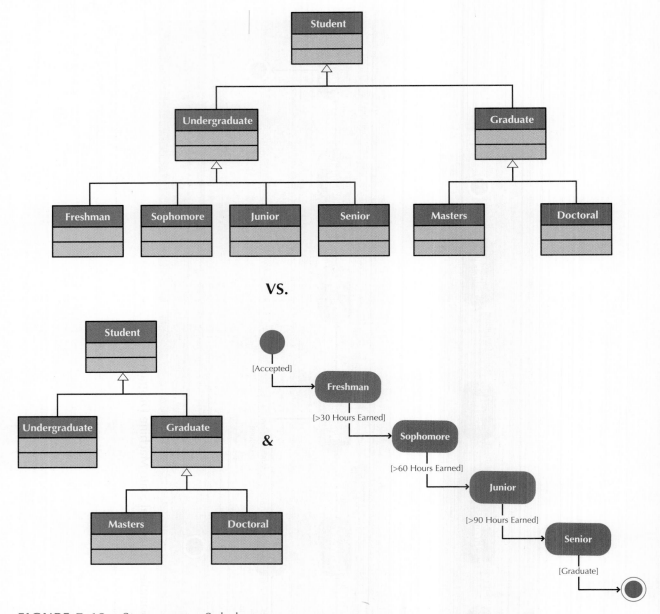

FIGURE 7-12 States versus Subclasses

object-oriented development is iterative and incremental. As such, as we progress to a "correct" model of the problem domain, we will make many mistakes.

Building a Behavioral State Machine

Behavioral state machines are drawn to depict a single class from a class diagram. Typically, the classes are very dynamic and complex, requiring a good understanding of their states over time and events triggering changes. You should examine your class diagram to identify which classes will need to undergo a complex series of state changes and draw a diagram for

FIGURE 7-13

Steps for Building
a Behavioral State
Machine

1. Set the context.
2. Identify the initial, final, and stable states of the object.
3. Determine the order in which the object will pass through the stable states.
4. Identify the events, actions, and guard conditions associated with the transitions.
5. Validate the behavioral state machine.

each of them. In this section, we describe a five-step process used to build a behavioral state machine (see Figure 7-13).[16] Like the other behavioral models, the first step in the process is to determine the context of the behavioral state machine, which is shown in the label of the frame of the diagram. The context of a behavioral state machine is usually a class. However, it also could be a set of classes, a subsystem, or an entire system.

The second step is to identify the various states that an object will have over its lifetime. This includes establishing the boundaries of the existence of an object by identifying the initial and final states of an object. We also must identify the stable states of an object. The information necessary to perform this is gleaned from reading the use-case descriptions, talking with users, and relying on the requirements-gathering techniques that you learned about in Chapter 4. An easy way to identify the states of an object is to write the steps of what happens to an object over time, from start to finish, similar to how the normal flow of events section of a use-case description would be created.

The third step is to determine the sequence of the states that an object will pass through during its lifetime. Using this sequence, the states are placed onto the behavioral state machine in a left-to-right order.

The fourth step is to identify the events, actions, and guard conditions associated with the transitions between the states of the object. The events are the *triggers* that cause an object to move from one state to the next state. In other words, an event causes an action to execute that changes the value(s) of an object's attribute(s) in a significant manner. The actions are typically operations contained within the object. Also, guard conditions can model a set of test conditions that must be met for the transition to occur. At this point in the process, the transitions are drawn between the relevant states and labeled with the event, action, or guard condition.

The fifth step is to validate the behavioral state machine by making sure that each state is reachable and that it is possible to leave all states except for final states. Obviously, if an identified state is not reachable, either a transition is missing or the state was identified in error. Furthermore, only final states can be a dead end from the perspective of an object-life cycle.[17]

YOUR

TURN

7-3 Drawing a Behavioral State Machine

You have been working with the system for the campus housing service that helps students find apartments. One of the dynamic classes in this system is probably the apartment class. Draw a behavioral state machine to show the various states that an apartment class transitions throughout its lifetime. Can you think of other classes that would make good candidates for a behavioral state machine?

---

[16] The approach described in this section is adapted from Booch, Rumbaugh, and Jacobson, *The Unified Modeling Language User Guide.*

[17] We describe validation in Chapter 8.

CRUD ANALYSIS

One useful technique for identifying potential collaborations is *CRUD analysis*.[18] CRUD analysis uses a *CRUD matrix*, in which each interaction among objects is labeled with a letter for the type of interaction: C for create, R for read or reference, U for update, and D for delete. In an object-oriented approach, a class/actor-by-class/actor matrix is used.[19] Each cell in the matrix represents the interaction between instances of the classes. For example, in Figure 7-1, an instance of the Receptionist actor creates an instance of the Appointment class. As such, assuming a Row:Column ordering, a C is placed in the cell aReceptionist: Appointment. Also, in Figure 7-1, an instance of the Receptionist actor references an instance of the Appointments. In this case, an R is placed in the aReceptionist:Appointments cell. Figure 7-14 shows the CRUD matrix based on the Make Appointment use case.

Unlike the interaction diagrams and behavioral state machines, a CRUD matrix is most useful as a systemswide representation. Once a CRUD matrix is completed for the entire system, the matrix can be scanned quickly to ensure that each and every class can be instantiated. Furthermore, each type of interaction can be validated for each class. For example, if a class represents only temporary objects, then the column in the matrix should have a D in it somewhere. Otherwise, the instances of the class will never be deleted. Because a data warehouse contains historical data, objects that are to be stored in one should not have any U or D entries in their associated columns. As such, CRUD analysis can be used as a way to partially validate the interactions among the objects in an object-oriented system. Finally, the more interactions among a set of classes, the more likely they should be clustered together in a collaboration. However, the number and type of interactions are only an estimate at this point in the development of the system. As such, care should be taken when using this technique to cluster classes together to identify collaborations.

CRUD analysis also can be used to identify complex objects. The more (C)reate, (U)pdate, or (D)elete entries in the column associated with a class, the more likely the instances of the class will have a complex life cycle. As such, these objects are candidates for state modeling with a behavioral state machine.

	Receptionist	PatientList	Patient	UnpaidBills	Appointments	Appointment
Receptionist		RU	CRUD	R	RU	CRUD
PatientList			R			
Patient						
UnpaidBills						
Appointments						R
Appointment						

FIGURE 7-14   CRUD Matrix for the Make Appointment Use Case

[18] CRUD analysis has typically been associated with structured analysis and design [see Alan Dennis, Barbara Haley Wixom and Roberta M. Roth, *Systems Analysis Design*, 3nd ed. (New York: Wiley, 2006)] and information engineering [see James Martin, *Information Engineering, Book II Planning and Analysis* (Englewood Cliffs, NJ: Prentice Hall, 1990)]. In our case, we have simply adapted it to object-oriented systems development. Specific details on collaborations are described in Chapter 8.

[19] Another useful but more detailed form of the CRUD matrix is a Class/Actor:Operation-by-Class/Actor:Operation matrix. For validation and verification purposes, this more detailed matrix is more useful. However, for our purposes at this point in our discussion, the Class/Actor-by-Class/Actor matrix is sufficient.

APPLYING THE CONCEPTS AT CD SELECTIONS

Because Alec and the team has now completed the functional and structural models for the Internet Sales System, he has decided that it was time to move on and begin to create the behavioral models. Alec understood that in some ways the behavioral models will allow them to complete their understanding of the problem that they are solving. As such, Alec and his team created sequence diagrams, communication diagrams, behavioral state machines, and a CRUD matrix. As in Chapter 6, the team created behavioral models for all the use cases and classes in the evolving system description. However, in the sections that follow, we see only the models associated with the Place Order use case and the Order class. The sections are organized in the same manner as the chapter: sequence diagrams, communication diagrams, behavioral state machine, and CRUD matrix.

Sequence Diagrams

To begin with Alec decided to follow the steps for building a sequence diagram listed in Figure 7-4. As such, Alec first needed to determine the context of the sequence diagram. He decided to use a scenario[20] from the Place Order use case that was created in Chapter 5 and illustrated in Figure 5-17. (Refer to the original use case for the details.) Figure 7-15 lists the normal flow of events, which contains the scenario that this sequence diagram describes.

The second step was to identify the objects that participated in the scenario being modeled. The classes associated with the Place Order use case are shown in Figure 6-18. For example, the classes used for the Place Order use case include Customer, CD, Marketing Information, Credit Card Clearance Center, Order, Order Item, Vendor, Search Request, CD List, Review, Artist Information, Sample Clip, Individual, and Organizational.

FIGURE 7-15

Normal Flow of Events of the Places Order Use Case

Normal Flow of Events:
1. Customer submits a search request to the system.
2. The System provides the Customer a list of recommended CDs.
3. The Customer chooses one of the CDs to find out additional information.
4. The System provides the Customer with basic information and reviews on the CD.
5. The Customer calls the Maintain Order use case.
6. The Customer iterates over 3 through 5 until done shopping.
7. The Customer executes the Checkout use case.
8. The Customer leaves the Web site.

---

[20] Remember, as stated previously, a scenario is a single executable path through a use case.

During the role playing of the *CRC cards*, Anne, one of the analysts assigned to the CD Selections Internet System development team, asked whether a Shopping Cart class should be included. She stated that every Internet sales site she had been to had a shopping cart that was used to build an order. However, the instance of the Shopping Cart class existed only until either an order was placed or the shopping cart was emptied. Based on this obvious oversight, both the Place Order use case and the class diagram will have to be modified. Brian, another analyst, pointed out that the CDs themselves were going to have to be associated with some type of searchable storage. Otherwise, it would be impossible to find the CD in which the customer was interested. However, Alec decided that the CD List class would suffice for both the searchable storage and a temporary instance that would be created as a result of a search request. Alec pointed out to the team that this process was fairly typical in object-oriented development. The more the development team learned about the requirements, the more the models (functional, structural, and behavioral) would evolve. Alec reminded them that the important thing to remember was that an object-oriented development process was incremental and it iterated over all of the models. As such, as they understood the problem better, the team would most likely have to make changes to the functional and structural models already completed.

Based on the team's current understanding of the Place Order use case, they decided that instances of the Search Request, CD List, CD, Marketing Material, Customer, Review, Artist Information, Sample Clip, and Shopping Cart classes would be required to describe this scenario. Furthermore, they realized that the Customer actor interacted with the scenario. To complete this step, the team laid out the objects on the sequence diagram by drawing them, from left to right, across the diagram.

The third step was to set the lifeline for each object. To do this, they drew a vertical dotted line below each of the objects (aSR, aCDL, CDs, aCD, MI, aR, AI, SC, aSC and anOrder) and the actor (aCustomer). They placed an X at the bottom of the lifelines for aCDL and aSC because they go away at the end of this process.

The fourth step was to add the messages to the diagram. By examining the steps in Figure 7-15, the team was able to determine the way in which the messages should be added to the diagram. Figure 7-16 shows the diagram they created. Notice how they did not include messages back to Customer in response to create SR and add CD. In these cases, the team assumed that aCustomer would receive response messages about the requested CD and inserted CD, respectively.

The fifth step was to add the execution occurrence to each object's and actor's lifeline. This was done by drawing a narrow rectangle box over top of the lifelines to represent when the objects (actors) are sending and receiving messages [e.g., in Figure 7-16 aCustomer is active during the entire process, whereas aSR is active only at the beginning of the process (the top of the diagram)].

Finally, the CD Selections team validated the diagram by ensuring that the diagram accurately and completely represented the scenario of the Place Order use case being modeled. Figure 7-16 portrays their completed sequence diagram.

YOUR TURN   7-5 CD Selections Internet Sales System

*I*n Your Turn 5-8, you completed the detailed use cases for the CD Selections Internet Sales System. In Your Turn 6-4, you completed the structural model. Based on the completed functional and structural models, create the sequence diagrams for the remaining scenarios of all of the use cases in Figure 5-18.

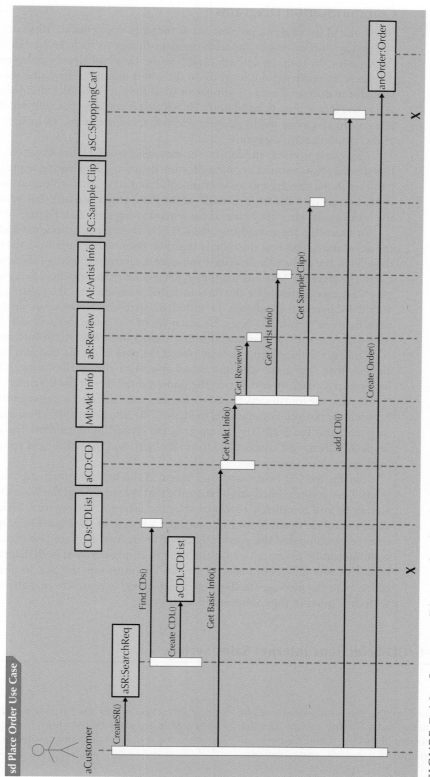

FIGURE 7-16   Sequence Diagram for the Places Order Use Case

259

Communication Diagrams

Brian, one of the analysts, pointed out to the team that sequence diagrams and communication diagrams essentially modeled the same things. As such, he felt that it was not worth the time for the team to do both. And because they had already completed the sequence diagrams, he really did not want to do the communication diagrams also. However, even though the diagrams are very similar in what they portray, Alec decided that it would be worth the team's time to build both. He felt that it could be possible that the different formats of the diagrams might uncover additional requirements. As such, the team also created communication diagrams.

Alec chose to create the communication diagrams by following the steps in Figure 7-8 that describe how to build a communication diagram. Like creating sequence diagrams, the first step in the process was to determine the context of the communication diagram. Alec chose to start with the same scenario of the Place Order use case that he and the team had used previously when they created the sequence diagrams (see Figure 7-16).

By executing the second step, the CD Selections team again identified the objects and the associations that link the objects together. Because they are using the same scenario as they did in the previously described sequence diagram, instances of the Search Request, CD List, CD, Marketing Material, Customer, Review, Artist Information, Sample Clip, and Shopping Cart classes should be the ones included. Also, because the Customer actor interacts with the scenario, it should also be included. Furthermore, the team identified the associations between the objects (e.g., the instances of CD are associated with instances of Mkt Info, which, in turn, are associated with instances of Review, Artist Info, and Sample Clip).

During the third step, the team placed the objects on the diagram based on the associations that they have with the other objects in the collaboration. This was done to increase in readability and, hence, the understandability of the diagram (see Figure 7-17).

During the fourth step, the CD Selections team added the messages to the associations between the objects. For example, in Figure 7-17, the Create SR() message is the first message sent, and the FindCDs() message is the second message sent. The aCustomer actor sends the Create SR() message to the aSR object, and the aSR object sends the FindCDs() message to the CDs object.

Finally, the CD Selections team executed the fifth and final step: validating the diagram. They accomplished this by ensuring that the scenario of the Place Order use case was accurately and completely represented by the diagram. See Figure 7-17 for the completed communication diagram for this particular scenario of the Place Order use case. Furthermore, they compared the previously created sequence diagram (see Figure 7-16) with the communication diagram (see Figure 7-17) to ensure that that both diagrams were equivalent. The only difference between the two diagrams was the ease in portraying the time ordering of the messages in the sequence diagram to represent how the objects interacted with each other in the communication diagram.

YOUR TURN   7-6 CD Selections Internet Sales System

In Your Turn 5-8, you completed the detailed use cases for the CD Selections Internet Sales System. In Your Turn 6-4, you completed the structural model. Based on the completed functional and structural models, create the communication diagrams for the remaining scenarios of all the use cases in Figure 5-18. Be sure to compare these diagrams to the sequence diagrams created in Your Turn 7-5.

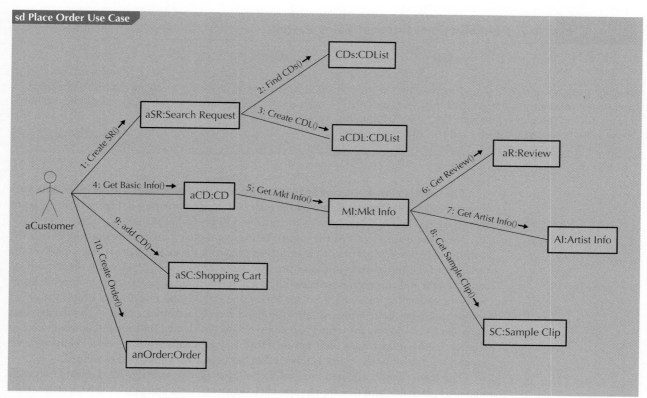

FIGURE 7-17   Communication Diagram for the Places Order Use Case

Behavioral State Machines

As in the previous example diagrams, we focus our attention only on the Place Order use case. Alec decided to follow the steps building a behavioral state machine (see Figure 7-13). As with the earlier diagrams, the first step was to determine the context for the behavioral state machine to be drawn. Upon reviewing the objects involved in the scenario described by the sequence diagram (see Figure 7-16) and the communication diagram (see Figure 7-17), Alec decided that the team should focus on the Order class.

The second step was to identify the various states that an order will go through during its lifetime. To enable the discovery of the initial, final, and stable states of an order, Alec and the development team interviewed a customer representative who dealt with processing customer orders on a regular basis. Based on this interview, the team uncovered the life of an order (see Figure 7-18) from start to finish, from an order's perspective.

FIGURE 7-18
The Life of an Order

1. The customer creates an order on the Web.
2. The customer submits the order once he or she is finished.
3. The credit authorization needs to be approved for the order to be accepted.
4. If denied, the order is returned to the customer for changes or deletion.
5. If accepted, the order is placed.
6. The order is shipped to the customer.
7. The customer receives the order.
8. The order is closed.

YOUR

TURN

7-7 CD Selections Internet Sales System

In Your Turn 7-5 and 7-6, you completed the sequence and communication diagrams for all the use-case scenarios for the CD Selections Internet Sales System. Based on these scenario-based diagrams and the structural model you created in Your Turn 6-4, create a set of behavioral state machines for each concrete class contained in the class diagram.

The third step was to determine the sequence of the states that an order object will pass through during its lifetime. Based on the order's lifecycle portrayed in Figure 7-18, the team identified and laid out the states of the order on the behavioral state machine.

Next, the team identified the events, actions, and guard conditions associated with the transitions between the states of the order. For example, the event Order is created moves the order from the Initial state to the In process state (see Figure 7-19). During the Processing state, a credit authorization is requested. The guard condition Authorization = Approved prevents the order from moving from the Processing state to the Placed state unless the credit authorization has been approved. Also, the guard condition Authorization = Denied prevents the order from moving from the Processing state to the Denied state unless the credit authorization has been denied. As such, between the two guard conditions, the order is stuck in the processing state until the credit authorization has been either approved or denied.

The team finally validated the Order's behavioral state machine by ensuring that each state was reachable and that it was possible to leave all states except for the final states. Furthermore, the team made sure that all states and transitions for the order had been modeled. At this point in time, one of the analysts on the team, Brian, suggested that possibly there were multiple types of orders being described in the behavioral state machine. Specifically, he thought that there were denied and accepted orders. Based on this discovery, he suggested that two new classes, for each subtype of order, be created. However, upon further review by the entire team, it was decided that adding these classes to the class diagram and modifying all the other diagrams to reflect this change would not add anything to the understanding of the problem. Therefore, it was decided not to add the classes. However, in many cases, modeling the states that an object will go through during its lifetime may, in fact, uncover additional useful subclasses. Figure 7-19 illustrates the behavioral state machine that the CD Selections team created for an order object.[21]

CRUD Matrix

As an attempt to tie the functional, structural, and behavioral models together, Alec decided to create a CRUD matrix. To accomplish this, Alec assigned Anne the task of creating the matrix. As in the previous examples, we have limited this example to the Place Order use case.

To begin with, Anne created a class-by-class matrix. She then placed a (C)reate, (R)ead, (U)pdate, or a (D)elete in each cell of the matrix that represented an interaction between instances of the classes. For example, in Figures 7-16 and 7-17, an instance of SearchReq created an instance of CDList. Also, in Figures 7-16 and 7-17, an instance of CD references an instance of MktInfo. In this case, an R was placed in the CD:MktInfo cell. Figure 7-20 shows the CRUD matrix that Anne created based on the Place Order use case.

---

[21] If the development team had been carefully reading our textbook, they would have seen that they could have reused the Order behavioral state machine in Figure 7-11.

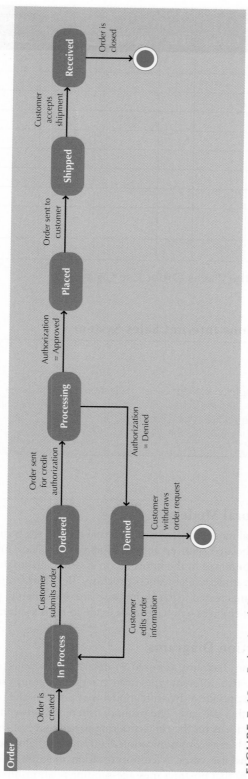

FIGURE 7-19 Behavioral State Machine for the Order Class

263

	Customer	SearchReq	CDList	CD	Mkt Info	Review	Artist Info	Sample Clip	Shopping Cart	Order
Customer		RU							U	C
SearchReq			CR							
CDList										
CD					R					
Mkt Info						U	U	U		
Review										
Artist Info										
Sample Clip										
Shopping Cart										
Order										

FIGURE 7-20   CRUD Matrix for the Places Order Use Case

YOUR TURN

7-8 CD Selections Internet Sales System

*I*n Your Turn 7-5, 7-6, and 7-7, you completed the sequence diagrams, the communication diagrams, and the behavioral state machines based on the detailed use cases (Your Turn 5-8) and the structural model (Your Turn 6-4) for the CD Selections Internet Sales System. Based on these results, complete the CRUD matrix begun by Anne (see Figure 7-20).

SUMMARY

Behavioral Models

Behavioral models describe the internal logic of the business processes described by the use cases associated with an information system. They provide a detailed view of how the objects contained in the information system collaborate to support the use cases that represent the underlying business processes. Behavioral models, like functional and structural models, are created using an iterative and incremental process. Based on this process, it is likely that changes will have to be made to both the functional and structural models.

Interaction Diagrams

Interaction diagrams are used to describe how objects collaborate to support use cases. There are two different types of interaction diagrams: sequence and communication. Both diagrams provide a dynamic model that portrays the interaction among the objects associated with a use case. The primary difference between the two diagrams is that sequence diagrams focus on the time-ordering or sequence of the messages being sent between the objects, whereas communication diagrams spotlight the collaborating nature of the objects supporting use cases. A communication diagram is essentially an object diagram (see Chapter 6) that portrays message-sending relationships instead of structural relationships.

Behavioral State Machine

The behavioral state machine shows the different states through which a single class passes during its life in response to events, along with responses and actions. A state is a set of values that describes an object at a specific point in time, and it represents a point in an object's life in which it satisfies some condition, performs some action, or waits for something to happen. An event is something that takes place at a certain point in time and changes a value(s) that describes an object, which, in turn, changes the object's state. As objects move from state to state, they undergo transitions.

When drawing a behavioral state machine, like the other behavioral diagrams, the first thing is to set the context of the diagram: system, subsystem, set of classes, or an individual class. Then, rectangles with rounded corners are placed on the model to represent the various states on which the context will take. Next, arrows are drawn between the rectangles to denote the transitions, and event labels are written above the arrows to describe the event that causes the transition to occur. Typically, behavioral state machines are used to portray the dynamic aspect of a complex class.

CRUD Analysis

CRUD analysis is a very useful approach to identifying potential collaborations among classes and to verifying and validating a system. It allows the analyst to see what type of interactions (create, read/reference, update, or delete) that the different types of objects have in the system in a very concise format (CRUD Matrix). CRUD analysis also supports the identification of the more complex objects that could benefit from state modeling using a behavioral state machine.

KEY TERMS

Action	CRUD matrix	Operation
Activity	Dynamic model	Operation call message
Actor	Event	Packages
Association	Execution occurrence	Return message
Attributes	Final state	Scenario
Behavior	Frame	Self-delegation
Behavior models	Generic sequence diagram	Sequence diagram
Behavioral state machines	Guard condition	State
Class	Initial state	State symbol
Class diagram	Instance	Temporary object
Collaboration	Instance sequence diagram	Transition
Communication diagram	Lifeline	Trigger
Condition	Message	Use case
CRC cards	Method	
CRUD analysis	Object	

QUESTIONS

1. How is behavioral modeling related to structural modeling?

2. Why is iteration important when creating a behavioral model?

3. What are the main building blocks for the sequence diagram? How are they represented on the model?

4. How does a use case relate to a sequence diagram? A communication diagram?

5. Describe the steps used to create a sequence diagram.
6. How do you show that a temporary object is to go out of existence on a sequence diagram?
7. Do lifelines always continue down the entire page of a sequence diagram? Explain.
8. Contrast the following sets of terms:
   state; behavior
   class; object
   action; activity
   use case; scenario
   method; message
9. Describe the steps used to create a communication diagram.
10. Describe the main building blocks for the communication diagram and how they are represented on the model.
11. How do you show the sequence of messages on a communication diagram?
12. How do you show the direction of a message on a communication diagram?
13. What is CRUD analysis and what is it used for?
14. What are the steps in building a behavioral state machine?
15. Describe the type of class that is best represented by a behavioral state machine. Give two examples of classes

that would be good candidates for a behavioral state machine.
16. What kinds of events can lead to state transitions on a behavioral state machine?
17. Are states always depicted using rounded rectangles on a behavioral state machine? Explain.
18. How are guard conditions shown on a behavioral state machine?
19. Identify the models that contain each of the following components:
   actor
   association
   class
   extends association
   final state
   guard condition
   initial state
   links
   message
   multiplicity
   object
   state
   transition
   update operation

EXERCISES

A. Think about sending a first-class letter to an international pen pal. Describe the process that the letter goes through to get from your initial creation of the letter to being read by your friend, from the letter's perspective. Draw a behavioral state machine that depicts the states that the letter moves through.
B. Create a sequence diagram for each of the following scenario descriptions for a video store system. A Video Store (AVS) runs a series of fairly standard video stores.
   1. Every customer must have a valid AVS customer card in order to rent a video. Customers rent videos for three days at a time. Every time a customer rents a video, the system must ensure that the customer does not have any overdue videos. If so, the overdue videos must be returned and an overdue fee paid before customer can rent more videos.
   2. If the customer has returned overdue videos but has not paid the overdue fee, the fee must be paid before new videos can be rented. If the customer is a premier customer, the first two overdue fees can be waived, and the customer can rent a video.

   3. Every morning, the store manager prints a report that lists overdue videos; if a video is two or more days overdue, the manager calls to remind the customer to return the video.
C. Create a communication diagram for the video system in Exercise B.
D. Perform a CRUD analysis for the current state of the video system in Exercise B.
E. Consider the video store that is described in Question B. Draw a behavioral state machine that describes the various states that a video goes through from the time it is placed on the shelf through the rental and return process.
F. Create a sequence diagram for each of the following scenario descriptions for a health club membership system.
   1. When members join the health club, they pay a fee for a certain length of time. The club wants to mail out reminder letters to members asking them to renew their memberships one month before their memberships expire. About half of the members do not renew their memberships. These members

are sent follow-up surveys to complete asking why they decided not to renew so that the club can learn how to increase retention. If the member did not renew because of cost, a special discount is offered to that customer. Typically, 25 percent of accounts are reactivated because of this offer.

2. Every time a member enters the club, an attendant takes his or her card and scans it to make sure the person is an active member. If the member is not active, the system presents the amount of money it costs to renew the membership. The customer is given the chance to pay the fee and use the club, and the system makes note of the reactivation of the account so that special attention can be given to this customer when the next round of renewal notices are dispensed.

G. Create a communication diagram for each of the health club membership scenarios described in Exercise F.

H. Perform a CRUD analysis for the current state of the health club membership system in Exercise F.

I. Picnics R Us (PRU) is a small catering firm with five employees. During a typical summer weekend, PRU caters fifteen picnics with twenty to fifty people each. The business has grown rapidly over the past year, and the owner wants to install a new computer system for managing the ordering and buying processes. PRU has a set of ten standard menus. When potential customers call, the receptionist describes the menus to them. If the customer decides to book a picnic, the receptionist records the customer information (e.g., name, address, phone number, etc.) and the information about the picnic (e.g., place, date, time, which one of the standard menus, total price) on a contract. The customer is then faxed a copy of the contract and must sign and return it along with a deposit (a credit-card number or check) before the picnic is officially booked. The remaining money is collected when the picnic is delivered. Sometimes, the customer wants something special (e.g., birthday cake). In this case, the receptionist takes the information and gives it to the owner, who determines the cost; the receptionist then calls the customer back with the price information. Sometimes the customer accepts the price; other times, the customer requests some changes that have to go back to the owner for a new cost estimate. Each week, the owner looks through the picnics scheduled for that weekend and orders the supplies (e.g., plates) and food (e.g., bread, chicken) needed to make them. The owner would like to use the system for marketing as well. It should be able to track how customers learned about PRU and identify repeat customers, so that PRU can mail special offers to them. The owner also wants to track the picnics on which PRU sent a contract but the customer did not sign the contract and actually book a picnic.

1. Create an activity diagram for the PRU system.
2. Identify the use cases for the PRU system.
3. Create the use-case diagram for the PRU system.
4. Create a class diagram for the PRU System.
5. Choose one use case and create sequence diagrams.
6. Create a communication diagram for the use case chosen in Question 5.
7. Create a behavioral state machine to depict one of the classes on the class diagram in Question 2.
8. Perform a CRUD analysis for the current state of the system's representation.

J. Draw a behavioral state machine that describes the various states that a travel authorization can have through its approval process. A travel authorization form is used in most companies to approve travel expenses for employees. Typically, an employee fills out a blank form and sends it to his or her boss for a signature. If the amount is fairly small (< $300), then the boss signs the form and routes it to accounts payable to be input into the accounting system. The system cuts a check that is sent to the employee for the right amount, and after the check is cashed, the form is filed away with the canceled check. If the check is not cashed within 90 days, the travel form expires. When the amount of the travel voucher is a large amount (> $300), then the boss signs the form and sends it to the CFO, along with a paragraph explaining the purpose of the travel; the CFO signs the form and passes it along to accounts payable. Of course, both the boss and the CFO can reject the travel authorization form if they do not feel that the expenses are reasonable. In this case, the employee can change the form to include more explanation or decide to pay the expenses.

K. Think about your school or local library and the processes involved in checking out books, signing up new borrowers, and sending out overdue notices from the library's perspective. Describe three use cases that represent these three functions.

1. Create a use-case diagram for the library system.
2. Create a class diagram for the library system.
3. Choose one use case and create sequence diagrams.
4. Create a communication diagram for the use case chosen in Question 3.
5. Create a behavioral state machine to depict one of the classes on the class diagram in Question 2.

L. Of-the-Month Club (OTMC) is an innovative young firm that sells memberships to people who have an interest in certain products. People pay membership fees for one year and each month receive a product by mail. For example, OTMC has a coffee-of-the-month club, which sends members one pound of special coffee each month. OTMC currently has six memberships (coffee, wine, beer, cigars, flowers, and computer games), each of which costs a different amount. Customers usually belong to just one, but some belong to two or more. When people join OTMC, the telephone operator records the name, mailing address, phone number, e-mail address, credit-card information, start date, and membership service(s) (e.g., coffee). Some customers request a double or triple membership (e.g., two pounds of coffee, three cases of beer). The computer game membership operates a bit differently from the others. In this case, the member must also select the type of game (action, arcade, fantasy/science-fiction, educational, etc.) and age level. OTMC is planning to greatly expand the number of memberships it offers (e.g., video games, movies, toys, cheese, fruit, vegetables), so the system needs to accommodate this future expansion. OTMC is also planning to offer three-month and six-month memberships.

1. Create an activity diagram for the OTMC system.
2. Identify the use cases for the OTMC system.
3. Create a use-case diagram for the OTMC system.
4. Create a class diagram for the OTMC System.
5. Choose one use case and create sequence diagrams.
6. Create a communication diagram for the use case chosen in Question 5.

7. Create a behavioral state machine to depict one of the classes on the class diagram in Question 2.
8. Perform a CRUD analysis for the current state of the system's representation.

M. Think about the system that handles student admissions at your university. The primary function of the system should be able to track a student from the request for information through the admissions process until the student is either admitted or rejected from attending the school. Write a use-case description that can describe an Admit Student use case.

1. Create a use-case diagram for the one use case. Assume that students of alumni are handled differently than other students. Also, a generic Update Student Information use case is available for your system to use.
2. Create a class diagram for the use case that you selected for Question 2. An admissions form includes the contents of the form, SAT information, and references. Additional information is captured about students of alumni, such as their parent's graduation year, contact information, and college major.
3. Choose one use case and create sequence diagrams. Assume that a temporary student object is used by the system to hold information about people before they send in an admission form. After the form is sent in, these people are considered students.
4. Create a communication diagram for the use case chosen in Question 3.
5. Create a behavioral state machine to depict a person as he or she moves through the admissions process.
6. Perform a CRUD analysis for the current state of the system's representation.

MINICASES

1. Refer to the use-case descriptions and use-case diagram you prepared for Professional and Scientific Staff Management (PSSM) for Minicase 2 in Chapter 5. PSSM was so satisfied with your work that they wanted the behavioral models created so that they could understand the interaction that would take place between the users and the system in greater detail.
   a. Create both sequence and a communication diagrams for each use case.
   b. Based on the sequence and communication diagrams, create CRC cards for each class identified and a class diagram.

   c. Perform a CRUD analysis for the current state of the system's representation.
2. Refer to the structural model that you created for Holiday Travel Vehicles for Minicase 1 in Chapter 6.
   a. Choose one of the more complex classes and create a behavioral state machine for it.
   b. Based on the structural model you created, role-play the CRC cards to develop a set of use cases that describe a typical sales process.
   c. Create both sequence and communication diagrams for the use case.
   d. Perform a CRUD analysis for the current state of the system's representation.

PART THREE

DESIGN

Whereas analysis modeling concentrated on the functional requirements of the evolving system, design modeling incorporates the nonfunctional requirements. That is, design modeling focuses on *how* the system will operate. First, the project team verifies and validates the analysis models (functional, structural, and behavioral). Next, a set of factored and partitioned analysis models are created. The class and method designs are illustrated using the class specifications (using CRC cards and class diagrams), contracts, and method specifications. Next, the data management layer is addressed by designing the actual database or file structure to be used for object persistence, and a set of classes that will map the class specifications into the object persistence format chosen. Concurrently, the team produces the user interface layer design using use scenarios, windows navigation diagrams, real use cases, interface standards, and user interface templates. The physical architecture layer design is created using deployment diagrams and hardware software specification. This collection of deliverables represents the system specification that is handed to the programming team for implementation.

Factored/Partitioned
Structural Model

Factored/Partitioned
Functional Model

Factored/Partitioned
Behavioral Model

Package
Diagrams

Alternative
Matrix

Class
Specifications

Contracts

Method
Specifications

Object
Persistence Design

Data Access & Manipula-
tion Class Design

User Interface
Design

Deployment
Diagrams

Hardware/Software
Specifications

CHAPTER 8

SYSTEM DESIGN

Object-oriented system development uses the requirements that were gathered during analysis to create a blueprint for the future system. A successful object-oriented design builds upon what was learned in earlier phases and leads to a smooth implementation by creating a clear, accurate plan of what needs to be done. This chapter describes the initial transition from analysis to design and presents three ways to approach the design for the new system.

OBJECTIVES

- Understand the verification and validation of the analysis models.
- Understand the transition from analysis to design.
- Understand the use of factoring, partitions, and layers.
- Be able to create package diagrams.
- Be familiar with the custom, packaged, and outsource design alternatives.
- Be able to create an alternative matrix.

CHAPTER OUTLINE

INTRODUCTION

The purpose of analysis is to figure out what the business needs are. The purpose of design is to decide how to build the system. The major activity that takes place during *design* is evolving the set of analysis representations into design representations.

Throughout design, the project team carefully considers the new system in respect to the current environment and systems that exist within the organization as a whole. Major considerations of the how of a system are environmental factors such as integrating with existing systems, converting data from legacy systems, and leveraging skills that exist in-house. Although the planning and analysis are undertaken to develop a possible system, the goal of design is to create a blueprint for a system that makes sense to implement.

An important initial part of design is to examine several design strategies and decide which will be used to build the system. Systems can be built from scratch, purchased and customized, or outsourced to others, and the project team needs to investigate the viability of each alternative. This decision influences the tasks that are to be accomplished during design.

At the same time, detailed design of the individual classes and methods that are used to map out the nuts and bolts of the system and how they are to be stored must still be completed. Techniques such as CRC cards, class diagrams, contract specification, method specification, and database design provide the final design details in preparation for the implementation phase, and they ensure that programmers have sufficient information to build the right system efficiently. These topics are covered in Chapters 9 and 10.

Design also includes activities such as designing the user interface, system inputs, and system outputs, which involve the ways that the user interacts with the system. Chapter 11 describes these three activities in detail, along with techniques such as storyboarding and prototyping, which help the project team design a system that meets the needs of its users and is satisfying to use.

Finally, physical architecture decisions are made regarding the hardware and software that will be purchased to support the new system and the way that the processing of the system will be organized. For example, the system can be organized so that its processing is centralized at one location, distributed, or both centralized and distributed, and each solution offers unique benefits and challenges to the project team. Because global issues and security will influence the implementation plans that are made, they need to be considered along with the system's technical architecture. Physical architecture, security, and global issues are described in Chapter 12.

The many steps of design are highly interrelated and, as with the steps in analysis, the analysts often go back and forth among them. For example, prototyping in the interface design step often uncovers additional information that is needed in the system. Alternatively, a system that is being designed for an organization that has centralized systems may require substantial hardware and software investments if the project team decides to change to a system in which all the processing is distributed.

In this chapter, we overview the processes that are used to evolve the analysis models into design models. But, before we move on into design, we really need to be sure that the current analysis models are consistent. As such, we next discuss an approach to verify and validate the analysis models. Afterward, we describe different higher-level constructs that can be used to evolve the analysis models into design models. Then, we introduce the use of packages and package diagrams as a means of representing the higher-level constructs used to evolve the models. Finally, we examine the three fundamental approaches to developing new systems: make, buy, or outsource.

| PRACTICAL | Avoiding Classic Design Mistakes |
| TIP | |

In Chapters 2 and 3, we discussed several classic mistakes and how to avoid them. Here, we summarize four classic mistakes in the design phase and discuss how to avoid them.

1. *Reducing design time:* If time is short, there is a temptation to reduce the time spent in "unproductive" activities such as design so that the team can jump into "productive" programming. This results in missing important details that have to be investigated later at a much higher time and cost (usually at least ten times higher).

   Solution: If time pressure is intense, use timeboxing to eliminate functionality or move it into future versions.

2. *Feature creep*: Even if you are successful at avoiding scope creep, about 25 percent of system requirements will still change. And, changes—big and small—can significantly increase time and cost.

   Solution: Ensure that all changes are vital and that the users are aware of the impact on cost and time. Try to move proposed changes into future versions.

3. *Silver bullet syndrome:* Analysts sometimes believe the marketing claims for some design tools that claim to solve all problems and magically reduce time and costs. No *one* tool or technique can eliminate overall time or costs by more than 25 percent (although some can reduce individual steps by this much).

   Solution: If a design tool has claims that appear too good to be true, just say no.

4. *Switching tools midproject:* Sometimes analysts switch to what appears to be a better tool during design in the hopes of saving time or costs. Usually, any benefits are outweighed by the need to learn the new tool. This also applies even to minor upgrades to current tools.

   Solution: Don't switch or upgrade unless there is a compelling need for *specific* features in the new tool, and then explicitly *increase* the schedule to include learning time.

Adapted from Steve McConnell, *Rapid Development* (Redmond, WA: Microsoft Press, 1966).

VERIFYING AND VALIDATING THE ANALYSIS MODELS[1]

Before we evolve our analysis representations into design representations, we need to verify and validate the current set of analysis models to ensure that they faithfully represent the problem domain under consideration. This includes testing the fidelity of each model; for example, we must be sure that the activity diagram(s), use-case descriptions, and use-case diagrams all describe the same functional requirements, and between the models, for instance, transitions on a behavioral state machine are associated with operations contained in a class diagram. Before we describe the specific tests to consider, we describe walkthroughs, a manual approach that supports verifying and validating the evolving models.[2] Afterward, we describe a set of rules that can be used for *verification* and *validation* of the analysis models.

Verification and Validation through Walkthroughs

A *walkthrough* is essentially a peer review of a product. In the case of the analysis models, a walkthrough is a review of the different models and diagrams created during analysis. This

---

[1] The material in this section has been adapted from E. Yourdon, *Modern Structured Analysis* (Englewood Cliffs, NJ: Prentice Hall, 1989). Verifying and validating are a type of testing. We also describe testing in Chapter 13.

[2] Even though many modern CASE tools can automate much of the verifying and validating of the analysis models, we feel that it is paramount that systems analysts understand the principles of verification and validation. Furthermore, some tools such as VISIO that supports UML diagramming are only diagramming tools. Regardless, the analyst is expected to perform all diagramming correctly.

review typically is completed by a team of individuals that comes from the analysis team, the design team, and the client. The purpose of an analysis walkthrough is to thoroughly *test* the fidelity of the analysis models to the functional requirements and to ensure that the models are consistent. That is, an analysis walkthrough uncovers *errors* or *faults* in the evolving specification. However, a walkthrough does not correct errors—it simply identifies them. Error correction is to be accomplished by the analysis team after the walkthrough is completed.

Walkthroughs are very interactive. As the presenter "walks" through the representation, members of the walkthrough team should ask questions regarding the representation. For example, if the presenter is walking through a class diagram, another member of the team could ask why certain attributes, operations, or associations were not included. The actual process of simply presenting the representation to a new set of eyes can uncover obvious misunderstandings and omissions. In many cases, the representation creator can get lost in the proverbial trees and not see the forest.[3] In fact, many times the act of walking through the representation causes a presenter to see the error himself or herself. For psychological reasons, hearing the representation helps the analyst to see the representation more completely.[4] As such, the representation creators should regularly do a walkthrough of the analysis models themselves by reading the representations out loud to themselves, regardless of how they think it may make them look.

There are specified roles that different members of the walkthrough team can play. The first is the *presenter* role. This should be played by the individual who is primarily responsible for the specific representation being reviewed. This individual presents the representation to the walkthrough team. The second role is *recorder,* or *scribe.* The recorder should be a member of the analysis team. This individual carefully takes the minutes of the meeting by recording all significant events that occur during the walkthrough. In particular, all errors that are uncovered must be documented so that the analysis team can address them. Another important role is to have someone who raises issues regarding maintenance of the representation. Yourdon refers to this individual as a *maintenance oracle.*[5] Due to the emphasis on reusability in object-oriented development, this role becomes particularly crucial. Finally, someone must be responsible for calling, setting up, and running the walkthrough meetings.

For a walkthrough to be successful, the members of the walkthrough team must be fully prepared. As such, all materials to be reviewed must be distributed with sufficient time for the team members to review them before the actual meeting. Furthermore, all team members should be expected to mark up the representations so that during the walkthrough meeting, all relevant issues can be discussed. Otherwise, the walkthrough will be both inefficient and ineffective. During the actual meeting, as the presenter is walking through the representation, the team members should point out any potential errors or misunderstandings. In many cases, the errors and misunderstandings are caused by invalid assumptions that would not be uncovered without the walkthrough.

One potential danger of walkthroughs is when management decides the results of uncovering errors in the representation are a reflection of an analyst's capability. This must be avoided at all costs. Otherwise, the underlying purpose of the walkthrough—to improve the fidelity of the representation—will be thwarted. Depending on the organization, it may be necessary to omit management from the walkthrough process. If not, the walkthrough process may break down into a slugfest to make some team members to look good by destroying the presenter. To say the least, this is obviously counterproductive.

---

[3] In fact, all joking aside, in many cases the developer is down at the knothole level and can't even see the tree, let alone the forest.

[4] This has to do with using different senses. Because our haptic senses are the most sensitive, "touching" the representation would be best. However, it is not clear how one can touch a use case or a class.

[5] See Appendix D of Yourdon, *Modern Structured Analysis.*

Functional Model Verification and Validation

In this book, we have suggested three different representations for the functional model: activity diagrams, use-case descriptions, and use-case diagrams (see Chapter 5). In this section, we describe a set of rules to ensure that these three representations are consistent among themselves.

First, when comparing an activity diagram to a use-case description, there should be at least one event recorded in the normal flow of events, subflows, or alternate/exceptional flows of the use-case description for each activity or action that is included on an activity diagram, and each event should be associated with an activity or action. For example, in Figure 5-2, there is an activity labeled Get Patient Information that seems to be associated with the first two events contained in the normal flow of events of the use-case description shown in Figure 5-5.

Second, all objects portrayed as an object node in an activity diagram must be mentioned in an event in the normal flow of events, subflows, or alternate/exceptional flows of the use-case description. For example, the activity diagram in Figure 5-2 portrays an Appt object, and the use-case description refers to a new appointment and changing an existing appointment.

Third, sequential order of the events in a use-case description should occur in the same sequential order of the activities contained in an activity diagram. For example in Figures 5-2 and 5-5, the events associated with the Get Patient Information activity (events 1 and 2) should occur before the events associated with the Make Payment Arrangements activity (event 4).

Fourth, when comparing a use-case description to a use-case diagram, there must be one and only one use-case description for each use case, and vice versa. For example, Figure 5-5 portrays the use-case description of the Make Appointment use case. However, the use-case diagrams shown in Figures 5-8, 5-9, and 5-10 are not totally consistent with the use-case description given in Figure 5-5. As such, either the use-case description needs to be modified by breaking it down into multiple use-case descriptions or the use-case diagram needs to be modified. Upon reviewing the two representations, it was decided to create a new use-case diagram (see Figure 8-1) that was consistent with the use-case

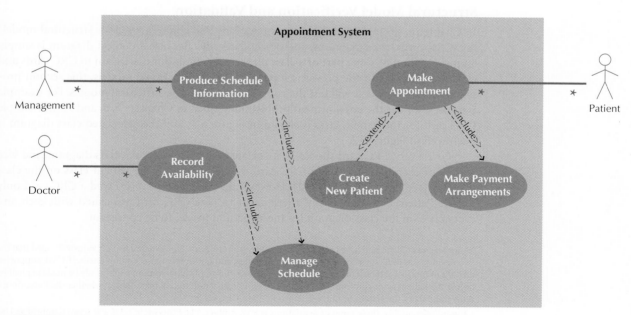

FIGURE 8-1   Modified Use-Case Diagram for the Appointment System

description given in Figure 5-5. However, there are still inconsistencies between the two representations: the use-case diagram has use cases portrayed on it, but there were no use-case descriptions associated with them. For presentation purposes, we have omitted those additional descriptions. As such, we can safely assume that they would go through the same testing procedure that the Make Appointment use case does.

Fifth, all actors listed in a use case description must be portrayed on the use-case diagram. Furthermore, each one must have an association link that connects it to the use case and must be listed with the association relationships in the use-case description. For example, the Patient actor is listed in the use-case description of the Make Appointment use case (see Figure 5-5), it is listed with the association relationships in the Make Appointment use-case description, and it is connected to the use case in the use-case diagram (see Figure 8-1).

Sixth, in some organizations, we should also include the stakeholders listed in the use-case description as actors in the use-case diagram. For example, there could have been an association between the Doctor actor and the Make Appointment use case (see Figures 5-5 and 8-1). However, in this case it was decided not to include this association because the Doctor never participates in the Make Appointment use case.[6]

Seventh, all other relationships listed in a use-case description (include, extend, and generalization) must be portrayed on a use-case diagram. For example, in Figure 5-5, there is an include relationship listed with the Make Payment Arrangements use case and in Figure 8-1, we see that it appears on the diagram between the two use cases.

Finally, there are many diagram-specific requirements that must be enforced. For example, in an activity diagram a decision node can be connected to activity or action nodes only with a control flow, and for every decision node there should be a matching merge node (see Chapter 5). Every type of node and flow has different restrictions. However, the complete restrictions for all the UML diagrams are beyond the scope of this text.[7] The concept map in Figure 8-2 portrays the associations among the functional models.

Structural Model Verification and Validation

In Chapter 6, we suggested three representations that could be used for structural modeling: CRC cards, class diagrams, and object diagrams. Because an object diagram is simply an instantiation of some part of a class diagram, we limit our discussion to CRC cards and class diagrams. As in the previous section regarding functional models, this section provides a set of rules that will test the consistency within the structural models. For example purposes, we use the appointment problem described in Chapters 5, 6, and 7. An example CRC card for the patient class is shown in Figure 6-1, and the associated class diagram is portrayed in Figure 6-2.

First, every CRC card should be associated with a class on the class diagram, and vice versa. For example, the Patient CRC card is associated with the Patient class on the class diagram (see Figures 6-1 and 6-2). However, in Chapter 6 we portrayed a CRC card only for the Patient class. Obviously, there needs to be a CRC card associated with each and every class on the class diagram for the structural model to be consistent.

---

[6] Another possibility could have been to include a Receptionist actor. However, we had previously decided that the Receptionist was in fact part of the Appointment System and not simply a user of the system. If UML supported the idea of internal actors, or actor-to-actor associations, this implicit association could easily be made explicit by having the Patient actor communicate with the Receptionist actor directly, regardless of whether the Receptionist actor was part of the system or not. See Footnote 19 in Chapter 5.

[7] A good reference for these types of restrictions is S.W. Ambler, *The Elements of UML 2.0 Style* (Cambridge, UK: Cambridge University Press, 2005).

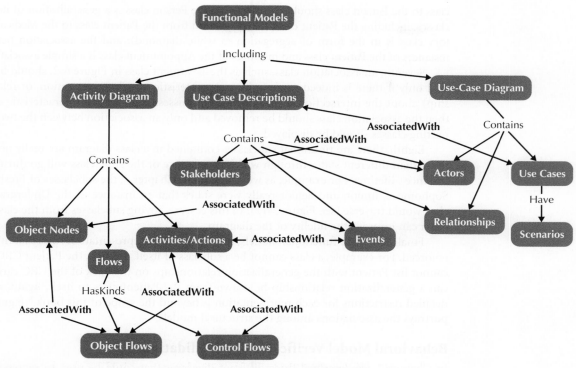

FIGURE 8-2   Interrelationships among Functional Models

Second, the responsibilities listed on the front of the CRC card must be included as operations in a class on a class diagram, and vice versa. For example, the make appointment responsibility on the patient CRC card also appears as the make appointment() operation in the Patient class on the class diagram.

Third, collaborators on the front of the CRC card imply some type of relationship on the back of the CRC card and some type of association that is connected to the associated class on the class diagram. For example, the appointment collaborator on the front of the CRC card also appears as an other association on the back of the CRC card and as an association on the class diagram that connects the Patient class with the Appointment class.

Fourth, attributes listed on the back of the CRC card must be included as attributes in a class on a class diagram, and vice versa. For example, the amount attribute on the Patient CRC card is included in the attribute list of the Patient class on the class diagram.

Fifth, the object type of the attributes listed on the back of the CRC card and with the attributes in the attribute list of the class on a class diagram implies an association from the class to the class of the object type. For example, technically speaking, the amount attribute implies an association with the double-object type. Simple types such as int and double are never shown on a class diagram. Furthermore, depending on the problem domain, object types such as Person, Address, or Date may not be explicitly shown either. However, if we know that messages are being sent to instances of those object types, we probably should include these implied associations as relationships.

Sixth, the relationships included on the back of the CRC card must be portrayed using the appropriate notation on the class diagram. For example in Figure 6-1, instances of the Patient class is *a-kind-of* Person, it has instances of the Medical History class as part of it, and it has an association with instances of the Appointment class. As such, the association from the Patient class to the Person class should represent that the Person class is a generalization of its subclasses, including the Patient class; the association from the Patient class to the Medical History class is in the form of aggregation (a white diamond); and the association between instances of the Patient class and instances of the Appointment class is a simple association.

Seventh, an association class, such as the Treatment class in Figure 6-2, should be created only if there is indeed some unique characteristic (attribute, operation, or relationship) about the intersection of the connecting classes. If no unique characteristic exists, then the association class should be removed and only an association between the two connecting classes should be displayed.

Eighth, there are times that subclasses contained in a class diagram are really nothing more than different states through which an instance of the superclass will go during the instance's lifetime. For example, as we discussed in Chapter 7, the subclasses of Freshman, Sophomore, Junior, and Senior really were states that an instance of the Undergraduate class would traverse (see Figure 7-12). In this case, removing subclasses from the class diagram can improve the fidelity of the diagram.

Finally, as in the functional models, there are specific representation rules that must be enforced. For example, a class cannot be a subclass of itself. As such, the Patient CRC card cannot list Patient with the generalization relationships on the back of the CRC card, nor can a generalization relationship be drawn from the Patient class to itself. Again, all the detailed restrictions for each representation is beyond the scope of this book.[8] Figure 8-3 portrays the associations among the structural models.

Behavioral Model Verification and Validation

In Chapter 7, we described three different diagrams that could be used to represent the behavioral model: the sequence diagram, the communication diagram, and the behavioral state machine. The sequence and communication diagrams modeled the interaction among instances of classes that worked together to support the business processes included in a system, whereas the behavioral state machine described the state changes through which an object traverses during its lifetime. We again use the appointment system described in Chapters 5, 6, and 7 and focus on Figures 7-1, 7-5, 7-9, and 7-14 to describe a set of rules that can be used to ensure that the behavioral model is internally consistent.

First, every actor and object included on a sequence diagram must be included as an actor and an object on a communication diagram, and vice versa. For example, in Figures 7-1 and 7-5, the aReceptionist actor and the Patients object appear on both diagrams.

Second, if there is a message on the sequence diagram, there must be an association on the communications diagram, and vice versa. For example, Figure 7-1 portrays a message being sent from the aReceptionist actor to the Patients object. As such, a matching association appears in the corresponding communication diagram (see Figure 7-5).

Third, every message that is included on a sequence diagram must appear as a message on an association in the corresponding communication diagram, and vice versa. For example, the LookUpPatient() message sent by the aReceptionist actor to the Patients object on the sequence diagram (see Figure 7-1) appears as a message on the association between the aReceptionist actor and the Patients object on the communication diagram (see Figure 7-5).

Fourth, if a guard condition appears on a message in the sequence diagram, there must be an equivalent guard condition on the corresponding communication diagram, and vice versa. For example, the message sent from the aReceptionist actor to the UnpaidBills object has a guard condition of [aPatient Exists] (see Figure 7-1). Figure 7-5 shows the matching guard condition included on the communication diagram.

---

[8] See Footnote 7.

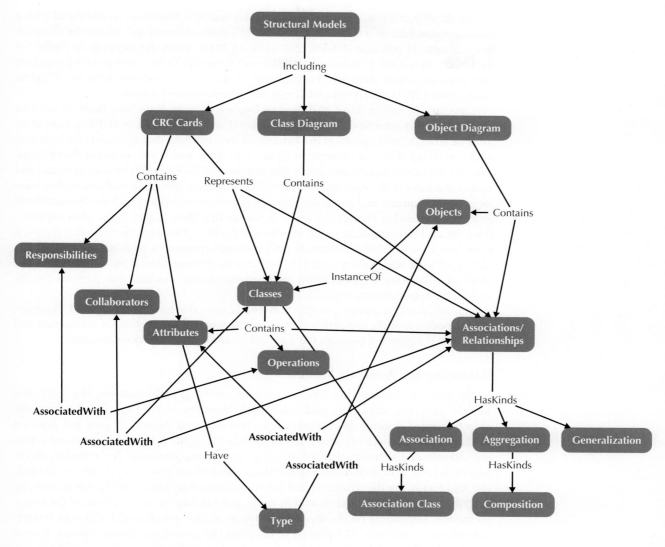

FIGURE 8-3   Interrelationships among Structural Models

Fifth, the sequence number included as part of a message label in a communications diagram implies the sequential order in which the message will be sent. As such, it must correspond to the top-down ordering of the messages being sent on the sequence diagram. For example, the LookUpPatient message sent from the aReceptionist actor to the Patients object on the sequence diagram (see Figure 7-1) is the second from the top of the diagram. As such, the LookUpPatient message sent from the aReceptionist actor to the Patients object on the communications diagram (see Figure 7-5) is labeled with the number 2.[9]

---

[9] There are more complicated numbering schemes that could be used. However, for our purposes, a simple sequential number is sufficient.

Sixth, all transitions contained in a behavior state machine must be associated with a message being sent on a sequence and communication diagram, and it must be classified as a (C)reate, (U)pdate, or (D)elete message in a CRUD matrix. For example, in Figure 7-9 the Checks In transition must be associated with a message in the corresponding sequence and communication diagrams. Furthermore, it should be associated with an (U)pdate entry in the CRUD matrix associated with the hospital patient system.

Seventh, all entries in a CRUD matrix imply a message being sent from an actor or object to another actor or object. If the entry is a (C)reate, (U)pdate, or (D)elete, then there must be an associated transition in a behavioral state machine that represents the instances of the receiving class. For example in Figure 7-14 the R and U entries in the Receptionist row and Appointments column imply that instances of the Receptionist actor will read and update instances of the Appointments class. As such, there should be read and update messages on the sequence and communication diagrams corresponding with the appointments processes. Reviewing Figures 7-1 and 7-5, we see that there is a message, MatchAppts(), from the aReceptionist actor to the Appointments object. However, based on this review, it is unclear as to whether the MatchAppts() message represents a read, update, or both. As such, additional analysis is required.[10] Furthermore, because there is an (U)pdate message involved, there must be a transition on a behavioral state machine that portrays the life cycle of an Appointments object.

Finally, there are many representation specific rules that have been proposed. However, as in the other models, these rules are beyond the scope of this section on verification and validation.[11] Figure 8-4 portrays the associations among the behavioral models.

Balancing the Analysis Models

In the previous sections, we have focused on verifying and validating the individual models: function, structural, and behavioral. However, as we have stated repeatedly, object-oriented systems development is incremental and iterative. Figure 8-5 portrays the fact that the object-oriented analysis models are highly interrelated. Next, we center our attention on ensuring that the different models are consistent. For example, do the functional and structural models agree? What about the functional and behavioral models? And finally, are the structural and behavioral models trustworthy? In this section, we describe a set of rules that are useful to verify and validate the intersections of the analysis models. Depending on the specific constructs of each actual model, different interrelationships are relevant. The process of ensuring the consistency among them is known as *balancing the models*.

Balancing Functional and Structural Models   To balance the functional and structural models, we must ensure that the two sets of models are consistent with each other. That is, the activity diagrams, use-case descriptions, and use-case diagrams must agree with the CRC cards and class diagrams that represent the evolving model of the problem domain. Figure 8-6 shows the interrelationships between the functional and structural models. By reviewing this figure, we uncover four sets of associations between the models. This gives us a place to begin balancing the functional and structural models.[12]

---

[10] We have delayed the description of designing operations/methods until Chapter 9. As such, the detailed information required to understand a specific message has not been created yet. However, in many cases, enough information will already have been created to validate many of the transitions in behavioral state machines and CRUD matrices.

[11] See Footnote 7.

[12] Role-playing the CRC cards (see Chapter 6) also can be very useful in verifying and validating the interrelationships among the functional and structural models.

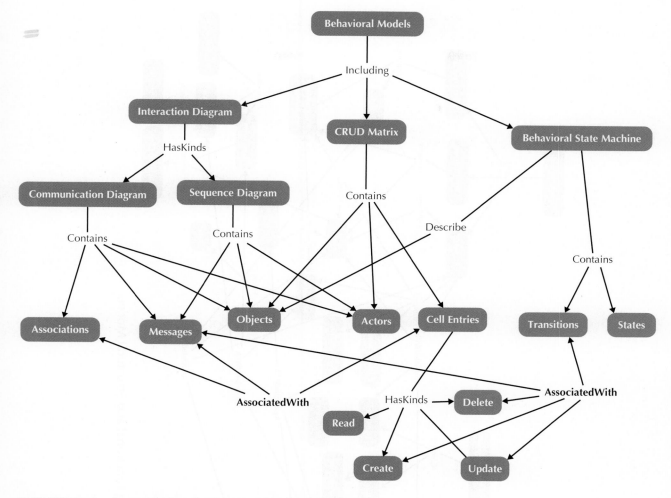

FIGURE 8-4   Interrelationships among Behavioral Models

First, each and every class on a class diagram and each and every CRC card must be associated with at least one use case, and vice versa. For example, the CRC card, portrayed in Figure 6-1, and its related class contained in the class diagram (see Figure 6-2) is associated with the Make Appointment use case described in Figure 5-5.

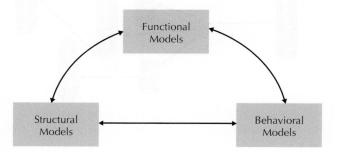

FIGURE 8-5
Object-Oriented
Analysis Models

FIGURE 8-6 Interrelationships among Functional and Structural Models

Second, every activity or action contained in an activity diagram (see Figure 5-2) and every event contained in a use-case description (see Figure 5-5) should be related to one or more responsibilities on a CRC card and one or more operations in a class on a class diagram and vice versa. For example, the Get Patient Information activity on the example activity diagram (see Figure 5-2) and the first two events on the use-case description (see Figure 5-5) are associated with the make appointment responsibility on the CRC card (see Figure 6-1) and the makeAppointment() operation in the Patient class on the class diagram (see Figure 6-2).

Third, every object node on an activity diagram must be associated with an instance of a class on a class diagram, (i.e., an object) and a CRC card, such as the Appt object node (see Figure 5-2) and the Appointment class (see Figure 6-2) or an attribute contained in a class and on a CRC card. However, in Figure 5-2, there is an additional object node, Appt Request Info, that does not seem to be related to any class in the class diagram portrayed in Figure 6-2. Thus, either the activity or class diagram is in error or the object node must represent an attribute. In this case, it does not seem to represent an attribute. As such, we could add a class to the class diagram that creates temporary objects associated with the object node on the activity diagram. However, it is unclear what operations, if any, would be associated with these temporary objects. Therefore, a better solution would be to delete the Appt Request Info object nodes from the activity diagram. In reality, this object node represented only a set of bundled attribute values, that is, data that would be used in the appointment system process (see Figure 8-7).

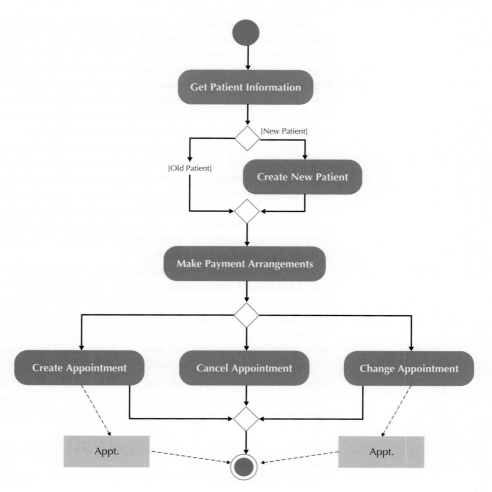

FIGURE 8-7
Updated Appointment
System Activity Diagram

Fourth, every attribute and association/aggregation relationships contained on a CRC card (and connected to a class on a class diagram) should be related to the subject or object of an event in a use-case description. For example, in Figure 5-5, the second event states: The Patient provides the Receptionist with their name and address. By reviewing the CRC card in Figure 6-1 and the class diagram in Figure 6-2, we see that the Patient class is a subclass of the Person class and, hence, inherits all the attributes, associations, and operations defined with the Person class, where name and address attributes are defined.

Balancing Functional and Behavioral Models   As in balancing the functional and structural models, we must ensure the consistency of the two sets of models. In this case, the activity diagrams, use-case descriptions, and use-case diagrams must agree with the sequence diagrams, communication diagrams, behavioral state machines, and CRUD matrix. Figure 8-8 portrays the interrelationships between the functional and behavioral models. Based on these interrelationships, we see that there are four areas with which we must be concerned.[13]

First, the sequence and communication diagrams must be associated with a use case on the use-case diagram and a use-case description. For example, the sequence diagram in Figure 7-1 and the communication diagram in Figure 7-5 are related to scenarios of the Make Appointment use case that appears in the use-case description in Figure 5-5 and the use-case diagram in Figure 8-1.

Second, actors on sequence diagrams, communication diagrams, and/or CRUD matrices must be associated with actors on the use-case diagram or referenced in the use-case description, and vice versa. For example, the aPatient actor in the sequence diagram in Figure 7-1, the communication diagram in Figure 7-5, and the CRUD matrix in Figure 7-14 appears in the use-case diagram in Figure 8-1 and the use-case description of Figure 5-5. However, the aReceptionist does not appear in the use-case diagram but is referenced in the events associated with the Make Appointment use-case description. In this case, the aReceptionist actor is obviously an internal actor, which cannot be portrayed on UML's use-case diagram.

Third, messages on sequence and communication diagrams, transitions on behavioral state machines, and entries in a CRUD matrix must be related to activities and actions on an activity diagram and events listed in a use-case description, and vice versa. For example, the CreateAppt() message on the sequence and communication diagrams (see Figures 7-1 and 7-5) is related to the CreateAppointment activity (see Figure 8-7) and the S-1: New Appointment subflow on the use-case description (see Figure 5-5). Furthermore, the C entry into the Receptionist Appointment cell of the CRUD matrix is also associated with these messages, activity, and subflow.

Fourth, all complex objects represented by an object node in an activity diagram must have a behavioral state machine that represents the object's lifecycle, and vice versa. As stated in Chapter 7, complex objects tend to be very dynamic and pass through a variety of states during their lifetimes. However, the Appointment object node (see Figure 8-7) seems to represent a fairly simple object. It seems only to be created, changed, and canceled. As such, in this case no behavioral state machine would be necessary.

Balancing Structural and Behavioral Models   To discover the interrelationships among the structural and behavioral models, we use the concept map in Figure 8-9. In this case, there are five areas in which we must ensure the consistency among the models.[14]

---

[13] Performing CRUD analysis (see Chapter 7) could also be useful in reviewing the intersections among the functional and behavioral models.

[14] Both role-playing (see Chapter 6) and CRUD analysis (see Chapter 7) also can be very useful in this undertaking.

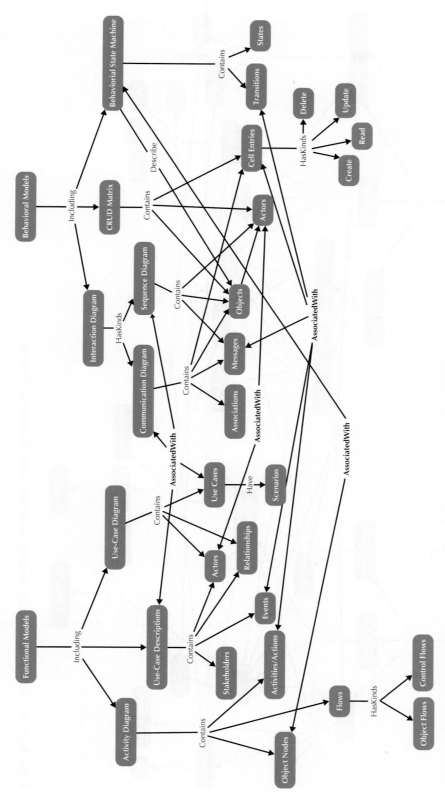

FIGURE 8-8 Interrelationships among Functional and Behavioral Models

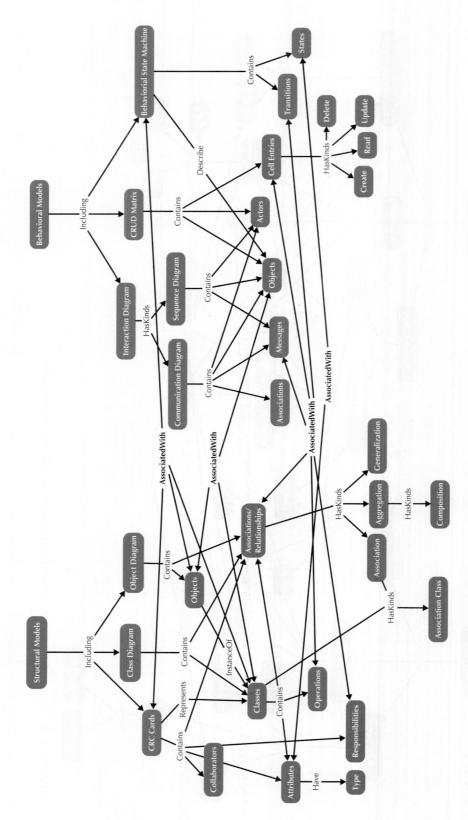

FIGURE 8-9   Interrelationships among Structural and Behavioral Models

First, objects that appear in a CRUD matrix must be associated with classes that are represented by CRC cards and appear on the class diagram, and vice versa. For example, the Patient class in the CRUD matrix in Figure 7-14 is associated with the CRC card in Figure 6-1 and the Patient class in the class diagram in Figure 6-2.

Second, because behavioral state machines represent the life cycle of complex objects, they must be associated with instances (objects) of classes on a class diagram and with a CRC card that represent the class of the instance. For example, the behavior state machine that describes an instance of an Order class in Figure 7-11 implies that an Order class exists on a related class diagram and that a CRC card exists for the related class.

Third, communication and sequence diagrams contain objects that must be an instantiation of a class that is represented by a CRC card and is located on a class diagram. For example, Figure 7-1 and Figure 7-5 have an anAppt object that is an instantiation of an Appointment class. As such, the Appointment class must exist in the class diagram (see Figure 6-2) and a CRC card should exist that describes it. However, there are several objects on the communication and sequence diagrams associated with classes that do not exist on the class diagram. These include a PatientsList class, a BillsList class, and ApptsList class. At this point in time, the analyst must decide to either modify the class diagram by adding these classes or rethink the communication and sequence diagrams. In this specific case, because we are moving on into design, it is better to add the classes to the class diagram (see Figure 8-10).

Fourth, messages contained on the sequence and communication diagrams, transitions on behavioral state machines, and cell entries on a CRUD matrix must be associated with responsibilities and associations on CRC cards and operations in classes and associations connected to the classes on class diagrams. For example, the CreateAppt() message on the sequence and communication diagrams (see Figures 7-1 and 7-5) relate to the makeAppointment operation of the Patient class and the schedules association between the Patient and Appointment classes on the class diagram (see Figure 6-2).

Fifth, the states in a behavioral state machine must be associated with different values of an attribute or set of attributes that describe an object. For example, in Figure 7-12 an instance of the Undergraduate class has different states based on the value of an attribute: HoursEarned.

Summary   Figure 8-11 portrays a concept map that is a complete picture of the interrelationships among the diagrams covered in Chapters 5, 6, and 7. It is obvious from the complexity of this figure that balancing all the functional, structural, and behavioral models is a very time-consuming, tedious, and difficult task. However, without paying this level of attention to the evolving models that represent the system, the models will not provide a sound foundation on which to design and build the system.

EVOLVING THE ANALYSIS MODELS INTO DESIGN MODELS

Now that we have successfully verified and validated our analysis models, we need to begin evolving them into appropriate design models. The purpose of the analysis models was to represent the underlying business problem domain as a set of collaborating objects. In other words, the analysis activities defined the functional requirements. To achieve this, the analysis activities ignored nonfunctional requirements such as performance and the system environment issues (e.g., distributed or centralized processing, user interface issues, and database issues). In contrast, the primary purpose of the design models is to increase the likelihood of successfully delivering a system that implements the functional requirements in a manner that is affordable and easily maintainable. Therefore, in system design, we address both the functional and nonfunctional requirements.

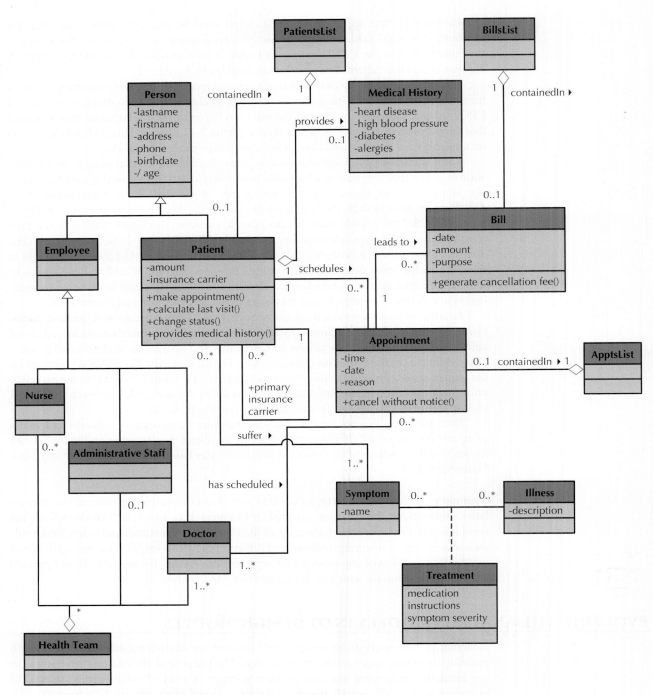

FIGURE 8-10   Updated Appointment System Class Diagram

From an object-oriented perspective, system design models simply refine the system analysis models by adding system environment (or solution domain) details to them and refining the problem domain information already contained in the analysis models. When evolving the analysis model into the design model, you should first carefully review the use cases and the current set of classes (their operations and attributes, and the relationships between them).

FIGURE 8-11 Interrelationships among Object-Oriented Analysis Models

Are all the classes necessary? Are there any missing classes? Are the classes fully defined? Are their any missing attributes or methods? Do the classes have any unnecessary attributes and methods? Is the current representation of the evolving system optimal? Obviously, if we have already verified and validated the analysis models, quite a bit of this has already taken place. Yet, object-oriented systems development is both incremental and iterative. Therefore, we must review the analysis models again. However this time, we begin looking at the models of the problem domain through a design lens. As such, we will make modifications to the problem domain models that will enhance the efficiency and effectiveness of the evolving system.

In the following sections, we introduce factoring, partitions and collaborations, and layers as a way to evolve problem domain-oriented analysis models into optimal solution domain-oriented design models. From an enhanced Unified Process perspective (see Figure 1-11), we are moving from the analysis workflow to the design workflow. Furthermore, we are moving further into the Elaboration phase and partially into the Construction phase.

Factoring

Factoring is the process of separating out a *module* into a stand-alone module in and of itself. The new module can be a new *class* or a new *method*. For example, when reviewing a set of classes, it may be discovered that they have a similar set of attributes and methods. As such, it may make sense to factor out the similarities into a separate class. Depending on whether the new class should be in a superclass relationship to the existing classes or not, the new class can be related to the existing classes through *generalization* (*a-kind-of*) or possibly through *aggregation* (*has-parts*) relationship. For example, using the appointment system example in the previous chapters (see Figure 6-2), if the Employee class had not been identified, we could possibly identify it at this stage by factoring out the similar methods and attributes from the Nurse, Administrative Staff, and Doctor classes. In this case, we would relate the new class (Employee) to the existing classes using the generalization (a-kind-of) relationship. Obviously, by extension we also could have created the Person class if it had not been previously identified.

Abstraction and *refinement* are two processes closely related to factoring. Abstraction deals with the creation of a higher-level idea from a set of ideas. Identifying the Employee class is an example of abstracting from a set of lower classes to a higher one. In some cases, the abstraction process will identify *abstract classes*, whereas in other situations, it will identify additional *concrete classes*.[15] The refinement process is the opposite of the abstraction process. In the appointment system example in the previous chapters (see Figure 6-2), we could identify additional subclasses of the Administrative Staff class, such as Receptionist, Secretary, and Bookkeeper. Of course we would add the new classes only if there were sufficient differences between them. Otherwise, the more general class, Administrative Staff, would suffice.

Partitions and Collaborations

Based on all the factoring, refining, and abstracting that can take place to the evolving system, the sheer size of the system representation can overload both the user and the developer. At this point in the evolution of the system, it may make sense to split the representation into a set of *partitions*. A partition is the object-oriented equivalent of a subsystem,[16] where a subsystem is a decomposition of a larger system into its component systems (e.g., an accounting

---

[15] See Chapter 6 for the differences between abstract and concrete classes.

[16] Some authors refer to partitions as subsystems [e.g., see R. Wirfs-Brock, B. Wilkerson, and L. Weiner, *Designing Object-Oriented Software* (Englewood Cliffs, NJ: Prentice Hall, 1990)], whereas others refer to them as layers [e.g., see I. Graham, *Migrating to Object Technology* (Reading, MA: Addison-Wesley, 1994)]. However, we have chosen to use the term partition [C. Larman, *Applying UML and Patterns: An Introduction to Object-Oriented Analysis and Design* (Englewood Cliffs, NJ: Prentice Hall, 1998)] to minimize confusion between subsystems in a traditional system development approach and layers associated with Rational's Unified Approach.

information system could be functionally decomposed into an accounts payable system, an account receivable system, a payroll system, etc.). From an object-oriented perspective, partitions are based on the pattern of activity (messages sent) among the objects in an object-oriented system. We describe an easy approach to model partitions and collaborations later in this chapter: packages and package diagrams.

A good place to look for potential partitions is the *collaborations* modeled in UML's communication diagrams (see Chapter 7). If you recall, one useful way to identify collaborations is to create a communication diagram for each use case. However, because an individual class may support multiple use cases, an individual class can participate in multiple use-case-based collaborations. In those cases where classes are supporting multiple use cases, the collaborations should be merged together. Furthermore, the class diagram should be reviewed to see how the different classes are related to one another. For example, if attributes of a class have complex object types, such as Person, Address, Department, etc., and these object types were not modeled as associations in the class diagram, we need to recognize these implied associations. As such, creating a diagram that combines the class diagram with the communication diagrams can be very useful to show to what degree the classes are coupled.[17] The greater the coupling between classes, the more likely the classes should be grouped together in a collaboration or partition. Finally, by looking at a CRUD matrix, we can use CRUD analysis (see Chapter 7) to identify potential classes on which to merge collaborations.

Depending on the complexity of the merged collaboration, it may be useful in decomposing the collaboration into multiple partitions. In this case, in addition to having collaborations between objects, it is possible to have collaborations among partitions. The general rule of thumb is the more messages sent between objects, the more likely the objects belong in the same partition. The fewer messages sent, the less likely the two objects belong together.

Another useful approach to identifying potential partitions is to model each collaboration between objects in terms of clients, servers, and contracts. A *client* is an instance of a *class* that sends a *message* to an instance of another class for a *method* to be executed; a *server* is the instance of a class that receives the message; and a *contract* is the specification that formalizes the interactions between the client and server objects (see Chapters 6 and 9). This approach allows the developer to build up potential partitions by looking at the contracts that have been specified between objects. In this case, the more contracts there are between objects, the more likely that the objects belong in the same partition. The fewer contracts, the less chance there is that the two classes do belong in the same partition.

Remember, the primary purpose of identifying collaborations and partitions is to determine which classes should be grouped together in design.

YOUR TURN   8-1 Campus Housing

You have been working with the system for the campus housing service over the previous three chapters. In Chapter 5, you were asked to create a set of use cases (Your Turn 5-5) and to create a use-case diagram (Your Turn 5-6). In Chapter 6, you created a structural model (CRC cards and class diagram) for the same situation (Your Turn 6-2). In Chapter 7, you created a sequence diagram (Your Turn 7-1) and a communication diagram (Your Turn 7-2) for one of the use cases you had identified in Chapter 5. Finally, you also created a behavioral state machine for the apartment class in Chapter 7 (Your Turn 7-3).

Based on the current set of functional, structural, and behavioral models portrayed in these diagrams, apply the abstraction and refinement processes to identify additional abstract and concrete classes that would be useful to include in the evolving system.

---

[17] We describe the concept of coupling in Chapter 9.

Layers	Examples	Relevant Chapters
Foundation	Date, Enumeration	8, 9
Problem Domain	Employee, Customer	5, 6, 7, 8, 9
Data Access and Management	DataInputStream, FileInputStream	9, 10
Human–Computer Interaction	Button, Panel	9, 11
Physical Architecture	ServerSocket, URLConnection	9, 12

FIGURE 8-12
Layers and
Sample Classes

Layers

Until this point in the development of our system, we have focused only on the problem domain; we have totally ignored the system environment (data management, user interface, and physical architecture). To successfully evolve the analysis model of the system into a design model of the system, we must add the system environment information. One useful way to do this, without overloading the developer, is to use *layers*. A layer represents an element of the software architecture of the evolving system. We have focused only on one layer in the evolving software architecture: the problem domain layer. There should be a layer for each of the different elements of the system environment (e.g., data management, user interface, physical architecture). Like partitions and collaborations, layers also can be portrayed using packages and package diagrams (see the following text).

The idea of separating the different elements of the architecture into separate layers can be traced back to the MVC architecture of *Smalltalk*.[18] When Smalltalk was first created,[19] the authors decided to separate the application logic from the logic of the user interface. In this manner, it was possible to easily develop different user interfaces that worked with the same application. To accomplish this, they created the *Model-View-Controller (MVC)* architecture, where *Models* implemented the application logic (problem domain) and *Views* and *Controllers* implemented the logic for the user interface. Views handled the output and Controllers handled the input. Because graphical user interfaces were first developed in the Smalltalk language, the MVC architecture served as the foundation for virtually all graphical user interfaces that have been developed today (including the Mac interfaces, the Windows family, and the various Unix-based GUI environments).

Based on Smalltalk's innovative MVC architecture, many different software layers have been proposed.[20] Based on these proposals, we suggest the following layers on which to base software architecture: foundation, problem domain, data management, human–computer interaction, and physical architecture (see Figure 8-12). Each layer limits the types of classes that can exist on it (e.g., only user interface classes may exist on the human–computer interaction layer). The following provides a short description of each layer.

[18] See S. Lewis, *The Art and Science of Smalltalk: An Introduction to Object-Oriented Programming Using Visual-Works* (Englewood Cliffs, NJ: Prentice Hall, 1995).

[19] Smalltalk was first invented in the early 1970s by a software-development research team at Xerox PARC. It introduced many new ideas into the area of programming languages (e.g., object orientation, windows-based user interfaces, reusable class library, and the idea of a development environment). In many ways, Smalltalk is the father (or mother) of all object-based and object-oriented languages, such as Visual Basic, C++, and Java.

[20] For example, Problem Domain, Human Interaction, Task Management, and Data Management (P. Coad and E. Yourdon, *Object-Oriented Design* [Englewood Cliffs, NJ: Yourdon Press, 1991)]; Domain, Application, and Interface (I. Graham, *Migrating to Object Technology* [Reading, MA: Addison-Wesley, 1994)]; Domain, Service, and Presentation [C. Larman, *Applying UML and Patterns: An Introduction to Object-Oriented Analysis and Design* (Englewood Cliffs, NJ: Prentice Hall, 1998)]; Business, View, and Access [A. Bahrami, *Object-Oriented Systems Development using the Unified Modeling Language* (New York: McGraw-Hill, 1999)]; Application-Specific, Application-General, Middleware, System-Software [I. Jacobson, G. Booch, and J. Rumbaugh, *The Unified Software Development Process* (Reading, MA: Addison-Wesley, 1999)]; and Foundation, Architecture, Business, and Application [M. Page-Jones, *Fundamentals of Object-Oriented Design in UML* (Reading, MA: Addison-Wesley, 2000)].

Foundation   The *foundation layer* is, in many ways, a very uninteresting layer. It contains classes that are necessary for any object-oriented application to exist. They include classes that represent fundamental data types (e.g., integers, real numbers, characters, strings), classes that represent fundamental data structures, sometimes referred to as container classes (e.g., lists, trees, graphs, sets, stacks, queues), and classes that represent useful abstractions, sometimes referred to as utility classes (e.g., date, time, money). Today, the classes found on this layer are typically included with the object-oriented development environments.

Problem Domain   The *problem-domain layer* is what we have focused our attention on up until now. At this stage of the development of our system, we will need to further detail the classes so that it will be possible to implement them in an effective and efficient manner. Many issues need to be addressed when designing classes, no matter on which layer they appear. For example, there are issues related to factoring, cohesion and coupling, connascence, encapsulation, proper use of inheritance and polymorphism, constraints, contract specification, and detailed method design. These issues are discussed in Chapter 9.

Data Management   The *data management layer* addresses the issues involving the persistence of the objects contained in the system. The types of classes that appear in this layer deal with how objects can be stored and retrieved. The classes contained in this layer allow the problem domain classes to be independent of the storage utilized and, hence, increase the portability of the evolving system. Some of the issues related to this layer include choice of the storage format (such as relational, object/relational, and object databases) and storage optimization (such as clustering and indexing). A complete description of all the issues related to the *data management layer* is beyond the scope of this book.[21] However, we do present the fundamentals in Chapter 10.

Human–Computer Interaction   The *human–computer interaction layer* contains classes associated with the View and Controller idea from Smalltalk. The primary purpose of this layer is to keep the specific user interface implementation separate from the problem domain classes. This increases the portability of the evolving system. Typical classes found on this layer include classes that can be used to represent buttons, windows, text fields, scroll bars, check boxes, drop-down lists, and many other classes that represent user interface elements.

   When it comes to designing the user interface for an application, there are many issues that must be addressed. For example, how important is consistency across different user interfaces, what about differing levels of user experience, how is the user expected to be able to navigate through the system, what about help systems and online manuals, what types of input elements should be included (e.g., text box, radio buttons, check boxes, sliders, drop-down list boxes, etc.), and what types of output elements should be included (e.g., text, tables, graphs, etc.)? Like the data management layer, a complete description of all the issues related to human–computer interaction is beyond the scope of this book.[22] However from the user's perspective, the user interface is the system. As such, we present the basic issues in user interface design in Chapter 11.

---

[21] There are many good database design books that are relevant to this layer; see, for example, F. R. McFadden, J. A. Hoffer, Mary B. Prescott, *Modern Database Management,* 4th ed. (Reading, MA: Addison-Wesley, 1998); M. Blaha and W. Premerlani, *Object-Oriented Modeling and Design for Database Applications* (Englewood Cliffs, NJ: Prentice Hall, 1998); and R. J. Muller, *Database Design for Smarties: Using UML for Data Modeling* (San Francisco: Morgan Kaufmann, 1999).

[22] One of the best books on user interface design is B. Schneiderman, *Designing the User Interface: Strategies for Effective Human Computer Interaction*, 3rd ed. (Reading, MA: Addison-Wesley, 1998).

Physical Architecture   The *physical architecture layer* addresses how the software will execute on specific computers and networks. As such, this layer includes classes that deal with communication between the software and the computer's operating system and the network. For example, classes that address how to interact with the various ports on a specific computer are included in this layer. This layer also includes classes that interact with so-called middleware applications, such as the OMG's *CORBA* and Microsoft's DCOM and .NET architectures that handle distributed objects.

Unlike the foundation layer, there are many design issues that must be addressed before choosing the appropriate set of classes for this layer. These design issues include the choice of a computing or network architecture (such as the various client-server architectures), the actual design of a network, hardware and server software specification, global/international issues (such as multilingual requirements), and security issues. Like the data management and human–computer interaction layers, a complete description of all the issues related to the physical architecture is beyond the scope of this book. However, we do present the basic issues in Chapter 12.

PACKAGES AND PACKAGE DIAGRAMS

In UML, collaborations, partitions, and layers can be represented by a higher-level construct: a package.[23] A *package* is a general construct that can be applied to any of the elements in UML models. In Chapter 5, we introduced the idea of packages as a way to group use cases together to make the use-case diagrams easier to read and to keep the models at a reasonable level of complexity. In Chapters 6 and 7, we did the same thing for class and communication diagrams, respectively. In this section we describe a *package diagram*: a diagram composed only of packages. A package diagram is effectively a class diagram that only shows packages.

The symbol for a package is similar to a tabbed folder (see Figure 8-13). Depending on where a package is used, packages can participate in different types of relationships. For example, in a class diagram, packages represent groupings of classes. Therefore, aggregation and association relationships are possible.

In a package diagram, it is useful to depict a new relationship, the *dependency relationship*. A dependency relationship is portrayed by a dashed arrow (see Figure 8-13). A dependency

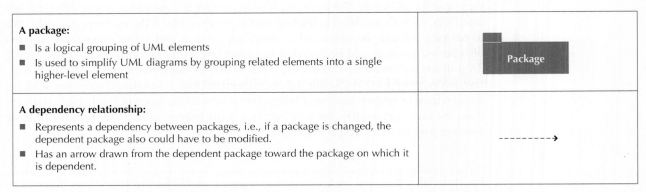

A package:	
■ Is a logical grouping of UML elements ■ Is used to simplify UML diagrams by grouping related elements into a single higher-level element	Package
A dependency relationship:	
■ Represents a dependency between packages, i.e., if a package is changed, the dependent package also could have to be modified. ■ Has an arrow drawn from the dependent package toward the package on which it is dependent.	---------→

FIGURE 8-13   Syntax for Package Diagram

[23] This discussion is based on material in Chapter 7 of M. Fowler with K. Scott, *UML Distilled: A Brief Guide to the Standard Object Modeling Language,* 3rd ed. (Reading, MA: Addison-Wesley, 2004).

FIGURE 8-14

Package Diagram of Dependency Relationships among Layers

relationship represents the fact that a modification dependency exists between two packages. That is, it is possible that a change in one package potentially could cause a change to be required in another package. Figure 8-14 portrays the dependencies among the different layers (foundation, problem domain, data management, human–computer interaction, and physical architecture). For example, if a change occurs in the problem domain layer, it most likely will cause changes to occur in the human–computer interaction, physical architecture, and data management layers. Notice that these layers "point to" the problem domain layer; as such, they are dependent on it. However, the reverse is not true.[24]

At the class level, there could be many causes for dependencies among classes. For example, if the protocol for a method is changed, then this causes the interface for all objects of this class to change. Therefore, all classes that have objects that send messages to the instances of the modified class may have to be modified. Capturing dependency relationships among the classes and packages helps the organization in maintaining object-oriented information systems.

As already stated, collaborations, partitions, and layers are modeled as packages in UML. Furthermore, collaborations are normally factored into a set of partitions, which are typically placed on a layer. In addition, partitions can be composed of other partitions. Also, it is possible to have classes in partitions, which are contained in another partition, which is placed on a layer. All these groupings are represented using packages in UML. Remember that a package is simply a generic, grouping construct used to simplify UML models through the use of composition.[25]

A simple package diagram, based on the appointment system example from the previous chapters, is shown in Figure 8-15. This diagram portrays only a very small portion of the entire system. In this case, we see that the Patient UI, Patient-DAM, and Patient Table

[24] A useful side effect of the dependencies among the layers is that the project manager can divide the project team up into separate subteams: one for each design layer. This is possible because each of the design layers is dependent on the problem domain layer, which has been the focus of analysis. In design, the team can gain some productivity-based efficiency by working on the different layer designs in parallel.

[25] For those familiar with traditional approaches, such as structured analysis and design, packages serve a similar purpose as the leveling and balancing processes used in data flow diagramming.

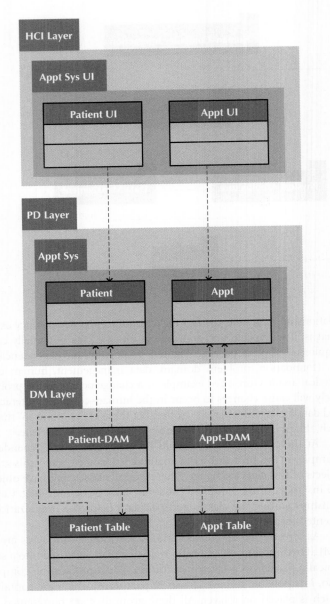

FIGURE 8-15

Partial Package
Diagram of the
Appointment System

classes depend on the Patient class. Furthermore, the Patient-DAM class depends on the Patient Table class. The same can be seen with the classes dealing with the actual appointments. By isolating the Problem Domain classes (such as the Patient and Appt classes) from the actual object persistence classes (such as the Patient Table and Appt Table classes) through the use of the intermediate Data Management classes (Patient-DAM and Appt-DAM classes), we isolate the Problem Domain classes from the actual storage medium.[26] This greatly simplifies the maintenance and increases the reusability of the Problem Domain classes. Of course, in a complete description of a real system, there would be many more dependencies.

[26] These issues are described in more detail in Chapter 10.

FIGURE 8-16

Steps for Identifying
Packages and Building
Package Diagram

1. Set the context.
2. Cluster classes together based on shared relationships.
3. Model clustered classes as a package.
4. Identify dependency relationships among packages.
5. Place dependency relationships between packages.

Identifying Packages and Creating Package Diagrams

In this section, we describe a simple five-step process to create package diagrams (see Figure 8-16). The first step is to set the context for the package diagram. Remember, packages can be used to model partitions and/or layers. Revisiting the appointment system again, let's set the context as the problem domain layer.

The second step is to cluster the classes together into partitions based on the relationships that the classes share. The relationships include generalization, aggregation, the various associations, and the message sending that takes place between the objects in the system. To identify the packages in the appointment system, we should look at the different analysis models [e.g., the class diagram (see Figure 6-2), the communication diagrams (see Figure 7-5)], and the CRUD matrix (see Figure 7-14). Classes in a generalization hierarchy should be kept together in a single partition.

The third step is to place the clustered classes together in a partition and model the partitions as packages. Figure 8-17 portrays five packages: PD Layer, Person Pkg, Patient Pkg, Appt Pkg, and Treatment Pkg.

The fourth step is to identify the dependency relationships among the packages. In this case, we review the relationships that cross the boundaries of the packages to uncover potential dependencies. In the appointment system, we see association relationships that connect the Person Pkg with the Appt Pkg (via the association between the Doctor class and the Appointment class), and the Patient Pkg, which is contained within the Person Pkg, with the Appt Pkg (via the association between the Patient and Appointment classes) and the Treatment Pkg (via the association between the Patient and Symptom classes).

The fifth step is to place the dependency relationships on the evolved package diagram. In the case of the Appointment system, there are dependency relationships between the Person Pkg and the Appt Pkg and the Person Pkg and the Treatment Pkg. To increase the understandability of the dependency relationships among the different packages, a pure package diagram that shows only the dependency relationships among the packages can be created (see Figure 8-18).

Verifying and Validating Package Diagrams

Like all the previous models, package diagrams need to be verified and validated. In this case, the package diagrams were derived primarily from the class diagram, the communications diagrams, and the CRUD matrix. Only two areas need to be reviewed.

YOUR TURN

8-2 Campus Housing

*B*ased on the factoring of the evolving system in Your Turn 8-1, identify a set of partitions for the Problem Domain layer and model them in a package diagram.

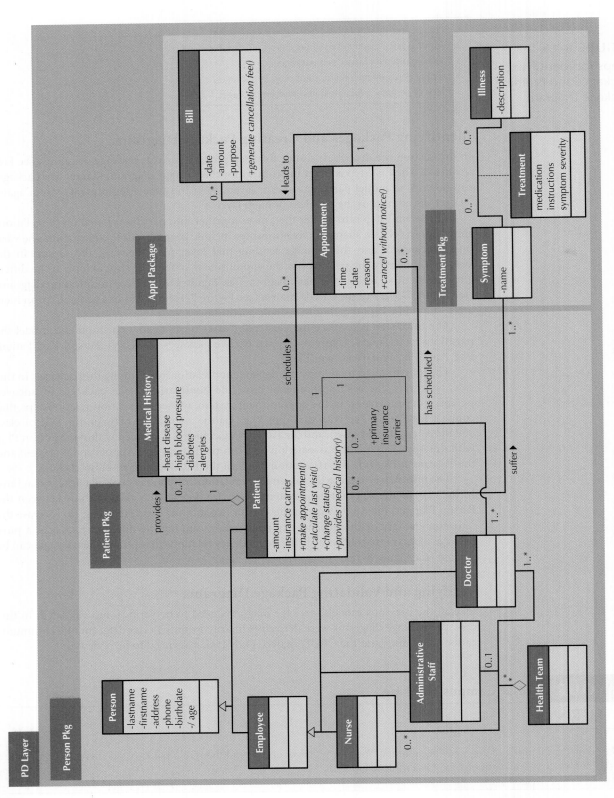

FIGURE 8-17 Package Diagram of the PD Layer for the Appointment System

FIGURE 8-18
Overview Package
Diagram of the PD
Layer for the
Appointment System

First, the identified packages must make sense from a problem domain point of view. For example, in the context of an appointment system, the packages in Figure 8-18 (Person, Patient, Appt, and Treatment) seem to be reasonable.

Second, all dependency relationships must be based on message-sending relationships on the communications diagram, associations on the class diagram, and cell entries in the CRUD matrix. In the case of the appointment system, the identified dependency relationships are reasonable (see Figures 6-2, 7-5, 7-14, and 8-18).

DESIGN STRATEGIES

Until now, we have assumed that the system will be built and implemented by the project team; however, there are actually three ways to approach the creation of a new system: developing a custom application in-house, buying a packaged system and customizing it, and relying on an external vendor, developer, or service provider to build the system. Each of these choices has its strengths and weaknesses, and each is more appropriate in different scenarios. The following sections describe each design choice in turn, and then we present criteria that you can use to select one of the three approaches for your project.

Custom Development

Many project teams assume that *custom development,* or building a new system from scratch, is the best way to create a system. For one thing, teams have complete control over the way the system looks and functions. Furthermore, custom development also allows developers to be flexible and creative in the way they solve business problems. Additionally, a custom application would be easier to change to include components that take advantage of current technologies that can support such strategic efforts.

Building a system in-house also builds technical skills and functional knowledge within the company. As developers work with business users, their understanding of the

CONCEPTS 8-A Building a Custom System—with Some Help

IN ACTION

I worked with a large financial institution in the southeast that suffered serious financial losses several years ago. A new CEO was brought in to change the strategy of the organization to being more "customer focused." The new direction was quite innovative, and it was determined that custom systems, including a data warehouse, would have to be built to support the new strategic efforts. The problem was that the company did not have the in-house skills for these kinds of custom projects.

The company now has one of the most successful data-warehouse implementations because of its willingness to use outside skills and its focus on project management. To supplement skills within the company, eight sets of external consultants, including hardware vendors, system integrators, and business strategists, were hired to take part and transfer critical skills to internal employees. An in-house project manager coordinated the data-warehouse implementation full-time, and her primary goals were to set clearly expectations, define responsibilities, and com-

municate the interdependencies that existed among the team members.

This company shows that successful custom development can be achieved, even when the company may not start off with the right skills in-house. But, this kind of project is not easy to pull off—it takes a talented project manager to keep the project moving along and to transition the skills to the right people over time.

—*Barbara Wixom*

Questions

1. What are the risks in building a custom system without the right technical expertise?
2. Why did the company select a project manager from within the organization?
3. Would it have been better to hire an external professional project manager to coordinate the project?

business grows and they become better able to align IS with strategies and needs. These same developers climb the technology learning curve so that future projects applying similar technology require much less effort.

Custom application development, however, also includes a dedicated effort that involves long hours and hard work. Many companies have a development staff that already is overcommitted to filling huge backlogs of systems requests, and they just do not have time for another project. Also, a variety of skills—technical, interpersonal, functional, project management, and modeling—all have to be in place for the project to move ahead smoothly. IS professionals, especially highly skilled individuals, are quite difficult to hire and retain.

The risks associated with building a system from the ground up can be quite high, and there is no guarantee that the project will succeed. Developers could be pulled away to work on other projects, technical obstacles could cause unexpected delays, and the business users could become impatient with a growing timeline.

Packaged Software

Many business needs are not unique, and because it makes little sense to reinvent the wheel, many organizations buy *packaged software* that has already been written rather than developing their own custom solution. In fact, there are thousands of commercially available software programs that have already been written to serve a multitude of purposes. Think about your own need for a word processor—did you ever consider writing your own word processing software? That would be very silly considering the number of good software packages available that are relatively inexpensive.

Similarly, most companies have needs that can be met quite well by packaged software, such as payroll or accounts receivable. It can be much more efficient to buy programs that have already been created, tested, and proven, and a packaged system can be bought and installed in a relatively short period of time, when compared with a custom system. Plus, packaged systems incorporate the expertise and experience of the vendor who created the software.

Packaged software can range from reusable components (such as ActiveX and Javabeans) to small, single-function tools (such as the shopping cart program) to huge, all-encompassing systems such as *enterprise resource planning (ERP)* applications that are installed to automate an entire business. Implementing ERP systems is a process in which large organizations spend millions of dollars installing packages by companies such as SAP, PeopleSoft, Oracle, and Baan and then change their businesses accordingly. Installing ERP software is much more difficult than installing small application packages because benefits can be harder to realize and problems are much more serious.

One problem is that companies buying packaged systems must accept the functionality that is provided by the system, and rarely is there a perfect fit. If the packaged system is large in scope, its implementation could mean a substantial change in the way the company does business. Letting technology drive the business can be a dangerous way to go.

Most packaged applications allow for *customization,* or the manipulation of system parameters to change the way certain features work. For example, the package might have a way to accept information about your company or the company logo that would then appear on input screens. Or, an accounting software package could offer a choice of various ways to handle cash flow or inventory control so that it can support the accounting practices in different organizations. If the amount of customization is not enough and the software package has a few features that don't quite work the way the company needs it to work, the project team can create workarounds.

A *workaround* is a custom-built add-on program that interfaces with the packaged application to handle special needs. It can be a nice way to create needed functionality that does not exist in the software package. But, workarounds should be a last resort for several reasons. First, workarounds are not supported by the vendor who supplied the packaged software, so upgrades to the main system may make the workaround ineffective. Also, if problems arise, vendors have a tendency to blame the workaround as the culprit and refuse to provide support.

Although choosing a packaged software system is simpler than custom development, it too can benefit from following a formal methodology, just as if a custom application were being built.

Systems integration refers to the process of building new systems by combining packaged software, existing legacy systems, and new software written to integrate these together. Many consulting firms specialize in systems integration, so it is not uncommon for companies to select the packaged software option and then outsource the integration of a variety of packages to a consulting firm. (Outsourcing is discussed in the next section).

The key challenge in systems integration is finding ways to integrate the data produced by the different packages and legacy systems. Integration often hinges on taking data produced by one package or system and reformatting it for use in another package or system. The project team starts by examining the data produced by and needed by the different packages or systems and identifying the transformations that must occur to move the data from one to the other. In many cases, this involves "fooling" the different packages or systems into thinking that the data were produced by an existing program module that the package or system expects to produce the data rather than the new package/system that is being integrated.

CONCEPTS

IN ACTION

8-B Collecting Data for the Long Term

*P*harmaceutical companies are generally heavily regulated companies. It may take years for a new drug to make it to the marketplace because of time for the development phase, highly monitored testing, and final approval by the Food and Drug Administration (FDA). Once in the marketplace, other companies can try to produce generic drugs that seem to be compatible with the name-brand drug.

Occasionally an approved drug, some years into its life span, gets scrutiny for higher-than-expected side effects. For example, a drug that is effective in lowering cholesterol may also cause the side effect of an increased chance for cataract growth that was not discovered during the initial testing and approval cycle. Data are collected on all aspects of clinical trials and from the marketplace, but some relationships are just harder to find.

Questions

1. Is there a particular system approach to being able to collect and analyze data from a mountain of data?
2. If you were building a strategic planning system for tracking a drug from proposal through development and testing and into the marketplace, how would you approach it?
3. What requirements might be necessary to build such a system?

A third approach is through the use of an *object wrapper.*[27] An object wrapper is essentially an object that "wraps around" a legacy system, enabling an object-oriented system to send messages to the legacy system. Effectively, object wrappers create an application program interface (API) to the legacy system. The creation of an object wrapper allows the protection of the corporation's investment in the legacy system.

Outsourcing

The design choice that requires the least amount of in-house resources is *outsourcing*—hiring an external vendor, developer, or service provider to create the system. Outsourcing has become quite popular in recent years. Some estimate that as many as 50 percent of companies with IT budgets of more than $5 million are currently outsourcing or evaluating the approach.

According to Wikipedia, "outsourcing involves transferring or sharing management control and/or decision-making of a business function to an outside supplier, which involves a degree of two-way information exchange, coordination and trust between the outsourcer and its client." From an IT perspective, IT outsourcing can include things such as (1) hiring consultants to solve a specific problem, (2) hiring contract programmers to implement a solution, (3) hiring a firm to manage the IT function and assets of a company, or (4) actually outsourcing the entire IT function to a separate firm. Today, through the use of Application Service Providers (ASPs) and Web Services technology, it is possible to use a pay-as-you-go approach for a software package.[28] Essentially, IT outsourcing involves hiring a third party to perform some IT function that traditionally would be performed in-house.

There can be great benefit to having someone else develop a company's system. They may be more experienced in the technology or have more resources, such as experienced programmers. Many companies embark upon outsourcing deals to reduce costs, whereas others see it as an opportunity to add value to the business.

---

[27] Ian Graham, *Object-Oriented Methods: Principles & Practice*, 3rd ed. (Reading, MA: Addison-Wesley, 2001).

[28] For an economic explanation of how this could work, see H. Baetjer, *Software as Capital: An Economic Perspective on Software Engineering* (Los Alamitos, CA: IEEE Computer Society Press, 1997).

For whatever reason, outsourcing can be a good alternative for a new system. However, it does not come without costs. If you decide to leave the creation of a new system in the hands of someone else, you could compromise confidential information or lose control over future development. In-house professionals are not benefiting from the skills that could be learned from the project; instead the expertise is transferred to the outside organization. Ultimately, important skills can walk right out the door at the end of the contract. Furthermore, when offshore outsourcing is being considered, we must also be cognizant of language issues, time-zone differences, and cultural differences (for example, acceptable business practices as understood in one country that may be unacceptable in another). All these concerns, if not dealt with properly, can prevail over any advantage that outsourcing or offshore outsourcing could realize.

However, most risks can be addressed if a company decides to outsource, but two are particularly important. First, the company must thoroughly assess the requirements for the project—a company should never outsource what is not understood. If rigorous planning and analysis has occurred, then the company should be well aware of its needs. Second, the company should carefully choose a vendor, developer, or service with a proven track record with the type of system and technology that its system needs.

Three primary types of contracts can be drawn to control the outsourcing deal. A *time-and-arrangements contract* is very flexible because a company agrees to pay for whatever time and expenses are needed to get the job done. Of course, this agreement could result in a large bill that exceeds initial estimates. This works best when the company and the outsourcer are unclear about what it is going to take to finish the job.

A company will pay no more than expected with a *fixed-price contract* because if the outsourcer exceeds the agreed-upon price, it will have to absorb the costs. Outsourcers are much more careful about defining requirements clearly up front, and there is little flexibility for change.

The type of contract gaining in popularity is the *value-added contract,* whereby the outsourcer reaps some percentage of the completed system's benefits. The company has very little risk in this case, but it must expect to share the wealth once the system is in place.

Creating fair contracts is an art because flexibility must be carefully balanced with clearly defined terms. Often needs change over time. As such, the contract should not be so specific and rigid that alterations can't be made. Think about how quickly technology like the World Wide Web changes. It is difficult to foresee how a project may evolve over a long period of time. Short-term contracts help leave room for reassessment if needs change or if relationships are not working out the way both parties expected. In all cases, the relationship with the outsourcer should be viewed as a partnership where both parties benefit and communicate openly.

Managing the outsourcing relationship is a full-time job. Thus, someone needs to be assigned full-time to manage the outsourcer, and the level of that person should be appropriate for the size of the job (a multimillion dollar outsourcing engagement should be handled by a high-level executive). Throughout the relationship, progress should be tracked and measured against predetermined goals. If a company does embark upon an outsourcing design strategy, it should be sure to get adequate information. Many books have been written that provide much more detailed information on the topic.[29] Figure 8-19 summarizes some guidelines for outsourcing.

[29] For more information on outsourcing, we recommend M. Lacity and R. Hirschheim, *Information Systems Outsourcing: Myths, Metaphors, and Realities* (New York, NY: Wiley, 1993); L. Willcocks and G. Fitzgerald, *A Business Guide to Outsourcing Information Technology* (London: Business Intelligence, 1994); E. Carmel, *Offshoring Information Technology: Sourcing and Outsourcing to a Global Workforce* (Cambridge, UK: Cambridge University Press, 2005); J. K Halvey and B. M. Melby, *Information Technology Outsourcing Transactions: Process, Strategies, and Contracts,* 2nd Ed. (Hoboken, NJ: Wiley, 2005); and T. L. Friedman, *The World is Flat: A Brief History of the Twenty-First Century, Updated and Expanded Edition* (New York: Farrar, Straus, and Giroux, 2006).

FIGURE 8-19
Outsourcing
Guidelines

Outsourcing Guidelines
• Keep the lines of communication open between you and your outsourcer.
• Define and stabilize requirements before signing a contact.
• View the outsourcing relationship as a partnership.
• Select the vendor, developer or service provider carefully.
• Assign a person to managing the relationship.
• Don't outsource what you don't understand.
• Emphasize flexible requirements, long-term relationships and short-term contracts.

CONCEPTS

IN ACTION

8-C EDS Value-Added Contract

*V*alue-added contracts can be quite rare—and very dramatic. They exist when a vendor is paid a percentage of revenue generated by the new system, which reduces the up-front fee, sometimes to zero. The City of Chicago and EDS (a large consulting and systems integration firm) agreed to reengineer the process by which the city collects the fines on 3.6 million parking tickets per year and thus signed a landmark deal of this type three years ago. At the time, because of clogged courts and administrative problems, the city collected on only about 25 percent of all tickets issued. It had a $60 million backlog of uncollected tickets.

Dallas-based EDS invested an estimated $25 million in consulting and new systems in exchange for the right to up to 26 percent of the uncollected fines, a base processing fee for new tickets, and software rights. To date, EDS has taken in well over $50 million on the deal, analysts say. The deal has come under some fire from various quarters as an example of an organization giving away too much in a risk-reward-sharing deal. City officials, however, counter that the city has pulled in about $45 million in previously uncollected fines and has improved its collection rate to 65 percent with little up-front investment.

Source: Jeff Moad. "Outsourcing? Go out on the limb together." pp. 58–61, Vol. 41, No. 2. *Datamation*, February 1, 1995.

Question

Do you think the City of Chicago got a good deal from this arrangement? Why or why not?

Selecting a Design Strategy

Each of the design strategies just discussed has its strengths and weaknesses, and no one strategy is inherently better than the others. Thus, it is important to understand the strengths and weaknesses of each strategy and when to use each. Figure 8-20 summarizes the characteristics of each strategy.

Business Need   If the business need for the system is common and technical solutions already exist in the marketplace that can meet the business need of the system, it makes little sense to build a custom application. Packaged systems are good alternatives for common business needs. A custom alternative must to be explored when the business need is unique or has special requirements. Usually, if the business need is not critical to the company, then outsourcing is the best choice—someone outside of the organization can be responsible for the application development.

In-house Experience   If in-house experience exists for all the functional and technical needs of the system, it will be easier to build a custom application than if these skills do not exist. A packaged system may be a better alternative for companies that do not have the

	Use Custom Development When...	Use a Packaged System When...	Use Outsourcing When...
Business Need	The business need is unique.	The business need is common.	The business need is not core to the business.
In-house Experience	In-house functional and technical experience exists.	In-house functional experience exists.	In-house functional or technical experience does not exist.
Project Skills	There is a desire to build in-house skills.	The skills are not strategic.	The decision to outsource is a strategic decision.
Project Management	The project has a highly skilled project manager and a proven methodology.	The project has a project manager who can coordinate vendor's efforts.	The project has a highly skilled project manager at the level of the organization that matches the scope of the outsourcing deal.
Time frame	The time frame is flexible.	The time frame is short.	The time frame is short or flexible.

FIGURE 8-20   Selecting a Design Strategy

technical skills to build the desired system. For example, a project team that does not have Web commerce technology skills may want to acquire a Web commerce package that can be installed without many changes. Outsourcing is a good way to bring in outside experience that is missing in-house so that skilled people are in charge of building a system.

Project Skills   The skills that are applied during projects are either technical (e.g., Java, SQL) or functional (e.g., electronic commerce), and different design alternatives are more viable, depending on how important the skills are to the company's strategy. For example, if certain functional and technical expertise that relate to Internet sales applications and Web commerce application development are important to an organization because it expects the Internet to play an important role in its sales over time, then it makes sense for the company to develop Web commerce applications in-house, using company employees, so that the skills can be developed and improved. On the other hand, some skills, such as network security, may be either beyond the technical expertise of employees or not of interest to the company's strategists—it is just an operational issue that needs to be addressed. In this case, packaged systems or outsourcing should be considered so internal employees can focus on other business-critical applications and skills.

Project Management   Custom applications require excellent project management and a proven methodology. There are so many things, such as funding obstacles, staffing holdups, and overly demanding business users, that can push a project offtrack. Therefore, the project team should choose to develop a custom application only if it is certain that the underlying coordination and control mechanisms will be in place. Packaged and outsourcing alternatives also need to be managed; however, they are more shielded from internal obstacles because the external parties have their own objectives and priorities (e.g., it may be easier for an outside contractor to say no to a user than it is for a person within the company). The latter alternatives typically have their own methodologies, which can benefit companies that do not have an appropriate methodology to use.

Time Frame   When time is a factor, the project team should probably start looking for a system that is already built and tested. In this way, the company will have a good idea of how long the package will take to put in place and what the final result will contain. The time frame for custom applications is hard to pin down, especially when you consider how

YOUR	8-3 Choose a Design Strategy
TURN	

Suppose that your university is interested in creating a new course registration system that can support Web-based registration. What should the university consider when determining whether to invest in a custom, packaged, or outsourced system solution?

many projects end up missing important deadlines. If a company must choose the custom development alternative and the time frame is very short, it should consider using techniques such as timeboxing to manage this problem. The time to produce a system using outsourcing really depends on the system and the outsourcer's resources. If a service provider has services in place that can be used to support the company's needs, then a business need could be implemented quickly. Otherwise, an outsourcing solution could take as long as a custom development initiative.

DEVELOPING THE ACTUAL DESIGN

Once the project team has a good understanding of how well each design strategy fits with the project's needs, it must begin to understand exactly *how* to implement these strategies. For example, what tools and technology would be used if a custom alternative were selected? What vendors make packaged systems that address the project needs? What service providers would be able to build this system if the application were outsourced? This information can be obtained from people working in the IS department and from recommendations by business users. Or, the project team can contact other companies with similar needs and investigate the types of systems that they have put in place. Vendors and consultants usually are willing to provide information about various tools and solutions in the form of brochures, product demonstrations, and information seminars. However, a company should be sure to validate the information it receives from vendors and consultants. After all, they are trying to make a sale. Therefore, they may stretch the capabilities of their tool by focusing on only the positive aspects of the tool while omitting the tool's drawbacks.

It is likely that the project team will identify several ways that a system could be constructed after weighing the specific design options. For example, the project team may have found three vendors that make packaged systems that potentially could meet the project's needs. Or, the team may be debating over whether to develop a system using Java as a development tool and the database management system from Oracle or to outsource the development effort to a consulting firm such as Accenture or American Management Systems. Each alternative will have pros and cons associated with it that need to be considered, and only one solution can be selected in the end.

Alternative Matrix

An *alternative matrix* can be used to organize the pros and cons of the design alternatives so that the best solution will be chosen in the end. This matrix is created using the same steps as the feasibility analysis, which was presented in Chapter 2. The only difference is that the alternative matrix combines several feasibility analyses into one matrix so that the alternatives can easily be compared. An alternative matrix is a grid that contains the technical, budget, and organizational feasibilities for each system candidate, pros and cons associated with adopting each solution, and other information that is helpful when making comparisons. Sometimes weights are provided for different parts of the matrix to show when some criteria are more important to the final decision.

To create the alternative matrix, draw a grid with the alternatives across the top and different criteria (e.g., feasibilities, pros, cons, and other miscellaneous criteria) along the side. Next, fill in the grid with detailed descriptions about each alternative. This becomes a useful document for discussion because it clearly presents the alternatives being reviewed and comparable characteristics for each one.

Suppose that a company is thinking about implementing a packaged financial system such as Oracle E-Business Financials or Microsoft's Dynamics GP, but there is not enough expertise in-house to be able to create a thorough alternative matrix. This situation is quite common—often the alternatives for a project are unfamiliar to the project team, so outside expertise is needed to provide information about the alternatives' criteria.

One helpful tool is the *request for proposals (RFP)*. An RFP is a document that solicits proposals to provide the alternative solutions from a vendor, developer, or service provider. Basically, an RFP explains the system that a company is trying to build and the criteria that it will use to select a system. Vendors then respond by describing what it would mean for them to be a part of the solution. They communicate the time, the cost, and exactly how their product or services will address the needs of the project.

There is no formal way to write an RFP, but it should include basic information such as the description of the desired system, any special technical needs or circumstances, evaluation criteria, instructions for how to respond, and the desired schedule. An RFP can be a very large document (i.e., hundreds of pages) because companies try to include as much detail as possible about their needs so that the respondent can be just as detailed in the solution that would be provided. Thus, RFPs typically are used for large projects rather than small ones because they take a lot of time to create; even more time and effort are needed for vendors, developers, and service providers to develop high-quality responses— only a project with a fairly large price tag would be worth the time and cost to develop a response for an RFP.

A less-effort-intensive tool is a *request for information (RFI)* that includes the same format as the RFP. The RFI is shorter and contains less-detailed information about a company's needs, and it requires general information from respondents that communicates the basic services that they can provide.

The final step, of course, is to decide which solution to design and implement. The decision should be made by a combination of business users and technical professionals after the issues involved with the different alternatives are well understood. Once the decision is finalized, design can continue as needed, based on the selected alternative.

YOUR TURN

8-4 Alternative Matrix

Suppose you have been assigned the task of selecting a CASE tool for your class to use for a semester project. Using the Web or other reference resources, select three CASE tools (e.g., ArgoUML, IBM's Rational Rose, or Visual Paradigm). Create an alternative matrix that can be used to compare the three software products so that a selection decision can be made.

APPLYING THE CONCEPTS AT CD SELECTIONS

In the previous installments of the CD Selections Case, the CD Selections Internet Sales System has been described:

- In Chapter 4, the functional and nonfunctional requirements for the system were given (see Figure 4-15).
- In Chapter 5, the functional model was given (see Figures 5-14 through 5-19).
- In Chapter 6, the structural model was given (see Figures 6-15 through 6-18).
- In Chapter 7 the behavioral models were developed for the Places Order use case and the Order class (Figures 7-15 through 7-19).

In this section of the case, we see how Alec and his team prepared to move from an analysis, or problem-domain, orientation to a design, or solution-domain, orientation. Following, we will see that to get ready for this transition, Alec and his team first create a package diagram to partition the problem-domain layer. Next, they go through a verification and validation of all of the analysis models; finally, they choose a design strategy to develop the actual design. As in the previous installments of the case, we deal only with the Places Order use case. However, object-oriented systems development is holistic. Therefore, to be complete, Alec and company had to finish all the previous Your Turns associated with the case.

Packages and Package Diagrams

At this point in the development of the Internet Sales System for CD Selections, Alec wanted to explicitly partition the evolving system. To do this, Alec decided to use packages to represent both the layers and partitions in each layer. Once he made this decision, he chose to follow the steps for identifying packages and building package diagrams in Figure 8-9. Because, at this point in the development, the team has been focusing only on analysis models, Alec decided that the team should concentrate on identifying only potential partitions on the Problem Domain Layer.

The second step, cluster classes together, was accomplished by reviewing the relationships among the different classes (see Figures 6-18, 7-17, and 7-20). Through this review process, the team saw that there were generalization, aggregation, various associations, and message-sending relationships. They also saw the entries in the CRUD matrix. Because they understood that classes in a generalization hierarchy should be kept together, they clustered the Customer, Individual, and Organization classes together to form a partition. Brian pointed out that it is also preferred to keep classes together that participate in aggregation relationships. Based on aggregation relationships, they clustered the Mkt Info, Review, Artist Info, and Sample Clip classes together in a partition. Based on the association relationship and the message-sending pattern of activity between the CD and Mkt Info classes, Anne suggested that they should be in the same partition. Furthermore, because the Vendor class was related only to the CD class, Alec suggested that they be placed in the same partition. Finally, the development team decided to place the Order and Order Item classes together and the Search Req and CD List classes together in their own partitions.

The third step was to model each of these partitions as packages. Figure 8-21 shows the classes being contained in their respective packages. Observe that the Credit-Card Clearance Center is not currently contained in any package.

Next, Alec quickly identified four associations among the different packages: the Customer Package and the Order Package, the Customer Package and the Search Package, the Order Package and the CD Package, and the Search Package and the CD Package. He also identified an association between the Credit-Card Clearance Center class and the Customer Package. Based on these associations, five dependency relationships were identified.

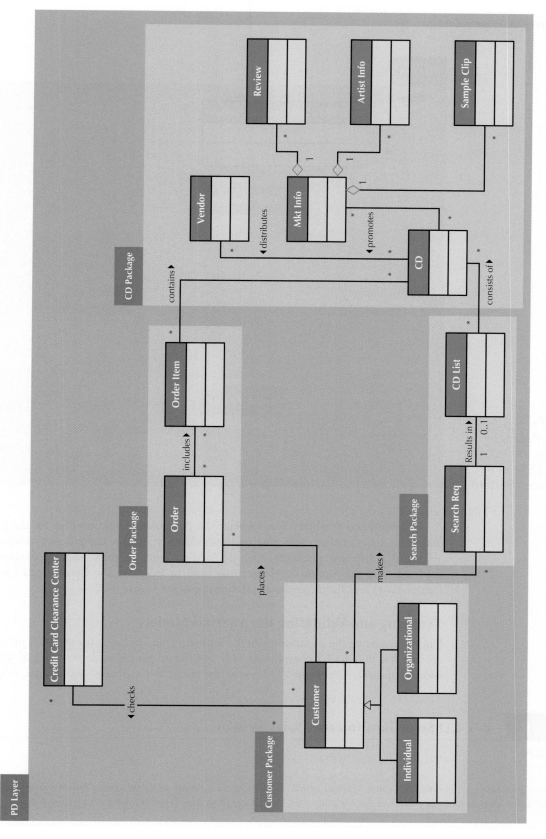

FIGURE 8-21 Package Diagram of the PD of CD Selections Internet Sales System

FIGURE 8-22
Overview Package
Diagram of the PD
Layer of CD Selections
Internet Sales System

The fifth and final step was to place the dependency relationships on the package diagram. Again, to increase the understandability of the dependency relationships among the different packages, Alec decided to create a pure package diagram that depicted only the highest-level packages (and in this case the Credit-Card Clearance Center class) and the dependency relationships (see Figure 8-22).

Verifying and Validating the Analysis Models

Upon completing the partitioning of the Problem Domain Layer, the team felt pretty good about what they had accomplished. However, based on his understanding of what was coming up next, Alec wanted to be sure that the analysis models—functional, structural,

YOUR TURN   8-5 CD Selections Internet Sales System

In previous Your Turns, you completed the functional, structural, and behavioral models for the CD Selections Internet Sales System. Based on these results, create more complete package diagrams based on those created by Alec.

| YOUR | 8-6 CD Selections Internet Sales System |
| TURN | |

At this point in time, you should go back through all of   using a walkthrough process.
the analysis models created and verifiy and validate them

behavioral—and the partitions made sense, so he decided that all the work to date needed to go through a verification and validation step. To say the least, the team was not all that excited about this. In fact, Brian pointed out that the team had been verifying and validating everything as they went along. As such, he argued that this would simply be a waste of time. But, Alec prevailed. He explained that in past projects, when they had not assured the quality of the Problem Domain Layer, teams had run into significant problems. These problems included the system not solving the right problem, significant cost overruns, and the system not being delivered on time. Because this team was relatively inexperienced with the technology that they were about to use, Alec told the team that there was not enough slack in the workplan to run into problems related to the analysis models. He suggested that the team perform a walkthrough with the model and that they ensure all the relationships among the diagrams were fully tested (see Figures 8-2, 8-3, 8-4, 8-6, 8-8, 8-9, and 8-10).

The good news was that the team had actually been very careful in developing the analysis models and they did not uncover any errors within the diagrams. Brian brought up his earlier point in an "I told you so" manner. However, Alec let him blow off a little steam and simply reminded the team that it was better to have done the verification and validation step now and not have to be sorry later. He pointed out that the other layers are mostly dependent on the Problem Domain Layer (see Figure 8-14) and any mistake not caught now could be very costly to catch later.

Developing the Actual Design

Once the team had verified and validated the analysis models, Alec had to decide on a design strategy. As he saw it, he had three different approaches that he could take with the new system: He could develop the entire system using development resources from CD Selections, he could buy a commercial Internet sales packaged software program (or a set of different packages and integrate them), or he could hire a consulting firm or service provider to create the system. Immediately, Alec ruled out the third option. Building Internet applications, especially sales systems, was important to the CD Selections' business strategy. By outsourcing the Internet Sales System, CD Selections would not develop Internet application development skills and business skills within the organization.

Instead, Alec decided that a custom development project using the company's standard Web development tools would be the best choice for CD Selections. In this way, the company would be developing critical technical and business skills in-house, and the project team would be able to have a high level of flexibility and control over the final product. Also, Alec wanted the new Internet Sales System to directly interface with the existing distribution system, and there was a chance that a packaged solution would not be able to integrate as well into the CD Selections environment.

There was one part of the project that potentially could be handled using packaged software: the shopping cart portion of the application. Alec realized that a multitude of

	Alternative 1: Shop-With-Me	Alternative 2: WebShop	Alternative 3: Shop-N-Go
Technical Feasibility	• Developed using C: very little C experience in-house • Orders sent to company using email files	• Developed using C and JAVA: would like to develop in-house JAVA skills • Flexible export features for passing order information to other systems	• Developed using JAVA: would like to develop in-house JAVA skills • Orders saved to a number of file formats
Economic Feasibility	• $150 initial charge	• $700 up front charge, no yearly fees	• $200/year
Organizational Feasibility	• Program used by other retail music companies	• Program used by other retail music companies	• Brand new application: few companies have experience with Shop-N-Go to date
Other Benefits	• Very simple to use	• Tom in IS support has had limited, but positive experience with this program • Easy to customize	
Other Limitations			• The interface is not easily customized

FIGURE 8-23   Alternative Matrix for Shopping Cart Program

programs had been written and were available (at low prices) to handle a customer's order transaction over the Web. These programs allow customers to select items for an order form, input credit-card and billing information, and finalize the order transaction. Alec believed that the project team should at least consider some of these packaged alternatives so that less time had to be spent writing a program that handled basic Web tasks; then more time could be devoted to innovative marketing ideas and custom interfaces with the distribution system.

To help better understand some of the shopping cart programs that were available in the market and how their adoption could benefit the project, Alec created an alternative matrix that compared three different shopping cart programs with each other (see Figure 8-23). Although all three alternatives had positive points, Alec saw Alternative B (Web-Shop) as the best alternative for handling the shopping cart functionality for the new Internet sales system. WebShop was written in JAVA, the tool that CD Selections selected as its standard Web development language; the expense was reasonable, with no hidden or recurring costs; and there was a person in-house who had some positive experience with the program. Alec made a note to look into acquiring WebShop as the shopping cart program for the Internet sales system.

SUMMARY

Design contains many steps that guide the project team through planning out exactly how a system needs to be constructed. The requirements that were identified and the models that were created during analysis serve as the primary inputs for the design activities. In object-oriented design, the primary activity is to evolve the analysis models into design models by optimizing the problem-domain information already contained in the analysis models and adding system environment details to them.

Verifying and Validating the Analysis Models

Before actually adding system environment details to the analysis models, the various representations need to be verified and validated. One very useful approach to test the fidelity of the representations is to perform a walkthrough in which developers walk through the representations by presenting the different models to members of the analysis team, members of the design team, and representatives of the client. The walkthrough must validate each model to be sure that the different representations within the model all agree with each other; for example, for the functional model, activity diagrams, use-case descriptions, and use-case diagrams must be consistent with each other. Furthermore, the different models (functional, structural, and behavioral) must be consistent. Finally, care must be taken during the walkthroughs to ensure that the presenter is not simply degraded and destroyed.

Evolving the Analysis Models into Design Models

When evolving the analysis models into design models, the analysis models: activity diagrams, use-case descriptions, use-case diagrams, CRC cards, class and object diagrams, sequence diagrams, communication diagrams, and behavioral state machines should first be carefully reviewed. During this review, factoring, refinement, and abstraction processes can be used to polish the current models. During this polishing, it is possible that the analysis models may become overly complex. If this occurs, then the models should be partitioned based on the interactivity (message sending) and relationships (generalization, aggregation, and association) shared among the classes. The more a class has in common with another class (i.e., the more relationships shared), the more likely they belong in the same partition.

The second thing to do to evolve the analysis mode is to add the system environment (physical architecture, user interface, and data access and management) information to the problem domain information already contained in the model. To accomplish this and to control the complexity of the models, layers are used. A layer represents an element of the software architecture of the system. We recommend five different layers to be used: foundation, physical architecture, human–computer interaction, data access and management, and problem domain. Each layer supports only certain types of classes (e.g., database manipulation classes would be allowed only on the data access and management layer).

Packages and Package Diagrams

A package is a general UML construct used to represent collaborations, partitions, and layers. Its primary purpose is to support the logical grouping of other UML constructs together (e.g., use cases and classes by the developer and user to simplify and increase the understandability of a UML diagram). There are instances in which a diagram that contains only packages is useful. A package diagram contains packages and dependency relationships. A dependency relationship represents the possibility of a modification dependency existing between two packages (i.e., changes in one package could cause changes in the dependent package).

Identifying packages and creating a package diagram is accomplished using a five-step process. The five steps can be summed up as setting the context, clustering similar classes, placing the clustered classes into a package, identifying dependency relationships among the packages, and placing the dependency relationship on the package diagram.

Design Strategies

During the design phase, the project team also needs to consider three approaches to creating the new system, including developing a custom application in-house, buying a packaged system and customizing it, and relying on an external vendor, developer, or system provider to build and/or support the system.

Custom development allows developers to be flexible and creative in the way they solve business problems, and it builds technical and functional knowledge within the organization. But, many companies have a development staff that is already overcommitted to filling huge backlogs of systems requests, and they just don't have time to devote to a project where a system is built from scratch. It can be much more efficient to buy programs that have been created, tested, and proven, and a packaged system can be bought and installed in a relatively short period of time as compared with a custom solution. Workarounds can be used to meet the needs that are not addressed by the packaged application.

The third design strategy is to outsource the project and pay an external vendor, developer, or service provider to create the system. It can be a good alternative to approaching the new system; however, it does not come without costs. However, if a company decides to leave the creation of a new system in the hands of someone else, the organization could compromise confidential information or lose control over future development.

Each of the design strategies discussed here has its strengths and weaknesses, and no one strategy is inherently better than the others. Thus, it is important to consider such issues as the uniqueness of business need for the system, the amount of in-house experience that is available to build the system, and the importance of the project skills to the company. Also, the existence of good project management and the amount of time available to develop the application play a role in the selection process.

Developing the Actual Design

Ultimately, the decision must be made regarding the specific type of system that needs to be designed. An alternative matrix can help make this decision by presenting feasibility information for several candidate solutions in a way in which they can be compared easily. Both the Request for Proposal and Request for Information are two ways to gather accurate information regarding the alternatives.

KEY TERMS

A-kind-of
Abstract classes
Abstraction
Aggregation
Alternative matrix
Balancing the models
Class
Client
Collaboration
Concrete classes
Contract
Controller
CORBA
Custom development
Customization
Data management layer
Dependency relationship
Design phase
Enterprise resource systems (ERP)
Errors

Factoring
Faults
Fixed-price contract
Foundation layer
Generalization
Has-parts
Human–computer interaction layer
Layer
Maintenance oracle
Message
Method
Model
Model-View-Controller (MVC)
Module
Object wrapper
Outsourcing
Package
Package diagram
Packaged software
Partition

Physical architecture layer
Presenter
Problem domain layer
Recorder
Refinement
Request for information (RFI)
Request for proposals (RFP)
Scribe
Server
Smalltalk
Systems integration
Test
Time-and-arrangements
   contract
Validation
Value-added contract
Verification
View
Walkthrough
Workaround

QUESTIONS

1. What are the relationships within the functional models, structural models, and the behavioral models that need to be tested?
2. What is the primary difference between an analysis model and a design model?
3. What is meant by balancing the models?
4. What are the interrelationships between the functional, structural, and the behavioral models that need to be tested?
5. What is a package? How are packages related to partitions and layers?
6. What are the five steps for identifying packages and creating package diagrams?
7. What needs to be verified and validated in package diagrams?
8. What is a partition? How does a partition relate to a collaboration?
9. What does factoring mean? How is it related to abstraction and refinement?
10. What is a layer? Name the different layers.
11. What is a dependency relationship? How do you identify them?
12. What is a walkthrough? How does it relate to verification and validation?
13. What are the different roles played during a walkthrough? What are their purposes?
14. What situations are most appropriate for a custom development design strategy?
15. What are some problems with using a packaged software approach to building a new system? How can these problems be addressed?
16. What are the differences between the time-and-arrangements, fixed-price, and value-added contracts for outsourcing?
17. How are the alternative matrix and feasibility analysis related?
18. When is outsourcing considered a good design strategy? When is it not appropriate?
19. What is an RFP? How is this different from an RFI?

EXERCISES

A. Perform a verification and validation walkthrough of the functional and structural models of the dentist office problem-domain layer described in exercises E, F, and G in Chapter 5 and exercise K in Chapter 6.
B. Based on the functional models you created for exercises E, F, and G in Chapter 5 and the structural model you created in exercise K in Chapter 6, draw a package diagram for the problem-domain layer for the dentist office system. Be sure to verify and validate the diagram.
C. Perform a verification and validation walkthrough of the functional and structural models of the real estate system problem-domain layer described in exercises H and I in Chapter 5 and Exercise L in Chapter 6.
D. Based on the functional models you created for exercises H and I in Chapter 5 and the structural model you created in exercise L in Chapter 6, draw a package diagram for the problem-domain layer for the real estate system. Be sure to verify and validate the diagram.
E. Perform a verification and validation walkthrough of the functional, structural, and behavioral models of the video store system problem-domain layer described in exercises L and M in Chapter 5, exercise N in Chapter 6, and exercises B, C, and D in Chapter 7.
F. Based on the functional models you created for exercises L and M in Chapter 5, the structural model you created in exercise N in Chapter 6, and the behavioral models you created in exercises B, C, and D in Chapter 7, draw a package diagram for the problem-domain layer for the video store system. Be sure to verify and validate the diagram.
G. Perform a verification and validation walkthrough of the functional, structural, and behavioral models of the health club membership system problem-domain layer described in exercises N and O in Chapter 5, exercise O in Chapter 6, and exercises F, G, and H in Chapter 7.
H. Based on the functional models you created for exercises N and O in Chapter 5, the structural model you created in exercise O in Chapter 6, and the behavioral models you created in exercises F, G, and H in Chapter 7, draw a package diagram for the problem domain layer for the health club membership system. Be sure to verify and validate the diagram.

I. Perform a verification and validation walkthrough of the functional, structural, and behavioral models of the catering system problem-domain layer described in exercises P and Q in Chapter 5, exercise P in Chapter 6, and exercise I in Chapter 7.

J. Based on the functional models you created for exercises P and Q in Chapter 5, the structural model you created in exercise P in Chapter 6, and the behavioral models you created in exercise I in Chapter 7, draw a package diagram for the problem-domain layer for the catering system. Be sure to verify and validate the diagram.

K. Perform a verification and validation walkthrough of the functional, structural, and behavioral models of the Of-the-Month Club system problem-domain layer described in exercises T and U in Chapter 5, exercise R in Chapter 6, and exercise L in Chapter 7.

L. Based on the functional models you created for exercises T and U in Chapter 5, the structural model you created in Exercise R in Chapter 6, and the behavioral models you created in exercise L in Chapter 7, draw a package diagram for the problem-domain layer for the system for the Of-the-Month Club. Be sure to verify and validate the diagram.

M. Perform a verification and validation walkthrough of the functional, structural, and behavioral models of the university library system problem-domain layer described in exercises R and S in Chapter 5, exercise Q in Chapter 6, and exercise K in Chapter 7.

N. Based on the functional models you created for exercises R and S in Chapter 5, the structural model you created in Exercise Q in Chapter 6, and the behavioral models you created in exercises K in Chapter 7, draw a package diagram for the problem domain layer for the university library system. Be sure to verify and validate the diagram.

O. Which design strategy would you recommend for the construction of the system in each exercise? Why?
   a. exercises A & B
   b. exercises C & D
   c. exercises E & F
   d. exercises G & H
   e. exercises I & J
   f. exercises K & L
   g. exercises M & N

P. Suppose you and your friends are starting a small business painting houses in the summertime. You need to buy a software package that handles the financial transactions of the business. Create an alternative matrix that compares three packaged systems (e.g., Quicken, MS Money, Quickbooks). Which alternative appears to be the best choice?

Q. Suppose you are leading a project that will implement a new course-enrollment system for your university. You are thinking about either using a packaged course-enrollment application or outsourcing the job to an external consultant. Create an outline for an RFP to which interested vendors and consultants could respond.

MINICASES

1. Refer to the analysis models that you created for professional and scientific staff management (PSSM) for minicase 2 in Chapter 5 and for minicase 1 in Chapter 7.
   a. Add packages to your use-case diagram to simplify it.
   b. Use the communication diagram to identify logical partitions in your class diagram. Add packages to the diagram to represent the partitions.

2. Refer to the analysis models that you created for Holiday Travel Vehicles for minicase 1 in Chapter 6 and for minicase 2 in Chapter 7.
   a. Add packages to your use-case diagram to simplify it.
   b. Use the communication diagram to identify logical partitions in your class diagram. Add packages to the diagram to represent the partitions.

3. Refer to the Singapore Importers Pte Ltd minicase in Chapter 6. After completing all the analysis models (both the as-is and to-be models), the general manager finally understood why it was important to understand the as-is system before delving into the development of the to-be system. However, you now tell him that the to-be models are only the problem-domain portion of the design. To say the least, he is now very confused. After explaining to him the advantages of using a layered approach to developing the system, he says, "I don't care about reusability or maintenance. I only want the system to be implemented as soon as possible. You IS people are always trying to pull a fast one on the users. Just get the system done!"

   What is your response to the general manager? Do you jump into the implementation he had insisted on? What do you do next?

4. Ting Ting Lim is the general manager of the Hong Kong office of Davies International, a British multi-national human resources firm. The firm began in the UK in the

early 1920s, and gradually opened offices in many British colonies around the world. Its information systems were largely developed at its UK head office. Over the last decade, there has been significant growth in business outside of the UK, and the Hong Kong office is now the firm's largest office. Offices in other countries such as India and Singapore have also grown dramatically.

Ms. Lim is reflecting on the client management software system that Davies International purchased five years ago. At that time, the firm had just gone through a growth spurt, and the combination of automated and manual procedures that managed client accounts had become unwieldy. Ms. Lim was in the committee that researched and selected the package currently used. She had learned of the software at a seminar. Initially, it worked fairly well for the firm. Some office procedures had to be changed to fit the package, but the office staff had expected this and were prepared for it.

Since that time, the Hong Kong office has continued to grow, not only through an expanding client base, but also from the acquisition of several smaller employment-related businesses in Hong Kong and an expansion onto the Chinese mainland. In a move to diversify its human resource management services, the firm's support staff has also expanded. Ms. Lim is particularly proud of the IS department the Hong Kong office has built up over the years. Using strong ties with a local university, an attractive compensation package, and a good working environment, the IS department is well-staffed with competent, innovative people, plus a steady stream of college interns that keeps the department vibrant. One of the IS teams pioneered the use of the Internet to offer services to a whole new Pankajet segment, a move that has proven to be successful and spread to Davies International offices overseas.

It seems clear that a major change is needed in the client management software, and Ms. Lim has already begun to budget for such a project. This software is a central part of Davies International operations, and Ms. Lim wants to be sure that a quality system is obtained this time. She knows that the vendor of their current system has made some revisions and additions to its product line. There are other software vendors who offer similar products. Ms. Lim is also considering the idea of appointing the Hong Kong office's IS department to develop a custom software application.

a. Outline the issues that Ms. Lim should consider that would support the development of a custom software application in-house.

b. Outline the issues that Ms. Lim should consider which would support the purchase of a software package.

c. Within the context of a systems development project, when should the decision of "make-versus-buy" be made? How should Ms. Lim proceed? Explain your answer.

CHAPTER 9

CLASS AND METHOD DESIGN

The most important step of the design phase is designing the individual classes and methods. Object-oriented systems can be quite complex, so analysts need to create instructions and guidelines for programmers that clearly describe what the system must do. This chapter presents a set of criteria, activities, and techniques used to design classes and methods. Together they are used to ensure the object-oriented design communicates how the system needs to be coded.

OBJECTIVES

- Become familiar with coupling, cohesion, and connascence.
- Be able to specify, restructure, and optimize object designs.
- Be able to identify the reuse of predefined classes, libraries, frameworks, and components.
- Be able to specify constraints and contracts.
- Be able to create a method specification.

CHAPTER OUTLINE

INTRODUCTION

WARNING: *This material may be hazardous to your mental stability*. Not really, but now that we have your attention, you must realize that this material is fairly technical in nature and that it is extremely important in today's "flat" world. Today, much of the actual implementation will be done in a different geographic location than where the analysis and design are performed. As such, we must ensure that the design is specified in a "correct" manner. We must make certain that there is no, or at least minimal, ambiguity in the design specification.

In today's flat world, the common language spoken among developers is very likely to be UML and some object-oriented language, such as Java, and not English. English has always been and always will be ambiguous. Furthermore, to what version of English do we refer? As both Oscar Wilde and George Bernard Shaw independently pointed out, the United States and England are divided by a common language. A simple, but relevant, example is the number of zeros in one billion. In American English, there are nine, but in British English there are twelve. Obviously, this could lead to problems when one is building financial information systems.

Practically speaking, Class and Method design is where all the work actually gets done during design. No matter which layer you are focusing on, the classes, which will be used to create the system objects, must be designed. Some people believe that with reusable class libraries and off-the-shelf components, this type of low-level, or detailed, design is a waste of time and that we should jump immediately into the "real" work: coding the system. However, if the past shows us anything, it shows that low-level, or detailed, design is critical despite the use of libraries and components. Detailed design is still very important for three reasons. First, with today's modern CASE tools, quite a bit of the actual code can be generated by the tool from the detailed design. Second, even preexisting classes and components needs to be understood, organized, and pieced together. Third, it is still common for the project team to have to write some code and produce original classes that support the application logic of the system.

Jumping right into coding will guarantee results that can be disastrous. For example, even though the use of layers can simplify the individual classes, they can increase the complexity of the interactions between them. As such, if the classes are not designed carefully, the resulting system can be very inefficient. Or worse, the instances of the classes (i.e., the objects) will not be capable of communicating with each other, which, of course, will result in the system not working properly.

Furthermore, in an object-oriented system, changes can take place at different levels of abstraction. These levels include variable, method, class/object, package,[1] library, and/or application/system levels (see Figure 9-1). The changes that take place at one level can impact other levels (e.g., changes to a class can affect the package level, which can affect both the system level and the library level, which in turn can cause changes back down at the class level). Finally, changes can occur at different levels at the same time.

The good news is that the detailed design of the individual classes and methods is fairly straightforward, and the interactions among the objects on the problem-domain layer have been designed, in some detail, during analysis (see Chapters 5–7). The other layers (data management, human–computer interaction, and system architecture), are highly dependent on the problem-domain layer (see Figure 8-7). Therefore, if the problem-domain classes are designed correctly, the design of the classes on the other layers will fall into place, relatively speaking.

That being said, it has been our experience that many project teams are much too quick at jumping into writing code for the classes without first designing them. Some of this has been caused by the fact that object-oriented systems analysis and design has evolved from object-oriented programming. Until recently there has been a general lack of accepted guidelines on how to design and develop effective object-oriented systems. However, with the acceptance of UML as a standard object notation, standardized approaches based on work of many object methodologists have begun to emerge.[2]

---

[1] A package is a group of collaborating objects. Other names for a package include cluster, partition, pattern, subject, and subsystem.

[2] For example, OPEN [I. Graham, B. Henderson-Seller, and H. Yanoussi, *The Open Process Specification* (Reading, MA: Addison-Wesley, 1997)], RUP [P. Kruchten, *The Rational Unified Process: An Introduction*, 2nd ed. (Reading, MA: Addison-Wesley, 2000)], and the Enhanced Unified Process (see Chapter 1).

FIGURE 9-1

Levels of Abstraction in Object-Oriented Systems

Source: Adapted from David P. Tegarden, Steven D. Sheetz, and David E. Monarchi, "A Software Complexity Model of Object-Oriented Systems," *Decision Support Systems* 13 (March 1995): 241–262.

In this chapter, we begin by reviewing the basic characteristics of object orientation. Next, we present a set of useful design criteria and activities that are applicable across any layer for class and method design. Finally, we present a set of techniques that are useful for design methods: contracts and method specifications.

Review of the Basic Characteristics of Object Orientation[3]

Object-oriented systems can be traced back to the Simula and the Smalltalk programming languages. However, until the increase in processor power and the decrease in processor cost that occurred in the 1980s, object-oriented approaches were not practical. Many of the specific details concerning the basic characteristics of object-orientation are language dependent; that is, each object-oriented programming language tends to implement some of the object-oriented basics in different ways. As such, we need to know which programming language is going to be used to implement the different aspects of the solution. Otherwise, the system may behave in a manner different than the analyst, designer, and client expect. Today, the C++, Java, and Visual Basic.Net programming languages tend to be the more predominant ones used. In this section, we review the basic characteristics of object orientation and point out where the language specific issues emerge.

Classes, Objects, Methods, and Messages

The basic building block of the system is the *object*. Objects are *instances* of *classes* that we use as templates to define objects. A class defines both the data and processes that each object contains. Each object has *attributes* that describe data about the object. Objects have *state*, which is defined by the value of its attributes and its relationships with other objects at a particular point in time. And, each object has *methods*, which specify what processes the object can perform. From our perspective, methods are used to implement the *operations* that specified the *behavior* of the objects (see Chapter 6). In order to get an object to perform a method (e.g., to delete itself), a *message* is sent to the object. A message is essentially a function or procedure call from one object to another object.

---

[3] For an introduction to the basic characteristics, see Appendix 1.

Encapsulation and Information Hiding

Encapsulation is the mechanism that combines the processes and data into a single object. *Information hiding* suggests only the information required to use an object be available outside the object; that is, information hiding is related to the *visibility* of the methods and attributes (see Chapter 6). Exactly how the object stores data or performs methods is not relevant, as long as the object functions correctly. All that is required to use an object are the set of methods and the messages needed to be sent to trigger them. The only communication between objects should be through an object's methods. The fact that we can use an object by sending a message that calls methods is the key to reusability because it shields the internal workings of the object from changes in the outside system, and it keeps the system from being affected when changes are made to an object.

Polymorphism and Dynamic Binding

Polymorphism means having the ability to take several forms. By supporting polymorphism, object-oriented systems can send the same message to a set of objects, which can be interpreted differently by different classes of objects. And, based on encapsulation and information hiding, an object does not have to be concerned with *how* something is done when using other objects. It simply sends a message to an object and that object determines how to interpret the message. This is accomplished through the use of dynamic binding.

Dynamic binding refers to the ability of object-oriented systems to defer the data typing of objects to run time. For example, imagine that you have an array of type employee that contains instances of hourly employees and salaried employees (see Figure 9-2). Both these types of employees implement a compute pay method. An object can send the message to each instance contained in the array to compute the pay for that individual instance. Depending on whether the instance is an hourly employee or a salaried employee, a different method will be executed. The specific method is chosen at run time. With this ability, individual classes are easier to understand. However, the specific level of support for polymorphism and dynamic binding is language specific. Most object-oriented programming languages support dynamic binding of methods, and some support dynamic binding of attributes. As such, it is important to know what object-oriented programming language is going to be used.

But polymorphism can be a double-edged sword. Through the use of dynamic binding, there is no way to know before run time which specific object will be asked to execute its method. In effect, there is a decision made by the system that is not coded anywhere.[4]

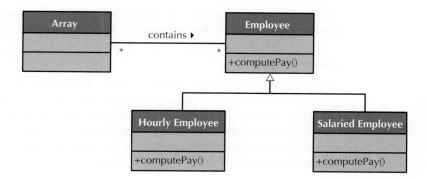

FIGURE 9-2
Polymorphism
Example

---

[4] From a practical perspective, there is an implied case statement. The system chooses the method based on the type of object being asked to execute it and the parameters passed as arguments to the method.

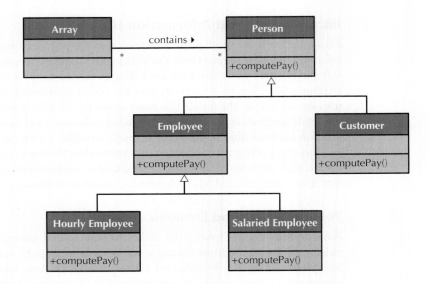

FIGURE 9-3

Polymorphism
Misuse Example

Furthermore, because all these decisions are made at run time, it is possible to send a message to an object that it does not understand (i.e., the object does not have a corresponding method). This can cause a run-time error that, if not programmed to handle correctly, can cause the system to abort.[5]

Finally, if the methods are not semantically consistent, the developer cannot assume that all methods with the same name will perform the same generic operation. For example, imagine that you have an array of type person that contains instances of employees and customers (see Figure 9-3). These both implement a compute pay method. An object can send the message to each instance contained in the array to execute the compute pay method for that individual instance. In the case of an instance of employee, the compute pay method computes the amount that the employee is owed by the firm, whereas the compute pay method associated with an instance of a customer computes the amount owed the firm by the customer. Depending on whether the instance is an employee or a customer, a different meaning is associated with the method. As such, the semantics of each method must be determined individually. This substantially increases the difficulty of understanding individual objects. The key to controlling the difficulty of understanding object-oriented systems when using polymorphism is to ensure that all methods with the same name implement that same generic operation (i.e., they are semantically consistent).

Inheritance

Inheritance allows developers to define classes incrementally by reusing classes defined previously as the basis for new classes. Although we could define each class separately, it might be simpler to define one general superclass that contains the data and methods needed by the subclasses and then have these classes inherit the properties of the superclass. Subclasses inherit the appropriate attributes and methods from the superclasses above them. Inheritance makes it simpler to define classes.

---

[5] In Java, these errors are referred to as exceptions that the system "throws" and must "catch." In other words, the programmer must correctly program the throw and catch or the Java virtual machine will abort. Again, each programming language can handle these situations in a unique manner.

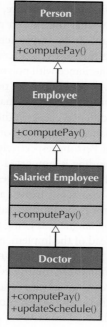

FIGURE 9-4
Redefinition and
Inheritance Conflict
Example

There have been many different types of inheritance mechanisms associated with object-oriented systems.[6] The most common inheritance mechanisms include different forms of single and multiple inheritance. *Single inheritance* allows a subclass to have only a single parent class. Currently, all object-oriented methodologies, databases, and programming languages permit extending the definition of the superclass through single inheritance.

Some object-oriented methodologies, databases, and programming languages allow a subclass to redefine some or all the attributes and/or methods of its superclass. With *redefinition* capabilities, it is possible to introduce an *inheritance conflict* [i.e., an attribute (or method) of a subclass with the same name as an attribute (or method) of a superclass]. For example in Figure 9-4, Doctor is a subclass of Employee. Both have methods named ComputePay(). This causes an inheritance conflict. Furthermore, when the definition of a superclass is modified, all its subclasses are affected. This may introduce additional inheritance conflicts in one (or more) of the superclass's subclasses. For example in Figure 9-4, Employee could be modified to include an additional method, UpdateSchedule(). This would add another inheritance conflict between Employee and Doctor. Therefore, developers must be aware of the effects of the modification not only in the superclass, but also in each subclass that inherits the modification.

Finally, through redefinition capabilities, it is possible for a programmer to arbitrarily cancel the inheritance of methods by placing stubs[7] in the subclass that will override the definition of the inherited method. If the cancellation of methods is necessary for the correct definition of the subclass, then it is likely that the subclass has been misclassified (i.e., it is inheriting from the wrong superclass).

As you can see, from a design perspective, inheritance conflicts and redefinition can cause all kinds of problems with interpreting the final design and implementation.[8] However, most inheritance conflicts are due to poor classification of the subclass in the inheritance hierarchy (the generalization a-kind-of semantics are violated), or the actual inheritance mechanism violates the encapsulation principle (i.e., subclasses are capable of directly addressing the attributes or methods of a superclass). To address these issues, Jim Rumbaugh, and his colleagues, suggested the following guidelines:[9]

- Do not redefine query operations.
- Methods that redefine inherited ones should restrict only the semantics of the inherited ones.
- The underlying semantics of the inherited method should never be changed.
- The signature (argument list) of the inherited method should never be changed.

However, many existing object-oriented programming languages violate these guidelines. When it comes to implementing the design, different object-oriented programming

---

[6] See, for example, M. Lenzerini, D. Nardi, and M. Simi, *Inheritance Hierarchies in Knowledge Representation and Programming Languages* (New York: Wiley, 1991).

[7] In this case, a stub is simply the minimal definition of a method to prevent syntax errors occurring.

[8] For more information, see Ronald J. Brachman, "I Lied about the Trees Or, Defaults and Definitions in Knowledge Representation," *AI Magazine* 5, no. 3 (Fall 1985): 80–93.

[9] J. Rumbaugh, M. Blaha, W. Premerlani, F. Eddy, and W. Lorensen, *Object-Oriented Modeling and Design* (Englewood Cliffs, NJ: Prentice Hall, 1991).

languages address inheritance conflicts differently. Therefore, it is important at this point in the development of the system to know what the chosen programming language supports. And, we must be sure that the design can be implemented as intended. Otherwise, the design needs to be modified before it is turned over to remotely located programmers.

When considering the interaction of inheritance with polymorphism and dynamic binding, object-oriented systems provide the developer with a very powerful, but dangerous, set of tools. Depending on the object-oriented programming language used, this interaction can allow the same object to be associated with different classes at different times. For example, an instance of Doctor can be treated as an instance of Employee or any of its direct and indirect superclasses, such as SalariedEmployee and Person, respectively (see Figure 9-4). Therefore, depending on whether static or dynamic binding is supported, the same object may execute different implementations of the same method at different times. Or, if the method is defined only with the SalariedEmployee class and it is currently treated as an instance of the Employee class, the instance may cause a run-time error to occur.[10] As stated previously, it is important to know what object-oriented programming language is going to be used so that these kinds of issues can be solved with the design, instead of the implementation, of the class.

With *multiple inheritance*, a subclass may inherit from more than one superclass. In this situation, the types of inheritance conflicts are multiplied. In addition to the possibility of having an inheritance conflict between the subclass and one (or more) of its superclasses, it is now possible to have conflicts between two (or more) superclasses. In this latter case, there are three different types of additional inheritance conflicts that can occur:

- Two inherited attributes (or methods) have the same name (spelling) and semantics.
- Two inherited attributes (or methods) have different names but identical semantics (i.e., they are *synonyms*).
- Two inherited attributes (or methods) have the same name but different semantics (i.e., they are *heteronyms, homographs,* or *homonyms*). This also violates the proper use of polymorphism.

For example, in Figure 9-5, Robot-Employee is a subclass of both Employee and Robot. In this case, Employee and Robot conflict with the attribute name. Which one should Robot-Employee inherit? Because they are the same, semantically speaking, does

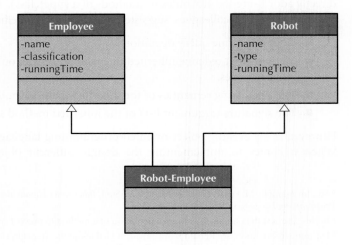

FIGURE 9-5

Additional Inheritance Conflicts with Multiple Inheritance

---

[10] This happens with novices quite regularly when using C++.

CONCEPTS

IN ACTION

9-A Inheritance Abuses

Meilir Page-Jones, through his consulting company, identified a set of abuses of inheritance. In some cases, these abuses led to lengthy and bloody disputes, gruesome implementations, and in one case, it led to the destruction of the development team. In all cases, the error was in not enforcing a generalization (a-kind-of) semantics. In one case, the inheritance hierarchy was inverted: BoardMember was a superclass of Manager, which was a superclass of Employee. However, in this case, an Employee is NOT a-kind-of Manager, which is NOT a-kind-of BoardMember. In fact, the opposite was true. However, if you think of an Organization Chart, a BoardMember is super to a Manager, which is super to an Employee. In another example, the client's firm

attempted to use inheritance to model a membership idea (e.g., Student is a member of a club). However, the club should have had an attribute that contained the student members. In the other examples, inheritance was used to implement an association relationship and an aggregation relationship.

Source: Meilir Page-Jones, *Fundamentals of Object-Oriented Design in UML* (Reading, MA: Addison-Wesley, 2000).

Question

As an analyst, how can you attempt to avoid these types of inheritance abuses?

it really matter? It is also possible that Employee and Robot could have a semantic conflict on the classification and type attributes if they have the same semantics. Practically speaking, the only way to prevent this situation is for the developer to catch it during the design of the subclass. Finally, what if the runningTime attributes have different semantics? In the case Employee objects, the runningTime attribute stores the employee's time running a mile, whereas the runningTime attribute for Robot objects stores the average time between checkups. Should Robot-Employee inherit both of them? It really depends on whether the robot employees can run the mile or not. With the potential for these additional types of conflicts, there is a risk of decreasing the understandability in an object-oriented system instead of increasing it through the use of multiple inheritance. As such, great care should be taken when using multiple inheritance.

DESIGN CRITERIA

When considering the design of an object-oriented system, a set of criteria exists that can be used to determine whether the design is a good one or a bad one. According to Coad and Yourdon,[11] "A good design is one that balances trade-offs to minimize the total cost of the system over its entire lifetime." These criteria include coupling, cohesion, and connascence.

Coupling

Coupling refers to how interdependent or interrelated the modules (classes, objects, and methods) are in a system. The higher the interdependency, the more likely changes in part of a design can cause changes to be required in other parts of the design. For object-oriented systems, Coad and Yourdon[12] identified two types of coupling to consider: interaction and inheritance.

Interaction coupling deals with the coupling among methods and objects through message passing. Lieberherr and Holland put forth the *law of Demeter* as a guideline to

[11] Peter Coad and Edward Yourdon, *Object-Oriented Design* (Englewood Cliffs, NJ: Yourdon Press, 1991), p. 128.
[12] Ibid.

minimize this type of coupling.[13] Essentially, the law minimizes the number of objects that can receive messages from a given object. The law states that an object should send messages only to one of the following:

- Itself (For example, in Figure 9-6a, Object1 can send Message1 to itself. In other words, a method associated with Object1 can use other methods associated with Object1.[14])
- An object that is contained in an attribute of the object or one of its superclasses (For example in Figure 9-6b, PO1 should be able to send messages using both its Customer and Date attributes.)
- An object that is passed as a parameter to the method (For example in Figure 9-6c, the aPatient instance sends the message RequestAppt(name, address) to the aReceptionist instance, which is allowed to send messages to the instances contained in the name and address parameters.)
- An object that is created by the method (For example in Figure 9-6c, the method RequestAppt associated with the aReceptionist instance creates an instance of the Appointment class. As such, the RequestAppt method is allowed to send messages to anAppt.)
- An object that is stored in a global variable[15]

In each case, interaction coupling is increased. For example, the coupling increases between the objects if the calling method passes attributes to the called method or if the calling method depends on the value being returned by the called method.

There are six types of interaction coupling, each falling on different parts of a good-to-bad continuum. They range from no direct coupling to content coupling. Figure 9-7 presents the different types of interaction coupling. In general, interaction coupling should be minimized. The one possible exception is that non-problem-domain classes must be coupled to their corresponding problem-domain classes. For example, a report object (on the human–computer interaction layer) that displays the contents of an employee object (on the problem-domain layer) will be dependent on the employee object. In this case, for optimization purposes, the report class may be even content or pathologically coupled to the employee class. However, problem-domain classes should never be coupled to non-problem-domain classes.

Inheritance coupling, as its name implies, deals with how tightly coupled the classes are in an inheritance hierarchy. Most authors tend to say simply that this type of coupling is desirable. However, depending on the issues raised previously with inheritance—inheritance conflicts, redefinition capabilities, and dynamic binding—a high level of inheritance coupling may not be a good thing. For example, in Figure 9-8, should Method2() defined in Subclass be allowed to call Method1() defined in Superclass? Or, should Method2() defined in Subclass refer to Atribute1 defined in Superclass? Or, even more confusing, assuming that Superclass is an abstract class, can a Method1() call Method2() or use Attribute2 defined in Subclass? Obviously, the first two examples have some intuitive sense. Using the properties of a superclass is the primary purpose of inheriting from it in the first place. On the other hand, the third example is somewhat counterintuitive. However, due to

[13] Karl J. Lieberherr and Ian M. Holland, "Assuring Good Style for Object-Oriented Programs," *IEEE Software*, 6, no. 5 (September, 1989): 38–48; and Karl J. Lieberherr, *Adaptive Object-Oriented Software: The Demeter Method with Propagation Patterns* (Boston, MA: PWS Publishing, 1996).

[14] Obviously, this is stating what is expected.

[15] From a design perspective, global variables should be avoided. Most pure object-oriented programming languages do not explicitly support global variables. As such, we do not address them any further.

(a)

(b)

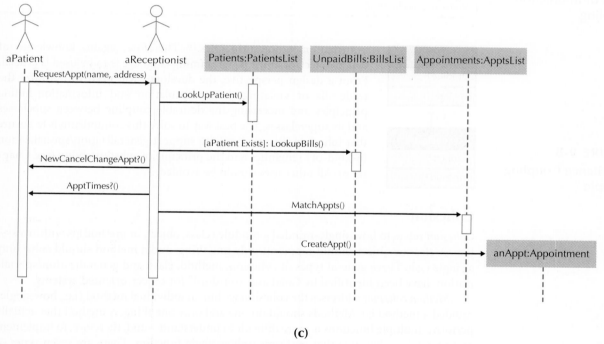

(c)

FIGURE 9-6 Interaction Coupling Examples

the way that different object-oriented programming languages support dynamic binding, polymorphism and inheritance, all these examples could be possible.

As Snyder has pointed out, most problems with inheritance involve the ability within the object-oriented programming languages to violate the encapsulation and

Level	Type	Description
Good	No Direct Coupling	The methods do not relate to one another; that is, they do not call one another.
	Data	The calling method passes a variable to the called method. If the variable is composite (i.e., an object), the entire object is used by the called method to perform its function.
	Stamp	The calling method passes a composite variable (i.e., an object) to the called method, but the called method only uses a portion of the object to perform its function.
	Control	The calling method passes a control variable whose value will control the execution of the called method.
	Common or Global	The methods refer to a "global data area" that is outside the individual objects.
Bad	Content or Pathological	A method of one object refers to the inside (hidden parts) of another object. This violates the principles of encapsulation and information hiding. However, C++ allows this to take place through the use of "friends."

Source: These types were adapted from Meilir Page-Jones, *The Practical Guide to Structured Systems Design,* 2nd ed. (Englewood Cliffs, NJ: Yardon Press, 1988); and Glenford Myers, *Composite/Structured Design (*New York: Van Nostrand Reinhold, 1978).

FIGURE 9-7
Types of Interaction Coupling

FIGURE 9-8
Inheritance Coupling Example

information-hiding principles.[16] Therefore, again, knowledge of which object-oriented programming language is to be used is crucial. From a design perspective, the developer will need to optimize the trade-offs of violating the encapsulation and information-hiding principles and increasing the desirable coupling between subclasses and its superclasses. The best way to solve this conundrum is to ensure that inheritance is used only to support generalization/specialization (a-kind-of) semantics and the principle of substitutability (see Chapter 6). All other uses should be avoided.

Cohesion

Cohesion refers to how single-minded a module (class, object, or method) is within a system. A class or object should represent only one thing, and a method should solve only a single task. Three general types of cohesion, method, class, and generalization/specialization, have been identified by Coad and Yourdon[17] for object-oriented systems.

Method cohesion addresses the cohesion within an individual method (i.e., how single-minded a method is). Methods should do one and only one thing. A method that actually performs multiple functions is more difficult to understand—and, therefore, to implement and maintain—than one that performs only a single function. There are seven types of method cohesion that have been identified (see Figure 9-9). They range from functional

---

[16] Alan Snyder, "Encapsulation and Inheritance in Object-Oriented Programming Languages," in N. Meyrowitz, ed., *OOPSLA '86 Conference Proceedings, ACM SigPlan Notices,* 21, no. 11 (November 1986); and Alan Snyder, "Inheritance and the Development of Encapsulated Software Components," in B. Shriver and P. Wegner, eds., *Research Directions in Object-Oriented Programming* (Cambridge, MA: MIT Press, 1987).

[17] Coad and Yourdon, *Object-Oriented Design.*

Level	Type	Description
Good	Functional	A method performs a single problem-related task (e.g., calculate current GPA).
	Sequential	The method combines two functions in which the output from the first one is used as the input to the second one (e.g., format and validate current GPA).
	Communicational	The method combines two functions that use the same attributes to execute (e.g., calculate current and cumulative GPA).
	Procedural	The method supports multiple weakly related functions. For example, the method could calculate student GPA, print student record, calculate cumulative GPA, and print cumulative GPA.
	Temporal or Classical	The method supports multiple related functions in time (e.g., initialize all attributes).
	Logical	The method supports multiple related functions, but the choice of the specific function is chosen based on a control variable that is passed into the method. For example, the called method could open a checking account, open a savings account, or calculate a loan, depending on the message that is send by its calling method.
Bad	Coincidental	The purpose of the method cannot be defined or it performs multiple functions that are unrelated to one another. For example, the method could update customer records, calculate loan payments, print exception reports, and analyze competitor pricing structure.

Source: These types were adapted from Page-Jones, *The Practical Guide to Structured Systems*, and Myers, *Composite/Structured Design.*

FIGURE 9-9
Types of Method Cohesion

cohesion (good) down to coincidental cohesion (bad). In general, method cohesion should be maximized.

Class cohesion is the level of cohesion among the attributes and methods of a class (i.e., how single-minded a class is). A class should represent only one thing, such as an employee, a department, or an order. All attributes and methods contained in a class should be required for the class to represent the thing. For example, an employee class should have attributes that deal with a social security number, last name, first name, middle initial, addresses, and benefits, but it should not have attributes such as door, engine, or hood. Furthermore, there should be no attributes or methods that are never used. In other words, a class should have only the attributes and methods necessary to fully define instances for the problem at hand. In this case, we have *ideal class cohesion*. Glenford Meyers suggested that a cohesive class[18] should have these attributes:

- It should contain multiple methods that are visible outside the class (i.e., a single-method class rarely makes sense).
- Each visible method performs only a single function (i.e., it has functional cohesion).

[18] We have adapted his informational-strength module criteria from structured design to object-oriented design. [see Glenford J. Myers, *Composite/Structured Design* (New York, NY: Van Nostrand Reinhold, 1978)].

Level	Type	Description
Good	Ideal	The class has none of the mixed cohesions.
	Mixed-Role	The class has one or more attributes that relate objects of the class to other objects on the same layer (e.g., the problem domain layer), but the attribute(s) have nothing to do with the underlying semantics of the class.
	Mixed-Domain	The class has one or more attributes that relate objects of the class to other objects on a different layer. As such, they have nothing to do with the underlying semantics of the thing that the class represents. In these cases, the offending attribute(s) belongs in another class located on one of the other layers. For example, a port attribute located in a problem domain class should be in a system architecture class that is related to the problem domain class.
Worse	Mixed-Instance	The class represents two different types of objects. The class should be decomposed into two separate classes. Typically, different instances only use a portion of the full definition of the class.

Source: Page-Jones, *Fundamentals of Object-Oriented Design in UML*.

FIGURE 9-10
Types of Class
Cohesion

- All methods reference only attributes or other methods defined within the class or one of its superclasses (i.e., if a method is going to send a message to another object, the remote object must be the value of one of the local object's attributes).[19]
- It should not have any control-flow couplings between its visible methods.

Page-Jones[20] has identified three less-than-desirable types of class cohesion: mixed-instance, mixed-domain, and mixed-role (see Figure 9-10). An individual class can have a mixture of any of the three types.

Generalization/specialization cohesion addresses the sensibility of the inheritance hierarchy. How are the classes in the inheritance hierarchy related? Are the classes related through a generalization/specialization (a-kind-of) semantics? Or, are they related via some association, aggregation, or membership type of relationship that was created for simple reuse purposes? Recall all the issues raised previously on the use of inheritance. For example, in Figure 9-11, the subclasses ClassRooms and Staff inherit from the superclass Department. Obviously, instances of the ClassRooms and Staff classes are not a-kind-of Department. However, in the early days of object-oriented programming, this use of inheritance was quite common. When a programmer saw that there were some common properties that a set of classes shared, the programmer would create an artificial abstraction that defined the commonalities. This was potentially useful in a reuse sense, but it turned out to cause many maintenance nightmares. In this case, instances of the ClassRooms and Staff classes are associated with or a-part-of an instance of Department. Today we know that highly cohesive inheritance hierarchies should support only the semantics of generalization and specialization (a-kind-of) and the principle of substitutability.

[19] This restricts message passing to only the first, second, and fourth conditions supported by the law of Demeter. For example, in Figure 9-6c, aReceptionist must have attributes associated with it that contains objects for Patients, Unpaid Bills, and Appointments. Furthermore, once anAppt is created, a Receptionist must have an attribute with anAppt as it's value to send any additional messages.

[20] See Meilir Page-Jones, *Fundamentals of Object-Oriented Design in UML* (Reading, MA: Addison-Wesley, 2000).

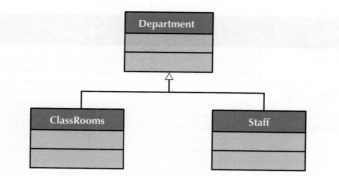

FIGURE 9-11

Generalization/
Specialization vs.
Inheritance Abuse

Connascence

Connascence[21] generalizes the ideas of cohesion and coupling, and it combines them with the arguments for encapsulation. To accomplish this, three levels of encapsulation have been identified. Level-0 encapsulation refers to the amount of encapsulation realized in an individual line of code, level-1 encapsulation is the level of encapsulation attained by combining lines of code into a method, and level-2 encapsulation is achieved by creating classes that contain both methods and attributes. Method cohesion and interaction coupling address primarily level-1 encapsulation. Class cohesion, generalization/specialization cohesion, and inheritance coupling address only level-2 encapsulation. Connascence, as a generalization of cohesion and coupling, addresses both level-1 and level-2 encapsulation.

But what exactly is connascence? Connascence literally means to be born together. From an object-oriented design perspective, it really means that two modules (classes or methods) are so intertwined that if you make a change in one, it is likely that a change in the other will be required. On the surface, this is very similar to coupling and, as such, should be minimized. However, when you combine it with the encapsulation levels, it is not quite as simple as that. In this case, we want to (1) minimize overall connascence by eliminating any unnecessary connascence throughout the system, (2) minimize connascence across any encapsulation boundaries, such as method boundaries and class boundaries, and (3) maximize connascence within any encapsulation boundary.

Based on these guidelines, a subclass should never directly access any hidden attribute or method of a superclass [i.e., a subclass should not have special rights to the properties of its superclass(es)]. If direct access to the nonvisible attributes and methods of a superclass by its subclass is allowed—and is permitted in most object-oriented programming languages—and a modification to the superclass is made, then due to the connascence between the subclass and its superclass, it is likely that a modification to the subclass also will be required.[22] In other words, the subclass has access to something across an encapsulation boundary (the class boundary between the subclass and the superclass). Practically speaking, you should maximize the cohesion (connascence) within an encapsulation boundary and minimize the coupling (connascence) between the encapsulation boundaries. There are many possible types of connascence. Figure 9-12 describes five of the types.

---

[21] See Meilir Page-Jones, "Comparing Techniques by Means of Encapsulation and Connascence," *Communications of the ACM* 35, no. 9 (September 1992): 147–151.

[22] Based on these guidelines, the use of the protected visibility, as supported in Java and C++, should be minimized if not avoided. Furthermore, so-called friends as defined in C++ also should be minimized or avoided. Due to the level of dependencies these language features create, any convenience afforded to a programmer is more than offset in potential design, understandability, and maintenance problems. As such, these features must be used with great caution and must be fully documented.

Type	Description
Name	If a method refers to an attribute, it is tied to the name of the attribute. If the attribute's name changes, the content of the method will have to change.
Type or Class	If a class has an attribute of type A, it is tied to the type of the attribute. If the type of the attribute changes, the attribute declaration will have to change.
Convention	A class has an attribute in which a range of values has a semantic meaning (e.g., account numbers whose values range from 1000 to 1999 are assets). If the range would change, then every method that used the attribute would have to be modified.
Algorithm	Two different methods of a class are dependent on the same algorithm to execute correctly (e.g., insert an element into an array and find an element in the same array). If the underlying algorithm would change, then the insert and find methods would also have to change.
Position	The order of the code in a method or the order of the arguments to a method is critical for the method to execute correctly. If either is wrong, then the method will, at least, not function correctly.

Source: Meilir Page-Jones, "Comparing Techniques by Means of Encapsulation and Connascence" and Meilir Page-Jones, *Fundamentals of Object-Oriented Design in UML.*

FIGURE 9-12
Types of Connascence

OBJECT DESIGN ACTIVITIES

The design activities for classes and methods are really an extension of the analysis and evolution activities presented previously (see Chapters 5 through 8). In this case, we expand the descriptions of the partitions, layers, and classes. Practically speaking, the expanded descriptions are created through the activities that take place during the detailed design of the classes and methods. The activities used to design classes and methods include additional specification of the current model, identifying opportunities for reuse, restructuring the design, optimizing the design, and, finally, mapping the problem-domain classes to an implementation language. Of course, any changes made to a class on one layer can cause the classes on the other layers that are coupled to it to be modified as well. The object design activities are described in this section.

Adding Specifications

At this point in the development of the system, it is crucial to review the current set of functional, structural, and behavioral models. First, we should ensure that the classes on the problem-domain layer are both necessary and sufficient to solve the underlying problem. To do this, we need to be sure that there are no missing attributes or methods and no extra or unused attributes or methods in each class. Furthermore, are there any missing or extra classes? If we have done our job well during analysis, there will be few, if any, attributes, methods, or classes to add to the models. And, it is unlikely that we have any extra attributes, methods, or classes to delete from the models. However, we still need to ensure that we have factored, abstracted, and refined the evolving models and created the relevant partitions and collaborations (see Chapter 8). We have mentioned this before, but we cannot overemphasize the importance of constantly reviewing the evolving system. Remember, it is always better to be safe than sorry.

Second, we need to finalize the visibility (hidden or visible) of the attributes and methods in each class. Depending on the object-oriented programming language used, this could be predetermined. [For example, in Smalltalk attributes are hidden and methods are visible. Other languages allow the programmer to set the visibility of each attribute or method. For example, in C++ and Java, you can set the visibility to private (hidden), public (visible), or protected (visible to subclasses, but not to other classes)].[23] By default, most object-oriented analysis and design approaches assume Smalltalk's approach.

Third, we need to decide on the signature of every method in every class. The *signature* of a method comprises three parts: the name of the method, the parameters or arguments that must be passed to the method, including their object type, and the type of value that the method will return to the calling method. The signature of a method is related to the method's *contract*.[24]

Fourth, we need to define any constraints that must be preserved by the objects (e.g., an attribute of an object that can have values only in a certain range). There are three different types of constraints: preconditions, postconditions, and invariants.[25] These are captured in the form of contracts (described later in this chapter) and assertions added to the CRC cards and class diagrams. We also must decide how to handle a violation of a constraint. Should the system simply abort? Should the system automatically undo the change that caused the violation? Should the system let the end user determine the approach to correct the violation? In other words, the designer must design the errors that the system is expected to handle. It is best not to leave these types of design decisions for the programmer to solve. Violations of a constraint are known as *exceptions* in languages such as C++ and Java.

Even though we have described these activities in the context of the problem-domain layer, they are also applicable to the other layers: data management (Chapter 10), human–computer interaction (Chapter 11), and physical architecture (Chapter 12).

Identifying Opportunities for Reuse

Previously, we looked at possibly utilizing reuse in our models in analysis through the use of *patterns* (see Chapter 6). In design, in addition to using analysis patterns, there are opportunities for using design patterns, frameworks, libraries, and components. The opportunities will vary, depending on which layer is being reviewed. For example, it is doubtful that a class library will be of much help on the problem-domain layer, but an appropriate class library could be of great help on the foundation layer. In this section, we describe the use of design patterns, frameworks, libraries, and components.

Like analysis patterns, design patterns are simply useful grouping of collaborating classes that provide a solution to a commonly occurring problem. The primary difference between analysis and design patterns is that design patterns are useful in solving "a general design problem in a particular context,"[26] whereas analysis patterns tended to aid in filling out a problem-domain representation. For example, a useful pattern is the Whole-Part pattern (see Figure 9-13a). The Whole-Part pattern explicitly supports the Aggregation and Composition relationships within the UML. Another useful design pattern is the Command Processor pattern (see Figure 9-13b). The primary purpose of the Command Processor pattern is to force the designer to explicitly separate the interface to an object (Command) from the actual execution of the operation (Supplier) behind the interface. Finally, some of the design patterns support different physical architectures (see Chapter 12). For example, the Forwarder-Receiver pattern (see Figure 9-13c) supports a peer-to-peer architecture. Many design patterns are available in C++ or Java source code.

---

[23] It is also possible to control visibility through packages and friends (see Footnote 21).

[24] Contracts were introduced in Chapter 6 and they are described in detail later in this chapter.

[25] Constraints are described in more detail later in this chapter.

[26] Erich Gamma, Richard Helm, Ralph Johnson, and John Vlissides, *Design Patterns: Elements of Reusable Object-Oriented Software* (Reading, MA: Addison-Wesley, 1995).

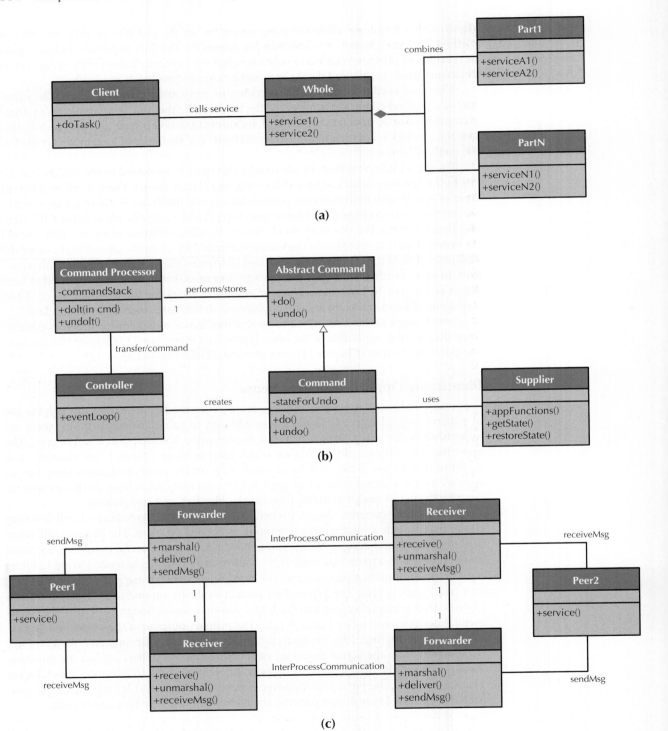

FIGURE 9-13 Sample Design Patterns

Source: These design patterns, along with many more, can be found in Frank Buschmann, Regine Meunier, Hans Rohnert, Peter Sommerlad, and Michael Stal, *Pattern-Oriented Software Architecture: A System of Patterns* (Chichester, UK: Wiley, 1996).

A *framework* is composed of a set of implemented classes that can be can be used as a basis for implementing an application. For example, there are frameworks available for CORBA and DCOM on which the implementation of part of the system architecture layer could be based. Most frameworks allow us to create subclasses to inherit from classes in the framework. There are object-persistence frameworks that can be purchased and used to add persistence to the problem-domain classes, which would be helpful on the data management layer. Of course, when inheriting from classes in a framework, we are creating a dependency (i.e., increasing the inheritance coupling from the subclass to the superclass). Therefore, if we use a framework and the vendor makes changes to the framework, we will have to at least recompile the system when we upgrade to the new version of the framework.

A *class library* is similar to a framework in that it typically has a set of implemented classes that were designed for reuse. However, frameworks tend to be more domain specific. In fact, frameworks may be built using a class library. A typical class library could be purchased to support numerical or statistical processing, file management (data management layer), or user interface development (human–computer interaction layer). In some cases, instances of classes contained in the class library can be created, and in other cases classes in the class library can be extended by creating subclasses based on them. As with frameworks, if we use inheritance to reuse the classes in the class library, we will run into all the issues dealing with inheritance coupling and connascence. If we directly instantiate classes in the class library, we will create a dependency between our object and the library object based on the signatures of the methods in the library object. This increases the interaction coupling between the class library object and our object.

A *component* is a self-contained, encapsulated piece of software that can be plugged into a system to provide a specific set of required functionalities. Today, there are many components available for purchase that have been implemented using *ActiveX* or *JavaBean* technologies. A component has a well-defined *API (application program interface)*. An API is essentially a set of method interfaces to the objects contained in the component. The internal workings of the component are hidden behind the API. Components can be implemented using class libraries and frameworks. However, components also can be used to implement frameworks. Unless the API changes between versions of the component, upgrading to a new version normally will require only linking the component back into the application. As such, recompilation typically is not required.

Which of these approaches should we use? It depends on what we are trying to build. In general, frameworks are used mostly to aid in developing objects on the system architecture, human–computer interaction, or data management layers; components are used primarily to simplify the development of objects on the problem-domain and human–computer interaction layers; and class libraries are used to develop frameworks and components and to support the foundation layer. Whichever of these reuse approaches you use, you must remember that reuse brings many potential benefits and possible problems. For example, the software has previously been verified and validated, which should reduce the amount of testing required for our system. However as stated before, if the software on which we are basing our system changes, then most likely, we will also have to change our system. Furthermore, if the software is from a third-party firm, we are creating a dependency from our firm (or our client's firm) to the third-party vendor. As such, we need to have some confidence that the vendor will be in business for a while.

Restructuring the Design

Once the individual classes and methods have been specified and the class libraries, frameworks, and components have been incorporated into the evolving design, we should use factoring to restructure the design. *Factoring* (Chapter 8) is the process of separating out

9-1 Campus Housing

In the previous chapters, you have been working on a system for the campus housing service. Based on the current set of functional, structural, and behavioral models that you have developed, are there potential opportunities of reuse in developing the system? Search the Web for potential patterns, class libraries, and components that could be useful in developing this system.

aspects of a method or class into a new method or class to simplify the overall design. For example, when reviewing a set of classes on a particular layer, we might discover that a subset of them shares a similar definition. In that case, it may be useful to factor out the similarities and create a new class. Based on the issues related to cohesion, coupling, and connascence, the new class may be related to the old classes via inheritance (generalization) or through an aggregation or association relationship.

Another process that is useful for restructuring the evolving design is *normalization*. Normalization is described in Chapter 10 in relation to relational databases. However, normalization can be useful at times to identify potential classes that are missing from the design. Also related to normalization is the requirement to implement the actual association and aggregation relationships as attributes. Virtually no object-oriented programming language differentiates between attributes and association and aggregation relationships. Therefore, all association and aggregation relationships must be converted to attributes in the classes. For example in Figure 9-14a, the Customer and State classes are associated with the Order class. Therefore, we need to add attributes to the Order class to reference the Customer and State classes. Furthermore, the Product-Order association class must be converted to a class (see Figure 9-14b).

Finally, all inheritance relationships should be challenged to ensure that they support only a generalization/specialization (a-kind-of) semantics. Otherwise, all the problems mentioned previously with inheritance coupling, class cohesion, and generalization/specialization cohesion will come to pass.

Optimizing the Design[27]

Up until now, we have focused our energy on developing an understandable design. With all the classes, patterns, collaborations, partitions, and layers designed and with all the class libraries, frameworks, and components included in the design, understandability has been, as it should have been, our primary focus. However, increasing the understandability of a design typically creates an inefficient design. Conversely, focusing on efficiency issues will deliver a design that is more difficult to understand. A good practical design will manage the inevitable trade-offs that must occur to create an acceptable system. In this section, we describe a set of simple optimizations that can be used to create a more efficient design.[28]

---

[27] The material contained in this section is based on James Rumbaugh, Michael Blaha, William Premerlani, Frederick Eddy, and William Lorensen, *Object-Oriented Modeling and Design* (Englewood Cliffs, NJ: Prentice Hall, 1991); and Bernd Brugge and Allen H. Dutoit, *Object-Oriented Software Engineering: Conquering Complex and Changing Systems* (Englewood Cliffs, NJ: Prentice Hall, 2000).

[28] The optimizations described here are only suggestions. In all cases, the decision to implement one or more of these optimizations really depends on the problem domain of the system and the environment on which the system will reside, that is, the data access and management layer (see Chapter 10), the human–computer interaction layer (see Chapter 11), and the physical architecture layer (see Chapter 12).

FIGURE 9-14   Converting Associations to Attributes

The first optimization to consider is to review the access paths between objects. In some cases, a message from one object to another may have a long path to traverse (i.e., it goes through many objects). If the path is long and the message is sent frequently, a redundant path should be considered. Adding an attribute to the calling object that will store a direct connection to the object at the end of the path can accomplish this.

A second optimization is to review each attribute of each class. Which methods use the attributes and which objects use the methods should be determined. If the only methods that use an attribute are read and update methods and only instances of a single class send messages to read and update the attribute, then the attribute may belong with the calling class instead of the called class. Moving the attribute to the calling class will substantially speed up the system.

A third optimization is to review the direct and indirect fan-out of each method. *Fan-out* refers to the number of messages sent by a method. The direct fan-out is the number of messages sent by the method itself, whereas the indirect fan-out also includes the number of messages sent by the methods called by the other methods in a message tree. If the fan-out of a method is high relative to the other methods in the system, the method should be optimized. One way to do this is to consider adding an index to the attributes used to send the messages to the objects in the message tree.

A fourth optimization is to look at the execution order of the statements in often-used methods. In some cases, it may be possible to rearrange some of the statements to be more efficient. For example, if it is known, based on the objects in the system, that a search routine can be narrowed by searching on one attribute before another one, then the search algorithm should be optimized by forcing it to always search in a predefined order.

A fifth optimization is to avoid recomputation by creating a *derived attribute* (or *active value*) (e.g., a total that will store the value of the computation). This is also known as caching computational results. It can be accomplished by adding a *trigger* to the attributes contained in the computation (i.e., those attributes on which the derived attribute is dependent). This would require a recomputation to take place only when one of the attributes that go into the computation is changed. Another approach is to simply mark the derived attribute for recomputation and delay the recomputation until the next time the derived attribute is accessed. This last approach delays the recomputation as long as possible. In this manner, a computation does not occur unless it must occur. Otherwise, every time a derived attribute needs to be accessed, a computation will be required.

A sixth optimization that should be considered deals with objects that participate in a one-to-one association, that is, they both must exist for either to exist. In this case, it may make sense, for efficiency purposes, to collapse the two defining classes into a single class.

YOUR TURN

9-2 Campus Housing

Assume that you are the project leader for the campus housing system that you have been developing over the previous chapters and that you just modified in Your Turn 9-1. However, as you review the current set of models, you realize that even though the models provide a rather complete description of the problem-domain layer, the evolving models have begun to become unmanageable.

And, as project leader, you also need to guarantee that the design will be efficient. Create a set of discussion points that you will use to explain to your development team the importance of optimizing the design before jumping into coding. Be sure to include an example of each optimization technique that can be used in the current set of models for the campus housing system.

However, this optimization may need to be reconsidered when storing the "fatter" object in a database. Depending on the type of object persistence used (see Chapter 10), it may actually be more efficient to keep the two classes separate. Or, it could make more sense for the two classes to be combined on the problem-domain layer but kept separate on the data access and management layer.

Mapping Problem-Domain Classes to Implementation Languages[29]

Up until this point in the development of the system, it has been assumed that the classes and methods in the models would be implemented directly in an object-oriented programming language. However, now it is important to map the current design to the capabilities of the programming language used. For example, if we have used multiple inheritance in your design but we are implementing in a language that supports only single inheritance, then the multiple inheritance must be factored out of the design. Or, if the implementation is to be done in an object-based language, one that does not support inheritance,[30] or a non-object-based language, such as C or Pascal, we must map the problem-domain objects to programming constructs that can be implemented using the chosen implementation environment. In this section, we describe a set of rules that can be used to do the necessary mapping.

Implementing Problem Domain Classes in a Single-Inheritance Language   The only issue associated with implementing problem-domain objects is the factoring out of any multiple inheritance—that is, the use of more than one superclass—used in the evolving design. For example, if you were to implement the solution in Java, Smalltalk, or Visual Basic.net, you must factor out any multiple inheritance. The easiest way to do this is to use the following rule:

RULE 1a: Convert the additional inheritance relationships to association relationships. The multiplicity of the new association from the subclass to the superclass should be 1..1. If the additional superclasses are concrete, that is, they can be instantiated themselves, then the multiplicity from the superclass to the subclass is 0..1. Otherwise, it is 1..1. Furthermore, an exclusive-or (XOR) constraint must be added between the associations. Finally, you must add appropriate methods to ensure that all information is still available to the original class.

OR

RULE 1b: Flatten the inheritance hierarchy by copying the attributes and methods of the additional superclass(es) down to all of the subclasses and remove the additional superclass from the design.[31]

Figure 9-15 demonstrates the application of these rules. Figure 9-15a portrays a simple example of multiple inheritance where Class 1 inherits from both SuperClass1 and SuperClass2, and Class 2 inherits from both SuperClass2 and SuperClass3. Assuming that SuperClass2 is

---

[29] The mapping rules presented in this section are based on material in Coad and Yourdon, *Object-Oriented Design*.

[30] In this case, we are talking about implementation inheritance, not the so-called interface inheritance. Interface inheritance supported by Visual Basic and Java supports only inheriting the requirements to implement certain methods, not any implementation. Java and Visual Basic.net also support single inheritance as described in this text, whereas Visual Basic 6 supports only interface inheritance.

[31] It is also a good idea to document this modification in the design so that in the future, modifications to the design can be maintained easily.

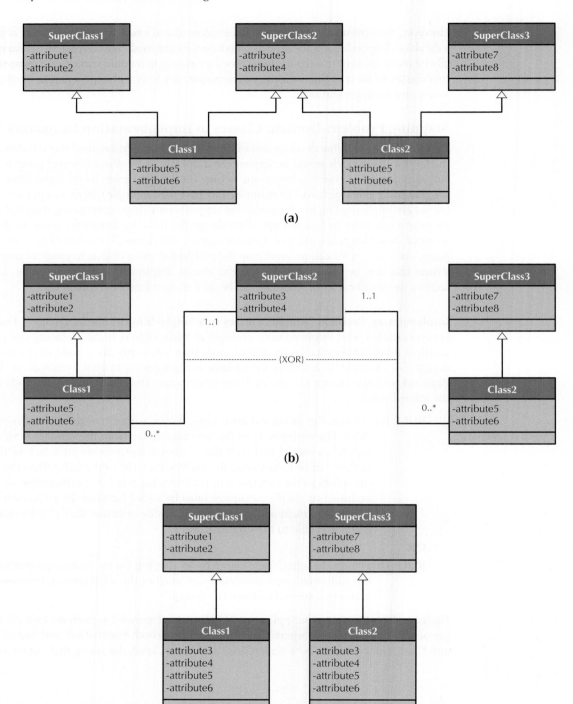

FIGURE 9-15   Factoring Out Multiple-Inheritance Effect for a Single-Inheritance Language

concrete, we apply Rule 1a to part a, and we end up with the diagram in part b, where we have added the association between Class1 and SuperClass2 and the association between Class2 and SuperClass2. Furthermore, the multiplicities have been added correctly, and the XOR constraint has been applied. If we apply Rule 1b to part a, we end up with the diagram in part c, where all the attributes of SuperClass2 have been copied down into Class1 and Class2. In this latter case, you may have to deal with the effects of inheritance conflicts (see earlier in the chapter).

The advantage of Rule 1a is that all problem-domain classes identified during analysis are preserved. This allows for maximum flexibility of maintenance of the design of the problem domain layer. However, Rule 1a increases the amount of message passing required in the system, and it has added processing requirements involving the XOR constraint, thus reducing the overall efficiency of the design. As such, our recommendation is to limit Rule 1a to be applied only when dealing with "extra" superclasses that are concrete because they have an independent existence in the problem domain. Use Rule 1b when they are abstract because they do not have an independent existence from the subclass.

Implementing Problem Domain Objects in a Object-Based Language   If we are going to implement our solution in an *object-based language* (i.e., a language that supports the creation of object but does not support implementation inheritance), we must factor out all uses of inheritance from the problem-domain class design. For example, if we were to implement our design in Visual Basic 6 or earlier, we would have to remove all uses of inheritance in the design. Applying the preceding rule to all superclasses will enable us to restructure our design without any inheritance.

Figure 9-16 demonstrates the application of the preceding rules. Figure 9-16a shows the same simple example of multiple inheritance portrayed in Figure 9-15, where Class1 inherits from both SuperClass1 and SuperClass2, and Class2 inherits from both SuperClass2 and SuperClass3. Assuming that SuperClass2 is concrete and that Super-Class1 and SuperClass3 are abstract, we apply Rule 1a to part a and we end up with the diagram in part b, where we have added the associations, the multiplicities, and the XOR constraint. If we apply Rule 1b to part a, we end up with the diagram in part c, where all the attributes of the superclasses have been copied down into Class1 and Class2. In this latter case, you may have to deal with the effects of inheritance conflicts (see earlier in the chapter).

Implementing Problem-Domain Objects in a Traditional Language   From a practical perspective, we are much better off implementing an object-oriented design in an object-oriented programming language, such as C++, Java, Smalltalk, or Visual Basic.net. However, implementing an object-oriented design in an object-based language, such as Visual Basic 6, is preferable to attempting to implement it in a traditional programming language, such as C or Cobol. Practically speaking, the gulf between an object-oriented design and a

YOUR TURN   **9-3 Dentist Office Appointment System**

In the previous chapters, we have been using a dentist office appointment system as an example. Assume that you now know that the system must be implemented in Visual Basic 6, which does not support implementation inheritance. As such, redraw the class diagram, factoring out the use of inheritance in the design by applying the above rules.

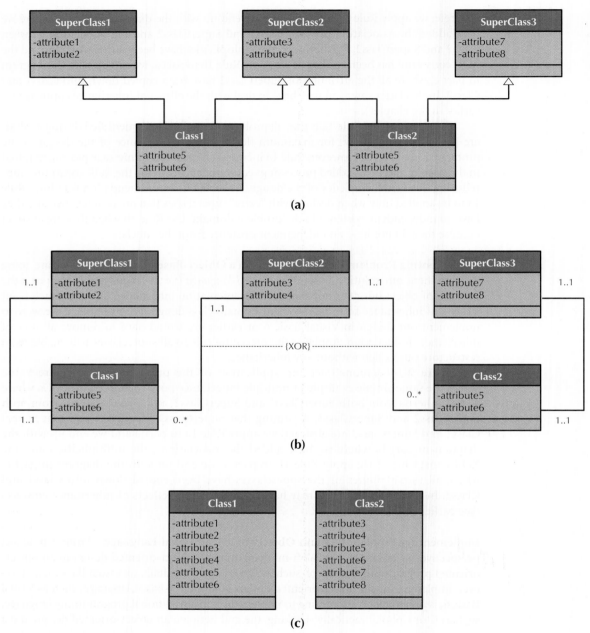

FIGURE 9-16   Factoring Out Multiple Inheritance Effect for an Object-Based Language

traditional programming language is simply too great for mere mortals to be able to cross. The best advice that we can give about implementing an object-oriented design in a traditional programming language is to run away as fast and as far as possible from the project. However, if we are brave (foolish?) enough to attempt this, we must realize that, in addition to factoring out inheritance from the design, we will have to factor out all uses of polymorphism, dynamic binding, encapsulation, and information hiding. To say the least, this is quite a bit of additional work to be accomplished. Furthermore, the manner in which we factor these object-oriented features out of the detailed design of the system tends to be language dependent. As such, this is beyond the scope of this text.

CONSTRAINTS AND CONTRACTS

Contracts were introduced in Chapter 6 in association with collaborations. A contract formalizes the interactions between the client and server objects, where a *client (consumer)* object is an instance of a class that sends a message to a *server (supplier)* object that executes one of its methods in response to the request. Contracts are modeled on the legal notion of a contract, where both parties, client and server objects, have obligations and rights. Practically speaking, a contract is a set of constraints and guarantees. If the constraints are met, then the server object will guarantee certain behavior.[32] Constraints can be written in either a natural language (e.g., English, *Structured English,* pseudocode) or a formal language (e.g., UML's Object Constraint Language[33]).

Types of Constraints

There are three different types of constraints typically captured in object-oriented design: preconditions, postconditions, and invariants.

Contracts are used primarily to establish the preconditions and postconditions for a method to be able to execute properly. A *precondition* is a constraint that must be met for a method to execute. For example, the parameters passed to a method must be valid for the method to execute. Otherwise, an exception should be raised. A *postcondition* is a constraint that must be met after the method executes, or the effect of the method execution must be undone. For example, the method can not make any of the attributes of the object take on an invalid value. In this case, an exception should be raised, and the effect of the method's execution should be undone.

Whereas preconditions and postconditions model the constraints on an individual method, *invariants* model constraints that must always be true for all instances of a class. Examples of invariants include domains or types of attributes, multiplicity of attributes, and the valid values of attributes. This includes the attributes that model association and aggregation relationships. For example, if an association relationship is required, an invariant should be created that will enforce it to have a valid value for the instance to exist. Invariants are normally attached to the class. As such, we can attach invariants to the CRC cards or class diagram by adding a set of assertions to them.

In Figure 9-17, the back of the CRC card constrains the attributes of an Order to specific types. For example, Order Number must be an unsigned long, and Customer must be an instance of the Customer class. Furthermore, additional invariants were added to four of the attributes. For example, Cust ID must not only be an unsigned long, but it also must have one and only one value [i.e., a multiplicity of (1..1)], and it must have the same value as the result of the GetCustID() message sent to the instance of Customer stored in the Customer attribute. Also shown is the constraint for an instance to exist, an instance of the Customer class, an instance of the State class, and at least one instance of the Product class must be associated with the Order object (see the

[32] The idea of using contracts in design evolved from the "Design by Contract" technique developed by Bertrand Meyer. See Bertrand Meyer, *Object-Oriented Software Construction* (Englewood Cliffs, NJ: Prentice Hall, 1988).
[33] See Jos Warmer and Anneke Kleppe, *The Object Constraint Language: Precise Modeling with UML* (Reading, MA: Addison-Wesley, 1999).

Front:

Class Name: Order	ID: 2	Type: Concrete, Domain
Description: An Individual that needs to receive or has received medical attention		Associated Use Cases: 3

Responsibilities	Collaborators
Calculate subtotal	
Calculate tax	
Calculate shipping	
Calculate total	

(a)

Back:

Attributes:

Order Number	(1..1)	(unsigned long)	
Date	(1..1)	(Date)	
Sub Total	(0..1)	(double)	
Tax	(0..1)	(double)	{Sub Total = sum (Product Order. GetExtension())}
Shipping	(0..1)	(double)	
Total	(0..1)	(double)	
Customer	(1..1)	(Customer)	
Cust ID	(1..1)	(unsigned long)	{Cust ID = Customer. GetCustID()}
State	(1..1)	(State)	
StateName	(1..1)	(String)	{State Name = State. GetState()}

Relationships:

Generalization (a-kind-of):

Aggregation (has-parts):

Other Associations: Product {1..*} Customer {1..1} State {1..1}

(b)

FIGURE 9-17
Invariants on a
CRC Card

Relationships section of the CRC card). Figure 9-18 portrays the same set of constraints on a class diagram. However, if all invariants are placed on a class diagram, the diagram will become very difficult to understand. As such, we recommend extending the CRC card to document the invariants instead of attaching them all to the class diagram.

YOUR TURN

9-4 Invariants

*U*sing the CRC card in Figure 9-17 and the class diagram in Figure 9-18 as guides, add invariants for the Customer class, the State class, the Product class, and the Product-Order Association to their respective CRC cards and the class diagram.

Questions

1. How easy is it to interpret the class diagram once all of the invariants were added?

2. Look at the class diagram in Figure 6-2 and the package diagram in Figure 8-21. What would they look like if the invariants were all attached to the diagrams? What would you recommend doing to avoid this situation?

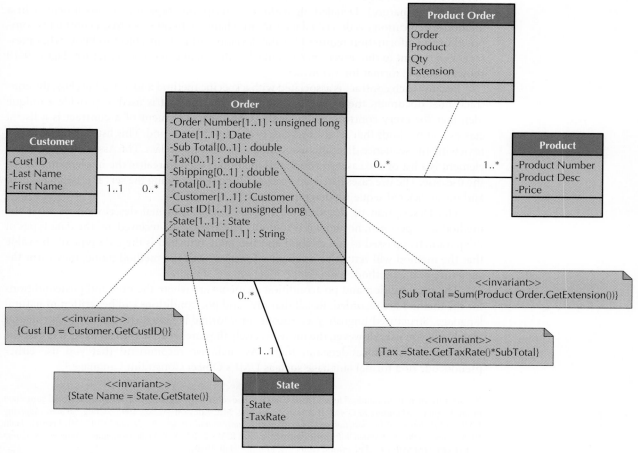

FIGURE 9-18 Invariants on a Class Diagram

Elements of a Contract

Contracts document the message passing that takes place between objects. Technically speaking, a contract should be created for each message sent and received by each object; one for each interaction. However, there would be quite a bit of duplication if this were done. In practice, a contract is created for each method that can receive messages from other objects (i.e., one for each visible method).

A contract should contain the information necessary for a programmer to understand what a method is to do (i.e., they are declarative in nature). This information includes the method name, class name, ID number, client objects, associated use cases, description, arguments received, type of data returned, and the pre- and postconditions.[34] Contracts do not have a detailed algorithmic description of how the method is to work (i.e., they are not procedural in nature). Detailed algorithmic descriptions typically are documented in a method specification (as described later in this chapter). In other words, a contract is composed of the information required for the developer of a client object to know what messages can be sent to the server objects and what the client can expect in return. Figure 9-19 shows a sample format for a contract.

Because each contract is associated with a specific method and a specific class, the contract must document them. The ID number of the contract is used to provide a unique identifier for every contract. The Clients (Consumers) element of a contract is a list of classes and methods that send a message to this specific method. This list is determined by reviewing the sequence diagrams associated with the server class. The Associated Use Cases element is a list of use cases in which this method is used to realize the implementation of the use case. The use cases listed here can be found by reviewing the server class's CRC card and the associated sequence diagrams.

The Description of Responsibilities provides an informal description of what the method is to perform, not how it is to do it. The arguments received are the data types of the parameters passed to the method and the value returned is the data type of the value that the method will return to its clients. Together with the method name, they form the signature of the method.

The precondition and postcondition elements are where the pre- and postconditions for the method are recorded. Recall that pre- and postconditions can be written in natural language, Structured English, pseudocode, or a formal language. It really does not matter which one you use. However, the more precisely they are written, the less likely it is that the programmer will misunderstand them. As such, we recommend that you use either pseudocode or a formal language such as UML's Object Constraint Language.[35]

[34] Currently, there is no standard format for a contract. The contract in Figure 9-19 is based on material contained in Ian Graham, *Migrating to Object Technology* (Reading, MA: Addison-Wesley, 1995); Craig Larman, *Applying UML and Patterns: An Introduction to Object-Oriented Analysis and Design* (Englewood Cliffs, NJ: Prentice Hall, 1998); Meyer, *Object-Oriented Software Construction;* and R. Wirfs-Brock, B. Wilkerson, and L. Wiener, *Designing Object-Oriented Software* (Englewood Cliffs, NJ: Prentice Hall, 1990).

[35] See Warmer and Kleppe, *The Object Constraint Language: Precise Modeling with UML.*

Method Name:	Class Name:		ID:
Clients (Consumers):			
Associated Use Cases:			
Description of Responsibilities:			
Arguments Received:			
Type of Value Returned:			
Pre-Conditions:			
Post-Conditions:			

FIGURE 9-19
Sample Contract Form

YOUR	9-6 Contract
TURN	

*U*sing the CRC card in Figure 9-17, the class diagram in Figure 9-18, and the sample contract format in Figure 9-19 as guides, create contracts for the calculate subtotal, calculate tax, calculate shipping, and calculate total methods.

METHOD SPECIFICATION

Once the analyst has communicated the big picture of how the system needs to be put together, he or she needs to describe the individual classes and methods in enough detail so that programmers can take over and begin writing code. Methods on the CRC cards, class diagram, and contracts are described using *method specifications*. Method specifications are written documents that include explicit instructions on how to write the code to implement the method. Typically, project team members write a specification for each method and then pass them all along to programmers, who write the code during implementation of the project. Specifications need to be very clear and easy to understand, or programmers will be slowed down trying to decipher vague or incomplete instructions.

There is no formal syntax for a method specification, so every organization uses its own format, often using a form like the one in Figure 9-20. Typical method specification forms contain four components that convey the information that programmers will need to write the appropriate code: general information, events, message passing, and algorithm specification.

Method Name:	Class Name:	ID:
Contract ID:	Programmer:	Date Due:

Programming Language:

☐ Visual Basic   ☐ Smalltalk   ☐ C++   ☐ Java

Triggers/Events:

Arguments Received: Data Type:	Notes:

Messages Sent & Arguments Passed: ClassName.MethodName:	Data Type:	Notes:

Arguments Returned: Data Type:	Notes:

Algorithm Specification:

Misc. Notes:

FIGURE 9-20
Method Specification
Form

General Information

The top of the form in Figure 9-20 contains general information, such as the name of the method, name of the class in which this implementation of the method will reside, ID number, Contract ID (which identifies the contract associated with this method implementation), programmer assigned, the date due, and the target programming language. This information is used to help manage the programming effort.

Events

The second section of the form is used to list the events that trigger the method. An *event* is a thing that happens or takes place. Clicking the mouse generates a mouse event, pressing a key generates a keystroke event—in fact, almost everything the user does generates an event to occur.

In the past, programmers used procedural programming languages (e.g., COBOL, C) that contained instructions that were implemented in a predefined order, as determined by the computer system, and users were not allowed to deviate from the order. Many programs today are *event driven* (e.g., programs written in languages such as Visual Basic, Smalltalk, C++, or Java), and event-driven programs include methods that are executed in response to an event initiated by the user, system, or another method. After initialization, the system waits for an event to occur. When it does, a method is fired that carries out the appropriate task, and then the system waits once again.

We have found that many programmers still use method specifications when programming in event-driven languages, and they include the event section on the form to capture when the method will be invoked. Other programmers have switched to other design tools that capture event-driven programming instructions, such as the behavioral state machine described in Chapter 7.

Message Passing

The next sections of the method specification describe the message passing to and from the method, which are identified on the sequence and collaboration diagrams. Programmers will need to understand what arguments are being passed into, from, and returned by the method because the arguments ultimately will translate into attributes and data structures within the actual method.

Algorithm Specification

Algorithm specifications can be written in Structured English, some form of pseudocode, or some type of formal language.[36] Structured English is simply a formal way of writing instructions that describe the steps of a process. Because it is the first step toward the implementation of the method, it looks much like a simple programming language. Structured English uses short sentences that clearly describe exactly what work is performed on what data. There are many versions of Structured English because there are no formal standards; each organization has its own type of Structured English.

Figure 9-21 shows some examples of commonly used Structured English statements. Action statements are simple statements that perform some action. An If statement controls actions that are performed under different conditions, and a For statement (or a While statement) performs some actions until some condition is reached. Finally, a Case statement is an advanced form of an If statement that has several mutually exclusive branches.

Pseudocode is a language that contains logical structures, including sequential statements, conditional statements, and iteration. It differs from Structured English in that pseudocode contains details that are programming specific, such as initialization instructions or linking, and it also is more extensive so that a programmer can write the module by mirroring the pseudocode instructions. In general, pseudocode is much more like real code, and its audience is the programmer, as opposed to the analyst. Its format is not as

---

[36] For our purposes, Structured English or pseudocode will suffice. However, there has been some work with the Catalysis, Fusion, and Syntropy methodologies to include formal languages, such as VDM and Z, into specifying object-oriented systems.

Common Statements	Example
Action Statement	Profits = Revenues – Expenses Generate Inventory-Report
If Statement	IF Customer Not in the Customer Object Store THEN Add Customer record to Customer Object Store ELSE Add Current-Sale to Customer's Total-Sales Update Customer record in Customer Object Store

For Statement	FOR all Customers in Customer Object Store DO
	    Generate a new line in the Customer-Report
	    Add Customer's Total-Sales to Report-Total

Case Statement	CASE
	    IF Income < 10,000: Marginal-tax-rate = 10 percent
	    IF Income < 20,000: Marginal-tax-rate = 20 percent
	    IF Income < 30,000: Marginal-tax-rate = 31 percent
	    IF Income < 40,000: Marginal-tax-rate = 35 percent
	    ELSE Marginal-Tax-Rate = 38 percent
	ENDCASE

FIGURE 9-21
Structured English

important as the information it conveys. Figure 9-22 shows a short example of pseudocode for a module that is responsible for getting CD information.

Writing good pseudocode can be difficult—imagine creating instructions that someone else can follow without having to ask for clarification or making a wrong assumption. For example, have you ever given a friend directions to your house, and they ended up getting lost? To you, the directions may have been very clear, but that is because of your personal assumptions. To you, the instruction take the first left turn may really mean take a left turn at the first stoplight. Someone else may have made the interpretation take a left turn at the first road, with or without a light. Therefore, when writing pseudocode, pay special attention to detail and readability.

If the algorithm of a method is complex, a tool that can be useful for algorithm specification is UML's *activity diagram* (see Chapter 5). Recall that activity diagrams can be used to specify any type of process. Obviously, an algorithm specification represents a process. However, due to the nature of object orientation, processes tend to be highly distributed over many little methods over many objects. As such, needing to use an activity diagram to specify the algorithm of a method may, in fact, hint at a problem in the design. For example, the method should be further decomposed or there could be missing classes.

The last section of the method specification provides space for other information that needs to be communicated to the programmer, such as calculations, special business rules, calls to subroutines or libraries, and other relevant issues. This also can point out changes or improvements that will be made to any of the other design documentation based on problems that the analyst detected during the specification process.[37]

```
(Get_CD_Info module)
    Accept (CD.Title) {Required}
    Accept (CD.Artist) {Required}
    Accept (CD.Category) {Required}
    Accept (CD.Length)
Return
```

FIGURE 9-22
Pseudocode

[37] Remember that the development process is very incremental and iterative in nature. Therefore, changes could be cascaded back to any point in the development process (e.g., to use-case descriptions, use-case diagrams, CRC cards, class diagrams, object diagrams, sequence diagrams, communication diagrams, behavioral state machines, and package diagrams).

YOUR	9-7 **Method Specification**
TURN	

*U*sing the CRC card in Figure 9-17, the class diagram in Figure 9-18, and the contract created in Your Turn 9-6 as guides, create method specifications for the calculate subtotal, calculate tax, calculate shipping, and calculate total methods.

APPLYING THE CONCEPTS AT CD SELECTIONS

Alec and his team began the detailed object design process by reviewing the class and package diagram for the problem-domain layer (see Figures 6-18, 8-14, and 8-15). Alec made it clear that the team should be aware of the cohesion, coupling, and connascence design criteria and to review the models with those in mind. Furthermore, he insisted that they look to see if there were any additional specifications necessary, any opportunities for reuse that could be exploited, and any further restructuring of the design. Alec assigned Anne to review all results and to look for any possible optimizations that could be implemented. Finally, since the implementation would be in Java, he asked Anne to also ensure that the design could be implemented in a single-inheritance-based language.

Upon their review, it was discovered that there were quite a few many-to-many (*..*) association relationships on the class diagram. Alec questioned whether this was a correct representation of the actual situation. Brian admitted that when they put together the class diagram, they had decided to model most of the associations as a many-to-many multiplicity, figuring that this could be easily fixed at a later point in time when they had more precise information. Alec also questioned why this issue was not addressed during the verification and validation step. However, he did not assign any blame at this point in time. Instead, because Brian was the team member most familiar with structural modeling and was the analyst in charge of the data management layer (see Chapter 10), Alec assigned him to evaluate the multiplicity of each association in the model and to restructure and optimize the evolving problem domain model.

Figure 9-23 shows the updated version of the class diagram. As you can see, Brian included both the lower and upper values of the multiplicity of the associations. He did this to remove any ambiguity about the associations. Because there is a one-to-one relationship between the CD class and the Mkt Info class, Brian considered merging them into a single class. However, he decided that they were sufficiently different to keep them separate. He reasoned this by assuming that not all customers would want to see all the Mkt Info associated with every CD.

Because Brian had already spent quite a bit of time on the classes in the CD package, upon reviewing the new, revised class diagram, Alec assigned it to Brian. The classes in the CD package were CD, Vendor, Mkt Info, Review, Artist Info, and Sample Clip. Because Anne was going to have to review all classes and packages from a more technical perspective, Alec decided to take on the Customer and Order packages himself (see Figures 8-14 and 8-15). However, because the Search package could become rather technical, he assigned it to Anne.

Next, Brian added invariants, preconditions, and postconditions to the classes and their methods. For example, Figure 9-24 portrays the back of the CRC card for the CD class. He decided to add only the invariant information to the CRC cards and not the class diagram in order to keep the class diagram as simple and as easy to understand as possible.

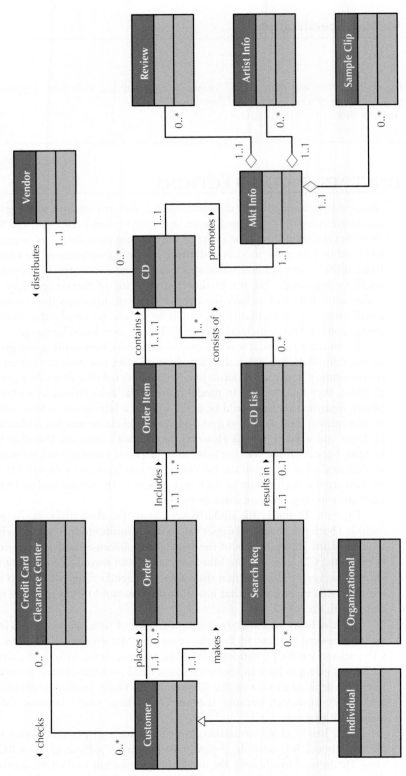

FIGURE 9-23 Revised CD Selections Internet Sales System Class Diagram (Places Order Use Case View)

Back:

Attributes:

CD Number	(1..1)	(unsigned long)
CD Name	(1..1)	(String)
Pub Date	(1..1)	(Date)
Artist Name	(1..1)	(String)
Artist Number	(1..1)	(unsigned long)
Vendor	(1..1)	(Vendor)
Vendor ID	(1..1)	(unsigned long) {Vendor ID = Vendor.GetVendorID()}

Relationships:

Generalization (a-kind-of): _____

Aggregation (has-parts): _____

Other Associations: Order Item {0..*} CD List {0..*} Vendor {1..1} Mkt Info {0..1}

FIGURE 9-24
Back of CD CRC Card

Notice the additional set of multiplicity, domain, and referential integrity invariants added to the attributes and relationships. During this process, he realized that not all CDs had marketing information associated with them. As such, he modified the multiplicity from the CD class to the Mkt Info class.

Furthermore, he needed to add contracts for each method. Figure 9-25 portrays the contract for the GetReview() method associated with the Mkt Info class. Notice that there is a precondition for this method to succeed—Review attribute, not Null.

Upon completing the CRC cards and contracts, Brian moved on to specifying the detailed design for each method. For example, the method specification for the GetReview() method is given in Figure 9-26. Brian developed this specification by reviewing the Places Order use case (see Figure 5-17), the sequence diagram (see Figure 7-16), and the contract (see Figure 9-25). Notice that Brian is enforcing the precondition on the contract by testing to see whether the Review attribute that contains the list of reviews contains a value or not. Because the method is to be implemented in Java, he has specified that an exception is to be thrown if there are no reviews. Finally, Brian updated the class diagram for the CD package (see Figure 9-27).

YOUR TURN

9-8 CD Selections Internet Sales System

At this point in time, you should play the role of Alec and complete the design of the problem domain layer by identifying all relevant invariants and placing them on the appropriate CRC cards and creating all necessary contracts and method specifications not included in the CD package.

Method Name: GetReview()	Class Name:		ID:
Clients (Consumers): CD Detailed Report			
Associated Use Cases: Places Order			
Description of Responsibilities: Return review objects for the Detailed Report Screen to display			
Arguments Received:			
Type of Value Returned: List of Review objects			
Preconditions: Review attribute not Null			
Postconditions:			

FIGURE 9-25
Get Review
Method Contract

SUMMARY

Review of the Basic Characteristics of Object-Oriented Systems

A class is a template on which objects can be instantiated. An object is a person, place, or thing about which we want to capture information. Each object has attributes and methods. The methods are executed by objects sending messages that trigger them. Encapsulation and information hiding allows an object to conceal its inner processes and data from the other objects. Polymorphism and dynamic binding allow a message to be interpreted differently by different kinds of objects. However, if polymorphism is not used in a semantically consistent manner, it can make an object design incomprehensible. Classes can be arranged in a hierarchical fashion in which subclasses inherit attributes and methods from superclasses to reduce the redundancy in development. However, through redefinition capabilities or multiple inheritance, inheritance conflicts can be introduced into the design.

Design Criteria

Coupling, cohesion, and connascence have been put forth as a set of criteria by which a design of an object-oriented system can be evaluated. Two types of coupling, interaction and inheritance, and three types of cohesion, method, class, and generalization/specialization, were described. Interaction coupling deals with the communication that takes place among the objects, and inheritance coupling deals with the innate dependencies in the use of inheritance in object-oriented systems. Method cohesion addresses how single-minded a method is. The fewer things a method does, the more cohesive it is. Class cohesion does the same for classes. A class should be a representation of one and only one thing. Generalization/specialization cohesion deals with the quality of an

Method Name: GetReview()	Class Name: Mkt Info		ID: 453
Contract ID: 89	Programmer: John Smith		Date Due: 7/7/06

Programming Language:

          ❑ **Visual Basic**   ❑ **Smalltalk**   ❑ **C++**   ☑ **Java**

Triggers/Events:

    Detail Button on Basic Report is pressed

Arguments Received: Data Type:	Notes:

Messages Sent & Arguments Passed: ClassName.MethodName:	Data Type:	Notes:

Argument Returned: Data Type:	Notes:
List	List of Review objects

Algorithm Specification:

    IF Review Not Null
           Return Review
    Else
        Throw Null Exception

Misc. Notes:

FIGURE 9-26
Create Review
Method Specification

inheritance hierarchy. Good inheritance hierarchies support only generalization and specialization (a-kind-of) semantics. Connascence generalizes coupling and cohesion and then combines them with the different levels of encapsulation. The general rule of thumb is to maximize the cohesion (connascence) within an encapsulation boundary and minimize the coupling (connascence) between the encapsulation boundaries.

FIGURE 9-27
Revised Package
Diagram for the CD
Package on the PD
Layer of the CD
Selections Internet
Sales System

Object Design Activities

There are five basic object design activities. First, additional specification is possible by carefully reviewing the models, deciding on the proper visibility of the attributes and methods, setting the signature for each method, and identifying any constraints associated with the classes or the classes' methods. Second, look for opportunities for reuse by reviewing the model and looking at possible patterns, class libraries, frameworks, and components that could be used to enhance the system. Third, restructure the model through the use of factoring and normalization. Be sure to take the programming language into consideration. It may be necessary to map the current design into the restricted capabilities of the language (e.g., the language supports only single inheritance). Also, be certain that the inheritance in your model supports only generalization/specialization (a-kind-of) semantics. Fourth, optimize the design. However, be careful in the process of optimizing the design. Optimizations typically decrease the understandability of the model. Fifth, map the problem domain classes to an implementation language.

Constraints and Contracts

There are three types of constraints associated with object-oriented design: invariants, preconditions, and postconditions. Invariants capture constraints that must always be true for all instances of a class (e.g., domains and values of attributes or multiplicity of relationships). Typically, invariants are attached to class diagrams and CRC cards. However, for clarity purposes, we suggest placing them only on the CRC cards.

Contracts formalize the interaction between objects (i.e., the message passing). As such, they include the pre- and postconditions that must be enforced for a method to execute properly. Contracts provide an approach to modeling the rights and obligations when client and server objects interact. From a practical perspective, all interactions that can take place between all possible client and server objects are not modeled on separate contracts. Instead, a single contract is drawn up for using each visible method of a server object.

Method Specification

A method specification is a written document that provides clear and explicit instructions on how a method is to behave. Without clear and unambiguous method specifications, critical design decisions will have to be made by programmers instead of by designers. Even though there is no standard format for a method specification, typically four types of information are captured. First, there is general information such as the name of the method, name of the class, contract ID, programmer assigned, the date due, and the target programming language. Second, due to the rise in popularity of GUI-based and event-driven systems, events also are captured. Third, the information that deals with the signature of the method, data received, data passed on to other methods, and data returned by the method is recorded. Finally, an unambiguous specification of the algorithm is given. The algorithm typically is modeled using Structured English, pseudocode, an activity diagram, or a formal language.

KEY TERMS

Active value
ActiveX
Activity diagram
API (application program interface)
Attribute
Behavior
Class
Class cohesion
Class library
Client
Cohesion
Component
Connascence
Constraint
Consumer
Contract
Coupling
Derived attribute
Design pattern
Dynamic binding
Encapsulation
Event

Event driven
Exceptions
Factoring
Fan-out
Framework
Generalization/specialization
    cohesion
Heteronyms
Homographs
Homonyms
Ideal class cohesion
Information hiding
Inheritance
Inheritance conflict
Inheritance coupling
Instance
Interaction coupling
Invariant
JavaBean
Law of Demeter
Message
Method

Method cohesion
Method specification
Multiple inheritance
Normalization
Object
Object-based language
Operations
Patterns
Polymorphism
Postcondition
Precondition
Pseudocode
Redefinition
Server
Signature
Single inheritance
State
Structured English
Supplier
Synonyms
Trigger
Visibility

QUESTIONS

1. What are the basic characteristics of object-oriented systems?
2. How important is it to know which object-oriented programming language is going to be used to implement the system?
3. What is the law of Demeter?
4. What are the six types of interaction coupling? Give one example of good interaction coupling and one example of bad interaction coupling.
5. What are the different ways to optimize an object system?
6. What is the typical downside of system optimization?
7. What are the four types of class cohesion? Give one example of each type.
8. What are the seven types of method cohesion? Give one example of good method cohesion and one example of bad method cohesion.
9. What the five types of connascence described in your text? Give one example of each type.

10. When designing a specific class, what types of additional specification for a class could be necessary?
11. What is an inheritance conflict? How does an inheritance conflict affect the design?
12. Give the guidelines to avoid problems with inheritance conflicts.
13. Define polymorphism. Give one example of a good use of polymorphism and one example of a bad use of polymorphism.
14. How do you specify a method's algorithm? Give an example of an algorithm specification for a compute pay method associated with an hourly employee class.
15. How are methods specified? Give an example of a method specification for a compute pay method associated with an hourly employee class.
16. Why is cancellation of methods a bad thing?
17. What is an invariant? How are invariants modeled in a design of a class? Give an example of an invariant for a hourly employee class.

18. How are factoring and normalization used in designing an object system?
19. What additional types of inheritance conflicts are there when using multiple inheritance?
20. What are constraints? What are the three different types of constraints?
21. What are patterns, frameworks, class libraries, and components? How are they used to enhance the evolving design of the system?
22. What are exceptions?
23. What is dynamic binding?
24. What is the purpose of a contract? How are contracts used?
25. Create a contract for a compute pay method associated with an hourly employee class.

EXERCISES

A. For the health club exercises in Chapter 5 (N, O), Chapter 6 (O), Chapter 7 (G, H), and Chapter 8 (G, H), choose one of the classes and create a set of invariants for attributes and relationships and add them to the CRC card for the class.

B. Choose one of the methods in the class that you chose for exercise A and create a contract and a method specification for it. Use Structured English for the algorithm specification.

C. For the Picnics R Us exercises in Chapter 5 (P, Q), Chapter 6 (P), Chapter 7 (I), and Chapter 8 (I, J), choose one of the classes and create a set of invariants for attributes and relationships and add them to the CRC card for the class.

D. Choose one of the methods in the class that you chose for exercise C and create a contract and a method specification for it. Use Structured English for the algorithm specification.

E. For the Of-the-Month Club exercises in Chapter 5 (T, U), Chapter 6 (R), Chapter 7 (L), and Chapter 8 (K, L), choose one of the classes and create a set of invariants for attributes and relationships and add them to the CRC card for the class.

F. Choose one of the methods in the class that you chose for exercise E and create a contract and a method spec-

ification for it. Use Structured English for the algorithm specification.

G. For the library exercises in Chapter 5 (R, S), Chapter 6 (Q), Chapter 7 (K), and Chapter 8 (M, N), choose one of the classes and create a set of invariants for attributes and relationships and add them to the CRC card for the class.

H. Choose one of the methods in the class that you chose for exercise G and create a contract and a method specification for it. Use Structured English for the algorithm specification.

I. Describe the difference in meaning between the following two class diagrams. Which is a better model? Why?

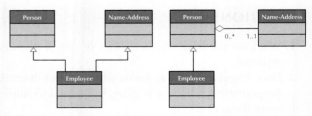

J. From a cohesion, coupling, and connascence perspective, is the following class diagram a good model? Why or why not?

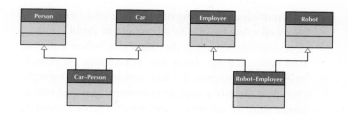

L. Create a set of inheritance conflicts for the two inheritance structures in the class diagrams of exercise K.

K. From a cohesion, coupling, and connascence perspective, are the following class diagrams good models? Why or why not?

MINICASES

1. You have been working with the Holiday Travel Vehicle problem for quite a while. You should go back and refresh your memory about the problem before attempting to solve this situation. Refer back to your solutions for Minicase 1 in Chapter 6 and Minicase 2 in Chapter 7.

   In the new system for Holiday Travel Vehicles, the system users follow a two-stage process to record complete information on all of the vehicles sold. When an RV or trailer first arrives at the company from the manufacturer, a clerk from the inventory department creates a new vehicle record for it in the computer system. The data entered at this time include basic descriptive information on the vehicle such as manufacturer, name, model, year, base cost, and freight charges. When the vehicle is sold, the new vehicle record is updated to reflect the final sales terms and the dealer-installed options added to the vehicle. This information is entered into the system at the time of sale when the salesperson completes the sales invoice.

   When it is time for the clerk to finalize the new vehicle record, the clerk will select a menu option from the system, which is called Finalize New Vehicle Record. The tasks involved in this process are described below.

   When the user selects the "Finalize New Vehicle Record" from the system menu, the user is immediately prompted for the serial number of the new vehicle. This serial number is used to retrieve the new vehicle record for the vehicle from system storage. If a record cannot be found, the serial number is probably invalid. The vehicle serial number is then used to retrieve the option records that describe the dealer-installed options that were added to the vehicle at the customer's request. There may be zero or more options. The cost of the option specified on the option record(s) is totaled. Then, the dealer cost is calculated using the vehicle's base cost, freight charge, and total option cost. The completed new vehicle record is passed back to the calling module.

   a. Update the structural model (CRC cards and class diagram) with this additional information.

   b. For each class in the structural model create a set of invariants for attributes and relationships and add them to the CRC cards for the classes.

   c. Choose one of the classes in the structural model. Create a contract for each method in that class. Be as complete as possible.

   d. Create a method specification for each method in the class you chose for question c. Use Structured English for the algorithm specification.

2. You are a team leader with a major IS outsourcing firm based in India. Bob Smith is the project manager of a joint major development project between your firm and a large US company. Bob has been in the software industry for 30 years and has always prided himself on his expertise in developing software. He is familiar with structured analysis and design as well as

information engineering of the 1980s and 1990s. He is also well-versed with the advantage of rapid application development. But the other day, when you and Bob were talking about the advantages of object-oriented approaches, he was perplexed. He thought that characteristics such as polymorphism and inheritance were an advantage for object-oriented systems. However, when you explained the problems with inheritance conflicts, redefinition capabilities, and the need for semantic consistency across different implementations of methods, he was ready to give up. You

then went on to explain the importance of contracts in maintaining the system. At this point in the conversation, he basically threw in the towel. As he walked off, you heard him say something like "I guess it's true, that it's hard to teach old dog new tricks."

Not wanting to upset a good client, you decided to write a short tutorial on object-oriented systems development for Bob. Create a detailed outline for the tutorial, using good design criteria such as coupling and cohesion.

CHAPTER 10

DATA BASE DESIGN

A project team designs the data management layer of a system using a four-step process: selecting the format of the storage, mapping the problem domain classes to the selected format, optimizing the storage to perform efficiently, and then designing the necessary data access and manipulation classes. This chapter first describes the different ways in which objects can be stored and several important characteristics that should be considered when choosing among object-persistence formats. Second, the chapter describes a problem-domain class to object-persistence format mapping process for the most important object-persistence formats. Third, because the most popular storage format today is the relational database, the chapter focuses on the optimization of relational databases from both storage and access perspectives. Fourth, the chapter describes the affect that nonfunctional requirements have on the data management layer. Last, the chapter describes how to design data access and manipulation classes.

OBJECTIVES

- Become familiar with several object-persistence formats.
- Be able to map problem domain objects to different object-persistence formats.
- Be able to apply the steps of normalization to a relational database.
- Be able to optimize a relational database for object storage and access.
- Become familiar with indexes for relational databases.
- Be able to estimate the size of a relational database.
- Understand the affect of nonfunctional requirements on the data management layer.
- Be able to design the data access and manipulation classes.

CHAPTER OUTLINE

INTRODUCTION

As explained in Chapter 8, the work done by any application can be divided into a set of layers. This chapter focuses on the *data management layer,* which includes both data access and manipulation logic, along with the actual design of the storage. The data storage component of the data management layer manages how data are stored and handled by the programs that run the system. This chapter describes how a project team designs the storage for objects (*object persistence*) using a four-step approach: selecting the format of the storage, mapping the problem-domain objects to the object-persistence format, optimizing the object-persistence format, and designing the data access and manipulation classes necessary to handle the communication between the system and the database.

Applications are of little use without the data that they support. How useful is a multimedia application that can't support images or sound? Why would someone log into a system to find information if it took him or her less time to locate the information manually? Design includes four steps to object-persistence design that decrease the chances of ending up with inefficient systems, long system-response times, and users who cannot get to the information that they need in the way that they need it—all of which can affect the success of the project.

The first part of this chapter describes a variety of storage formats and explains how to select the appropriate one for your application. From a practical perspective, there are four basic types of formats that can be used to store objects for application systems: files (sequential and random), object-oriented databases, object-relational databases, or relational databases.[1] Each type has certain characteristics that make it more appropriate for some types of systems over others.

Once the object-persistence format is selected to support the system, the problem-domain objects need to drive the design of the actual object storage. Then the object storage needs to be designed to optimize its processing efficiency, which is the focus of the next part of the chapter. One of the leading complaints by end users is that the final system is too slow, so to avoid such complaints project team members must allow time during design to carefully make sure that the file or database performs as fast as possible. At the same time, the team must keep hardware costs down by minimizing the storage space that the application will require. The goals of maximizing access to the objects and minimizing the amount of space taken to store objects can conflict, and designing object persistence efficiency usually requires trade-offs.

Finally, it is necessary to design a set of data access and manipulation classes to ensure the independence of the problem-domain classes from the storage format. The data access and manipulation classes handle all communication with the database. In this manner, the problem domain is decoupled from the object storage allowing the object storage to be changed without impacting the problem domain classes.

OBJECT-PERSISTENCE FORMATS

There are four main types of object-persistence formats: files (sequential and random access), object-oriented databases, object-relational databases, and relational databases. *Files* are electronic lists of data that have been optimized to perform a particular transaction. For

---

[1] There are other types of files, such as relative, indexed sequential, and multi-indexed sequential, and databases, such as hierarchical, network, and multidimensional. However, these formats typically are not used for object persistence.

Order Number	Date	Cust ID	Last Name	First Name	Amount	Tax	Total	Prior Customer	Payment Type
234	11/23/00	2242	DeBerry	Ann	$ 90.00	$5.85	$ 95.85	Y	MC
235	11/23/00	9500	Chin	April	$ 12.00	$0.60	$ 12.60	Y	VISA
236	11/23/00	1556	Fracken	Chris	$ 50.00	$2.50	$ 52.50	N	VISA
237	11/23/00	2242	DeBerry	Ann	$ 75.00	$4.88	$ 79.88	Y	AMEX
238	11/23/00	2242	DeBerry	Ann	$ 60.00	$3.90	$ 63.90	Y	MC
239	11/23/00	1035	Black	John	$ 90.00	$4.50	$ 94.50	Y	AMEX
240	11/23/00	9501	Kaplan	Bruce	$ 50.00	$2.50	$ 52.50	N	VISA
241	11/23/00	1123	Williams	Mary	$120.00	$9.60	$129.60	N	MC
242	11/24/00	9500	Chin	April	$ 60.00	$3.00	$ 63.00	Y	VISA
243	11/24/00	4254	Bailey	Ryan	$ 90.00	$4.50	$ 94.50	Y	VISA
244	11/24/00	9500	Chin	April	$ 24.00	$1.20	$ 25.20	Y	VISA
245	11/24/00	2242	DeBerry	Ann	$ 12.00	$0.78	$ 12.78	Y	AMEX
246	11/24/00	4254	Bailey	Ryan	$ 20.00	$1.00	$ 21.00	Y	MC
247	11/24/00	2241	Jones	Chris	$ 50.00	$2.50	$ 52.50	N	VISA
248	11/24/00	4254	Bailey	Ryan	$ 12.00	$0.60	$ 12.60	Y	AMEX
249	11/24/00	5927	Lee	Diane	$ 50.00	$2.50	$ 52.50	N	AMEX
250	11/24/00	2242	DeBerry	Ann	$ 12.00	$0.78	$ 12.78	Y	MC
251	11/24/00	9500	Chin	April	$ 15.00	$0.75	$ 15.75	Y	MC
252	11/24/00	2242	DeBerry	Ann	$132.00	$8.58	$140.58	Y	MC
253	11/24/00	2242	DeBerry	Ann	$ 72.00	$4.68	$ 76.68	Y	AMEX

FIGURE 10-1   Customer Order File

example, Figure 10-1 shows a customer order file with information about customers' orders, in the form in which it is used, so that the information can be accessed and processed quickly by the system.

A *database* is a collection of groupings of information, each of which is related to each other in some way (e.g., through common fields). Logical groupings of information could include such categories as customer data, information about an order, product information, and so on. A *database management system (DBMS)* is software that creates and manipulates these databases (see Figure 10-2 for a relational database example). Such *end user DBMSs* as Microsoft Access support small-scale databases that are used to enhance personal productivity, whereas *enterprise DBMSs,* such as DB2, Jasmine, and Oracle, can manage huge volumes of data and support applications that run an entire company. An end user DBMS is significantly less expensive and easier for novice users to use than its enterprise counterpart, but it does not have the features or capabilities that are necessary to support mission-critical or large-scale systems.

In the next sections we describe sequential and random access files, relational databases, object-relational databases, and object-oriented databases that can be used to handle a system's object persistence requirements. Finally, we describe a set of characteristics on which the different formats can be compared.

Sequential and Random Access Files

From a practical perspective, most object-oriented programming languages support sequential and random access files as part of the language.[2] In this section, we describe

[2] For example, see the FileInputStream, FileOutputStream, and RandomAccessFile classes in the java.io package.

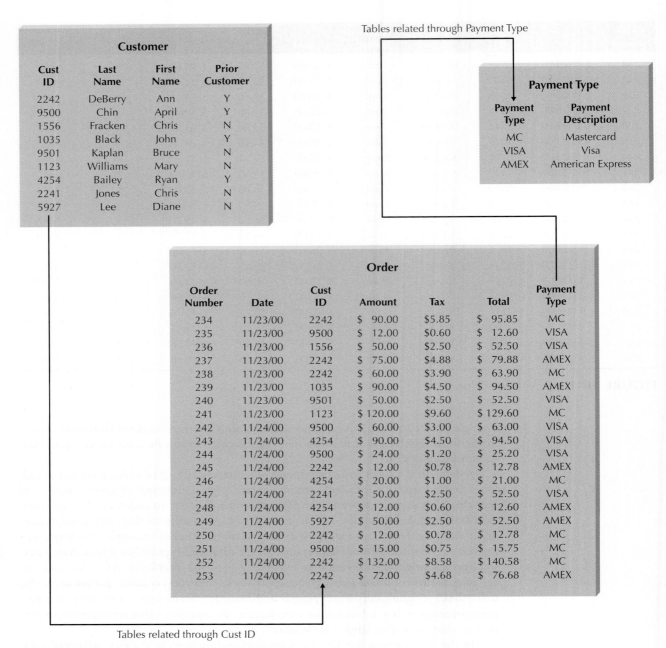

Tables related through Payment Type

Customer

Cust ID	Last Name	First Name	Prior Customer
2242	DeBerry	Ann	Y
9500	Chin	April	Y
1556	Fracken	Chris	N
1035	Black	John	Y
9501	Kaplan	Bruce	N
1123	Williams	Mary	N
4254	Bailey	Ryan	Y
2241	Jones	Chris	N
5927	Lee	Diane	N

Payment Type

Payment Type	Payment Description
MC	Mastercard
VISA	Visa
AMEX	American Express

Order

Order Number	Date	Cust ID	Amount	Tax	Total	Payment Type
234	11/23/00	2242	$ 90.00	$5.85	$ 95.85	MC
235	11/23/00	9500	$ 12.00	$0.60	$ 12.60	VISA
236	11/23/00	1556	$ 50.00	$2.50	$ 52.50	VISA
237	11/23/00	2242	$ 75.00	$4.88	$ 79.88	AMEX
238	11/23/00	2242	$ 60.00	$3.90	$ 63.90	MC
239	11/23/00	1035	$ 90.00	$4.50	$ 94.50	AMEX
240	11/23/00	9501	$ 50.00	$2.50	$ 52.50	VISA
241	11/23/00	1123	$ 120.00	$9.60	$ 129.60	MC
242	11/24/00	9500	$ 60.00	$3.00	$ 63.00	VISA
243	11/24/00	4254	$ 90.00	$4.50	$ 94.50	VISA
244	11/24/00	9500	$ 24.00	$1.20	$ 25.20	VISA
245	11/24/00	2242	$ 12.00	$0.78	$ 12.78	AMEX
246	11/24/00	4254	$ 20.00	$1.00	$ 21.00	MC
247	11/24/00	2241	$ 50.00	$2.50	$ 52.50	VISA
248	11/24/00	4254	$ 12.00	$0.60	$ 12.60	AMEX
249	11/24/00	5927	$ 50.00	$2.50	$ 52.50	AMEX
250	11/24/00	2242	$ 12.00	$0.78	$ 12.78	MC
251	11/24/00	9500	$ 15.00	$0.75	$ 15.75	MC
252	11/24/00	2242	$ 132.00	$8.58	$ 140.58	MC
253	11/24/00	2242	$ 72.00	$4.68	$ 76.68	AMEX

Tables related through Cust ID

FIGURE 10-2   Customer Order Database

what sequential access and random access files are.[3] We also describe how sequential access and random access files are used to support an application. For example, they can be used to support master files, look-up files, transaction files, audit files, and history files.

Sequential access files allow only sequential file operations to be performed (e.g., read, write, and search). Sequential access files are very efficient for sequential operations, such as report writing. However, for random operations, such as finding or updating a specific

[3] For a more complete coverage of issues related to the design of files, see Owen Hanson, *Design of Computer Data Files* (Rockville, MD: Computer Science Press, 1982).

object, they are very inefficient. On the average, 50 percent of the contents of a sequential access file will have to be searched through before finding the specific object of interest in the file. They come in two flavors: ordered and unordered.

An *unordered sequential access file* is basically an electronic list of information stored on disk. Unordered files are organized serially (i.e., the order of the file is the order in which the objects are written to the file). Typically, new objects simply are added to the file's end.

Ordered sequential access files are placed into a specific sorted order (e.g., in ascending order by customer number). However, there is overhead associated with keeping files in a particular sorted order. The file designer can keep the file in sorted order by always creating a new file each time a delete or addition occurs, or he or she can keep track of the sorted order via the use of a *pointer*, which is information about the location of the related record. A pointer is placed at the end of each record, and it "points" to the next record in a series or set. The underlying data/file structure in this case is a *linked list*.[4]

Random access files allow only random or direct file operations to be performed. This type of file is optimized for random operations, such as finding and updating a specific object. Random access files typically give a faster response time to find and update operations than any other type of file. However, because they do not support sequential processing, applications such as report writing are very inefficient. The various methods to implement random access files are beyond the scope of this book.[5]

There are times when it is necessary to be able to process files in both a sequential and random manner. One simple way to do this is to use a sequential file that contains a list of the keys (the field in which the file is to be kept in sorted order) and a random access file for the actual objects. This will minimize the cost of additions and deletions to a sequential file while allowing the random file to be processed sequentially by simply passing the key to the random file to retrieve each object in sequential order. It also allows for fast random processing to occur by using only the random access file, thus optimizing the overall cost of file processing. However, if a file of objects needs to be processed in both a random and sequential manner, the developer should look toward using a database (relational, object relational, or object oriented) instead.

There are many different application types of files—for example, master files, lookup files, transaction files, audit files, and history files. *Master files* store core information that is important to the business and, more specifically, to the application, such as order information or customer mailing information. They usually are kept for long periods of time, and new records are appended to the end of the file as new orders or new customers are captured by the system. If changes need to be made to existing records, programs must be written to update the old information.

Lookup files contain static values, such as a list of valid zip codes or the names of the U.S. states. Typically, the list is used for validation. For example, if a customer's mailing address is entered into a master file, the state name is validated against a lookup file that contains U.S. states to make sure that the operator entered the value correctly.

A *transaction file* holds information that can be used to update a master file. The transaction file can be destroyed after changes are added, or the file may be saved in case the transactions need to be accessed again in the future. Customer address changes, for one, would be stored in a transaction file until a program is run that updates the customer address master file with the new information.

---

[4] For more information on various data structures see Ellis Horowitz and Sartaj Sahni, *Fundamentals of Data Structures* (Rockville, MD: Computer Science Press, 1982) and Michael T. Goodrich and Roberto Tamassia, *Data Structures and Algorithms in Java* (New York: Wiley, 1998).

[5] For a more detailed look at the underlying data and file structures of the different types of files, see Mary E. S. Loomis, *Data Management and File Structures,* 2nd ed. (Englewood Cliffs, NJ: Prentice Hall, 1989); and Michael J. Folk and Bill Zoellick, *File Structures: A Conceptual Toolkit* (Reading, MA: Addison-Wesley, 1987).

For control purposes, a company might need to store information about how data change over time. For example, as human resource clerks change employee salaries in a human resource system, the system should record the person who made the changes to the salary amount, the date, and the actual change that was made. An *audit file* records before and after images of data as they are altered so that an audit can be performed if the integrity of the data is questioned.

Sometimes files become so large that they are unwieldy, and much of the information in the file is no longer used. The *history file* (or archive file) stores past transactions (e.g., old customers, past orders) that are no longer needed by system users. Typically the file is stored off-line; yet it can be accessed on an as-needed basis. Other files, such as master files, can then be streamlined to include only active or very recent information.

Relational Databases

A relational database is the most popular kind of database for application development today. A relational database is based on collections of tables with each table having a *primary key*—a field or fields whose values are unique for every row of the table. The tables are related to each other by placing the primary key from one table into the related table as a *foreign key* (see Figure 10-3). Most *relational database management systems (RDBMS)* support *referential integrity,* or the idea of ensuring that values linking the tables together through the primary and foreign keys are valid and correctly synchronized. For example, if an order-entry clerk using the tables in Figure 10-3 attempted to add order 254 for customer number 1111, he or she would have made a mistake because no customer exists in the Customer table with that number. If the RDBMS supported referential integrity, it would check the customer numbers in the Customer table; discover that the number 1111 is invalid; and return an error to the entry clerk. The clerk would then go back to the original order form and recheck the customer information. Can you imagine the problems that would occur if the RDBMS let the entry clerk add the order with the wrong information? There would be no way to track down the name of the customer for order 254.

Tables have a set number of columns and a variable number of rows that contain occurrences of data. *Structured query language (SQL)* is the standard language for accessing the data in the tables, and it operates on complete tables, as opposed to the individual rows in the tables. Thus, a query written in SQL is applied to all the rows in a table all at once, which is different than a lot of programming languages, which manipulate data row by row. When queries must include information from more than one table, the tables first are *joined* together based on their primary key and foreign key relationships and treated as if they were one large table. Examples of RDBMS software are Microsoft Access, Oracle, DB2, and Microsoft SQL Server.

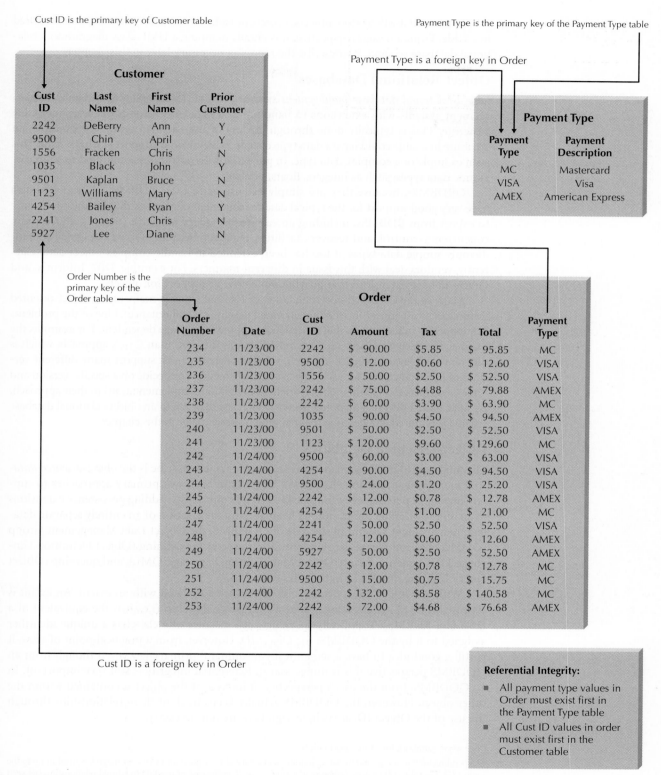

Cust ID is the primary key of Customer table

Payment Type is the primary key of the Payment Type table

Payment Type is a foreign key in Order

Customer

Cust ID	Last Name	First Name	Prior Customer
2242	DeBerry	Ann	Y
9500	Chin	April	Y
1556	Fracken	Chris	N
1035	Black	John	Y
9501	Kaplan	Bruce	N
1123	Williams	Mary	N
4254	Bailey	Ryan	Y
2241	Jones	Chris	N
5927	Lee	Diane	N

Payment Type

Payment Type	Payment Description
MC	Mastercard
VISA	Visa
AMEX	American Express

Order Number is the primary key of the Order table

Order

Order Number	Date	Cust ID	Amount	Tax	Total	Payment Type
234	11/23/00	2242	$ 90.00	$5.85	$ 95.85	MC
235	11/23/00	9500	$ 12.00	$0.60	$ 12.60	VISA
236	11/23/00	1556	$ 50.00	$2.50	$ 52.50	VISA
237	11/23/00	2242	$ 75.00	$4.88	$ 79.88	AMEX
238	11/23/00	2242	$ 60.00	$3.90	$ 63.90	MC
239	11/23/00	1035	$ 90.00	$4.50	$ 94.50	AMEX
240	11/23/00	9501	$ 50.00	$2.50	$ 52.50	VISA
241	11/23/00	1123	$ 120.00	$9.60	$ 129.60	MC
242	11/24/00	9500	$ 60.00	$3.00	$ 63.00	VISA
243	11/24/00	4254	$ 90.00	$4.50	$ 94.50	VISA
244	11/24/00	9500	$ 24.00	$1.20	$ 25.20	VISA
245	11/24/00	2242	$ 12.00	$0.78	$ 12.78	AMEX
246	11/24/00	4254	$ 20.00	$1.00	$ 21.00	MC
247	11/24/00	2241	$ 50.00	$2.50	$ 52.50	VISA
248	11/24/00	4254	$ 12.00	$0.60	$ 12.60	AMEX
249	11/24/00	5927	$ 50.00	$2.50	$ 52.50	AMEX
250	11/24/00	2242	$ 12.00	$0.78	$ 12.78	MC
251	11/24/00	9500	$ 15.00	$0.75	$ 15.75	MC
252	11/24/00	2242	$ 132.00	$8.58	$ 140.58	MC
253	11/24/00	2242	$ 72.00	$4.68	$ 76.68	AMEX

Cust ID is a foreign key in Order

Referential Integrity:

- All payment type values in Order must exist first in the Payment Type table
- All Cust ID values in order must exist first in the Customer table

FIGURE 10-3   Relational Database

To use a RDBMS to store objects, objects must be converted so that they can be stored in a table. From a design perspective, this entails mapping a UML class diagram to a relational database schema. We describe the mapping necessary later in this chapter.

Object-Relational Databases

Object-relational database management systems (ORDBMS) are relational database management systems with extensions to handle the storage of objects in the relational table structure. This is typically done through the use of user-defined types. For example, an attribute in a table could have a data type of map, which would support storing a map. This is an example of a complex data type. In pure RDBMSs, attributes are limited to simple or atomic data types, such as integers, floats, or chars.

ORDBMSs, because they are simply extensions to their RDBMS counterparts, also have very good support for the typical data management operations that business has come to expect from RDBMSs, including an easy-to-use query language (SQL), authorization, concurrency-control, and recovery facilities. However, because SQL was designed to handle only simple data types, it too has been extended to handle complex object data. Currently, vendors deal with this issue in different manners. For example, DB2, Informix, and Oracle all have extensions that provide some level of support for objects.

Many of the ORDBMSs on the market still do not support many of the object-oriented features that can appear in an object-oriented design (e.g., inheritance). One of the problems in supporting inheritance is that inheritance support is language dependent. For example, the manner in which Smalltalk supports inheritance is different than C++'s approach, which is different than Java's approach. As such, vendors currently must support many different versions of inheritance, one for each object-oriented language, or decide on a specific version and force developers to map their object-oriented design (and implementation) to their approach. As such, like RDBMSs, a mapping from a UML class diagram to an object-relational database schema is required. We describe the mapping necessary later in this chapter.

Object-Oriented Databases

The final type of database management system that we describe is the *object-oriented database management systems (OODBMS)*. There have been two primary approaches to supporting object persistence within the OODBMS community: adding persistence extensions to an object-oriented programming language and the creation of an entirely separate database management system. In the case of OODBMS, the Object Data Management Group (ODMG) has completed the standardization process for defining (Object Definition Language, ODL), manipulating (Object Manipulating Language, OML), and querying (Object Query Language, OQL) objects in an OODBMS.[6]

With an OODBMS, collections of objects are associated with an extent. An *extent* is simply the set of instances associated with a particular class (i.e., it is the equivalent of a table in a RDBMS). Technically speaking, each instance of a class has a unique identifier assigned to it by the OODBMS: the *Object ID*. However, from a practical point of view, it is still a good idea to have a semantically meaningful primary key (even though from an OODBMS perspective this is unnecessary). Referential integrity is still very important. In an OODBMS, from the user's perspective, it looks as if the object is contained within the other object. However, the OODBMS actually keeps track of these relationships through the use of the Object ID; as such, foreign keys are not necessary.[7]

---

[6] See www.odmg.org for more information.

[7] Depending on the storage and updating requirements, it usually is a good idea to use a foreign key in addition to the Object ID. The Object ID has no semantic meaning. As such, in the case of needing to rebuild relationships between objects, Object IDs are difficult to validate. Foreign keys, by contrast, should have some meaning outside of the DBMS.

OODBMSs provide support for some form of inheritance. However, as already discussed, inheritance tends to be language dependent. Currently, most OODBMSs are tied closely to either a particular *object-oriented programming language (OOPL)* or a set of OOPLs. Most OODBMSs originally supported either Smalltalk or C++. Today, many of the commercially available OODBMSs provide support for C++, Java, and Smalltalk.

OODBMSs also support the idea of *repeating groups (fields)* or *multivalued attributes*. These are supported through the use of *attribute sets* and *relationships sets*. RDBMSs do not explicitly allow for multivalued attributes or repeating groups. This is considered to be a violation of the first normal form (discussed later in this chapter) for relational databases. Some ORDBMSs do support repeating groups and multivalued attributes.

Up until recently, OODBMSs have mainly been used to support multimedia applications or systems that involve complex data (e.g., graphics, video, sound). Application areas, such as computer-aided design and manufacturing (CAD/CAM), financial services, geographic information systems, health care, telecommunications, and transportation, have been the most receptive to OODBMSs. They are also becoming popular technologies for supporting electronic commerce, online catalogs, and large Web multimedia applications. Examples of pure OODBMSs include Gemstone, Jasmine, O2, Objectivity, ObjectStore, POET, and Versant.

Although pure OODBMS exist, most organizations currently invest in *ORDBMS* technology. The market for OODBMS is expected to grow, but its ORDBMS and RDBMS counterparts dwarf it. One reason for this situation is that there are many more experienced developers and tools in the RDBMS arena. Furthermore, relational users find that using an OODBMS comes with a fairly steep learning curve.

SELECTING AN OBJECT-PERSISTENCE FORMAT

Each of the file and database storage formats that have been presented has its strengths and weaknesses, and no one format is inherently better than the others are. In fact, sometimes a project team will choose multiple formats (e.g., a relational database for one, a file for another, and an object-oriented database for a third). Thus, it is important to understand the strengths and weaknesses of each format and when to use each one. Figure 10-4 presents a summary of the characteristics of each and the characteristics that can help identify when each type of format is more appropriate.

Major Strengths and Weaknesses   The major strengths of files include that some support for sequential and random access files is normally part of an OOPL, files can be designed to be very efficient, and they are a good alternative for temporary or short-term storage. However, all file manipulation must be done through the OOPL. Files do not have any form of access control beyond that of the underlying operating system. Finally, in most cases, if files are used for permanent storage, redundant data most likely will result. This can cause many update anomalies.

RDBMSs bring with them proven commercial technology. They are the leaders in the DBMS market. Furthermore, they can handle very diverse data needs. However, they cannot handle complex data types, such as images. Therefore, all objects must be converted to a form that can be stored in tables composed of atomic or simple data. They provide no support for object orientation. This lack of support causes an *impedance mismatch* between the objects contained in the OOPL and the data stored in the tables. An impedance mismatch refers to the amount of work done by both the developer and DBMS and the potential information loss that can occur when converting objects to a form that can be stored in tables.

	Sequential and Random Access Files	Relational DBMS	Object Relational DBMS	Object-Oriented DBMS
Major Strengths	Usually part of an object-oriented programming language Files can be designed for fast performance Good for short-term data storage	Leader in the database market Can handle diverse data needs	Based on established, proven technology, e.g., SQL Able to handle complex data	Able to handle complex data Direct support for object orientation
Major Weaknesses	Redundant data Data must be updated using programs, i.e., no manipulation or query language No access control	Cannot handle complex data No support for object orientation Impedance mismatch between tables and objects	Limited support for object orientation Impedance mismatch between tables and objects	Technology is still maturing Skills are hard to find
Data Types Supported	Simple and Complex	Simple	Simple and Complex	Simple and Complex
Types of Application Systems Supported	Transaction processing	Transaction processing and decision making	Transaction processing and decision making	Transaction processing and decision making
Existing Storage Formats	Organization dependent	Organization dependent	Organization dependent	Organization dependent
Future Needs	Poor future prospects	Good future prospects	Good future prospects	Good future prospects

FIGURE 10-4   Comparison of Object-Persistence Formats

Because ORDBMSs are typically object-oriented extensions to RDBMSs, they inherit the strengths of RDBMSs. They are based on established technologies, such as SQL, and unlike their predecessors, they can handle complex data types. However, they provide only limited support for object orientation. The level of support varies among the vendors; therefore, ORDBMSs also suffer from the impedance mismatch problem.

OODBMSs support complex data types and have the advantage of directly supporting object orientation. Therefore, they do not suffer from the impedance mismatch that the previous DBMSs do. Even though the ODMG has released version 3.0 of its set of standards, the OODBMS community is still maturing. As such, this technology may still be too risky for some firms. The other major problems with OODBMS are the lack of skilled labor and the perceived steep learning curve of the RDBMS community.

Data Types Supported   The first issue is the type of data that will need to be stored in the system. Most applications need to store simple data types, such as text, dates, and numbers, and all files and DBMSs are equipped to handle this kind of data. The best choice for simple data storage, however, is usually the RDBMS because the technology has matured over time and has continuously improved to handle simple data very effectively.

Increasingly, applications are incorporating complex data, such as video, images, or audio. ORDBMSs or OODBMSs are best able to handle data of this type. Complex data stored as objects can be manipulated much faster than with other storage formats.

Type of Application System   There are many different kinds of application systems that can be developed. *Transaction-processing systems* are designed to accept and process many simultaneous requests (e.g., order entry, distribution, payroll). In transaction-processing systems, the data are continuously updated by a large number of users, and the queries that are asked of the systems typically are predefined or targeted at a small subset of records (e.g., List the orders that were backordered today or What products did customer #1234 order on May 12, 2001?).

Another set of application systems is the set designed to support decision making, such as *decision support systems (DSS), management information systems (MIS), executive information systems (EIS)*, and *expert systems (ES)*. These decision-making support systems are built to support users who need to examine large amounts of read-only historical data. The questions that they ask are often ad hoc, and they include hundreds or thousands of records at a time (e.g., List all customers in the West region who purchased a product costing more than $500 at least three times. or What products had increased sales in the summer months that have not been classified as summer merchandise?).

Transaction-processing systems and DSSs thus have very different data storage needs. Transaction-processing systems need data storage formats that are tuned for a lot of data updates and fast retrieval of predefined, specific questions. Files, relational databases, object-relational databases, and object-oriented databases can all support these kinds of requirements. By contrast, systems to support decision making are usually only reading data (not updating it), often in ad hoc ways. The best choices for these systems usually are RDBMSs because these formats can be configured specially for needs that may be unclear and less apt to change the data.

Existing Storage Formats   The storage format should be selected primarily on the basis of the kind of data and application system being developed. However, project teams should consider the existing storage formats in the organization when making design decisions. In this way, they can better understand the technical skills that already exist and how steep the learning curve will be when the storage format is adopted. For example, a company that is familiar with RDBMS will have little problem adopting a relational database for the project, whereas an OODBMS may require substantial developer training. In the latter situation, the project team may have to plan for more time to integrate the object-oriented database with the company's relational systems or possibly consider moving toward an ORDBMS solution.

Future Needs   Not only should a project team consider the storage technology within the company, but it should also be aware of current trends and technologies that are being used by other organizations. A large number of installations of a specific type of storage format suggest that skills and products are available to support the format. Therefore, the selection of that format is safe. For example, it would probably be easier and less expensive to find RDBMS expertise when implementing a system than to find help with an OODBMS.

Other Miscellaneous Criteria   Other criteria that should be considered include cost, licensing issues, concurrency control, ease of use, security and access controls, version management, storage management, lock management, query management, language bindings, and APIs. We also should consider performance issues, such as cache management, insertion, deletion, retrieval, and updating of complex objects. Finally, the level of support for object orientation (such as objects, single inheritance, multiple inheritance, polymorphism, encapsulation and information hiding, methods, multivalued attributes, repeating groups) is critical.

10-2 Donation-Tracking System

A major public university graduates approximately 10,000 students per year, and the development office has decided to build a Web-based system that solicits and tracks donations from the university's large alumni body. Ultimately, the development officers hope to use the information in the system to better understand the alumni giving patterns so that they can improve giving rates.

Question

1. What kind of system is this? Does it have characteristics of more than one?
2. What different kinds of data will this system use?
3. On the basis of your answers, what kind of data storage format(s) do you recommend for this system?

MAPPING PROBLEM-DOMAIN OBJECTS TO OBJECT-PERSISTENCE FORMATS[8]

As described in the previous section, there are many different formats from which to choose to support object persistence. Each of the different formats can have some conversion requirements. Regardless of the object-persistence format chosen, we suggest supporting primary keys and foreign keys by adding them to the problem domain classes at this point in time. However, this does imply some additional processing required. The developer has to set the value for the foreign key when adding the relationship to an object. In some cases, this overhead may be too costly. In those cases, this suggestion should be ignored. In the remainder of this section, we describe how to map the problem-domain classes to the different object-persistence formats. From a practical perspective, file formats are used mostly for temporary storage. As such, we do not consider them further.

We also recommend that the data management functionality specifics, such as retrieval and updating of data from the object storage, be included only in classes contained in the Data Management layer. This will ensure that the Data Management classes are dependent on the *Problem-Domain classes* and not the other way around. Furthermore, this allows the design of problem-domain classes to be independent of any specific object-persistence environment, thus increasing their portability and their potential for reuse. Like our previous recommendation, this one also implies additional processing. However, the increased portability and potential for reuse realized should more than compensate for the additional processing required.

Mapping Problem-Domain Objects to an OODBMS Format

If we support object persistence with an OODBMS, the mappings between the problem-domain objects and the OODBMS tend to be fairly straightforward. As a starting point, we suggest that each concrete problem-domain class should have a corresponding object-persistence class in the OODBMS. Furthermore, there will be a data access and management (DAM) class that contains the functionality required to manage the interaction between the object-persistence class and the problem-domain layer. For example, using

---

[8] The rules presented in this section are based on material in Ali Bahrami, *Object-Oriented Systems Development using the Unified Modeling Language* (New York: McGraw-Hill, 1999); Michael Blaha and William Premerlani, *Object-Oriented Modeling and Design for Database Applications* (Upper Saddle River, NJ: Prentice Hall, 1998); Akmal B. Chaudri and Roberto Zicari, *Succeeding with Object Databases: A Practical Look at Today's Implementations with Java and XML* (New York: Wiley, 2001); Peter Coad and Edward Yourdon, *Object-Oriented Design* (Upper Saddle River, NJ: Yourdon Press, 1991); and Paul R. Read, Jr., *Developing Applications with Java and UML* (Boston: Addison-Wesley, 2002).

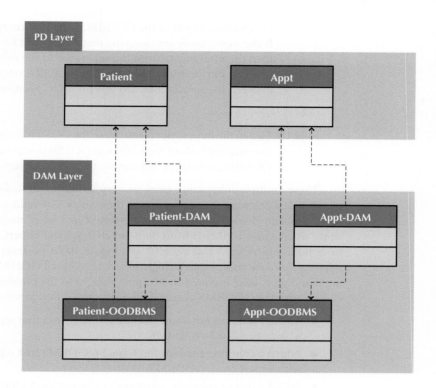

FIGURE 10-5
Appointment System
Problem-Domain and
DAM Layers

the appointment system example from the previous chapters, the Patient class is associated with an OODBMS class (see Figure 10-5). The Patient class essentially will be unchanged from analysis. The Patient-OODBMS class will be a new class that is dependent on the Patient class, whereas the Patient-DAM class will be a new class that depends on both the Patient class and the Patient-OODBMS class. The Patient-DAM class must be able to read from and write to the OODBMS. Otherwise, it will not be able to store and retrieve instances of the Patient class. Even though this does add overhead to the installation of the system, it allows the Problem-Domain class to be independent of the OODBMS being used. As such, if at a latter point in time another OODBMS or object-persistence format is adopted, only the DAM classes will have to be modified. This approach increases both the portability and the potential for reuse of the problem-domain classes.

Even though we are implementing the DAM layer using an OODBMS, a mapping from the problem-domain layer to the OODBMS classes in the data access and management layer may be required, depending on the level of support of inheritance in the OODBMS and the level of inheritance used in the problem-domain classes. If multiple inheritance is used in the problem domain but not supported by the OODBMS, then the multiple inheritance must be factored out of the OODBMS classes. For each case of multiple inheritance (i.e., more than one superclass), the following rules can be used to factor out the multiple inheritance effects in the design of the OODBMS classes.

Rule 1a: Add a column(s) to the OODBMS class(es) that represents the subclass(es) that will contain an Object ID of the instance stored in the OODBMS class that represents the "additional" superclass(es). This is similar in concept to a foreign key in an RDBMS. The multiplicity of this new association from the subclass to the "superclass" should be 1..1. Add a column(s) to the OODBMS class(es) that represents the superclass(es) that will contain an Object ID of the instance stored in the OODBMS class that represents the subclass(es). If the superclasses are concrete, that is, they can be instantiated themselves, then the multiplicity from the superclass to the subclass is 0..*, otherwise, it is 1..1. Furthermore, an exclusive-or (XOR) constraint must be added between the associations. Do this for each "additional" superclass.

OR

Rule 1b: Flatten the inheritance hierarchy of the OODBMS classes by copying the attributes and methods of the additional OODBMS superclass(es) down to all of the OODBMS subclasses and remove the additional superclass from the design.[9]

These multiple-inheritance rules are very similar to those described in Chapter 9. Figure 10-6 demonstrates the application of these rules. The right side of the figure portrays a set of problem-domain classes that are involved in a set of multiple-inheritance relationships, where Class 1 inherits from both SuperClass1 and SuperClass2, and Class 2 inherits from both SuperClass2 and SuperClass3. Figure 10-6a portrays the mapping of multiple inheritance relationships into a single inheritance-based OODBMS using Rule 1a. Assuming that SuperClass2 is concrete, we apply Rule 1a to the problem-domain classes, and we end up with the OODBMS classes on the left side of Part a, where we have:

- Added a column (attribute) to Class1-OODBMS that represents an association with SuperClass2-OODBMS;
- Added a column (attribute) to Class2-OODBMS that represents an association with SuperClass2-OODBMS;
- Added a pair of columns (attributes) to SuperClass2-OODBMS that represents an association with Class1-OODBMS and Class2-OODBMS and for completeness sake;
- Added associations between Class2-OODBMS and SuperClass2-OODBMS and Class2-OODBMS and SuperClass2-OODBMS that have the correct multiplicities and the XOR constraint explicitly shown.

We also display the dependency relationships from the OODBMS classes to the problem-domain classes. Furthermore, we illustrate the fact that the association between Class1-OODBMS and SuperClass2-OODBMS and the association between Class2-OODBMS and SuperClass2-OODBMS are based on the original factored-out inheritance relationships in the problem-domain classes by showing dependency relationships from the associations to the inheritance relationships.

On the other hand, if we apply Rule 1b to map the problem-domain classes to a single-inheritance-based OODBMS, we end up with the mapping in Figure 10-6b, where all the attributes of SuperClass2 have been copied into the Class1-OODBMS and Class2-OODBMS classes. In this latter case, you may have to deal with the effects of inheritance conflicts (see Chapter 9).

The advantage of Rule 1a is that all problem-domain classes identified during analysis are preserved in the database. This allows for maximum flexibility of maintenance of the design of the data management layer. However, Rule 1a increases the amount of message passing required in the system, and it has added processing requirements involving the XOR constraint, thus reducing the overall efficiency of the design. As such, our recommendation is to limit Rule 1a to be applied only when dealing with "extra" superclasses that are concrete because they have an independent existence in the problem domain. Use Rule 1b when they are abstract because they do not have an independent existence from the subclass.

[9] It is also a good idea to document this modification in the design so that in the future, modifications to the design can be easily maintained.

Problem-Domain Classes

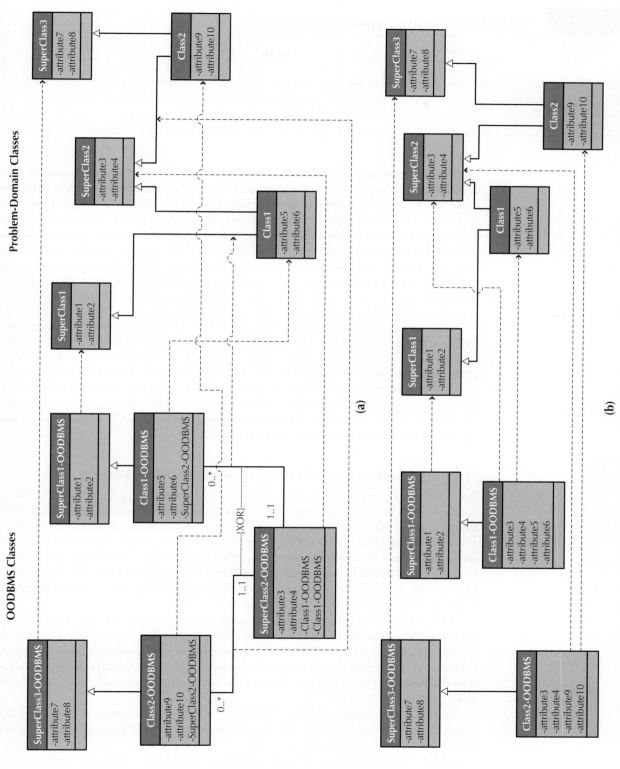

OODBMS Classes

(a)

(b)

FIGURE 10-6 Mapping Problem-Domain Objects to Single Inheritance–Based OODBMS

375

10-3 Dentist Office Appointment System

*I*n the previous chapters, we have been using a dentist office appointment system as an example. Assume that you now know that the OODBMS that will be used to support the system will support only single inheritance. Using a class diagram, draw the design for the database.

In either case, additional processing will be required. In the first case, cascading of deletes will work, not only from the individual object to all its elements, but also from the superclass instances to all the subclass instances. Most OODBMSs do not support this type of deletion. However, to enforce the referential integrity of the system, it must be done. In the second case, there will be a lot of copying and pasting of the structure of the superclass to the subclasses. In the case that a modification of the structure of the superclass is required; then the modification must be cascaded to all of the subclasses. Again, most OODBMSs do not support this type of modification cascading. Therefore, the developer must address it. However, multiple inheritance is rare in most business problems. As such, in most situations the preceding rules will never be necessary.

When instantiating problem-domain objects from OODBMS objects, additional processing will also be required. The additional processing will be in the retrieval of the OODBMS objects and taking their elements to create a problem-domain object. Also, when storing the problem-domain object, the conversion to a set of OODBMS objects is required. Basically speaking, anytime that an interaction takes place between the OODBMS and the system, if multiple inheritance is involved and the OODBMS supports only single inheritance, a conversion between the two formats will be required. This conversion is the purpose of the data access and management classes.

Mapping Problem Domain Objects to an ORDBMS Format

If we support object persistence with an ORDBMS, then the mapping from the problem domain objects to the data management objects is much more involved. Depending on the level of support for object orientation, different mapping rules are necessary. For our purposes, we assume that the ORDBMS supports Object IDs, multivalued attributes, and stored procedures. However, we assume that the ORDBMS does not provide any support for inheritance. Based on these assumptions, Figure 10-7 lists a set of rules that can be used to design the mapping from the problem-domain objects to the tables of the ORDBMS-based data access and management layer.

First, all concrete problem-domain classes must be mapped to the tables in the ORDBMS. For example in Figure 10-8, the Patient class has been mapped to Patient ORDBMS table. Notice that the Person class has also been mapped to an ORDBMS table. Even though the Person class is abstract, this mapping was done because in the complete class diagram (see Figure 7-2), the Person class had multiple direct subclasses (Employee and Patient).

Second, single-valued attributes should be mapped to columns in the ORDBMS tables. Again, referring to Figure 10-8, we see that the amount attribute of the Patient class has been included in the Patient Table class.

Third, depending on the level of support of stored procedures, the methods and derived attributes should be mapped either to stored procedures or program modules.

Rule 1: Map all concrete problem-domain classes to the ORDBMS tables. Also, if an abstract problem domain class has multiple direct subclasses, map the abstract class to an ORDBMS table.

Rule 2: Map single-valued attributes to columns of the ORDBMS tables.

Rule 3: Map methods and derived attributes to stored procedures or to program modules.

Rule 4: Map single-valued aggregation and association relationships to a column that can store an Object ID. Do this for both sides of the relationship.

Rule 5: Map multivalued attributes to a column that can contain a set of values.

Rule 6: Map repeating groups of attributes to a new table and create a one-to-many association from the original table to the new one.

Rule 7: Map multivalued aggregation and association relationships to a column that can store a set of Object IDs. Do this for both sides of the relationship.

Rule 8: For aggregation and association relationships of mixed type (one-to-many or many-to-one), on the single-valued side (1..1 or 0..1) of the relationship, add a column that can store a set of Object IDs. The values contained in this new column will be the Object IDs from the instances of the class on the multivalued side. On the multivalued side (1..* or 0..*), add a column that can store a single Object ID that will contain the value of the instance of the class on the single-valued side.

For generalization/inheritance relationships:

Rule 9a: Add a column(s) to the table(s) that represents the subclass(es) that will contain an Object ID of the instance stored in the table that represents the superclass. This is similar in concept to a foreign key in an RDBMS. The multiplicity of this new association from the subclass to the "superclass" should be 1..1. Add a column(s) to the table(s) that represents the superclass(es) that will contain an Object ID of the instance stored in the table that represents the subclass(es). If the superclasses are concrete, that is, they can be instantiated themselves, then the multiplicity from the superclass to the subclass is 0..*, otherwise, it is 1..1. Furthermore, an exclusive-or (XOR) constraint must be added between the associations. Do this for each superclass.

OR

Rule 9b: Flatten the inheritance hierarchy by copying the superclass attributes down to all of the subclasses and remove the superclass from the design.[10]

[10] Again, it is also a good idea to document this modification in the design so that in the future, modifications to the design can be maintained easily.

FIGURE 10-7   Mapping Problem-Domain Objects to ORDBMS Schema

Fourth, single-valued (one-to-one) aggregation and association relationships should be mapped to a column that can store an Object ID. This should be done for both sides of the relationship.

Fifth, multivalued attributes should be mapped to columns that can contain a set of values. For example in Figure 10-8, the insurance carrier attribute in the Patient class may contain multiple values because a patient may have more than one insurance carrier. As such, in the Patient table, a multiplicity has been added to the insurance carrier attribute to portray this fact.

The sixth mapping rule addresses repeating groups of attributes in a problem-domain object. In this case, the repeating group of attributes should be used to create a new table in the ORDBMS. Furthermore, it may imply a missing class in the problem-domain layer. Normally, when a set of attributes repeat together as a group, it implies a new class. Finally, we must create a one-to-many association from the original table to the new one.

The seventh rule supports mapping multivalued (many-to-many) aggregation and association relationships to columns that can store a set of Object IDs. Basically, this is a combination of the fourth and fifth rules. Like the fourth rule, this should be done for both sides of the relationships. For example in Figure 10-8, the Symptom table has a multivalued attribute (Patients) that can contain multiple Object IDs to Patient Table objects, and Patient table has a multivalued attribute (Symptoms) that can contain multiple Object IDs to Symptom Table objects.

The eighth rule combines the intentions of Rules 4 and 7. In this case, the rule maps one-to-many and many-to-one relationships. On the single-valued side (1..1 or 0..1) of the

FIGURE 10-8   Mapping Problem-Domain Objects to ORDBMS Schema Example

relationship, a column that can store a set of Object IDs from the table on the multivalued side (1..* or 0..*) of the relationship should be added. On the multivalued side, a column should be added to the table that can store an Object ID from an instance stored in the table on the single-valued side of the relationship. For example, in Figure 10-8, the Patient table has a multivalued attribute (Appts) that can contain multiple Object IDs to Appt Table objects, whereas the Appt table has a single-valued attribute (Patient) that can contain an Object ID to a Patient Table object.

10-4 Dentist Office Appointment System

*I*n Your Turn 10-3, you created a design for the database assuming that the database would be implemented using an OODBMS that supported only single inheritance. In this case, assume that you now know that an ORDBMS will be used and that it does not support any inheritance. However, it does support Object IDs, multivalued attributes, and stored procedures. Using a class diagram, draw the design for the database.

The ninth, and final, rule deals with the lack of support for generalization and inheritance. In this case, there are two different approaches. These approaches are virtually identical to the rules described with the preceding OODBMS object-persistence formats. For example in Figure 10-8, the Patient table contains an attribute (Person) that can contain an Object ID for a Person Table object, and the Person table contains an attribute (SubClassObjects) that contain an Object ID for an object, in this case, stored in the Patient table. In the other case, the inheritance hierarchy is flattened.

Of course, there will be additional processing required anytime that an interaction takes place between the database and the system. Every time an object must be created, retrieved from the database, updated, or deleted, the ORDBMS object(s) must be converted to the problem-domain object, or vice versa. Again, this is the purpose of the data access and manipulation classes. The only other choice is to modify the problem-domain objects. However, such a modification can cause problems between the problem-domain layer and the physical architecture and human–computer interface layers. Generally speaking, the cost of conversion between the ORDBMS and the problem-domain layer will be more than offset by the savings in development time associated with (1) the interaction between the problem-domain and physical architecture and human–computer interaction layers, and (2) the ease of maintenance of a semantically clean problem-domain layer. In the long run, due to the conversions necessary, the development and production cost of using an OODBMS may be less than the development and production cost implementing the object persistence in an ORDBMS.

Mapping Problem Domain Objects to an RDBMS Format

If we support object persistence with an RDBMS, then the mapping from the problem-domain objects to the RDBMS tables is similar to the mapping to an ORDBMS. However, the assumptions made for an ORDBMS are no longer valid. Figure 10-9 lists a set of rules that can be used to design the mapping from the problem-domain objects to the RDBMS-based data access and management layer tables.

The first four rules are basically the same set of rules used to map problem-domain objects to ORDBMS-based data management objects. First, all concrete problem-domain classes must be mapped to tables in the RDBMS. Second, single-valued attributes should be mapped to columns in the RDBMS table. Third, methods should be mapped to either stored procedures or program modules, depending on the complexity of the method. Fourth, single-valued (one-to-one) aggregation and association relationships are mapped to columns that can store the foreign keys of the related tables. This should be done for both sides of the relationship. For example in Figure 10-10, we needed to include tables in the RDBMS for the Person, Patient, Symptom, and Appt classes.

The fifth rule addresses multivalued attributes and repeating groups of attributes in a problem-domain object. In these cases, the attributes should be used to create new tables in the RDBMS. Furthermore, as in the ORDBMS mappings, repeating groups of attributes may

Rule 1: Map all concrete-problem domain classes to the RDBMS tables. Also, if an abstract problem-domain class has multiple direct subclasses, map the abstract class to a RDBMS table.

Rule 2: Map single-valued attributes to columns of the tables.

Rule 3: Map methods to stored procedures or to program modules.

Rule 4: Map single-valued aggregation and association relationships to a column that can store the key of the related table, i.e., add a foreign key to the table. Do this for both sides of the relationship.

Rule 5: Map multivalued attributes and repeating groups to new tables and create a one-to-many association from the original table to the new ones.

Rule 6: Map multivalued aggregation and association relationships to a new associative table that relates the two original tables together. Copy the primary key from both original tables to the new associative table, i.e., add foreign keys to the table.

Rule 7: For aggregation and association relationships of mixed type, copy the primary key from the single-valued side (1..1 or 0..1) of the relationship to a new column in the table on the multivalued side (1..* or 0..*) of the relationship that can store the key of the related table, i.e., add a foreign key to the table on the multivalued side of the relationship.

For generalization/inheritance relationships:

Rule 8a: Ensure that the primary key of the subclass instance is the same as the primary key of the superclass. The multiplicity of this new association from the subclass to the "superclass" should be 1..1. If the superclasses are concrete, that is, they can be instantiated themselves, then the multiplicity from the superclass to the subclass is 0..*, otherwise, it is 1..1. Furthermore, an exclusive-or (XOR) constraint must be added between the associations. Do this for each superclass.

OR

Rule 8b: Flatten the inheritance hierarchy by copying the superclass attributes down to all of the subclasses and remove the superclass from the design.[11]

[11] Again, it is also a good idea to document this modification in the design so that in the future, modifications to the design can be maintained easily.

FIGURE 10-9   Mapping Problem-Domain Objects to RDBMS Schema

imply missing classes in the problem-domain layer. As such, a new problem-domain class may be required. Finally, we should create a one-to-many or zero-to-many association from the original table to the new one. For example, in Figure 10-10, we needed to create a new table for insurance carrier because it was possible for a patient to have more than one insurance carrier.

The sixth rule supports mapping multivalued (many-to-many) aggregation and association relationships to a new table that relates the two original tables. In this case, the new table should contain foreign keys back to the original tables. For example, in Figure 10-10, we needed to create a new table that represents the suffer association between the Patient and Symptom problem-domain classes.

The seventh rule addresses one-to-many and many-to-one relationships. With these types of relationships, the multivalued side (0..* or 1..*) should be mapped to a column in its table that can store a foreign key back to the single-valued side (0..1 or 1..1). It is possible that we have already taken care of this situation because we earlier recommended inclusion of both primary and foreign key attributes in the problem-domain classes. In the case of Figure 10-10, we had already added the primary key from the Patient class to the Appt class as a foreign key (see personNumber). However, in the case of the reflexive relationship, primary insurance carrier, associated with the Patient class, we need to add a new attribute (primaryInsuranceCarrier) to be able to store the relationship.

The eighth, and final, rule deals with the lack of support for generalization and inheritance. As in the case of an ORDBMS, there are two different approaches. These approaches are virtually identical to the rules described with OODBMS and ORDBMS object-persistence formats given earlier. The first approach is to add a column to each table that represents a subclass for each of the concrete superclasses of the subclass. Essentially, this ensures that the primary key of the subclass is the same as the primary key for the superclass. If we had previously added the primary and foreign keys to the problem-domain objects, as we recommended, then we do not have to do anything else. The primary keys of the tables will be used to rejoin the instances stored in the tables that represent each of the pieces of the

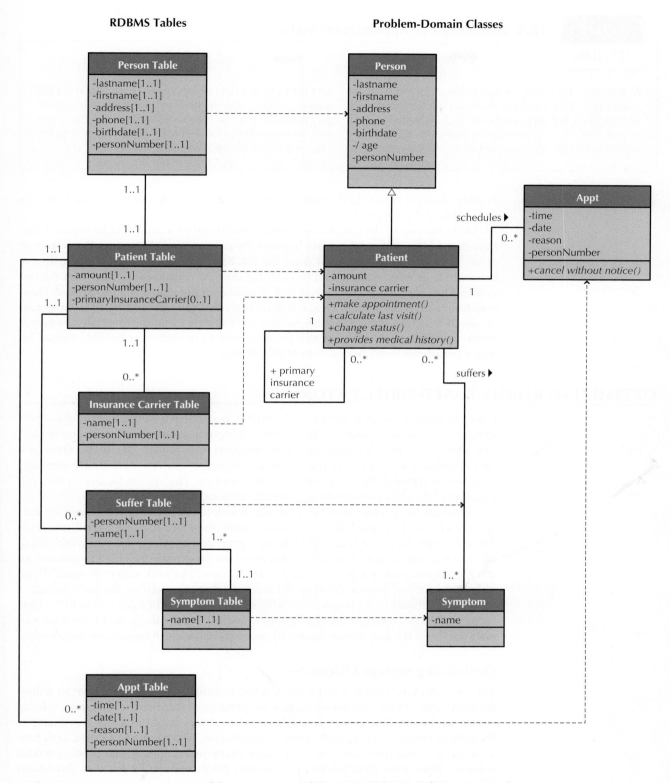

FIGURE 10-10   Mapping Problem-Domain Objects to RDBMS Schema Example

problem-domain object. Conversely, the inheritance hierarchy can be flattened and the rules (Rules 1–7) reapplied.

As in the case of the ORDBMS approach, additional processing will be required anytime that an interaction takes place between the database and the system. Every time an object must be created, retrieved from the database, updated, or deleted, the mapping between the problem domain and the RDBMS must be used to convert between the two different formats. In this case, a great deal of additional processing will be required. However, from a practical point of view, it is more likely that you will use a RDBMS for storage of objects than the other approaches because RDBMSs are by far the most popular format in the marketplace. As such, we will focus on how to optimize the RDBMS format for object persistence in the next section of this chapter.

OPTIMIZING RDBMS-BASED OBJECT STORAGE

Once the object-persistence format is selected, the second step is to optimize the object persistence for processing efficiency. The methods of optimization will vary based on the format that you select; however, the basic concepts will remain the same. Once you understand how to optimize a particular type of object persistence, you will have some idea as to how to approach the optimization of other formats. This section focuses on the optimization of the most popular storage format: relational databases.

There are two primary dimensions in which to optimize a relational database: for storage efficiency and for speed of access. Unfortunately, these two goals often conflict because the best design for access speed may take up a great deal of storage space as compared to other, less-speedy designs. This section describes how to optimize the object persistence for storage efficiency using a process called normalization. The next section presents design techniques, such as denormalization and indexing, that can speed up the performance of the system. Ultimately, the project team will go through a series of trade-offs until the ideal balance of the two optimization dimensions is reached. Finally, the project team must estimate the size of the data storage needed to ensure there is enough capacity on the server(s).

Optimizing Storage Efficiency

The most efficient tables in a relational database in terms of storage space have no redundant data and very few null values because the presence of these suggest that space is being wasted (and more data to store means higher data storage hardware costs). For example, the table in Figure 10-11 repeats customer information, such as name and state, each time a customer places an order, and it contains many null values in the product-related columns. These nulls occur whenever a customer places an order for less than three items (the maximum number on an order).

Sample Records:

Order

- Order Number : unsigned long
- Date : Date
- Cust ID : unsigned long
- Last Name : String
- First Name : String
- State : String
- Tax Rate : float
- Product 1 Number : unsigned long
- Product 1 Desc. : String
- Product 1 Price : double
- Product 1 Qty. : unsigned long
- Product 2 Number : unsigned long
- Product 2 Desc. : String
- Product 2 Price : double
- Product 2 Qty. : unsigned long
- Product 3 Number : unsigned long
- Product 3 Desc. : String
- Product 3 Price : double
- Product 3 Qty. : unsigned long

Order Number	Date	Cust ID	Last Name	First Name	State	Tax Rate	Prod. 1 Number	Prod. 1 Desc.	Prod. 1 Price	Prod. 1 Qty.	Prod. 2 Number	Prod. 2 Desc.	Prod. 2 Price	Prod. 2 Qty.	Prod. 2 Number
239	11/23/00	1035	Black	John	MD	0.05	555	Cheese Tray	$45.00	2					
260	11/24/00	1035	Black	John	MD	0.05	444	Wine Gift Pack	$60.00	1					
273	11/27/00	1035	Black	John	MD	0.05	222	Bottle Opener	$12.00	1					
241	11/23/00	1123	Williams	Mary	CA	0.08	444	Wine Gift Pack	$60.00	2					
262	11/24/00	1123	Williams	Mary	CA	0.08	222	Bottle Opener	$12.00	2					
287	11/27/00	1123	Williams	Mary	CA	0.08	222	Bottle Opener	$12.00	2					
290	11/30/00	1123	Williams	Mary	CA	0.08	555	Cheese Tray	$45.00	3					
234	11/23/00	2242	DeBerry	Ann	DC	0.065	555	Cheese Tray	$45.00	2					
237	11/23/00	2242	DeBerry	Ann	DC	0.065	111	Wine Guide	$15.00	1					
238	11/23/00	2242	DeBerry	Ann	DC	0.065	444	Wine Gift Pack	$60.00	1	444	Wine Gift Pack	$60.00	1	
245	11/24/00	2242	DeBerry	Ann	DC	0.065	222	Bottle Opener	$12.00	1					
250	11/24/00	2242	DeBerry	Ann	DC	0.065	222	Bottle Opener	$12.00	1					
252	11/24/00	2242	DeBerry	Ann	DC	0.065	222	Bottle Opener	$12.00	1	444	Wine Gift Pack	$60.00	2	
253	11/24/00	2242	DeBerry	Ann	DC	0.065	222	Bottle Opener	$12.00	1	444	Wine Gift Pack	$60.00	1	
297	11/30/00	2242	DeBerry	Ann	DC	0.065	333	Jams & Jellies	$20.00	2					
243	11/24/00	4254	Bailey	Ryan	MD	0.05	555	Cheese Tray	$45.00	2					
246	11/24/00	4254	Bailey	Ryan	MD	0.05	333	Jams & Jellies	$20.00	3					
248	11/24/00	4254	Bailey	Ryan	MD	0.05	222	Bottle Opener	$12.00	1	333	Jams & Jellies	$20.00	2	
235	11/23/00	9500	Chin	April	KS	0.05	222	Bottle Opener	$12.00	1					
242	11/23/00	9500	Chin	April	KS	0.05	333	Jams & Jellies	$20.00	3					
244	11/24/00	9500	Chin	April	KS	0.05	222	Bottle Opener	$12.00	2					
251	11/24/00	9500	Chin	April	KS	0.05	111	Wine Guide	$15.00	2					111

Redundant Data

Null Cells

FIGURE 10-11 Optimizing Storage

In addition to wasting space, redundancy and null values also allow more room for error and increase the likelihood that problems will arise with the integrity of the data. What if customer 1035 moved from Maryland to Georgia? In the case of Figure 10-11, a program must be written to ensure that all instances of that customer are updated to show GA as the new state of residence. If some of the instances are overlooked, then the table will contain an update anomaly, whereby some of the records contain the correctly updated value for state and other records contain the old information.

Nulls threaten data integrity because they are difficult to interpret. A blank value in the Order table's product fields could mean (1) the customer did not want more than one or two products on his or her order, (2) the operator forgot to enter in all three products on the order, or (3) the customer canceled part of the order and the products were deleted by the operator. It is impossible to be sure of the actual meaning of the nulls.

For both these reasons—wasted storage space and data integrity threats—project teams should remove redundancy and nulls from the table. During design, the class diagram is used to examine the design of the RDBMS tables (e.g., see Figure 10-10) and optimize it for storage efficiency. If you follow the modeling instructions and guidelines that were presented in Chapter 6, you will have little trouble creating a design that is highly optimized in this way because a well-formed logical data model does not contain redundancy or many null values.

Sometimes, however, a project team needs to start with a model that was poorly constructed or with one that was created for files or a nonrelational type of format. In these cases, the project team should follow a series of steps that serve to check the model for storage efficiency. These steps make up a process called normalization.[12]

Normalization is a process whereby a series of rules are applied to the RDBMS tables to determine how well formed they are (see Figure 10-12). These rules help analysts identify tables that are not represented correctly. Here, we describe three normalization rules that are applied regularly in practice. Figure 10-11 shows a model in 0 Normal Form, which is an unnormalized model before the normalization rules have been applied.

A model is in *first normal form (1NF)* if it does not lead to multivalued fields, fields that allow a set of values to be stored, or repeating fields, which are fields that repeat within a table to capture multiple values. The rule for 1NF says that all tables must contain the same number of columns (i.e., fields) and that all the columns must contain a single value. Notice that the model in Figure 10-11 violates 1NF because it causes product number, description, price, and quantity to repeat three times for each order in the table. The resulting table has many records that contain nulls in the product-related columns, and orders are limited to three products because there is no room to store information for more.

A much more efficient design (and one that conforms to 1NF) leads to a separate table to hold the repeating information; to do this, we create a separate table on the model to capture product order information. A zero-to-many relationship would then exist between the two tables. As shown in Figure 10-13, the new design eliminates nulls from the Order table and supports an unlimited number of products that can be associated with an order.

Second normal form (2NF) requires first that the data model is in 1NF and second that the data model leads to tables containing fields that depend on a *whole* primary key. This means that the primary key value for each record can determine the value for all the other

---

[12] Normalization also can be performed on the problem-domain layer. However, the normalization process should be used on the problem-domain layer only to uncover missing classes. Otherwise, optimizations that have nothing to do with the semantics of the problem domain can creep into the problem-domain layer.

0 Normal Form

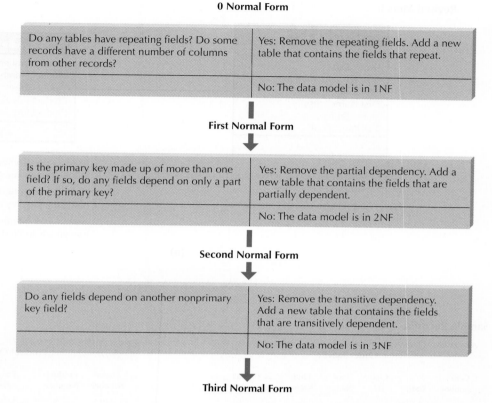

Do any tables have repeating fields? Do some records have a different number of columns from other records?	Yes: Remove the repeating fields. Add a new table that contains the fields that repeat.
	No: The data model is in 1NF

First Normal Form

Is the primary key made up of more than one field? If so, do any fields depend on only a part of the primary key?	Yes: Remove the partial dependency. Add a new table that contains the fields that are partially dependent.
	No: The data model is in 2NF

Second Normal Form

Do any fields depend on another nonprimary key field?	Yes: Remove the transitive dependency. Add a new table that contains the fields that are transitively dependent.
	No: The data model is in 3NF

Third Normal Form

FIGURE 10-12
The Steps of
Normalization

fields in the record. Sometimes fields depend on only part of the primary key (i.e., *partial dependency*), and these fields belong in another table.

For example, in the new Product Order table that was created in Figure 10-13, the primary key is a combination of the order number and product number, but the product description and price attributes are dependent only upon product number. In other words, by knowing product number, we can identify the product description and price. However, knowledge of the order number and product number is required to identify the quantity. To rectify this violation of 2NF, a table is created to store product information, and the description and price attributes are moved into the new table. Now, product description is stored only once for each instance of a product number as opposed to many times (every time a product is placed on an order).

A second violation of 2NF occurs in the Order table: customer first name and last name depend only upon the customer ID, not the whole key (Cust ID and Order number). As a result, every time the customer ID appears in the Order table, the names also appear. A much more economical way of storing the data is to create a Customer table with the Customer ID as the primary key and the other customer-related fields (i.e., last name and first name) listed only once within the appropriate record. Figure 10-14 illustrates how the model would look when placed in 2NF.

Third normal form (3NF) occurs when a model is in both 1NF and, 2NF and, in the resulting tables, none of the fields depend on nonprimary key fields (i.e., *transitive dependency*). Figure 10-14 contains a violation of 3NF: the tax rate on the order depends upon

Revised Model:

Order

-Order Number : unsigned long
-Date : Date
-Cust ID : unsigned long
-Last Name : String
-First Name : String
-State : String
-Tax Rate : float

0..*   1..*

Product Order

-Order Number : unsigned long
-Product Number : String
-Product Desc : String
-Product Price : double
-Product Qty : unsigned long

Note: Order Number will serve as part of the primary key of Order

Note: Cust ID also will serve as part of the primary key of Order

Note: Order Number will serve as part of the primary key of Product Order

Note: Product Number will serve as part of the primary key of Product Order

Note: Order Number also will serve as a foreign key in Product Order

(a)

Order 237 has 2 products

Sample Records:

Order Table

Order Number	Date	Cust ID	Last Name	First Name	State	Tax Rate
239	11/23/00	1035	Black	John	MD	0.05
260	11/24/00	1035	Black	John	MD	0.05
273	11/27/00	1035	Black	John	MD	0.05
241	11/23/00	1123	Williams	Mary	CA	0.08
262	11/24/00	1123	Williams	Mary	CA	0.08
287	11/27/00	1123	Williams	Mary	CA	0.08
290	11/30/00	1123	Williams	Mary	CA	0.08
234	11/23/00	2242	DeBerry	Ann	DC	0.065
237	11/23/00	2242	DeBerry	Ann	DC	0.065
238	11/23/00	2242	DeBerry	Ann	DC	0.065
245	11/24/00	2242	DeBerry	Ann	DC	0.065
250	11/24/00	2242	DeBerry	Ann	DC	0.065
252	11/24/00	2242	DeBerry	Ann	DC	0.065
253	11/24/00	2242	DeBerry	Ann	DC	0.065
297	11/30/00	2242	DeBerry	Ann	DC	0.065
243	11/24/00	4254	Bailey	Ryan	MD	0.05
246	11/24/00	4254	Bailey	Ryan	MD	0.05
248	11/24/00	4254	Bailey	Ryan	MD	0.05
235	11/23/00	9500	Chin	April	KS	0.05
242	11/23/00	9500	Chin	April	KS	0.05
244	11/24/00	9500	Chin	April	KS	0.05
251	11/24/00	9500	Chin	April	KS	0.05

Product Order Table

Order Number	Product Number	Product Desc	Product Price	Product Qty
239	555	Cheese Tray	$45.00	2
260	444	Wine Gift Pack	$60.00	1
273	222	Bottle Opener	$12.00	1
241	444	Wine Gift Pack	$60.00	2
262	222	Bottle Opener	$12.00	2
287	222	Bottle Opener	$12.00	2
290	555	Cheese Tray	$45.00	3
234	555	Cheese Tray	$45.00	2
237	111	Wine Guide	$15.00	1
237	444	Wine Gift Pack	$60.00	1
238	444	Wine Gift Pack	$60.00	1
245	222	Bottle Opener	$12.00	1
250	222	Bottle Opener	$12.00	1
252	222	Bottle Opener	$12.00	1
252	444	Wine Gift Pack	$60.00	2
253	222	Bottle Opener	$12.00	1
253	444	Wine Gift Pack	$60.00	1
297	333	Jams & Jellies	$20.00	2
243	555	Cheese Tray	$45.00	2
246	333	Jams & Jellies	$20.00	3
248	222	Bottle Opener	$12.00	1
248	333	Jams & Jellies	$20.00	2
248	111	Wine Guide	$15.00	1
235	222	Bottle Opener	$12.00	1
242	333	Jams & Jellies	$20.00	3
244	222	Bottle Opener	$12.00	2
251	111	Wine Guide	$15.00	2

Order 248 has 3 products

(b)

FIGURE 10-13   1NF: Remove Repeating Fields

Product Order
-Order Number : unsigned long
-Product Number : unsigned long
-Qty : unsigned long

Customer
-Cust ID : unsigned long
-Last Name : String
-First Name : String

Order
-Order Number : unsigned long
-Date : Date
-Cust ID : unsigned long
-State : String
-Tax Rate : float

Product
-Product Number : unsigned long
-Product Desc : String
-Price : double

1..1   0..*

0..*   1..*

Note: Cust ID will serve as the primary key of Customer.

Note: Order Number will serve as the primary key of Order.

Note: Cust ID will serve as a foreign key in Order.

Note: Order Number will serve as part of the primary key of Product Order.

Note: Order Number also will serve as a foreign key in Product Order.

Note: Product Number will serve as part of the primary key in Product Order.

Note: Product Number also will serve as a foreign key in Product Order.

Note: Product Number will serve as part of the primary key of Product Order.

Sample Records:

Customer Table

Cust ID	Last Name	First Name
1035	Black	John
1123	Williams	Mary
2242	DeBerry	Ann
4254	Bailey	Ryan
9500	Chin	April

Last Name and First Name was moved to the Customer table to eliminate redundancy

Order Table

Order Number	Date	Cust ID	State
239	11/23/00	1035	MD
260	11/24/00	1035	MD
273	11/27/00	1035	MD
241	11/23/00	1123	CA
262	11/24/00	1123	CA
287	11/27/00	1123	CA
290	11/30/00	1123	CA
234	11/23/00	2242	DC
237	11/23/00	2242	DC
238	11/23/00	2242	DC
245	11/24/00	2242	DC
250	11/24/00	2242	DC
252	11/24/00	2242	DC
253	11/24/00	2242	DC
297	11/30/00	2242	DC
243	11/24/00	4254	MD
246	11/24/00	4254	MD
248	11/24/00	4254	MD
235	11/23/00	9500	KS
242	11/23/00	9500	KS
244	11/24/00	9500	KS
251	11/24/00	9500	KS

Product Order Table

Order Number	Product Number	Product Qty
239	555	2
260	444	1
273	222	1
241	444	2
262	222	2
287	222	2
290	555	3
234	555	2
237	111	1
237	444	1
238	444	1
245	222	1
250	222	1
252	222	1
252	444	2
253	222	1
253	444	1
297	333	2
243	555	2
246	333	3
248	222	1
248	333	2
248	111	1
235	222	1
242	333	3
244	222	2
251	111	2

Product Table

Product Number	Product Desc	Product Price
111	Wine Guide	$15.00
222	Bottle Opener	$12.00
333	Jams & Jellies	$20.00
444	Wine Gift Pack	$60.00
555	Cheese Tray	$45.00

Product Desc and Price was moved to the Product table to eliminate redundancy

FIGURE 10-14   2NF Partial Dependencies Removed

YOUR TURN

10-6 Normalizing a Student Activity File

*S*uppose that you have been asked to build a system that tracks student involvement in activities around campus. You have been given a file with information that needs to be imported into the system, and the file contains the following fields:

student social security number
student last name

student first name
student advisor name
student advisor phone
activity 1 code
activity 1 description
activity 1 start date
activity 2 code

activity 2 description
activity 2 start date
activity 3 code
activity 3 description
activity 3 start date

Normalize the file. Show how the logical data model would change at each step.

the state to which the order is being sent. The solution involves creating another table that contains state abbreviations serving as the primary key and the tax rate as a regular field. Figure 10-15 presents the end results of applying the steps of normalization to the original model from Figure 10-11.

Optimizing Data Access Speed

After you have optimized the design of the object storage for efficiency, the end result is that data are spread out across a number of tables. When data from multiple tables need

FIGURE 10-15 3NF Normalized Field

to be accessed or queried, the tables must be first joined together. For example, before a user can print out a list of the customer names associated with orders, first the Customer and Order tables need to be joined, based on the customer number field (see Figure 10-15). Only then can both the order and customer information be included in the query's output. Joins can take a lot of time, especially if the tables are large or if many tables are involved.

Consider a system that stores information about 10,000 different products, 25,000 customers, and 100,000 orders, each averaging three products per order. If an analyst wanted to investigate whether there were regional differences in music preferences, he or she would need to combine all the tables to be able to look at products that have been ordered while knowing the state of the customers placing the orders. A query of this information would result in a huge table with 300,000 rows (i.e., the number of products that have been ordered) and 11 columns (the total number of columns from all of the tables combined).

The project team can use several techniques to try to speed up access to the data, including denormalization, clustering, and indexing.

Denormalization   After the object storage is optimized, the project team may decide to denormalize, or add redundancy back into the design. *Denormalization* reduces the number of joins that need to be performed in a query, thus speeding up access. Figure 10-16 shows a denormalized model for customer orders. The customer last name was added back into the Order table because the project team learned during analysis that queries about orders usually require the customer last name field. Instead of joining the Order table repeatedly to the Customer table, the system now needs to access only the Order table because it contains all of the relevant information.

Denormalization should be applied sparingly for the reasons described in the previous section, but it is ideal in situations in which information is queried frequently, yet updated rarely. There are three cases in which you may rely upon denormalization to reduce joins and improve performance. First, denormalization can be applied in the case of look-up tables, which are tables that contain descriptions of values (e.g., a table of product descriptions or a table of payment types). Because descriptions of codes rarely change, it may be more efficient to include the description along with its respective code in the main table to eliminate the need to join the look-up table each time a query is performed (see Figure 10-17a).

FIGURE 10-16
Denormalized Physical
Data Model

Notice that the payment description field
appears in both Payment Type and Order.

(a) Look-up Table

FIGURE 10-17

Denormalization
Situations (FK = foreign
key; PK = primary key)

Notice that the shipment state and shipment method
are included in both Shipment and Order.

(b) One-to-one Relationship

Second, one-to-one relationships are good candidates for denormalization. Although logically two tables should be separated, from a practical standpoint the information from both tables may regularly be accessed together. Think about an order and its shipping information. Logically, it may make sense to separate the attributes related to shipping into a separate table, but as a result the queries regarding shipping will probably always need a join to the Order table. If the project team finds that certain shipping information, such as state and shipping method, are needed when orders are accessed, they may decide to combine the tables or include some shipping attributes in the Order table (see Figure 10-17b).

Third, at times it will be more efficient to include a parent entity's attributes in its child entity on the physical data model. For example, consider the Customer and Order tables in Figure 10-16, which share a one-to-many relationship, with Customer as the parent and Order as the child. If queries regarding orders continuously require customer information, the most popular customer fields can be placed in Order to reduce the required joins to the Customer table, as was done with Customer Last Name.

Clustering Speed of access also is influenced by the way that the data are retrieved. Think about going shopping in a grocery store. If you have a list of items to buy but you

YOUR

TURN

10-7 Denormalizing a Student Activity File

Consider the logical data model that you created for Your Turn 10-3. Examine the model and describe possible opportunities for denormalization. How would you change the physical data model for this file, and what are the benefits of your changes?

are unfamiliar with the store's layout, then you need to walk down every aisle to make sure that you don't miss anything from your list. Likewise, if records are arranged on a hard disk in no particular order (or in an order that is irrelevant to your data needs), then any query of the records results in a *table scan* in which the DBMS has to access every row in the table before retrieving the result set. Table scans are the most inefficient of data retrieval methods.

One way to improve access speed is to reduce the number of times that the storage medium needs to be accessed during a transaction. One method is to *cluster* records together physically so that similar records are stored close together. With *intrafile clustering*, like records in the table are stored together in some way, such as in order by primary key or, in the case of a grocery store, by item type. Thus, whenever a query looks for records, it can go directly to the right spot on the hard disk (or other storage medium) because it knows in what order the records are stored, just as we can walk directly to the bread aisle to pick up a loaf of bread. *Interfile clustering* combines records from more than one table that typically are retrieved together. For example, if customer information is usually accessed with the related order information, then the records from the two tables may be physically stored in a way that preserves the customer-order relationship. Returning to the grocery store scenario, an interfile cluster would be similar to storing peanut butter, jelly, and bread next to each other in the same aisle because they are usually purchased together, not because they are similar types of items. Of course, each table can have only one clustering strategy because the records can be arranged physically in only one way.

Indexing   A familiar time saver is an index located in the back of a textbook, which points directly to the page or pages that contain a topic of interest. Think of how long it would take to find all the times that relational database appears in this textbook without the index to rely on! An *index* in data storage is like an index in the back of a textbook; it is a minitable that contains values from one or more columns in a table and the location of the values within the table. Instead of paging through the entire textbook, we can move directly to the right pages and get the information we need. Indexes are one of the most important ways to improve database performance. Whenever there are performance problems, the first place to look is an index.

A query can use an index to find the locations of only those records that are included in the query answer, and a table can have an unlimited number of indexes. Figure 10-18 shows an index that orders records by payment type. A query that searches for all the customers who used American Express can use this index to find the locations of the records that contain American Express as the payment type without having to scan the entire Order table.

Project teams can make indexes perform even faster by placing them into the main memory of the data storage hardware. Retrieving information from memory is much faster than from another storage medium, such as a hard disk—think about how much faster it is

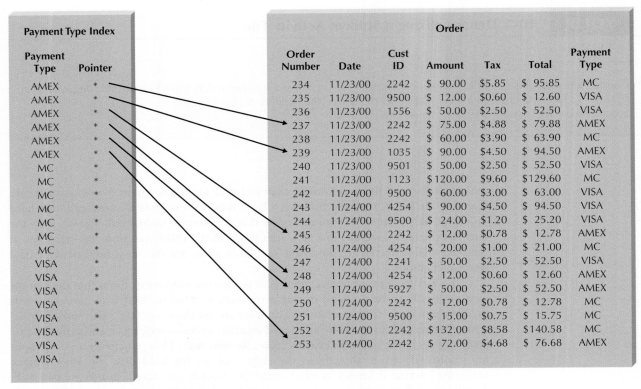

FIGURE 10-18   Payment Type Index

to retrieve a memorized phone number versus one that must be looked up in a phone book. Similarly, when a database has an index in memory, it can locate records very, very quickly.

Of course, indexes require overhead in that they take up space on the storage medium. Also, they need to be updated as records in tables are inserted, deleted, or changed. Thus, although queries lead to faster access to the data, they slow down the update process. In general, we should create indexes sparingly for transaction systems or systems that require

---

CONCEPTS IN ACTION   **10-A Mail-Order Index**

A Virginia-based mail-order company sends out approximately 25 million catalogs each year using a Customer table with 10 million names. Although the primary key of the Customer table is customer identification number, the table also contains an index of Customer Last Name. Most people who call to place orders remember their last name, not their customer identification number, so this index is used frequently.

An employee of the company explained that indexes are critical to reasonable response times. A fairly complicated query was written to locate customers by the state in which they lived, and it took more than three weeks to return an answer. A customer state index was created, and that same query provided a response in twenty minutes: that's 1,512 times faster!

Question

As an analyst, how can you make sure that the proper indexes have been put in place so that users are not waiting for weeks to receive the answers to their questions?

FIGURE 10-19
Guidelines for
Creating Indexes

Use indexes sparingly for transaction systems.

Use many indexes to increase response times in decision support systems.

For each table, create a unique index that is based on the primary key.

For each table, create an index that is based on the foreign key to improve the performance of joins.

Create an index for fields that are used frequently for grouping, sorting, or criteria.

a lot of updates, but we should apply indexes generously when designing systems for decision support (see Figure 10-19).

Estimating Data Storage Size

Even if we have denormalized our physical data model, clustered records, and created indexes appropriately, the system will perform poorly if the database server cannot handle its volume of data. Therefore, one last way to plan for good performance is to apply *volumetrics,* which means estimating the amount of data that the hardware will need to support. You can incorporate your estimates into the database server hardware specification to make sure that the database hardware is sufficient for the project's needs. The size of the database is based on the amount of *raw data* in the tables and the *overhead* requirements of the DBMS. To estimate size you will need to have a good understanding of the initial size of your database as well as its expected growth rate over time.

Raw data refers to all the data that are stored within the tables of the database, and it is calculated based on a bottom-up approach. First, write down the estimated average width for each column (field) in the table and sum the values for a total record size (see Figure 10-20). For example, if a variable-width LAST NAME column is assigned a width of 20 characters, you can enter 13 as the average character width of the column. In Figure 10-20, the estimated record size is 49.

Next, calculate the overhead for the table as a percentage of each record. Overhead includes the room needed by the DBMS to support such functions as administrative actions and indexes, and it should be assigned based on past experience, recommendations from technology vendors, or parameters that are built into software that was written to calculate volumetrics. For example, your DBMS vendor may recommend that you allocate 30 percent of the records' raw data size for overhead storage space, creating a total record size of 63.7 in the Figure 10-20 example.

Field	Average Size
Order Number	8
Date	7
Cust ID	4
Last Name	13
First Name	9
State	2
Amount	4
Tax Rate	2
Record Size	49
Overhead	30%
Total Record Size	63.7
Initial Table Size	50,000
Initial Table Volume	3,185,000
Growth Rate/Month	1,000
Table Volume @ 3 years	5,478,200

FIGURE 10-20
Calculating
Volumetrics

Finally, record the number of initial records that will be loaded into the table, as well as the expected growth per month. This information should have been collected during analysis. According to Figure 10-20, the initial space required by the first table is 3,185,000, and future sizes can be project based on the growth figure. These steps are repeated for each table to get a total size for the entire database.

Many CASE tools will provide you with database-size information based on how you set up the object persistence, and they will calculate volumetrics estimates automatically. Ultimately, the size of the database needs to be shared with the design team so that the proper technology can be put in place to support the system's data and potential performance problems can be addressed long before they affect the success of the system.

10-B Return on Investment from Virtualization—A Hard Factor to Determine

Many companies are undergoing server virtualization. This is the concept of putting multiple virtual servers onto one physical device. The payoffs can be significant: fewer servers, less electricity, less generated heat, less air conditioning, less infrastructure and administration costs; increased flexibility; less physical presence (i.e., smaller server rooms), faster maintenance of servers, and more. There are (of course) costs—such as licensing the virtualization software, labor costs in establishing the virtual servers onto a physical device, and labor costs in updating tables and access. But, determining the Return on Investment can be a challenge. Some companies have lost money on server virtualization, but most would say they have gained a positive return on investment with virtualization but have not really quantified the results.

Questions

1. How might a company really determine the return on investment for server virtualization?
2. Would server virtualization impact the amount of data storage required? Why or why not?
3. Is this a project that a systems analyst might be involved in? Why or why not?

NONFUNCTIONAL REQUIREMENTS AND DATA MANAGEMENT LAYER DESIGN[11]

Recall that nonfunctional requirements refer to behavioral properties that the system must have. These properties include issues related to performance, security, ease of use, operational environment, and reliability. In this text, we have grouped nonfunctional requirements into four categories: operational, performance, security, and cultural and political requirements. We describe each of these in relation to the DAM layer.

The *operational requirements* for the DAM layer include issues that deal with the technology being used to support object persistence. However, the choice of the *hardware and operating system* may limit the choice of the technology and format of the object persistence available. For example, if the system is to be deployed on a Macintosh-based network that is running the Mac OS, the choices of object persistence is fairly limited, for example, FileMaker. This, in turn, will determine which set of the mapping rules described earlier will have to be used. Another operational requirement could be the ability to import and export data using XML. Again, this could limit the object stores under consideration.

The primary *performance requirements* that affect the data access and management layer are speed and capacity. As described before, depending on the anticipated—and, afterwards, actual—usage patterns of the objects being stored, different indexing and caching approaches may be necessary. When considering distributing objects over a network, there are speed considerations that may cause objects to be replicated on different nodes in the network. As such, multiple copies of the same object may be stored in different locations on the network. This raises the issue of update anomalies described before in conjunction with normalization. Also, depending on the estimated size and growth of the system, different DBMSs may need to be considered. An additional requirement that may impact the design of the DAM layer deals with the availability of the objects being stored. It may make sense to limit the availability to different objects based on the time of day. For example, one class of users may be allowed to access a set of objects only from 8 to 12 in the morning and a second set of users may be able to access them only from 1 to 5 in the afternoon. Through the DBMS, these types of restrictions could be set.

[11] Because the vast majority of nonfunctional requirements impact the physical architecture layer, we provide additional details in Chapter 12.

The *security requirements* deal primarily with access controls, encryption, and backup. Through a modern DBMS, different types of access can be set (e.g., Read, Update, or Delete) granting access only to those users (or class of users) who have been authorized. Furthermore, *access control* can be set to guarantee that only users with "administrator" privileges are allowed to modify the object storage schema or access controls. Encryption requirements on this layer deal with whether the object should be stored in an encrypted format or not. Even though encrypted objects are more secure than unencrypted objects, the process of encrypting and decrypting the objects will slow down the system. Depending on the physical architecture being used, the cost of encryption may be negligible. For example, if we plan on encrypting the objects before transmitting them over a network, there may be no additional cost of storing them in the encrypted format. Backup requirements deal with ensuring that the objects are routinely copied and stored in case the object store becomes corrupted or unusable. Having a backup copy made on a periodic basis and storing the updates that have occurred since the last backup copy was made ensures that the updates are not lost and the object store can be reconstituted by running the copies of the updates against the backup copy to create a new current copy.

The only *political and cultural requirements* that can impact the DAM layer deals with how the detailed format of the data is to be stored. For example, in what format should a date be stored? Or, how many characters should be allocated for a last name field that is part of an Employee object? There could be a corporate IT bias toward different hardware and software platforms. If so, as mentioned before, this could limit the type of object store available.

DESIGNING DATA ACCESS AND MANIPULATION CLASSES

The final step in developing the data management layer is to design the *data access and manipulation classes* that act as a translator between the object-persistence and the problem-domain objects. As such, they should always be capable of at least reading and writing both the object-persistence and problem-domain objects. As described earlier and in Chapter 8, the object-persistence classes are derived from the concrete problem-domain classes, whereas the data access and manipulation classes depend on both the object-persistence and problem-domain classes.

Depending on the application, a simple rule to follow is that there should be one data access and manipulation class for each concrete problem domain class. In some cases, it might make sense to create data access and manipulation classes associated with the human–computer interaction classes (see Chapter 11). However, this creates a dependency from the data management layer to the human–computer interaction layer. Adding this additional complexity to the design of the system normally is not recommended.

Returning to the ORDBMS solution for the Appointment system example (see Figure 10-8), we see that we have four problem-domain classes and four ORDBMS tables. Following the previous rule, the DAM classes are rather simple. They have to support only a one-to-one translation between the concrete problem-domain classes and the ORDBMS tables (see Figure 10-21). Because the Person problem-domain class is an abstract class, only three data access and manipulation classes are required: Patient-DAM, Symptom-DAM, and Appt-DAM. However, the process to create an instance of the Patient problem-domain class can be fairly complicated. The Patient-DAM class may have to be able to retrieve information from all four ORDBMS tables. To accomplish this, the Patient-DAM class retrieves the information from the Patient table. Using the Object-IDs stored in the attribute values associated with the Person, Appts, and Symptoms attributes, the remaining information required to create an instance of Patient is easily retrieved by the Patient-DAM class.

FIGURE 10-21 Managing Problem-Domain Objects to ORDBMS using DAM Classes

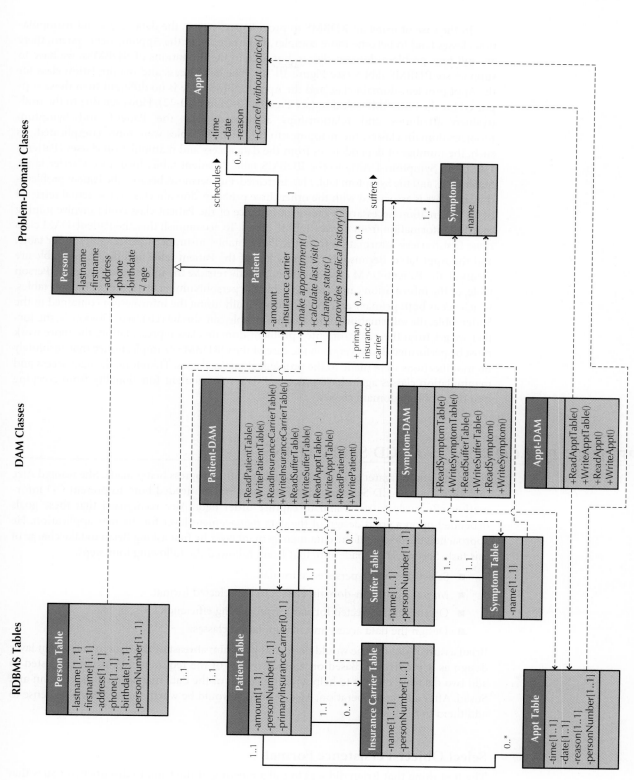

Problem-Domain Classes

DAM Classes

RDBMS Tables

FIGURE 10-22 Mapping Problem-Domain Objects to RDBMS using DAM Classes

In the case of using an RDBMS to provide persistence, the data access and manipulation classes tend to become more complex. For example, in the Appointment system, there are still four problem-domain classes, but, due to the limitations of RDBMSs, we have to support six RDBMS tables (see Figure 10-10). The data access and manipulation class for the Appt problem-domain class and the Appt RDBMS table is no different than those supported for the ORDBMS solution (see Figures 10-21 and 10-22). However, due to the multivalued attributes and relationships associated with the Patient and Symptom problem-domain classes, the mappings to the RDBMS tables were more complicated. As such, the number of dependencies from the data access and manipulation classes (Patient-DAM and Symptom-DAM) to the RDBMS tables (Patient table, Insurance Carrier table, Suffer table, and the Symptom table) has increased. Furthermore, because the Patient problem-domain class is associated with the other three problem-domain classes, the actual retrieval of all information necessary to create an instance of the Patient class could involve joining together information from all six RDBMS tables. To accomplish this, the Patient-DAM class must first retrieve information from the Patient table, Insurance Carrier table, Suffer table, and the Appt table. Because the primary key of the Patient table and the Person table are identical, the Patient-DAM class can either directly retrieve the information from the Person table, or the information can be joined using the personNumber attributes of the two tables, which act as both primary and foreign keys. Finally, using the information contained in the Suffer table, the information in the Symptom table can also be retrieved. Obviously, the further we get from the object-oriented problem-domain class representation, the more work must be performed. However, as in the case of the ORDBMS example, notice that absolutely no modifications were made to the problem-domain classes. Therefore, the data access and manipulation classes again have prevented data management functionality from creeping into the problem-domain classes.

APPLYING THE CONCEPTS AT CD SELECTIONS

Alec met with Margaret to keep up to date as to the progress being made. She wanted Alec to realize that the CD Selections Internet Sales System needed both to present CD information effectively to users and to capture order data. Alec recognized that these goals depended upon a good design of the data management layer for the new application. He approached the design of the data management layer by first asking Brian to take charge of this task. Second, Alec made sure that Brian followed the following four steps:

- Select the object-persistence format.
- Map the problem-domain classes to the selected format.
- Optimize the selected format for processing efficiency.
- Design the data access and manipulation classes.

Brian assured Alec that he would do so and keep him abreast of the data management layer design as it progressed. Based on a quick review of the requirements, Brian requested to add two database experts to help with the design of the data management layer: John and Susan. After a little deliberation, Alec decided it would be worth the additional expense to add them to the team.

Select Object-Persistence Format

The first thing that Brian did was to call a meeting of the team to discuss two issues that would drive the object-persistence format selection: What kind of objects would be in

Data	Type	Use	Suggested Format
Customer information	Simple (mostly text)	Transactions	Relational
Order Information	Simple (text and numbers)	Transactions	Relational
Marketing Information	Both simple and complex (eventually the system will contain audio clips, video, etc.)	Transactions	Object add-on?
Information that will be exchanged with the Distribution System	Simple text, formatted specifically for importing into the Distribution System	Transactions	Transaction file
Temporary Information	The Web component will likely need to hold information for temporary periods of time. (e.g., the shopping card will store order information before the order is actually placed)	Transactions	Transaction file

FIGURE 10-23

Types of Data in Internet Sales System

the system and how they would be used. Using a whiteboard, they listed their ideas (see Figure 10-23). The project team agreed that the bulk of the data in the system would be the text and numbers that are exchanged with Web users regarding customers and orders. A relational database would be able to handle the data effectively, and the technology would be well received at CD Selections because relational technology is already in place throughout the organization.

However, they realized that relational technology was not optimized to handle complex data, such as the images, sound, and video that the marketing facet of the system ultimately will require. Alec asked Brian to look into relational databases that offered object add-on products (i.e., an RDBMS that could become an ORDBMS). It might be possible for the team to invest in a RDBMS foundation and then upgrade to an ORDBMS version of the same product. However, in the meantime, Alec decided to store sample clips using a random file. This way, they could still deliver the system as envisioned while keeping the technology requirements reasonable.

The team noted that it also must design two transaction files to handle the interface with the distribution system and the Web shopping cart program. The Internet sales system would regularly download order information to the distribution system using a transaction file containing all the required information for that system. Also, the team must design the file that stores temporary order information on the Web server as customers shop through the Web site. The file would contain the fields that ultimately would be transferred to an order object.

Of course, Alec realized that other data needs might arise over time, but he felt confident that the major data issues were identified (e.g., the capability to handle complex data) and that the design of the data management layer would be based on the proper storage technologies.

Map Problem-Domain Objects to Object-Persistence Format

Based on the decision to use an RDBMS and a random file to store the problem-domain objects, Brian created an object-persistence design. To begin, Brian first reviewed the current class and package diagrams for the evolving Internet Sales System (see Figures 8-21, 8-22, 9-23, and 9-27). Focusing on Figures 9-23 and 9-27, Brian began applying the appropriate mapping rules (see Figure 10-9). Based on Rule 1, Brian identified twelve problem domain classes that needed to have their objects stored; as such, Brian created eleven RDBMS tables and one file to represent these objects. These included Credit-Card Clearance Center table, Customer table, Individual table, Organizational table, Order table, Order Item table, CD table, Vendor table, Mkt Info table, Review table, Artist Info table, and Sample Clip File. He also created a set of tentative primary keys for each of the tables and the file. Based on the fact that the objects in the Search Package (see Figures 8-21, 8-22, and 9-23), Search Req, and CD List were only temporary, Brian decided that there was no real need to address them at this point in the design.

Using Rule 4, Brian identified the need for the CD table and the Mkt Info table to both have each other's primary key stored as a foreign key to the other table. Upon further reflection, Brian reasoned that because an actual instance of marketing information was going to be associated with only a single instance of CD, and vice versa, he could have merged the two tables. However, the team had earlier decided to keep them separate, so he decided to simply use the primary key of the CD table as the primary key of the Mkt Info table.

While reviewing the current set of attributes for each of the tables, John suggested that the development had left out the idea of a CD containing a set of tracks. As such, they added a multivalued attribute, tracks, to the CD problem-domain class. However, Brian then pointed out that when they applied Rule 5 to the CD class, they really needed to factor the tracks attribute out as a separate table. Furthermore, as Brian, John, and Susan discussed the track attribute further, it was decided to include it as a problem-domain class as well.

Next, the data management layer team applied Rule 6 to the evolving object-persistence design. In doing this, Susan pointed out that the checks relationship between the Customer and Credit-Card Clearance Center problem-domain classes was a multivalued association. Therefore, it needed its own table in the relational database. Not to be upstaged by Susan, John immediately pointed out that Rule 7 was applicable to eight associations: Customer places Order, Order includes Order Item, Order Item contains CD, Vendor distributes CD, Mkt Info promotes CD, CD contains Tracks, and the three aggregation associations with the Mkt Info class (Review, Artist Info, and Sample Clip). As such, quite a few primary keys had to be copied into the relevant tables as foreign keys (e.g., the primary key of the Customer table had to be copied to the Order table). Can you identify the others?

Finally, Susan suggested the solution to the inheritance problem because RDBMSs do not support inheritance. She pointed out that when applying Rule 8a to the Customer superclass and the Individual and Organizational subclasses, the primary key of the Customer table also had to be copied to the tables representing the subclasses. Furthermore, she pointed out that an XOR condition existed between the two subclasses.

Based on all the suggestions and hard work accomplished by the data management layer team, Brian was able to create a design of the object persistence for the Internet Sales System (see Figure 10-24).

Optimize Object Persistence and Estimate Its Size

Upon completing the object-persistence design, Brian requested a meeting with the development team to walk through the design.[12] After the walkthrough, Alec asked Brian to stay behind to discuss the data management layer model. Now that the team had a good idea of the type of object-persistence formats that would be used, they were ready for the third step: optimizing the design for performance efficiency. Because Brian was the analyst in

[12] It seems that Brian finally got the importance of verifying and validating everything.

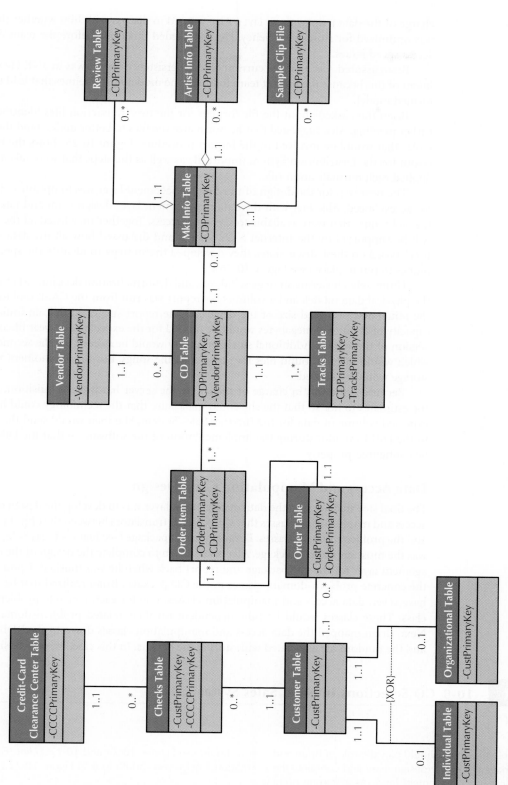

FIGURE 10-24 Internet Sales System Object-Persistence Design

401

charge of the data management layer, Alec wanted to discuss with him whether the model was optimized for storage efficiency. He also needed this done before the team discussed access speed issues.

Brian assured Alec that the current object-persistence model was in 3NF. He was confident of this because the project team followed the modeling guidelines that lead to a well-formed model.

Brian then asked about the file formats for the two transaction files identified in the earlier meeting. Alec suggested that he normalize the files to better understand the various tables that would be involved in the import procedure. Figure 10-25 shows the initial file layout for the Distribution System import file as well as the steps that were taken as Brian applied each normalization rule.

The next step for the design of the data management layer was to optimize the design for access speed. Alec met with the data management layer design team and talked about the techniques that were available to speed up access. Together they listed all the data that will be supported by the Internet Sales System and discussed how all the data would be used. Based on these discussions, they developed the strategy to identify the specific techniques to put in place (see Figure 10-26).

Ultimately, clustering strategies, indexes, and denormalization decisions were applied to the physical data model, and a volumetrics report was run from the CASE tool to estimate the initial and projected size of the database. The report suggested that an initial storage capacity of about 450 megabytes would be needed for the expected one-year life of the first version of the system. Additional storage capacity would be needed for the second version, which would include sound files for samples of the songs, but for the moment not much storage would be needed.

Because Anne was in charge of managing the server hardware acquisition, Alec gave the estimates to her so that the she could make sure that the technology could handle the expected volume of data for the Internet Sales System. She then would send the estimates to the DBMS vendor during the implementation of the software so that the DBMS could be configured properly.

Data Access and Manipulation Class Design

The final step in designing the data management layer was to develop the design of the data access and manipulation classes that would act as translators between the object persistence and the problem domain classes. Because the CD package (see Figures 8-21, 8-22, and 9-27) was the most important package, Alec asked Brian to complete the design of the data management layer for the CD package and report back when he was finished. Upon reviewing the concrete problem-domain classes in the CD package, Brian realized that he needed to have seven data access and manipulation classes, one for each concrete problem-domain class. These classes would be fully dependent on their related problem-domain classes. Next, Brian mapped the data access and manipulation classes down to the RDBMS tables and the random file associated with storing the objects. In this case, there were six RDBMS

YOUR **TURN** **10-8 CD Selections Internet Sales System**

At this point in time, you should play the role of Brian and his data management layer design team and complete the design of the data management layer. As a starting point you should use Figures 10-24 and 10-27. (*Hint:* For some additional help, you should look at Figure 10-22.)

Remove repeating groups of items and place them in a separate Item entity.

1NF:

There is one partial dependencies in the above data model because Item qty depends on the whole primary key, whereas Item Desc and Item Price is dependent only on Item No.

2NF:

Remove transitive dependencies. Customer home address depends upon Cust FName and Cust LName (and these do not serve as the identifier for the Order entity); therefore, a Customer entity is added to contain customer information. Order then contains only order information and necessary foreign keys.

3NF:

FIGURE 10-25   Internet Sales System Import File Normalization Process

Target	Comments	Suggestions to Improve Data Access Speed
All tables	Basic table manipulation	• Investigate if records should be clustered physically by primary key • Create indexes for primary keys • Create indexes for foreign key fields
All tables	Sorts and Grouping	• Create indexes for fields that are frequently sorted or grouped
CD information	Users will need to search CD information by title, artist, and category	• Create indexes for CD title, artist, and category
Order Information	Operators should be able to locate information about a particular customer's order	• Create an index in the Order table for orders by customer name
Entire Physical Model	Investigate denormalization opportunities for all fields that are not updated very often	• Investigate one-to-one relationships • Investigate look-up tables • Investigate one-to-many relationships

FIGURE 10-26
Internet Sales
System Performance

tables and one random file. Again, the data access and manipulation classes were dependent on the object-persistence format. Figure 10-27 depicts the data management layer and problem-domain layer for the CD package of the Internet sales system.

SUMMARY

Object-Persistence Formats

There are four basic types of object-persistence formats: files (sequential and random access), object-oriented databases, object-relational databases, and relational databases. Files are electronic lists of data that have been optimized to perform a particular transaction. There are two different access methods (sequential and random), and there are five different application types: master, look-up, transaction, audit, and history. Master files typically are kept for long periods of time because they store important business information, such as order information or customer mailing information. Look-up files contain static values that are used to validate fields in the master files, and transaction files temporarily hold information that will be used for a master file update. An audit file records before-and-after images of data as they are altered so that an audit can be performed if the integrity of the data is questioned. Finally, the history file stores past transactions (e.g., old customers, past orders) that are no longer needed by the system.

A database is a collection of information groupings related to each other in some way, and a DBMS (database management system) is software that creates and manipulates databases. There are three types of databases that are likely to be encountered during a project: relational, object relational, and object oriented. The relational database is the most popular kind of database for application development today. It is based on collections of tables that are related to each other through common fields, known as foreign keys. Object-relational databases are relational databases that have extensions that provide limited support for object orientation. The extensions typically include some support for the storage of objects in the relational table structure. Object-oriented databases come in two flavors: full-blown DBMS products and extensions to an object-oriented programming language. Both approaches typically fully support object orientation.

Problem-Domain Classes

DAM Classes

Object Persistence

FIGURE 10-27 Data Management and Problem-Domain Layer Design for the CD Package of the Internet Sales System

The application's data should drive the storage format decision. Relational databases support simple data types very effectively, whereas object databases are best for complex data. The type of system also should be considered when choosing among data storage formats (e.g., relational databases have matured to support transactional systems). Although less critical to the format selection decision, the project team needs to consider what technology exists within the organization and the kind of technology likely to be used in the future.

Mapping Problem-Domain Objects to Object-Persistence Formats

There are many different approaches to support object persistence. Each of the different formats for object persistence has some conversion requirements. The complexity of the conversion requirements increases the further the format is from an object-oriented format. An OODBMS has the fewest conversion requirements, whereas an RDBMS tends to have the most. No matter what format is chosen, all data access and management functionality should be kept out of the problem-domain classes to minimize the maintenance requirements of the system and to maximize the portability and reusability of the problem domain classes.

Optimizing RDBMS-Based Object Storage

There are two primary dimensions by which to optimize a relational database: storage efficiency and speed of access. The most efficient relational database tables in terms of data storage are those that have no redundant data and very few null values. Normalization is the process whereby a series of rules are applied to the data management layer to determine how well formed it is. A model is in first normal form (1NF) if it does not lead to repeating fields, which are fields that repeat within a table to capture multiple values. Second normal form (2NF) requires that all tables be in 1NF and lead to fields whose values are dependent on the *whole* primary key. Third normal form (3NF) occurs when a model is in both 1NF and 2NF and none of the resulting fields in the tables depend on nonprimary key fields (i.e., transitive dependency). With each violation, additional tables should be created to remove the repeating fields or improper dependencies from the existing tables.

Once we have optimized the design of the object persistence for storage efficiency, the data may be spread out across a number of tables. To improve speed, the project team may decide to denormalize—add redundancy back into—the design. Denormalization reduces the number of joins that need to be performed in a query, thus speeding up data access. Denormalization is best in situations where data are accessed frequently and updated rarely. There are three modeling situations that are good candidates for denormalization: lookup tables, entities that share one-to-one relationships, and entities that share one-to-many relationships. In all three cases, attributes from one entity are moved or repeated in another entity to reduce the joins that must occur while accessing the database.

Clustering occurs when similar records are stored close together on the storage medium to speed up retrieval. In intrafile clustering, similar records in the table are stored together in some way, such as in sequence. Interfile clustering combines records from more than one table that typically are retrieved together. Indexes also can be created to improve the access speed of a system. An index is a minitable that contains values from one or more columns in a table and tells where the values can be found. Instead of performing a table scan, which is the most inefficient way to retrieve data from a table, an index points directly to the records that match the requirements of a query.

Finally, the speed of the system can be improved if the right hardware is purchased to support it. Analysts can use volumetrics to estimate the current and future size of the database and then share these numbers with the people who are responsible for buying and configuring the database hardware.

Nonfunctional Requirements and Data Management Layer Design

Nonfunctional requirements can impact the design of the data access and management layer. Operational requirements can potentially limit the viability of different object-persistence formats. Performance requirements can cause various indexing and caching approaches to be considered. Furthermore, performance requirements can create situations where denormalization must be considered. The security requirements can cause the design to include varying types of access to be controlled and the possibility of using different encryption algorithms to make the data more difficult to utilize by unauthorized users. Finally, political and cultural concerns can influence the design of certain attributes and objects.

Designing Data Access and Manipulation Classes

Once the object persistence has been designed, a translation layer between the problem-domain classes and the object persistence should be created. The translation layer is implemented through data access and manipulation classes. In this manner, any changes to the object-persistence format chosen will require changes only to the data access and manipulation classes. The problem-domain classes will be completely isolated from the changes.

KEY TERMS

Access control
Attribute sets
Audit file
Cluster
Data access and manipulation classes
Data management layer
Database
Database management
    system (DBMS)
Decision support systems (DSS)
Denormalization
End user DBMS
Enterprise DBMS
Executive information systems (EIS)
Expert system (ES)
Extent
File
First normal form (1NF)
Foreign key
Hardware and operating system
History file
Impedance mismatch
Index

Interfile clustering
Intrafile clustering
Join
Linked list
Lookup file
Management information
    system (MIS)
Master file
Multivalued attributes (fields)
Normalization
Object ID
Object-oriented database
    management system (OODBMS)
Object-oriented programming
    language (OOPL)
Object persistence
Object-relational database
    management system (ORDBMS)
Operational requirements
Ordered sequential access file
Overhead
Partial dependency
Performance requirements

Pointer
Political and cultural requirements
Primary key
Problem-domain classes
Random access files
Raw data
Referential integrity
Relational database management
    system (RDBMS)
Relationship sets
Repeating groups (fields)
Second normal form (2NF)
Security requirements
Sequential access files
Structured query language (SQL)
Table scan
Third normal form (3NF)
Transaction file
Transaction-processing system
Transitive dependency
Unordered sequential access file
Update anomaly
Volumetrics

QUESTIONS

1. How are a file and a database different from each other?
2. Name five types of files and describe the primary purpose of each type.
3. What are the differences between sequential and random access files?
4. What is the most popular kind of database today? Provide three examples of products that are based on this database technology.
5. What is the difference between an end-user database and an enterprise database? Provide an example of each one.
6. Why should you consider the storage formats that already exist in an organization when deciding upon a storage format for a new system?
7. Describe the four steps in object-persistence design.
8. What are the factors in determining the type of object-persistence format that should be adopted for a system? Why are these factors so important?
9. Why is it important to understand the initial and projected size of a database during design?
10. Describe what should be considered when estimating the size of a database.
11. What is referential integrity and how is it implemented in an RDBMS?
12. List some of the differences between an ORDBMS and an RDBMS.
13. When implementing the object persistence in an ORDBMS, what types of issues must you address?
14. When implementing the object persistence in an RDBMS, what types of issues must you address?
15. What is the purpose of normalization?
16. How does a model meet the requirements of third normal form?
17. Describe three situations that can be good candidates for denormalization.
18. What are the advantages of using an ORDBMS over an OODBMS?
19. List some of the differences between an ORDBMS and an OODBMS.
20. Describe several techniques that can improve performance of a database.
21. What are the two dimensions in which to optimize a relational database?
22. What are the advantages of using an OODBMS over an RDBMS?
23. What are the advantages of using an OODBMS over an ORDBMS?
24. What is the primary purpose of the data access and manipulation classes?
25. Why should the data access and manipulation classes be dependent on the associated problem-domain classes instead of the other way around?
26. What are some of the nonfunctional requirements that can influence the design of the data management layer?
27. What are the key issues in deciding between using perfectly normalized databases and denormalized databases?
28. Name three ways in which null values can be interpreted in a relational database. Why is this problematic?
29. What is the difference between interfile and intrafile clustering? Why are they used?
30. What is an index and how can it improve the performance of a system?

EXERCISES

A. Using the Web, review one of the following products. What are the main features and functions of the software? In what companies has the DBMS been implemented, and for what purposes? According to the information that you found, what are three strengths and weaknesses of the product?
   1. Relational DBMS
   2. Object-relational DBMS
   3. Object-oriented DBMS

B. Using the Web or other resources, identify a product that can be classified as an end-user database and a product that can be classified as an enterprise database. How are the products described and marketed? What kinds of applications and users do they support? In what kinds of situations would an organization choose to implement an end-user database over an enterprise database?

C. Visit a commercial Web site (e.g., Amazon.com). If files were being used to store the data supporting the application, what types of files would be needed? What access type would be required? What data would they contain?

D. You have been given a file that contains the following fields relating to CD information. Using the steps of normalization, create a model that represents this file in third normal form. The fields include:

Musical group name	CD title 2
Musicians in group	CD title 3
Date group was formed	CD 1 length
Group's agent	CD 2 length
CD title 1	CD 3 length

Assumptions:

- Musicians in group contains a list of the members of the people in the musical group.
- Musical groups can have more than one CD, so both group name and CD title are needed to uniquely identify a particular CD.

E. Jim Smith's dealership sells Fords, Hondas, and Toyotas. The dealership keeps information about each car manufacturer with whom they deal so that they can get in touch with them easily. The dealership also keeps information about the models of cars that they carry from each manufacturer. They keep information like list price, the price the dealership paid to obtain the model, and the model name and series (e.g., Honda Civic LX). They also keep information about all sales that they have made (for instance, they will record a buyer's name, the car bought, and the amount paid for the car). In order to contact the buyers in the future, contact information is also kept (e.g., address, phone number). Create a class diagram for this situation. Apply the rules of normalization to the class diagram to check the diagram for processing efficiency.

F. Describe how you would denormalize the model that you created in exercise E. Draw the new class diagram based on your suggested changes. How would performance be affected by your suggestions?

G. Examine the model that you created in exercise F. Develop a clustering and indexing strategy for this

model. Describe how your strategy will improve the performance of the database.

H. Calculate the size of the database that you created in exercise F. Provide size estimates for the initial size of the database as well as for the database in one year's time. Assume that the dealership sells ten models of cars from each manufacturer to approximately 20,000 customers a year. The system will be set up initially with one year's worth of data.

I. For the health club exercises in Chapter 5 (N, O), Chapter 6 (O), Chapter 7 (G, H), Chapter 8 (G, H), and Chapter 9 (A):

1. Apply the rules of normalization to the class diagram to check the diagram for processing efficiency.
2. Develop a clustering and indexing strategy for this model. Describe how your strategy will improve the performance of the database.

J. For the Picnics R Us exercises in Chapter 5 (P, Q), Chapter 6 (P), Chapter 7 (I), Chapter 8 (I, J), and Chapter 9 (C):

1. Apply the rules of normalization to the class diagram to check the diagram for processing efficiency.
2. Develop a clustering and indexing strategy for this model. Describe how your strategy will improve the performance of the database.

K. For the Of-the-Month Club exercises in Chapter 5 (T, U), Chapter 6 (R), Chapter 7 (L), Chapter 8 (K, L), and Chapter 9 (E):

1. Apply the rules of normalization to the class diagram to check the diagram for processing efficiency.
2. Develop a clustering and indexing strategy for this model. Describe how your strategy will improve the performance of the database.

L. For the library exercises in Chapter 5 (R, S), Chapter 6 (Q), Chapter 7 (K), Chapter 8 (M, N), and Chapter 9 (G):

1. Apply the rules of normalization to the class diagram to check the diagram for processing efficiency.
2. Develop a clustering and indexing strategy for this model. Describe how your strategy will improve the performance of the database.

MINICASES

1. Refer to the Professional and Scientific Staff Management (PSSM) minicase in Chapters 5, 7, and 8.

  a. Apply the rules of normalization to the class diagram to check the diagram for processing efficiency.

  b. Develop a clustering and indexing strategy for this model. Describe how your strategy will improve the performance of the database.

2. In the new system under development for Holiday Travel Vehicles, seven tables will be implemented in the new relational database. These tables are: New Vehicle, Trade-in Vehicle, Sales Invoice, Customer, Salesperson, Installed Option, and Option. The expected average record size for these tables and the initial record count per table are given here.

Table Name	Average Record Size	Initial Table Size (records)
New Vehicle	65 characters	10,000
Trade-in Vehicle	48 characters	7,500
Sales Invoice	76 characters	16,000
Customer	61 characters	13,000
Salesperson	34 characters	100
Installed Option	16 characters	25,000
Option	28 characters	500

Perform a volumetrics analysis for the Holiday Travel Vehicle system. Assume that the DBMS that will be used to implement the system requires 35 percent overhead to be factored into the estimates. Also, assume a growth rate for the company of 10 percent per year. The systems development team wants to ensure that adequate hardware is obtained for the next three years.

3. The system development team at Hoffmann Baer, a small Swiss textile firm, is developing a new customer order entry system. In the process of designing the new system, the team has identified the following class and its attributes:

```
Inventory Order
Order Number (PK)
Order Date
Customer Name
Street Address
City
Canton-State
Postal Code
Country
Customer Type
Sales District Number
1 to 22 occurrences of:
    Item Name
    Quantity Ordered
    Item In
    Quantity Shipped
    Item Out
    Quantity Received
```

a. State the rule that is applied to place a class in first normal form (1NF). Revise the above class diagram so that it is in 1NF.
b. State the rule that is applied to place a class in second normal form (2NF). Create a class diagram using the class and attributes described (if necessary) to place it in 2NF.
c. State the rule that is applied to place a class in third normal form (3NF). Revise the class diagram to place it in 3NF.
d. When planning for the physical design of this database, can you identify any likely situations where the project team may chose to denormalize the class diagram? After going through the work of normalizing, why should this be considered?

CHAPTER 11

USER INTERFACE DESIGN

A user interface is the part of the system with which the users interact. It includes the screen displays that provide navigation through the system, the screens and forms that capture data, and the reports that the system produces (whether on paper, on the screen, or via some other media). This chapter introduces the basic principles and processes of interface design and discusses how to design the interface structure and standards, navigation design, input design, and output design. The chapter also describes the affect of the nonfunctional requirements on designing the human-computer interaction layer.

OBJECTIVES

- Understand several fundamental user interface design principles.
- Understand the process of user interface design.
- Understand how to design the user interface structure.
- Understand how to design the user interface standards.
- Understand commonly used principles and techniques for navigation design.
- Understand commonly used principles and techniques for input design.
- Understand commonly used principles and techniques for output design.
- Be able to design a user interface.
- Understand the affect of nonfunctional requirements on the human-computer interaction layer.

CHAPTER OUTLINE

Introduction
Principles for User Interface Design
   Layout
   Content Awareness
   Aesthetics
   User Experience
   Consistency
   Minimizing User Effort
User Interface Design Process
   Use Scenario Development
   Interface Structure Design
   Interface Standards Design
   Interface Design Prototyping
   Interface Evaluation

Navigation Design
   Basic Principles
   Types of Navigation Controls
   Messages
   Navigation Design
     Documentation
Input Design
   Basic Principles
   Types of Inputs
   Input Validation
Output Design
   Basic Principles
   Types of Outputs
   Media

INTRODUCTION

Interface design is the process of defining how a system will interact with external entities (e.g., customers, suppliers, other systems). In this chapter we focus on the design of *user interfaces,* but it is also important to remember that there are sometimes *system interfaces,* which exchange information with other systems. System interfaces are typically designed as part of a systems integration effort. They are defined in general terms as part of the physical architecture and data management layers.

The human–computer interface layer defines the way in which the users will interact with the system and the nature of the inputs and outputs that the system accepts and produces. The user interface includes three fundamental parts. The first is the *navigation mechanism,* the way in which the user gives instructions to the system and tells it what to do (e.g., buttons, menus). The second is the *input mechanism,* the way in which the system captures information (e.g., forms for adding new customers). The third is the *output mechanism,* the way in which the system provides information to the user or to other systems (e.g., reports, Web pages). Each of these is conceptually different, but they are closely intertwined. All computer displays contain navigation mechanisms, and most contain input and output mechanisms. Therefore, navigation design, input design, and output design are tightly coupled.

First, this chapter introduces several fundamental user interface design principles. Second, it provides an overview of the design process for the human–computer interaction layer. Third, the chapter provides an overview of the navigation, input, and output components that are used in interface design. This chapter focuses on the design of Web-based interfaces and *graphical user interfaces (GUI)* that use windows, menus, icons and a mouse (e.g., Windows, Macintosh).[1] Although text-based interfaces are still commonly used on mainframes and Unix systems, GUI interfaces are probably the most common type of interfaces that we will use, with the possible exception of printed reports.[2]

PRINCIPLES FOR USER INTERFACE DESIGN

In many ways, user interface design is an art. The goal is to make the interface pleasing to the eye and simple to use while minimizing the effort the users need to accomplish their work. The system is never an end in itself; it is merely a means to accomplish the business of the organization.

---

[1] Many people attribute the origin of GUI interfaces to Apple or Microsoft. Some people know that Microsoft copied from Apple, which, in turn, "borrowed" the whole idea from a system developed at the Xerox Palo Alto Research Center (PARC) in the 1970s. Very few know that the Xerox system was based on a system developed by Doug Englebart of Stanford that was first demonstrated at the Western Computer Conference in 1968. Around the same time, Doug also invented the mouse, desktop video conferencing, groupware, and host of other things we now take for granted. Doug is a legend in the computer science community and has won too many awards to count but is relatively unknown by the general public.

[2] A good book on GUI design is Susan Fowler, *GUI Design Handbook* (New York: McGraw-Hill, 1998).

Principle	Description
Layout	The interface should be a series of areas on the screen that are used consistently for different purposes—for example, a top area for commands and navigation, a middle area for information to be input or output, and a bottom area for status information.
Content Awareness	Users should always be aware of where they are in the system and what information is being displayed.
Aesthetics	Interfaces should be functional and inviting to users through careful use of white space, colors, and fonts. There is often a trade-off between including enough white space to make the interface look pleasing without losing so much space that important information does not fit on the screen.
User Experience	Although ease of use and ease of learning often lead to similar design decisions, there is sometimes a trade-off between the two. Novice users or infrequent users of software will prefer ease of learning, whereas frequent users will prefer ease of use.
Consistency	Consistency in interface design enables users to predict what will happen before they perform a function. It is one of the most important elements in ease of learning, ease of use, and aesthetics.
Minimal User Effort	The interface should be simple to use. Most designers plan on having no more than three mouse clicks from the starting menu until users perform work.

FIGURE 11-1
Principles of User
Interface Design

We have found that the greatest problem facing experienced designers is using space effectively. Simply put, there often is too much information that needs to be presented on a screen or report or form than will fit comfortably. Analysts must balance the need for simplicity and pleasant appearance against the need to present the information across multiple pages or screens, which decreases simplicity. In this section, we discuss some fundamental interface design principles, which are common for navigation design, input design, and output design[3] (see Figure 11-1).

Layout

The first element of design is the basic *layout* of the *screen, form,* or *report.* Most software designed for personal computers follows the standard Windows or Macintosh approach for screen design. The screen is divided into three boxes. The top box is the navigation area, through which the user issues commands to navigate through the system. The bottom box is the status area, which displays information about what the user is doing. The middle—and largest—box is used to display reports and present forms for data entry.

In many cases (particularly on the Web), multiple layout areas are used. Figure 11-2 shows a screen with five navigation areas, each of which is organized to provide different functions and navigation within different parts of the system. The top area provides the standard Internet Explorer navigation and command controls that change the contents of the entire system. The navigation area on the left edge navigates between sections and changes all content to its right. The other two navigation areas at the top and bottom of the page provide other ways to navigate between sections. The content in the middle of the page displays the results (i.e., a report on a book) and also provides additional navigation within the page about this book.

---

[3] A good book on the design of interfaces is Susan Weinschenk, Pamela Jamar, and Sarah Yeo, *GUI Design Essentials* (New York: Wiley, 1997).

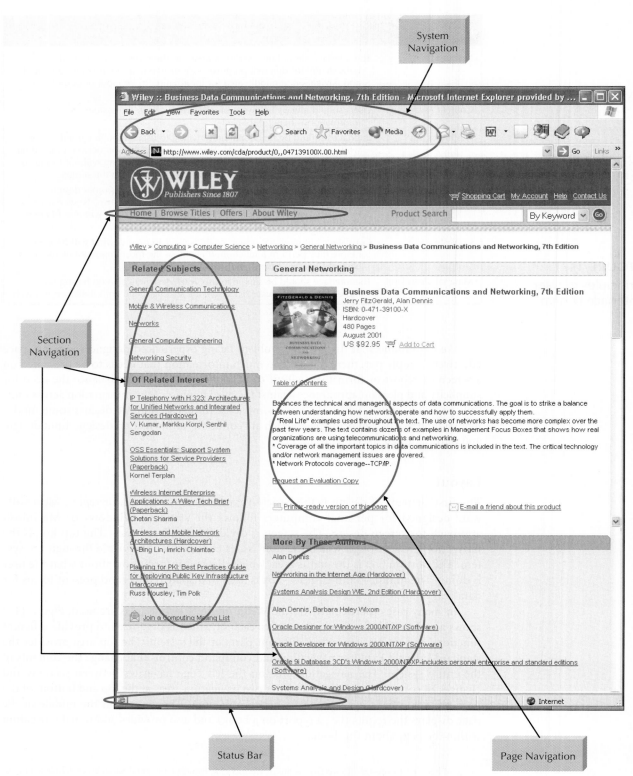

FIGURE 11-2   Layout with Multiple Navigation Areas

This use of multiple layout areas for navigation also applies to inputs and outputs. Data areas on reports and forms are often subdivided into subareas, each of which is used for a different type of information. These areas are almost always rectangular in shape, although sometimes space constraints will require odd shapes. Nonetheless, the margins on the edges of the screen should be consistent. Each of the areas within the report or form is designed to hold different information. For example, on an order form (or order report), one part may be used for customer information (e.g., name, address), one part, for information about the order in general (e.g., date, payment information), and one part, for the order details (e.g., how many units of which items at what price each). Each area is self-contained so that information in one area does not run into another.

The areas and information within areas should have a natural intuitive flow to minimize the users' movement from one area to the next. People in westernized nations (e.g., United States, Canada, Mexico) tend to read top-to-bottom, left-to-right, so related information should be placed so it is used in this order (e.g., address lines, followed by city, state/province, and then zip code/postal code). Sometimes the sequence is in chronological order, or from the general to the specific, or from most frequently to least frequently used. In any event, before the areas are placed on a form or report, the analyst should have a clear understanding of what arrangement makes the most sense for how the form or report will be used. The flow between sections should also be consistent, whether horizontal or vertical (see Figure 11-3). Ideally, the areas will remain consistent in size, shape, and placement for the forms used to enter information (whether paper or on-screen) and the reports used to present it.

Content Awareness

Content awareness refers to the ability of an interface to make the user aware of the information it contains with the least amount of effort on the user's part. All parts of the interface, whether navigation, input, or output, should provide as much content awareness as possible, but it is particularly important for forms or reports that are used quickly or irregularly (e.g., a Web site).

Content awareness applies to the interface in general. All interfaces should have titles (on the screen frame, for example). Menus should show where we are and, if possible, where we came from to get there. For example, in Figure 11-2, the line below the top section navigation area shows that the user has browsed from the Wiley home page to the computer science section, to the networking section, to the general networking section, and then to this specific book.

Content awareness also applies to the areas within forms and reports. All areas should be clear and well defined (with titles if space permits) so that it is difficult for the user to become confused about the information in any area. Then users can quickly locate the part of the form or report that is likely to contain the information they need. Sometimes the areas are marked using lines, colors, or headings (e.g., the section area in Figure 11-2), whereas in other cases the areas are only implied (e.g., the page controls at the bottom of Figure 11-2).

Content awareness also applies to the *fields* within each area. Fields are the individual elements of data that are input or output. The *field labels* that identify the fields on the interface should be short and specific—objectives that often conflict. There should be no uncertainty about the format of information within fields, whether for entry or display. For example a date of 10/5/07 is different depending on whether you are in the United States (October 5, 2007) or in Canada (May 10, 2007). Any fields for which there is the possibility of uncertainty or multiple interpretations should provide explicit explanations.

Patient Information

Patient Name:

First Name: [   ]

Last Name: [   ]

Address:

Street: [   ]

City: [   ]

State/Province: [   ]

Zip Code/Postal Code: [   ]

Home phone: [   ]

Office phone: [   ]

Cell phone: [   ]

Referring Doctor:

First Name: [   ]

Last Name: [   ]

Street: [   ]

City: [   ]

State/Province: [   ]

Zip Code/Postal Code: [   ]

Office phone: [   ]

(a) Vertical Flow

Patient Information

Patient Name:

First Name: [   ]   Last Name: [   ]

Street: [   ]   City: [   ]   State/Province: [   ]   Zip Code/Postal Code: [   ]

Home Phone: [   ]   Office Phone: [   ]   Cell Phone: [   ]

Referring Doctor:

First Name: [   ]   Last Name: [   ]

Street: [   ]   City: [   ]   State/Province: [   ]   Zip Code/Postal Code: [   ]

Office Phone: [   ]

(b) Horizontal Flow

FIGURE 11-3   Flow Between Interface Sections

Content awareness also applies to the information that a form or report contains. In general, all forms and reports should contain a preparation date (i.e., the date printed or the data completed) so that the age of information is obvious. Likewise, all printed forms and software should provide version numbers so that users, analysts, and programmers can identify outdated materials.

Figure 11-4 shows a form from the University of Georgia. This form illustrates the logical grouping of fields into areas with an explicit box (top left), as well as an implied area with no box (lower left). The address fields within the address area follow a clear, natural order. Field labels are short where possible (see the top left) but long where more information is needed to prevent misinterpretation (see the bottom left).

Aesthetics

Aesthetics refers to designing interfaces that are pleasing to the eye. Interfaces do not have to be works of art, but they do need to be functional and inviting to use. In most cases, less is more, meaning that a simple, minimalist design is the best.

Space is usually at a premium on forms and reports, and often there is the temptation to squeeze as much information as possible onto a page or a screen. Unfortunately, this can make a form or report so unpleasant that users do not want to use it. In general, all forms and reports need a minimum amount of *white space* that is intentionally left blank.

What was your first reaction when you looked at Figure 11-4? This is the most unpleasant form at the University of Georgia, according to staff members. Its *density* is too high; it has too much information packed into a too-small space with too little white space. Although it may be efficient to save paper by using one page, not two, it is not effective for many users.

In general, novice or infrequent users of an interface, whether on a screen or on paper, prefer interfaces with low density, often one with a density of less than 50 percent (i.e., less than 50 percent of the interface occupied by information). More experienced users prefer higher densities, sometimes approaching 90 percent occupied because they know where information is located and high densities reduce the amount of physical movement through the interface. We suspect the form in Figure 11-4 was designed for the experienced staff in the personnel office, who use it daily, rather than for the clerical staff in academic departments with less personnel experience, who use the form only a few times a year.

The design of text is equally important. As a general rule, all text should be in the same font and about the same size. Fonts should be no less than 8 points in size, but 10 points is often preferred, particularly if the interface will be used by older people. Changes in font and size are used to indicate changes in the type of information that is presented (e.g., headings, status indicators). In general, italics and underlining should be avoided because they make text harder to read.

Serif fonts (i.e., those having letters with serifs, or tails, such as Times Roman, or the font you are reading right now) are the most readable for printed reports, particularly for small letters. Sans serif fonts (i.e., those without serifs, such as Helvetica or Arial, like those used for the chapter titles in this book) are the most readable for computer screens and are often used for headings in printed reports. Never use all capital letters, except possibly for titles.

Color and patterns should be used carefully and sparingly and only when they serve a purpose. (About 10 percent of men are color blind, so the improper use of color can impair their ability to read information.) A quick trip around the Web will demonstrate the problems caused by indiscriminate use of colors and patterns. Remember, the goal is pleasant readability, not art; color and patterns should be use to strengthen the message, not overwhelm it. Color is best used to separate and categorize items, such as showing the difference between headings and regular text, or to highlight important information. Therefore, colors with high contrast should be used (e.g., black and white). In general, black text on a

EMPLOYEE PERSONNEL REPORT

UNIVERSITY OF GEORGIA

PAY TYPE

ACTION MO DA YR

DOCUMENT NO.	PAGE	DATE	FY	DEPARTMENT PHONE	COLLEGE OR DIVISION

DEPARTMENT/PROJECT

PRI DEPT | HIGH DEGREE | INSTITUTION | YEAR

UGA EMPLOYMENT HISTORY
☐ (C) CURRENT ☐ (P) PREVIOUS
DATE

SOC.SEC.NUM. | LAST NAME | FIRST NAME/INITIAL | MIDDLE INITIAL/NAME | SUF

UGA % TIME

STREET OR ROUTE NO. (LINE 1) | NON-WORK PHONE | BIRTH DATE | SPOUSE'S NAME | CHAIR

☐ (1) REGULAR   ☐ (3) TEMPORARY
☐ (2) UGA STUDENT   ☐ (4) NR-ALIEN
☐ (E) EXEMPT   ☐ (N) NON-EXEMPT   ☐ (T) TIPPED
☐ (M) MALE   ☐ (S) SINGLE   ☐ (Y) FACULTY-RANK
☐ (F) FEMALE   ☐ (M) MARRIED   ☐ (N) NON-FACULTY

STREET OR ROUTE NO. (LINE 2) | UNIVERSITY PHONE | CITIZEN OF | 1-9 VISA COUNTY

☐ (1) WHITE   ☐ (3) ORIENTAL/ASIAN   ☐ (5) HISPANIC
☐ (2) BLACK   ☐ (4) AMERICAN INDIAN ☐ (6) MULTIRACIAL
                                            ☐ (9)

CITY | STATE | ZIP + 4 | UNIVERSITY BUILDING NAME | BLDG.NO/FLOOR/ROOM

COOP. EXT. EMPLOYEES ONLY
UGA SALARY
COUNTY MONEY
TOTAL

PAYROLL PAYMENT DISTRIBUTION
☐ (1) SEND TO DEPT (DIST CODE)
☐ (2) DIRECT DEPOSIT(SEND PR105 TO PAYROLL
☐ (3) PICK UP AT PAYROLL WINDOW

FOR PAYROLL DEPT USE ONLY
FED EXM	STATE EXM	OASDI	RETIRE	COUNTY MONEY (PER PAY PERIOD)
		HI	EIC	

GDCP
GDCP

TRX	HOME DEPT	SHORT TITLE	POSN NO.	APPT. BEGIN MO DA YR HR	APPT. END MO DA YR HR	JOBCLASS CODE	POSITION TITLE	POS % TIME	C N	FULL TIME ANNUAL SALARY	S C	SUPPLEMENT AMOUNT

MO DA YR HR ... MO DA YR HR

PAYROLL AUTHORIZATION
TRX	HOME DEPT	SHORT TITLE	POSN NO	ACCOUNT	FISCAL YEAR	BUDGET	EFT

FROM
THRU
AMOUNT PER PAY PERIOD OR HOURLY RATE

TOTALS

☐ (A) NEW UGA EMPLOYEE   ☐ (B) LATERAL TRANSFER   ☐ (C) PROMOTION
☐ (D) REPLACEMENT POSN-NAME OF LAST INCUMBENT
☐ (E) APPOINTMENT TO NEW POSITION
☐ (F) CHANGE % TIME EMPLOYED FROM ___ TO ___
☐ (G) CONTINUATION WITHIN EXISTING BUDGET POSITION
☐ (H) REVISE DISTRIBUTION OF SALARY
☐ (I) TRANSFER FROM DEPT ___ TO ___
☐ (J) CHANGE PAY TYPE FROM ___ TO ___

☐ (K) CHANGE TITLE FROM ___ TO ___
☐ (L) CHANGE NAME FROM ___ TO ___
☐ (M) CHANGE SSN FROM ___ TO ___
☐ (N) LEAVE W/O PAY FROM ___ TO ___
☐ (O) CHG COUNTY $ FROM ___ TO ___
☐ (P) TERMINATION-REASON
☐ (Q) OTHER (SPECIFY)

REMARKS

DEPARTMENT HEAD	DATE	VICE PRESIDENT	DATE	BUDGET REVIEW	DATE	BUDGET OFFICE	DATE
DEAN/DIRECTOR	DATE	FACULTY RECORDS	DATE	CONTRACTS & GRANTS	DATE	PERSONNEL	DATE

FIGURE 11-4 Form Example

white background is the most readable, with blue on red the least readable. (Most experts agree that background patterns on Web pages should be avoided). Color has been shown to affect emotion, with red provoking intense emotion (e.g., anger) and blue provoking lowered emotions (e.g., drowsiness).

User Experience

User experience can essentially be broken down into two levels: those with experience and those without. Interfaces should be designed for both types of users. Novice users usually are most concerned with *ease of learning*—how quickly they can learn new systems. Expert users are usually most concerned with *ease of use*—how quickly they can use the system once they have learned how to use it. Often these two are complementary and lead to similar design decisions, but sometimes there are trade-offs. Novices, for example, often prefer menus, which show all available system functions, because these promote ease of learning. Experts, on the other hand, sometimes prefer fewer menus organized around the most commonly used functions.

Systems that will end up being used by many people on a daily basis are more likely to have a majority of expert users (e.g., order entry systems). Although interfaces should try to balance ease of use and ease of learning, these types of systems should put more emphasis on ease of use rather than ease of learning. Users should be able to access the commonly used functions quickly, with few keystrokes or a small number of menu selections.

In many other systems (e.g., decision support systems), most people will remain occasional users for the lifetime of the system. In this case, greater emphasis may be placed on ease of learning rather than ease of use.

Ease of use and ease of learning often go hand-in-hand—but sometimes they don't. Research shows that expert and novice users have different requirements and behavior patterns in some cases. For example, novices virtually never look at the bottom area of a screen that presents status information, whereas experts refer to the status bar when they need information. Most systems should be designed to support frequent users, except for systems designed to be used infrequently or when many new users or occasional users are expected (e.g., the Web). Likewise, systems that contain functionality that is used only occasionally must contain a highly intuitive interface or an interface that contains clear explicit guidance regarding its use.

The balance of quick access to commonly used and well-known functions and guidance through new and less-well-known functions is challenging to the interface designer, and this balance often requires elegant solutions. Microsoft Office, for example, addresses this issue through the use of the "show-me" functions, which demonstrate the menus and buttons for specific functions. These features remain in the background until they are needed by novice users (or even experienced users when they use an unfamiliar part of the system).

Consistency

Consistency in design is probably the single most important factor in making a system simple to use because it enables users to predict what will happen. When interfaces are consistent, users can interact with one part of the system and then know how to interact with the rest, aside, of course, from elements unique to those parts. Consistency usually refers to the interface within one computer system, so that all parts of the same system work in the same way. Ideally, the system should also be consistent with other computer systems in the organization and with commercial software that is used (e.g., Windows). For example, many users are familiar with the Web, so the use of Web-like interfaces can reduce the amount of learning required by the user. In this way, the user can reuse Web knowledge, thus significantly reducing the learning curve for a new system. Many software development tools support consistent system interfaces by providing standard interface objects (e.g., list boxes, pull-down menus, and radio buttons).

Consistency occurs at many different levels. Consistency in the *navigation controls* conveys how actions in the system should be performed. For example, using the same icon or command to change an item clearly communicates how changes are made throughout the system. Consistency in terminology is also important. This refers to using the same words for elements on forms and reports (e.g., not customer in one place and client in another). We also believe that consistency in report and form design is important, although a recent study suggests that being *too* consistent can cause problems.[4] When reports and forms are very similar except for very minor changes in titles, users sometimes mistakenly use the wrong form and either enter incorrect data or misinterpret its information. The implication for design is to make the reports and forms similar but give them some distinctive elements (e.g., color, size of titles) that enable users to immediately detect differences.

Minimizing User Effort

Finally, interfaces should be designed to minimize the amount of effort needed to accomplish tasks. This means using the fewest possible mouse clicks or keystrokes to move from one part of the system to another. Most interface designers follow the *three-clicks rule:* users should be able to go from the start or main menu of a system to the information or action they want in no more than three mouse clicks or three keystrokes.

USER INTERFACE DESIGN PROCESS

User interface design is a five-step process that is iterative—analysts often move back and forth between steps, rather than proceeding sequentially from step 1 to step 5 (see Figure 11-5). First, the analysts examine the *use cases* (see Chapter 5) and *sequence diagrams* (see Chapter 7) developed in analysis and interview users to develop *use scenarios* that describe commonly employed patterns of actions the users will perform so the interface enables users to quickly and smoothly perform these scenarios. Second, the analysts develop the *windows navigation diagram (WND)* that defines the basic structure of the interface. These diagrams show all the interfaces (e.g., screens, forms, and reports) in the system and how they are connected. Third, the analysts design *interface standards,* which are the basic design elements on which interfaces in the system are based. Fourth, the analysts create an *interface design prototype* for each of the individual interfaces in the system, such as navigation controls (including the conversion of the *essential use cases* to *real use cases*), input screens, output screens, forms (including preprinted paper forms), and reports. Finally, the individual interfaces are subjected to *interface evaluation* to determine if they are satisfactory and how they can be improved.

---

[4] John Satzinger and Lorne Olfman, "User Interface Consistency Across End-User Application: The Effects of Mental Models," *Journal of Management Information Systems* (Spring 1998): 167–193.

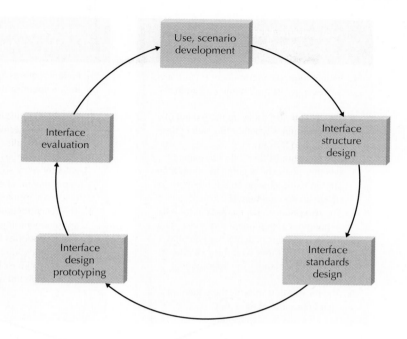

FIGURE 11-5
User Interface
Design Process

Interface evaluations almost always identify improvements, so the interface design process is repeated in a cyclical process until no new improvements are identified. In practice, most analysts interact closely with the users during the interface design process so that users have many chances to see the interface as it evolves, rather than waiting for one overall interface evaluation at the end of the interface design process. It is better for all concerned (both analysts and users) if changes are identified sooner rather than later. For example, if the interface structure or standards need improvements, it is better to identify changes before most of the screens that use the standards have been designed.[5]

Use Scenario Development

A use scenario is an outline of the steps that the users perform to accomplish some part of their work. A use scenario is one path through an essential use case. For example, Figure 8-1 shows the use-case diagram for the Appointment System. This figure shows that the Create New Patient use case is distinct from the Make Payment Arrangements use case. We model these two use cases separately because they represent separate processes that are included in the Make Appointment use case.

The use-case diagram was designed to model all possible uses of the system—its complete functionality or all possible paths through the use case at a fairly high level of abstraction. In one use scenario, a patient will make a request with the receptionist regarding an appointment with the dentist. The receptionist will look up the patient and check to see if the patient has any bills to be paid. The receptionist then will ask the patient whether he or she wants to set up a new appointment, cancel an existing appointment, or to change an existing appointment. If the patient wants to create an new appointment, the receptionist

---

[5] A good source for more information on user interface evaluation is Deborah Hix and H. Rex Hartson, *Developing User Interfaces, Ensuring Usability Through Product & Process* (New York: Wiley, 1993).

Use scenario: Existing Patient Makes New Appointment	Use scenario: Existing Patient Cancels Appointment
1. Patient requests appointment (1) and gives the receptionist their name and address (2).	1. Patient requests appointment (1) and gives the receptionist their name and address (2).
2. The receptionist looks up the patient (3) and determines whether the patient has any unpaid bills (4).	2. The receptionist looks up the patient (3) and determines whether the patient has any unpaid bills (4).
3. The receptionist then asks the patient whether he or she is going to set up a new appointment, change an appointment, or delete an appointment (5).	3. The receptionist then asks the patient whether he or she is going to set up a new appointment, change an appointment, or delete an appointment (5).
4. The receptionist asks the patient for a list of potential appointment times (S-1, 1).	4. The receptionist asks the patient for the appointment time to be canceled (S-2, 1).
5. The receptionist matches the potential appointment times with the available times and schedules the appointment (S-1, 2).	5. The receptionist finds and deletes the appointment (S-2, 2).
6. The receptionist informs the patient of their appointment time (6).	6. The receptionist informs the patient that their appointment time was canceled (6).

The numbers in parentheses refer to specific events in the essential use case.

FIGURE 11-6   Use Scenarios

asks the patient for some suggested appointment times, which the receptionist matches against potential times available. The receptionist finally creates a new appointment (see Figures 7-1 and 7-5).

In another use scenario, a patient simply wants to cancel an appointment. In this case, the receptionist looks up the patient and checks to see if the patient has any bills to be paid. The receptionist then asks the patient for the time of the appointment to be canceled. Finally, the receptionist deletes the appointment.

Use scenarios are presented in a simple narrative description that is tied to the essential use cases developed during analysis (see Chapter 5). Figure 11-6 shows the two use scenarios just described. The key point with using use cases for interface design is *not* to document all possible use scenarios within a use case. The goal is to document two or three of the most common use scenarios so the interface can be designed to enable the most common uses to be performed simply and easily.

YOUR TURN   11-2 Use Scenario Development for the Web

Visit the Web home page for your university and navigate through several of its Web pages. Develop two use   scenarios for it.

YOUR
TURN

11-3 Use Scenario Development for an ATM

Suppose you have been charged with the task of   Develop two use scenarios for it.
redesigning the interface for the ATM at your local bank.

Interface Structure Design

The interface structure defines the basic components of the interface and how they work together to provide functionality to users. A WND[6] is used to show how all the screens, forms, and reports used by the system are related and how the user moves from one to another. Most systems have several WNDs, one for each major part of the system.

A WND is very similar to a behavioral state machine (see Chapter 7), in that they both model state changes. A behavioral state machine typically models the *state* changes of an object, where a WND models the state changes of the user interface. In a WND, each state in which the user interface may be is represented as a box. Furthermore, a box typically corresponds to a user interface component, such as a *window,* form, *button,* or report. For example, in Figure 11-7, there are five separate states: Client Menu, Find Client Form, Add Client Form, Client List, and Client Information Report.

Transitions are modeled as either a single-headed or double-headed arrow. A single-headed arrow indicates that a return to the calling state is not required, whereas a double-headed arrow represents a required return. For example in Figure 11-7, the transition from the Client Menu state to the Find Client Form state does not require a return. The arrows are labeled with the action that causes the user interface to move from one state to another. For example, in Figure 11-7, to move from the Client Menu state to the Find Client Form state, the user must click Find Client Button on the Client Menu.

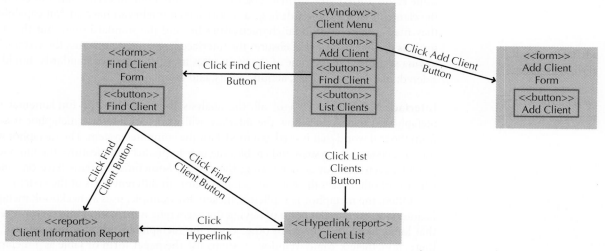

FIGURE 11-7   Sample WND

---

[6] A WND is actually an adaptation of the behavioral state machine and object diagrams [see Meilir Page-Jones, *Fundamentals of Object-Oriented Design in UML* (New York: Dorset House, 2000)].

Suppose you have been charged with the task of redesigning the interface for the ATM at your local bank. Design an interface structure design using a WND that shows how a user would navigate among the screens.

The last item to be described in a WND is the *stereotype*. A stereotype is modeled as a text item enclosed within guillemets (<< >>) or angle brackets (<< >>). The stereotype represents the type of user interface component of a box on the diagram. For example, the Client Menu is a window, whereas Find Client Form is a form.

The basic structure of an interface follows the basic structure of the business process itself, as defined in the use cases and behavioral model. The analyst starts with the essential use cases and develops the fundamental flow of control of the system as it moves from object to object. The analyst then examines the use scenarios to see how well the WND supports them. Quite often, the use scenarios identify paths through the WND that are more complicated then they should be. The analyst then reworks the WND to simplify the ability of the interface to support the use scenarios, sometimes by making major changes to the menu structure, sometimes by adding shortcuts.

Interface Standards Design

Interface standards are the basic design elements that are common across the individual screens, forms, and reports within the system. Depending on the application, there may be several sets of interface standards for different parts of the system (e.g., one for Web screens, one for paper reports, one for input forms). For example, the part of the system used by data entry operators may mirror other data entry applications in the company, whereas a Web interface for displaying information from the same system may adhere to some standardized Web format. Likewise, each individual interface may not contain all of the elements in the standards (e.g., a report screen might not have an edit capability), and they may contain additional characteristics beyond the standard ones, but the standards serve as the touchstone that ensures the interfaces are consistent across the system. The following sections discuss some of the main areas in which interface standards should be considered: metaphors, objects, actions, icons, and templates.

Interface Metaphor First of all, the analysts must develop the fundamental interface metaphor(s) that defines how the interface will work. An interface metaphor is a concept from the real world that is used as a model for the computer system. The metaphor helps the user understand the system and enables the user to predict what features the interface might provide, even without actually using the system. Sometimes systems have one metaphor, whereas in other cases there are several metaphors in different parts of the system.

Often, the metaphor is explicit. Quicken, for example, uses a checkbook metaphor for its interface, even to the point of having the users type information into an on-screen form that looks like a real check. In other cases, the metaphor is implicit or unstated, but it is there, nonetheless. Many Windows systems use the paper form or table as a metaphor.

In some cases, the metaphor is so obvious that it requires no thought. For example, most online stores use a shopping cart metaphor to temporarily store the items that the customer is considering purchasing. In other cases, a metaphor is hard to identify. In general, it is better not to force a metaphor that really doesn't fit a system, because an ill-fitting metaphor will confuse users by promoting incorrect assumptions.

Interface Objects   The template specifies the names that the interface will use for the major *interface objects*, the fundamental building blocks of the system, such as the classes. In many cases, the object names are straightforward, such as calling the shopping cart the "shopping cart." In other cases, it is not so simple. For example, Amazon.com sells much more than books. In some cases the user may not know whether he or she is looking for a book, CD, DVD, or VHS tape. In those cases, the user can use a catchall search item: Amazon.com. In the case that the user knows the type of item that he or she wants to buy, the user can limit the search by specifying more specific types of search items, such as book, CD, DVD, or VHS. Obviously, the object names should be easily understood and help promote the interface metaphor.

In general, in cases of disagreements between the users and the analysts over names, whether for objects or actions (discussed later), the users should win. A more understandable name always beats a more precise or more accurate one.

Interface Actions   The template also specifies the navigation and command language style (e.g., menus) and grammar (e.g., object-action order; see the navigation design section later in this chapter). It gives names to the most commonly used *interface actions* in the navigation design (e.g., buy versus purchase or modify versus change).

Interface Icons   The interface objects and actions and also their status (e.g., deleted or overdrawn) may be represented by *interface icons*. Icons are pictures that will appear on command buttons as well as in reports and forms to highlight important information. Icon design is very challenging because it means developing a simple picture less than half the size of a postage stamp that needs to convey an often-complex meaning. The simplest and best approach is to simply adopt icons developed by others (e.g., a blank page to indicate create a new file, a diskette to indicate save). This has the advantage of quick icon development, and the icons may already be well understood by users because they have seen them in other software.

Commands are actions that are especially difficult to represent with icons because they are in motion, not static. Many icons have become well known from widespread use, but icons are not as well understood as first believed. Use of icons can sometimes cause more confusion than insight. [For example, did you know that a picture of a sweeping broom (paintbrush?) in Microsoft Word means format painter?] Icon meanings become clearer with use, but sometimes a picture is not worth even one word; when in doubt, use a word, not a picture.

Interface Templates   An *interface template* defines the general appearance of all screens in the information system and the paper-based forms and reports that are used. The template design, for example, specifies the basic layout of the screens [e.g., where the navigation area(s), status area, and form/report area(s) will be placed] and the color scheme(s) that will be applied. It defines whether windows will replace one another on the screen or will cascade over the top of each other. The template defines a standard placement and order for common interface actions (e.g., File Edit View rather than File View Edit). In short, the template draws together the other major interface design elements: metaphors, objects, actions, and icons.

YOUR
TURN

11-5 Interface Standards Development

Suppose you have been charged with the task of redesigning the interface for the ATM at your local bank. Develop an interface standard that includes metaphors, objects, actions, icons, and a template.

Interface Design Prototyping

An interface design prototype is a mock-up or a simulation of a computer screen, form, or report. A prototype is prepared for each interface in the system to show the users and the programmers how the system will perform. In the "old days," an interface design prototype was usually specified on a paper form that showed what would be displayed on each part of the screen. Paper forms are still used today, but more and more interface design prototypes are being built using computer tools instead of paper. The four most common approaches to interface design prototyping are storyboards, windows layout diagrams, HTML prototypes, and language prototypes.

Storyboard   At its simplest, an interface design prototype is a paper-based *storyboard*. The storyboard shows hand-drawn pictures of what the screens will look like and how they flow from one screen to another, in the same way a storyboard for a cartoon shows how the action will flow from one scene to the next (see Figure 11-8). Storyboards are the simplest technique because all they require is paper (often a flip chart) and a pen—and someone with some artistic ability.

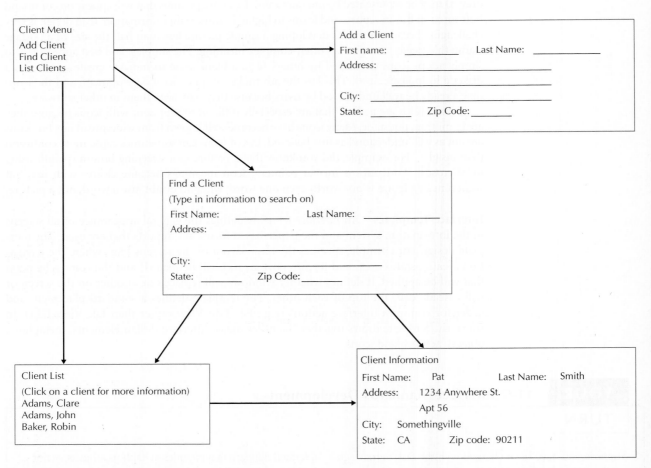

FIGURE 11-8   Sample Storyboard

FIGURE 11-9
Sample Windows
Layout Diagram

Windows Layout Diagram   A slight step up from a storyboard is a *windows layout diagram*. From our perspective, a windows layout diagram is a storyboard that more closely resembles the actual user interface that the user will gradually receive. Typically, it is created using a tool such as Visio's Windows User Interface Template. Using this type of tool, the designer can quickly drag and drop the user interface components onto the canvas to lay out the design of the user interface. For example, in Figure 11-9, an equivalent diagram for the Add a Client window for the storyboard is portrayed. Furthermore, by combining the windows layout diagrams with the windows navigation diagram, the designer can work effectively with a set of users to design both the look and feel of the evolving system without having to actually implement anything.

HTML Prototype   One of the most common types of interface design prototypes used today is the *HTML prototype.* As the name suggests, an HTML prototype is built using Web pages created in HTML (hypertext markup language). The designer uses HTML to create a series of Web pages, which show the fundamental parts of the system. The users can interact with the pages by clicking on buttons and entering pretend data into forms (but because there is no system behind the pages, the data are never processed). The pages are linked together so that as the user clicks on buttons, the requested part of the system appears. HTML prototypes are superior to storyboards in that they enable users to interact with the system and gain a better sense of how to navigate among the different screens. However, HTML has limitations—the screens shown in HTML will never appear exactly like the real screens in the system (unless, of course, the real system will be a Web system in HTML).

Language Prototype   A *language prototype* is an interface design prototype built using the actual language or tool that will be used to build the system. Language prototypes are designed in the same ways as HTML prototypes (they enable the user to move from screen to screen but perform no real processing). For example, in Visual Basic, it is possible to create and view screens without actually attaching program code to the screens. Language prototypes take longer to develop than storyboards or HTML prototypes but have the distinct advantage of showing *exactly* what the screens will look like. The user does not have to guess about the shape or position of the elements on the screen.

CONCEPTS

IN ACTION

11-A Interface Design Prototypes for a DSS Application

I was involved in the development of several DSSs while working as a consultant. On one project, a future user was frustrated because he could not imagine what a DSS looked like and how one would be used. He was a key user, but the project team had a difficult time involving him in the project because of his frustration. The team used SQLWindows (one of the most popular development tools at the time) to create a language prototype that demonstrated the future system's appearance, proposed menu system, and screens (with fields, but no processing).

The team was amazed at the user's response to the prototype. He appreciated being given a context with which to visualize the DSS, and he soon began to recommend improvements to the design and flow of the system and to identify some important information that was overlooked during analysis. Ultimately, the user became one of the strongest supporters of the system, and the project team felt sure that the prototype would lead to a much better product in the end.

—Barbara Wixom

Questions

1. Why do you think the team chose to use a language prototype rather than a storyboard or HTML prototype?
2. What trade-offs were involved in the decision?

Selecting the Appropriate Techniques   Projects often use a combination of different interface design prototyping techniques for different parts of the system. Storyboarding is the fastest and least expensive but provides the least amount of detail. Windows layout diagrams provide more of a feel that the user will experience, while remaining fairly inexpensive to develop. HTML prototyping is useful for testing the basic design and navigation (see the following) of the user interface. Language prototyping is the slowest, most expensive, and most detailed approach. Therefore, storyboarding is used for parts of the system in which the interface is well understood and when more expensive prototypes are thought to be unnecessary. However, in most cases it is probably worth the additional cost of developing windows layout diagrams in addition to storyboards. HTML prototypes and language prototypes are used for parts of the system that are critical, yet not well understood.

Interface Evaluation[7]

The objective of interface evaluation is to understand how to improve the interface design before the system is complete. Most interface designers intentionally or unintentionally design an interface that meets their personal preferences, which might or might not match the preferences of the users. The key message, therefore, is to have as many people as possible evaluate the interface, and the more users the better. Most experts recommend involving at least ten potential users in the evaluation process.

Many organizations save interface evaluation for the very last step in the systems development before the system is installed. Ideally, however, interface evaluation should be performed while the system is being designed—before it is built—so that any major design problems can be identified and corrected before the time and cost of programming has been spent on a weak design. It is not uncommon for the system to undergo one or two major changes after the users see the first interface design prototype because they identify problems that are overlooked by the project team.

---

[7] Verifying and validation approaches, in general, were described in Chapter 8. Also, further approaches to testing the evolving system are described in Chapter 13. In this section, we describe approaches that have been customized to the human–computer interaction layer.

As with interface design prototyping, interface evaluation can take many different forms, each with different costs and different amounts of detail. Four common approaches are heuristic evaluation, walkthrough evaluation, interactive evaluation, and formal usability testing. As with interface design prototyping, the different parts of a system can be evaluated using different techniques.

Heuristic Evaluation A *heuristic evaluation* examines the interface by comparing it to a set of heuristics or principles for interface design. The project team develops a checklist of interface design principles—from the list at the start of this chapter, for example, as well as the list of principles in the navigation, input, and output design sections later in this chapter. At least three members of the project team then individually work through the interface design prototype examining each and every interface to ensure it satisfies each design principle on a formal checklist. After each has gone through the prototype separately, they meet as a team to discuss their evaluations and identify specific improvements that are required.

Walkthrough Evaluation An interface design *walkthrough evaluation* is a meeting conducted with the users who will ultimately have to operate the system. The project team presents the prototype to the users and walks them through the various parts of the interface. The project team shows the storyboard and windows layout diagrams or actually demonstrates the HTML or language prototype and explains how the interface will be used. The users identify improvements to each of the interfaces that are presented.

Interactive Evaluation With an *interactive evaluation,* the users themselves actually work with the HTML or language prototype in a one-person session with member(s) of the project team (an interactive evaluation cannot be used with a storyboard or windows layout diagrams). As the user works with the prototype (often by going through the use scenarios, using the real use cases described later in this chapter, or just navigating at will through the system), he or she tells the project team member(s) what he or she likes and doesn't like and what additional information or functionality is needed. As the user interacts with the prototype, team member(s) record the cases when he or she appears to be unsure of what to do, makes mistakes, or misinterprets the meaning of an interface component. If the pattern of uncertainty, mistakes, or misinterpretations reoccurs across several of the users participating in the evaluation, it is a clear indication that those parts of the interface need improvement.

Formal Usability Testing Formal *usability testing* is commonly done with commercial software products and products developed by large organizations that will be widely used through the organization. As the name suggests, it a very formal—almost scientific—process that can be used only with language prototypes (and systems that have been completely built awaiting installation or shipping).[8] As with interactive evaluation, usability testing is done in one-person sessions in which a user works directly with the software. However, it is typically done in a special lab equipped with video cameras and special software that records each and every keystroke and mouse operation so they can be replayed to understand exactly what the user did.

The user is given a specific set of tasks to accomplish (usually the use scenarios) and after some initial instructions, the project team's members are not permitted to interact with the user to provide assistance. The user must work with the software without help, which can be hard on the users if they become confused with the system. It is critical that users understand that the goal is to test the interface, not their abilities, and if they are unable to complete the task, the interface—not the user—has failed the test.

---

[8] A good source for usability testing is Jakob Nielsen, and Robert Mack (eds.), *Usability Inspection Methods* (New York: Wiley, 1994). See also www.useit.com/papers.

11-6 Prototyping and Evaluation

*S*uppose you have been charged with the task of redesign-ing the interface for the ATM at your local bank. What type of prototyping and interface evaluation approach would you recommend? Why?

Formal usability testing is very expensive, because each one-user session can take one to two days to analyze due to the volume of detail collected in the computer logs and video-tapes. Sessions typically last one to two hours. Most usability testing involves five to ten users, because if there are fewer than five users, the results depend too much on the specific individual users who participated, and more than ten users are often too expensive to justify (unless a large commercial software developer is involved).

NAVIGATION DESIGN

The navigation component of the interface enables the user to enter commands to navigate through the system and perform actions to enter and review information it contains. The navigation component also presents messages to the user about the success or failure of his or her actions. The goal of the navigation system is to make the system as simple as possi-ble to use. A good navigation component is one the user never really notices. It simply functions the way the user expects, and thus the user gives it little thought.

Basic Principles

One of the hardest things about using a computer system is learning how to manipulate the navigation controls to make the system do what you want. Analysts usually need to assume that users have not read the manual, have not attended training, and do not have external help readily at hand. All controls should be clear and understandable and placed in an intu-itive location on the screen. Ideally, the controls should anticipate what the user will do and simplify his or her efforts. For example, many setup programs are designed so that for a typical installation, the user can simply keep pressing the Next button.

Prevent Mistakes   The first principle of designing navigation controls is to prevent the user from making mistakes. A mistake costs time and creates frustration. Worse still, a series of mistakes can cause the user to discard the system. Mistakes can be reduced by labeling commands and actions appropriately and by limiting choices. Too many choices can confuse the user, particularly when they are similar and hard to describe in the short space available on the screen. When there are many similar choices on a menu, consider creating a second menu level or a series of options for basic commands.

Never display a command that cannot be used. For example, many Windows applica-tions gray out commands that cannot be used; they are displayed on pull-down menus in a very light-colored font, but they cannot be selected. This shows that they are available but cannot be used in the current context. It also keeps all menu items in the same place.

When the user is about to perform a critical function that is difficult or impossible to undo (e.g., deleting a file), it is important to confirm the action with the user (and make sure the selection was not made by mistake). This is usually done by having the user respond to a confirmation message, which explains what the user has requested and asks the user to confirm that this action is correct.

Simplify Recovery from Mistakes   No matter what the system designer does, users will make mistakes. The system should make it as easy as possible to correct these errors. Ideally, the system will have an Undo button that makes mistakes easy to override; however, writing the software for such buttons can be very complicated.

Use Consistent Grammar Order   One of the most fundamental decisions is the *grammar order.* Most commands require the user to specify an object (e.g., file, record, word), and the action to be performed on that object (e.g., copy, delete). The interface can require the user to first choose the object and then the action (an *object-action order*) or first choose the action and then the object (an *action-object order*). Most Windows applications use an object-action grammar order (e.g., think about copying a block of text in your word processor).

The grammar order should be consistent throughout the system, both at the data element level as well as at the overall menu level. Experts debate about the advantages of one approach over the other, but because most users are familiar with the object-action order, most systems today are designed using that approach.

Types of Navigation Controls

There are two traditional hardware devices that can be used to control the user interface: the keyboard and a pointing device such as a mouse, trackball, or touch screen. In recent years, voice recognition systems have made an appearance, but they are not yet common. There are three basic software approaches for defining user commands: languages, menus, and direct manipulation.

Languages   With a *command language,* the user enters commands using a special language developed for the computer system (e.g., DOS and SQL both use command languages). Command languages sometimes provide greater flexibility than other approaches because the user can combine language elements in ways not predetermined by developers. However, they put a greater burden on users because users must learn syntax and type commands rather than select from a well-defined, limited number of choices. Systems today use command languages sparingly, except in cases where there is an extremely large number of command combinations that make it impractical to try to build all combinations into a menu (e.g., SQL queries for databases).

Natural language interfaces are designed to understand the user's own language (e.g., English, French, Spanish). These interfaces attempt to interpret what the user means, and often they present back to the user a list of interpretations from which to choose. An example of the use of natural language is Microsoft's Office Assistant, which enables users to ask free-form questions for help.

Menus   The most common type of navigation system today is the *menu.* A menu presents a user with a list of choices, each of which can be selected. Menus are easier to learn than languages because a limited number of available commands are presented to the user in an organized fashion. Clicking on an item with a pointing device or pressing a key that matches the menu choice (e.g., a function key) takes very little effort. Therefore, menus are usually preferred to languages.

Menu design needs to be done with care because the submenus behind a main menu are hidden from users until they click on the menu item. It is better to make menus broad and shallow (i.e., each menu containing many items with only one or two layers of menus) rather than narrow and deep (i.e., each menu containing only a few items, but each leading to three or more layers of menus). A broad and shallow menu presents the user with the most information initially so that he or she can see many options and requires only a few mouse clicks or keystrokes to perform an action. A narrow and deep menu makes users hunt for items hidden behind menu items and requires many more clicks or keystrokes to perform an action.

Research suggests that in an ideal world, any one menu should contain no more than eight items, and it should take no more than two mouse clicks or keystrokes from any menu to perform an action (or three from the main menu that starts a system).[9] However, analysts sometimes must break this guideline in the design of complex systems. In this case, menu items are often grouped together and separated by a horizontal line (see Figure 11-10). Often menu items have *hot keys* that enable experienced users to quickly invoke a command with keystrokes in lieu of a menu choice (e.g., Ctrl-F in Word invokes the Find command or Alt-F opens the file menu).

Menus should put together like items so that the user can intuitively guess what each menu contains. Most designers recommend grouping menu items by interface objects (e.g., customers, purchase orders, inventory) rather than by interface actions (e.g., new, update, format), so that all actions pertaining to one object are in one menu, all actions for another object are in a different menu, and so. However, this is highly dependent on the specific interface. As Figure 11-10 shows, Microsoft Word groups menu items by interface objects (e.g., File, Table, Window) *and* by interface actions (e.g., Edit, Insert, Format) on the same menu. Some of the more common types of menus include *menu bars, drop-down menus, pop-up menus, tab menus, toolbars,* and *image maps* (see Figure 11-11).

Direct Manipulation   With *direct manipulation,* the user enters commands by working directly with interface objects. For example, users can change the size of objects in Microsoft PowerPoint by clicking on them and moving their sides, or they can move files in Windows Explorer by dragging the filenames from one folder to another. Direct manipulation can be simple, but it suffers from two problems. First, users familiar with language- or menu-based interfaces don't always expect it. Second, not all commands are intuitive. [How do you copy (not move) files in Windows explorer? On the Macintosh, why does moving a folder to the trash delete the file if it is on the hard disk, but eject the diskette if the file is on a diskette?]

Messages

Messages are the way in which the system responds to a user and informs him or her of the status of the interaction. There are many different types of messages, such as *error messages, confirmation messages, acknowledgment messages, delay messages,* and *help messages* (see Figure 11-12). In general, messages should be clear, concise, and complete, which are sometimes conflicting objectives. All messages should be grammatically correct and free of jargon and abbreviations (unless they are the users' jargon and abbreviations). Avoid negatives because they can be confusing (e.g., replace Are you sure you do not want to continue? with Do you want to quit?). Likewise, avoid humor, because it wears off quickly after the same message appears dozens of times.

YOUR TURN   **11-7 Design a Navigation System**

Design a navigation system for a system into which users must enter information about customers, products, and orders. For all three, users will want to change, delete, find one specific record, and list all records.

---

[9] Kent L. Norman, *The Psychology of Menu Selection* (Norwood NJ.: Ablex Publishing Corp., 1991).

FIGURE 11-10   Common Types of Menus

Type of Menu	When to Use	Notes
Menu bar List of commands at the top of the screen; always on-screen	Main menu for system	Use the same organization as the operating system and other packages (e.g., File, Edit, View). Menu items are always one word, never two. Menu items lead to other menus rather than perform action. Never allow users to select actions they can't perform (instead, use grayed-out items).
Drop-down menu Menu that drops down immediately below another menu; disappears after one use	Second-level menu, often from menu bar	Menu items are often multiple words. Avoid abbreviations. Menu items perform action or lead to another cascading drop-down menu, pop-up menu, or tab menu.
Pop-up menu Menu that pops up and floats over the screen; disappears after one use	As a shortcut to commands for experienced users	Pop-up menus often (not always) invoked by a right click in Windows-based systems. These menus are often overlooked by novice users, so usually they should duplicate functionality provided in other menus.
Tab menu Multipage menu with one tab for each page that pops up and floats over the screen; remains on-screen until closed	When user needs to change several settings or perform several related commands	Menu items should be short to fit on the tab label. Avoid more than one row of tabs, because clicking on a tab to open it can change the order of the tabs and in virtually no other case does selecting from a menu rearrange the menu itself.
Tool bar Menu of buttons (often with icons) that remains on-screen until closed	As a shortcut to commands for experienced users	All buttons on the same tool bar should be the same size. If the labels vary dramatically in size, then use two different sizes (small and large). Buttons with icons should have a tool tip, an area that displays a text phrase explaining the button when the user pauses the mouse over it.
Image map Graphic image in which certain areas are linked to actions or other menus	Only when the graphic image adds meaning to the menu	The image should convey meaning to show which parts perform action when clicked. Tool tips can be helpful.

FIGURE 11-11 Types of Menus

Messages should require the user to acknowledge them (by clicking, for example), rather than being displayed for a few seconds and then disappearing. The exceptions are messages that inform the user of delays in processing, which should disappear once the delay has passed. In general, messages are text, but sometimes standard icons are used. For example, Windows displays an hourglass when the system is busy.

All messages should be carefully crafted, but error messages and help messages require particular care. Messages (and especially error messages) should always explain the problem in polite, succinct terms (e.g., what the user did incorrectly) and explain corrective action as clearly and as explicitly as possible so the user knows exactly what needs to be done. In the case of complicated errors, the error message should display what the user entered, suggest probable causes for the error, and propose possible user responses. When in doubt, provide either more information than the user needs or the ability to get additional information. Error messages should provide a message number. Message numbers are not intended for users, but their presence makes it simpler for help desks and customer support lines to identify problems and help users because many messages use similar wording.

Type of Messages	When to Use	Notes
Error message Informs the user that he or she has attempted to do something to which the system cannot respond	When user does something that is not permitted or not possible	Always explain the reason and suggest corrective action. Traditionally, error messages have been accompanied by a beep, but many applications now omit it or permit users to remove it.
Confirmation message Asks users to confirm that they really want to perform the action they have selected	When user selects a potentially dangerous choice, such as deleting a file	Always explain the cause and suggest possible action. Often include several choices other than OK and cancel.
Acknowledgment message Informs the user that the system has accomplished what it was asked to do	Seldom or never. Users quickly become annoyed with all the unnecessary mouse clicks	Acknowledgment messages are typically included because novice users often like to be reassured that an action has taken place. The best approach is to provide acknowledgment information without a separate message on which the user must click. For example, if the user is viewing items in a list and adds one, then the updated list on the screen showing the added item is sufficient acknowledgment.
Delay message Informs the user that the computer system is working properly	When an activity takes more than seven seconds	Should permit the user to cancel the operation in case he or she does not want to wait for its completion. Should provide some indication of how long the delay may last.
Help message Provides additional information about the system and its components	In all systems	Help information is organized by table of contents and/or keyword search. Context-sensitive help provides information that is dependent on what the user was doing when help was requested. Help messages and online documentation are discussed in Chapter 13.

FIGURE 11-12 Types of Messages

Navigation Design Documentation

The design of the navigation for a system is done through the use of WNDs and real use cases. Real use cases are derived from the essential use cases (see Chapter 5), use scenarios, and WNDs. Recall that an essential use case is one that describes only the minimum essential issues necessary to understand the required functionality. A real use case describes a specific set of steps that a user performs to use a specific part of a system. As such, real use cases are implementation dependent (i.e., they are detailed descriptions of how to use the system once it is implemented).

To evolve an essential use case into a real use case, two changes must be made. First, the use-case type must be changed from essential to real. Second, all events must be specified in terms of the actual user interface. Therefore, the normal flow of events, subflows, and alternate/exceptional flows must be modified. The normal flow of events, subflows, and alternate/exceptional flows for the real use case associated with the storyboard user interface prototype given in Figure 11-8 is shown in Figure 11-13. For example, step 2 of the normal flow of events states that "The System provides the Sales Rep with the Main Menu for the System," which allows the Sales Rep to interact with the Maintain Client List aspect of the system.

CONCEPTS

IN ACTION

11-B Complex Electrical Systems

*S*ystems integration across platforms and companies grows more complex with time. In a case study from Florida in 2008, an electrical company's real-time system detected a minor problem in the power grid and shut down the entire system—plunging more than 2 million people into the dark. The system experts place the blame on a substation software system that detected the minor fluctuation but had the ability to immediately shut down the entire system. Although there may be times where such a rapid response is vital (such as nuclear disasters in Chernobyl, Ukraine, and on Three

Mile Island), this was a case where such response was not warranted.

Questions

1. Because software controls substation operations, how might a systems analyst approach this as a systems project?
2. Are there special considerations that a systems analyst needs to think about when dealing with real-time systems?

INPUT DESIGN

Inputs facilitate the entry of data into the computer system, whether highly structured data, such as order information (e.g., item numbers, quantities, costs) or unstructured information (e.g., comments). Input design means designing the screens used to enter the information as well as any forms on which users write or type information (e.g., timecards, expense claims).

Basic Principles

The goal of the input mechanism is to simply and easily capture accurate information for the system. The fundamental principles for input design reflect the nature of the inputs (whether batch or online) and ways to simplify their collection.

Online versus Batch Processing   There are two general formats for entering inputs into a computer system: online processing and batch processing. With *online processing* (sometimes called *transaction processing*), each input item (e.g., a customer order, a purchase order) is entered into the system individually, usually at the same time as the event or transaction prompting the input. For example, when you take a book out from the library, buy an item at the store, or make an airline reservation, the computer system that supports that process uses online processing to immediately record the transaction in the appropriate database(s). Online processing is most commonly used when it is important to have *real-time information* about the business process. For example, when you reserve an airline seat, the seat is no longer available for someone else to use.

With *batch processing*, all the inputs collected over some time period are gathered together and entered into the system at one time in a batch. Some business processes naturally generate information in batches. For example, most hourly payrolls are done using batch processing because time cards are gathered together in batches and processed at once. Batch processing is also used for transaction processing systems that do not require real-time information. For example, most stores send sales information to district offices so that new replacement inventory can be ordered. This information can be sent in real-time as it is captured in the store so that the district offices are aware within a second or two that a product is sold. If stores do not need this up-to-the-second real-time data,

Use-Case Name: *Maintain Client List*		ID: *12*	Importance Level: *High*

Primary Actor: *Sales Rep*	Use-Case Type: *Detail, Real*

Stakeholders and Interests: *Sales Rep - wants to add, find or list clients*

Brief Description: *This use case describes how sales representatives can search and maintain the client list.*

Trigger: *Patient calls and asks for a new appointment or asks to cancel or change an existing appointment.*

Type: *External*

Relationships:
    Association: *Sales Rep*
    Include:
    Extend:
    Generalization:

Normal Flow of Events:

1. The Sales Rep starts up the system.
2. The System provides the Sales Rep with the Main Menu for the System.
3. The System asks Sales Rep if he or she would like to Add a client, Find an existing Client, or to List all existing clients.
    If the Sales Rep wants to add a client, they click on the Add Client Link and execute S-1: New Client.
    If the Sales Rep wants to find a client, they click on the Find Client Link and execute S-2: Find Client.
    If the Sales Rep wants to list all clients, they click on the List Client Link and execute S-3: List Clients.
4. The System returns the Sales Rep to the Main Menu of the System.

Subflows:

S-1: New Client
    1. The System asks the Sales Rep for relevant information.
    2. The Sales Rep types in the relevant information into the Form
    3. The Sales Rep submits the information to the System.
S-2: Find Client
    1. The System asks the Sales Rep for the search information.
    2. The Sales Rep types in the search information into the Form
    3. The Sales Rep submits the information to the System.
    4. If the System finds a single client that meets the search information, the System produces a client Information report and returns the Sales Rep to the Main Menu of the System
    Else If the System finds a list of clients that meet the search information, the System executes S-3: List Clients.
S-3: List Clients
    1. If this Subflow is executed from Step 3
        The System creates a List of All clients
    Else
        The System creates a List of clients that matched the S-2: Find Client search criteria.
    2. The Sales Rep selects a client.
    3. The System produces a client Information report.

Alternate/Exceptional Flows:
    S-2 4a.   The System produces an Error Message.

FIGURE 11-13
Real Use-Case
Example

they will collect sales data throughout the day and transmit it every evening in a batch to the district office. This batching simplifies the data communications process and often saves in communications costs, but it does mean that inventories are not accurate in real time but rather are accurate only at the end of the day after the batch has been processed.

Capture Data at the Source   Perhaps the most important principle of input design is to capture the data in an electronic format at its original source or as close to the original source as possible. In the early days of computing, computer systems replaced traditional manual systems that operated on paper forms. As these business processes were automated, many of the original paper forms remained, either because no one thought to replace them or because it was too expensive to do so. Instead, the business process continued to contain manual forms that were taken to the computer center in batches to be typed into the computer system by a *data entry operator.*

Many business processes still operate this way today. For example, most organizations have expense claim forms that are completed by hand and submitted to an accounting department, which approves them and enters them into the system in batches. There are three problems with this approach. First, it is expensive because it duplicates work (the form is filled out twice, once by hand, once by keyboard.)[10] Second, it increases processing time because the paper forms must be physically moved through the process. Third, it increases the cost and probability of error, because it separates the entry from the processing of information; someone may misread the handwriting on the input form, data may be entered incorrectly, or the original input may contain an error that invalidates the information.

Most transaction-processing systems today are designed to capture data at its source. *Source data automation* refers to using special hardware devices to automatically capture data without requiring anyone to type it. Stores commonly use *bar-code readers* that automatically scan products and enter data directly into the computer system. No intermediate formats such as paper forms are used. Similar technologies include *optical character recognition,* which can read printed numbers and text (e.g., on checks), *magnetic stripe readers,* which can read information encoded on a stripe of magnetic material similar to a diskette (e.g., credit cards), and *smart cards,* which contain microprocessors, memory chips, and batteries (much like credit-card-sized calculators). As well as reducing the time and cost of data entry, these systems reduce errors because they are far less likely to capture data incorrectly. Today, portable computers and scanners allow data to be captured at the source even when in mobile settings (e.g., air courier deliveries, use of rental cars).

These automatic systems are not capable of collecting a lot of information, so the next-best option is to capture data immediately from the source using a trained entry operator. Many airline and hotel reservations, loan applications, and catalog orders are recorded directly into a computer system while the customer provides the operator with answers to questions. Some systems eliminate the operator altogether and allow users to enter their own data. For example, several universities [e.g., the Massachusetts Institute of Technology (MIT)] no longer accept paper-based applications for admissions; all applications are typed by students into electronic forms.

The forms for capturing information (on a screen, on paper, etc.) should support the data source. That is, the order of the information on the form should match the natural flow of information from the data source, and data-entry forms should match paper forms used to initially capture the data.

Minimize Keystrokes   Another important principle is to minimize keystrokes. Keystrokes cost time and money, whether they are performed by a customer, user, or trained data-entry operator. The system should never ask for information that can be obtained in another way (e.g., by retrieving it from a database or by performing a calculation). Likewise, a system should not require a user to type information that can be selected from a list; selecting reduces errors and speeds entry.

[10] Or in the case of the University of Georgia, three times: first by hand on an expense form, a second time when it is typed onto a new form for the "official" submission because the accounting department refuses handwritten forms, and, finally, when it is typed into the accounting computer system.

11-8 Career Services

Suppose you are designing the new interface for a career services system at your university that accepts student résumés and presents them in a standard format to recruiters. Describe how you could incorporate the basic principles of input design into your interface design. Remember to include the use of online versus batch data input, the capture of information, and plans to minimize keystrokes.

In many cases, some fields have values that often recur. These frequent values should be used as the *default value* for the field so that the user can simply accept the value and not have to retype it time and time again. Examples of default values are the current date, the area code held by the majority of a company's customers, and a billing address, which is based on the customer's residence. Most systems permit changes to default values to handle data-entry exceptions as they occur.

Types of Inputs

Each data item that has to be input is linked to a field on the form into which its value is typed. Each field also has a field label, which is the text beside, above, or below the field that tells the user what type of information belongs in the field. Often the field label is similar to the name of the data element, but they do not have to have identical words. In some cases, a field will display a template over the entry box to show the user exactly how data should be typed. There are many different types of inputs, in the same way that there are many different types of fields (see Figure 11-14).

Text   As the name suggests, a *text box* is used to enter text. Text boxes can be defined to have a fixed length or can be scrollable and can accept a virtually unlimited amount of text. In either case, boxes can contain single or multiple lines of textual information. We never use a text box if we can use a selection box.

Text boxes should have field labels placed to the *left* of the entry area, their size clearly delimited by a box (or a set of underlines in a non-GUI interface). If there are multiple text boxes, their field labels and the left edges of their entry boxes should be aligned. Text boxes should permit standard GUI functions, such as cut, copy, and paste.

Numbers   A *number box* is used to enter numbers. Some software can automatically format numbers as they are entered, so that 3452478 becomes $34,524.78. Dates are a special form of numbers that sometimes have their own type of number box. Never use a number box if you can use a selection box.

Selection Box   A *selection box* enables the user to select a value from a predefined list. The items in the list should be arranged in some meaningful order, such as alphabetical for long lists or in order of most frequently used. The default selection value should be chosen with care. A selection box can be initialized as unselected. However, it is better to start with the most commonly used item already selected.

There are six commonly used types of selection boxes: *check boxes, radio buttons, on-screen list boxes, drop-down list boxes, combo boxes,* and *sliders* (see Figure 11-15). The choice among the types of text selection boxes generally comes down to one of screen space and the number of choices the user can select. If screen space is limited and only one item can be selected, then

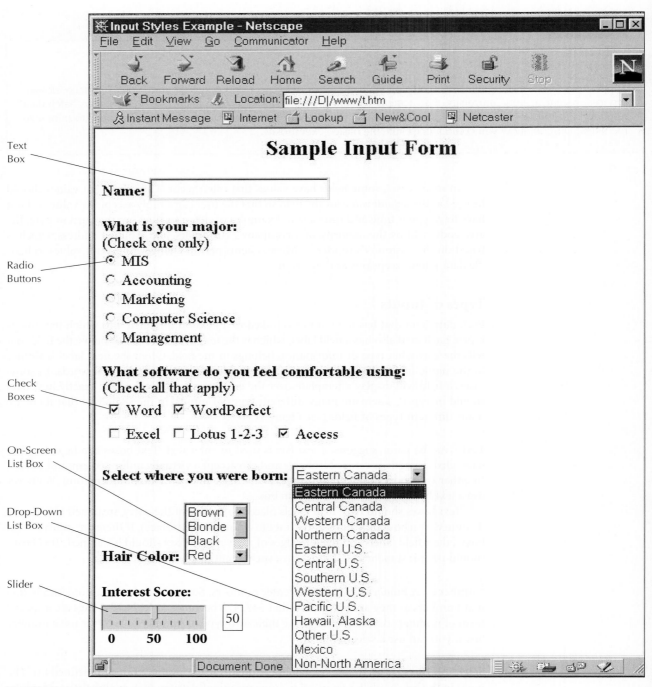

FIGURE 11-14   Types of Input Boxes

a drop-down list box is the best choice, because not all list items need to be displayed on the screen. If screen space is limited but the user can select multiple items, an on-screen list box that displays only a few items can be used. Check boxes (for multiple selections) and radio buttons (for single selections) both require all list items to be displayed at all times, thus requiring more screen space, but because they display all choices, they are often simpler for novice users.

Type of Box	When to Use	Notes
Check box Presents a complete list of choices, each with a square box in front	When several items can be selected from a list of items	Check boxes are not mutually exclusive. Do not use negatives for box labels. Check box labels should be placed in some logical order, such as that defined by the business process, or failing that, alphabetically or most commonly used first. Use no more than ten check boxes for any particular set of options. If you need more boxes, group them into subcategories.
Radio button Presents a complete list of mutually exclusive choices, each with a circle in front	When only one item can be selected from a set of mutually exclusive items	Use no more than six radio buttons in any one list; if you need more, use a drop-down list box. If there are only two options, one check box is usually preferred to two radio buttons, unless the options are not clear. Avoid placing radio buttons close to check boxes to prevent confusion between different selection lists.
On-screen list box Presents a list of choices in a box	Seldom or never—only if there is insufficient room for check boxes or radio buttons	This type of box can permit only one item to be selected (in which case it is an ugly version of radio buttons). This type of box can also permit many items to be selected (in which case it is an ugly version of check boxes), but users often fail to realize they can choose multiple items. This type of box permits the list of items to be scrolled, thus reducing the amount of screen space needed.
Drop-down list box Displays selected item in one-line box that opens to reveal list of choices	When there is insufficient room to display all choices	This type of box acts like radio buttons but is more compact. This type of box hides choices from users until it is opened, which can decrease ease of use; conversely, because it shelters novice users from seldom-used choices, it can improve ease of use. This type of box simplifies design if the number of choices is unclear, because it takes only one line when closed.
Combo box A special type of drop-down list box that permits user to type as well as scroll the list	Shortcut for experienced users	This type of box acts like drop-down list but is faster for experienced users when the list of items is long.
Slider Graphic scale with a sliding pointer to select a number	Entering an approximate numeric value from a large continuous scale	The slider makes it difficult for the user to select a precise number. Some sliders also include a number box to enable the user to enter a specific number.

FIGURE 11-15   Types of Selection Boxes

Input Validation

All data entered into the system need to be validated to ensure their accuracy. Input *validation* (also called *edit checks*) can take many forms. Ideally, computer systems should not accept data that fail any important validation check to prevent invalid information from entering the system. However, this can be very difficult, and invalid data often slip by data-entry operators and the users providing the information. It is up to the system to identify invalid data and either make changes or notify someone who can resolve the information problem.

There are six different types of validation checks: *completeness check, format check, range check, check digit check, consistency check,* and *database check* (see Figure 11-16). Every system should use at least one validation check on all entered data and, ideally, will perform all appropriate checks where possible.

11-9 Career Services

Consider a Web form that a student would use to input student information and résumé information into a career services application at your university. First, sketch out how this form would look and identify the fields that the form would include. What types of validity checks would you use to make sure that the correct information is entered into the system?

Type of Validation	When to Use	Notes
Completeness check Ensures all required data have been entered	When several fields must be entered before the form can be processed	If required information is missing, the form is returned to the user unprocessed.
Format check Ensures data are of the right type (e.g., numeric) and in the right format (e.g., month, day, year)	When fields are numeric or contain coded data	Ideally, numeric fields should not permit users to type text data, but if this is not possible, the entered data must be checked to ensure it is numeric. Some fields use special codes or formats (e.g., license plates with three letters and three numbers) that must be checked.
Range check Ensures numeric data are within correct minimum and maximum values	With all numeric data, if possible	A range check permits only numbers between correct values. Such a system can also be used to screen data for "reasonableness"—e.g., rejecting birthdates prior to 1880 because people do not live to be a great deal over 100 years old (most likely, 1980 was intended).
Check digit check Check digits are added to numeric codes	When numeric codes are used	Check digits are numbers added to a code as a way of enabling the system to quickly validate correctness. For example, U.S. Social Security numbers and Canadian Social Insurance numbers assign only eight of the nine digits in the number. The ninth number—the check digit—is calculated using a mathematical formula from the first eight numbers. When the identification number is typed into a computer system, the system uses the formula and compares the result with the check digit. If the numbers don't match, then an error has occurred.
Consistency checks Ensure combinations of data are valid	When data are related	Data fields are often related. For example, someone's birth year should precede the year in which he or she was married. Although it is impossible for the system to know which data are incorrect, it can report the error to the user for correction.
Database checks Compare data against a database (or file) to ensure they are correct	When data are available to be checked	Data are compared against information in a database (or file) to ensure they are correct. For example, before an identification number is accepted, the database is queried to ensure that the number is valid. Because database checks are more expensive than the other types of checks (they require the system to do more work), most systems perform the other checks first and perform database checks only after the data have passed the previous checks.

FIGURE 11-16   Types of Input Validation

OUTPUT DESIGN

Outputs are the reports that the system produces, whether on the screen, on paper, or in other media, such as the Web. Outputs are perhaps the most visible part of any system because a primary reason for using an information system is to access the information that it produces.

Basic Principles

The goal of the output mechanism is to present information to users so they can accurately understand it with the least effort. The fundamental principles for output design reflect how the outputs are used and ways to make it simpler for users to understand them.

Understand Report Usage   The first principle in designing reports is to understand how they are used. Reports can be used for many different purposes. In some cases—but not very often—reports are read cover to cover because all information is needed. In most cases, reports are used to identify specific items or used as references to find information, so the order in which items are sorted on the report or grouped within categories is critical. This is particularly important for the design of electronic or Web-based reports. Web reports that are intended to be read from start to finish should be presented in one long scrollable page, whereas reports that are used primarily to find specific information should be broken into multiple pages, each with a separate link. Page numbers and the date on which the report was prepared are also important for reference reports.

The frequency of the report may also play an important role in its design and distribution. *Real-time reports* provide data that are accurate to the second or minute at which they were produced (e.g., stock market quotes). *Batch reports* are those that report historical information that may be months, days, or hours old, and they often provide additional information beyond the reported information (e.g., totals, summaries, historical averages).

There are no inherent advantages to real-time reports over batch reports. The only advantages lie in the time value of the information. If the information in a report is time critical (e.g., stock prices, air-traffic control information), then real time reports have value. This is particularly important because real-time reports are often expensive to produce; unless they offer some clear business value, they may not be worth the extra cost.

Manage Information Load   Most managers get too much information, not too little (i.e., the *information load* that the manager must deal with is too great). The goal of a well-designed report is to provide all the information needed to support the task for which it was designed. This does not mean that the report needs to provide all the information *available* on the subject—just what the users decide they need in order to perform their jobs. In some cases, this may result in the production of several different reports on the same topics for the same users because they are used in different ways. This is not a bad design.

For users in westernized countries, the most important information should always be presented first in the top-left corner of the screen or paper report. Information should be provided in a format that is usable without modification. The user should not need to re-sort the report's information, highlight critical information to find it more easily amid a mass of data, or perform additional mathematical calculations.

Minimize Bias   No analyst sets out to design a biased report. The problem with bias is that it can be very subtle; analysts can introduce it unintentionally. *Bias* can be introduced by the way in which lists of data are sorted because entries that appear first in a list may receive more attention than those later in the list. Data are often sorted in alphabetical order, making those entries starting with the letter *A* more prominent. Data can be sorted in chronological order (or reverse chronological order), placing more emphasis on older

Unbiased graph with scale starting at 0

(a)

Biased graph with scale starting at 90

(b)

FIGURE 11-17
Bias in Graphs

(or most recent) entries. Data may be sorted by numeric value, placing more emphasis on higher or lower values. For example, consider a monthly sales report by state. Should the report be listed in alphabetical order by state name, in descending order by the amount sold, or in some other order (e.g., geographic region)? There are no easy answers to this, except to say that the order of presentation should match the way in which the information is used.

Graphical displays and reports can present particularly challenging design issues.[11] The scale on the axes in graphs is particularly subject to bias. For most types of graphs, the scale should always begin at zero; otherwise, comparisons among values can be misleading. For example, in Figure 11-17, have sales increased by very much since 1993? The numbers in both charts are the same, but the visual images the two present are quite different. A glance at Figure 11-17a would suggest only minor changes, whereas a glance at Figure 11-17b might suggest that there have been some significant increases. In fact, sales have increased by a total of 15 percent over five years, or 3 percent per year. Figure 11-17a presents the

[11] Some of the best books on the design of charts and graphical displays are by Edward R. Tufte, published by Graphics Press in Cheshire, Connecticut: *The Visual Display of Quantitative Information, Envisioning Information,* and *Visual Explanations.* Another good book is by William Cleveland, *Visualizing Data,* (Summit, NJ: Hobart Press, 1993).

11-10 Finding Bias

*R*ead through recent copies of a newspaper or popular press magazine, such as *Time, Newsweek,* or *BusinessWeek,* and find four graphs. Are any of these biased? if so, why?

most accurate picture; Figure 11-17b is biased because the scale starts very close to the lowest value in the graph and misleads the eye into inferring that there have been major changes (i.e., more than doubling from two lines in 1993 to five lines in 1998). Figure 11-17b is the default graph produced by Microsoft Excel.

Types of Outputs

There are many different types of reports, such as *detail reports, summary reports, exception reports, turnaround documents,* and *graphs* (see Figure 11-18). Classifying reports is challenging because many reports have characteristics of several different types. For example, some detail reports also produce summary totals, making them summary reports.

Media

There are many different types of media used to produce reports. The two dominant media in use today are paper and electronic. Paper is the more traditional medium. For almost as long as there have been human organizations, there have been reports on paper or similar media (e.g., papyrus, stone). Paper is permanent, easy to use, and accessible in most situations. It also is highly portable, at least for short reports.

11-C Selecting the Wrong Students

I helped a university department develop a small DSS to analyze and rank students who applied to a specialized program. Some of the information was numeric and could easily be processed directly by the system (e.g., grade-point average, standardized test scores). Other information required the faculty to make subjective judgments among the students (e.g., extracurricular activities, work experience). The users entered their evaluations of the subjective information via several data analysis screens in which the students were listed in alphabetical order.

To make the system easier to use, it was designed so that the reports listing the results of the analysis were also presented in alphabetical order by student name rather than in order from the highest-ranked student to the lowest-ranked student. In a series of tests prior to installation, the users selected the wrong students to admit in 20 percent of the cases. They assumed, wrongly, that the students listed first were the highest-ranked students and simply selected the first students on the list for admission. Neither the title on the report nor the fact that all the students' names were in alphabetical order made users realize that they had read the report incorrectly.

—Alan Dennis

Question

This system was biased because users assumed that the list of students implied ranking. Suppose that you are an analyst charged with minimizing bias in this application. Where else may you find bias in the application? How would you eliminate it?

Type of Reports	When to Use	Notes
Detail report Lists detailed information about all the items requested	When user needs full information about the items	This report is usually produced only in response to a query about items matching some criteria. This report is usually read cover to cover to aid understanding of one or more items in depth.
Summary report Lists summary information about all items	When user needs brief information on many items	This report is usually produced only in response to a query about items matching some criteria, but it can be a complete database. This report is usually read for the purpose of comparing several items to each other. The order in which items are sorted is important.
Turnaround document Outputs that "turn around" and become inputs	When a user (often a customer) needs to return an output to be processed	Turnaround documents are a special type of report that are both outputs and inputs. For example, most bills sent to consumers (e.g., credit-card bills) provide information about the total amount owed and also contain a form that consumers fill in and return with payment.
Graphs Charts used in addition to and instead of tables of numbers	When users need to compare data among several items	Well-done graphs help users compare two or more items or understand how one has changed over time. Graphs are poor at helping users recognize precise numeric values and should be replaced by or combined with tables when precision is important. Bar charts tend to be better than tables of numbers or other types of charts when it comes to comparing values between items (but avoid three-dimensional charts that make comparisons difficult). Line charts make it easier to compare values over time, whereas scatter charts make it easier to find clusters or unusual data. Pie charts show proportions or the relative shares of a whole.

FIGURE 11-18 Types of Reports

Paper also has several rather significant drawbacks. It is inflexible. Once the report is printed, it cannot be sorted or reformatted to present a different view of the information. Likewise, if the information on the report changes, the entire report must be reprinted. Paper reports are expensive, are hard to duplicate, and require considerable supplies (paper, ink) and storage space. Paper reports also are hard to move long distances quickly (e.g., from a head office in Toronto to a regional office in Bermuda).

Many organizations are, therefore, moving to electronic production of reports, whereby reports are "printed" but stored in electronic format on file servers or Web servers so that users can easily access them. Often the reports are available in more predesigned formats than their paper-based counterparts because the cost of producing and storing different formats is minimal. Electronic reports also can be produced on demand as needed, and they enable the user to more easily search for certain words. Furthermore, electronic reports can provide a means of supporting ad hoc reports, where users customize the contents of the report at the time the report is generated. Some users still print the electronic report on their own printers, but the reduced cost of electronic delivery over distance and the ease of enabling more users to access the reports than when they were only in paper form usually offsets the cost of local printing.

11-D Cutting Paper to Save Money

*O*ne of the Fortune 500 firms with which I have worked had an eighteen-story office building for its world headquarters. It devoted two full floors of this building to nothing more than storing "current" paper reports (a separate warehouse was maintained outside the city for archived reports such as tax documents). Imagine the annual cost of office space in the headquarters building tied up in these paper reports. Now imagine how a staff member would gain access to the reports, and you can quickly understand the driving force behind electronic reports, even if most users end up printing them. Within one year of switching to electronic reports

(for as many reports as practical), the paper report storage area was reduced to one small storage room.

—Alan Dennis

Questions

1. What types of reports are most suited to electronic format?
2. What types of reports are less suited to electronic reports?

NONFUNCTIONAL REQUIREMENTS AND HUMAN–COMPUTER INTERACTION LAYER DESIGN[12]

The human–computer interaction layer is heavily influenced by nonfunctional requirements. Earlier in this chapter, we dealt with issues such as user interface layout, content awareness, aesthetics, user experience, and consistency. None of these have anything to do with the functional requirements of the system. However, if they are ignored, the system can prove to be unusable. Like the data management layer, there are four primary types of nonfunctional requirements that can be important in designing the human–computer interaction layer: operational, performance, security, and cultural/political requirements.

Operational requirements, such as choice of hardware and software platforms, influence the design of the human–computer interaction layer. For example, something as simple as the number of buttons on a mouse (one, two, three, or more) changes the interaction that the user will experience. Other operational nonfunctional requirements that can influence the design of the human–computer interaction layer include system integration and portability. In these cases, a Web-based solution may be required, which can affect the design; not all features of a user interface can be implemented efficiently and/or effectively on the Web. This can cause additional user interface design to occur.

Performance requirements, over time, have become less of an issue for this layer. However, speed requirements are still paramount. Most users do not care for hitting return or clicking the mouse and having to take a coffee break while they are waiting for the system to respond. As such, efficiency issues must be still addressed. Depending on the user interface toolkit used, different user interface components may be required. Furthermore, the interaction of the human–computer interaction layer with the other layers must be considered. For example, if the system response is slow, incorporating more-efficient data structures with the problem domain layer, including indexes in the tables with the data management layer, and/or replicating objects across the physical architecture layer could be required.

Security requirements impacting the human–computer interaction layer deal primarily with the access controls implemented to protect the objects from unauthorized access. Most of these controls will be enforced through the DBMS on the data management layer and the

---

[12] Because the vast majority of nonfunctional requirements impact the physical architecture layer, we provide additional details in Chapter 12.

operating system on the physical architecture layer. However, the human–computer interaction layer design must include appropriate log-on controls and the possibility of encryption.

Cultural and political requirements that can impact the design of the human–computer interaction layer include multilingual requirements and unstated norms. The multilingual requirements include simple concerns such as the number of characters to allocate for different fields. For example, Spanish can take between 20 and 30 percent more characters to display the equivalent message when compared to English. Furthermore, there are many different alphabets. Also, translating a message from one language to another is still not a very automatable process.[13] Unstated norm requirements include things such as having the date displayed in the appropriate format (MM/DD/YYYY versus DD/MM/YYYY). Another concern is the layout of the user interface. Because English is read left to right and top to bottom, we tend to design the layout of the user interface in a similar manner. But not all languages are read in this order. Finally, the use of color can be problematic. Different cultures attach different meaning to different colors. For example the implification of the colors black and white depends on the culture that is viewing the user interface. In many western cultures, black implies bad and white implies good. However, in certain eastern cultures, the opposite is true. For a system to be truly useful in a global environment, the user interface must be customizable to address local cultural requirements.

APPLYING THE CONCEPTS AT CD SELECTIONS

In the CD Selections Internet Sales System case, there are three different high-level use cases (see Figure 5-19): Maintain CD Information, Place Order, and Maintain CD Marketing Information. There are also six additional use cases, Maintain Order, Checkout, Create New Customer, Place InStore Hold, Place Special Order, and Fill Mail Order, associated with the Place Order use case. To keep the complexity of the current example under control, in this section we focus only on the Place Order, Maintain Order, and Checkout use cases.

Use Scenario Development

The first step in the interface design process was to develop the key use scenarios for the Internet Sales System. Because Alec assigned the human–computer interaction layer design to himself, he began by examining the essential use cases (see Figure 5-18) and thinking about the types of users and how they would interact with the system. To begin with, Alec identified two use scenarios: the browsing shopper and the hurry-up shopper (see Figure 11-6).[14] Alec also thought of several other use scenarios for the Web site in general, but he omitted them because they were not relevant to the Internet sales portion. Likewise, he thought of several use scenarios that did not lead to sales (e.g., fans looking for information about their favorite artists and albums) and omitted them as well.

Interface Structure Design

Next, Alec created a WND for the Web system. He began with the Place Order, Maintain Order, and Checkout essential use cases to ensure that all functionality defined for the system was included in the WND. Figure 11-19 shows the WND for the Web portion of the

[13] For example, type the text "I would like my steak cooked rare into babel fish" (http://babelfish.yahoo.com/) and translate it to Russian and back to English. You will get back "I wanted would be my rare welded [steykom] done" not exactly the most useful translation.

[14] Of course, it may be necessary to modify the original essential use cases in light of these new subtypes of customer. Furthermore, the structural and behavioral models may have to be modified. Remember that object-oriented systems analysis and design is incremental and iterative; as such, additional requirements can be uncovered at any time.

Internet Sales System. The system started with a home page that contained the main menu for the sales system. Based on the essential use cases, Alec identified four basic operations that he felt made sense to support on the main menu: search the CD catalog, search by music category, review the contents of the shopping cart, and actually place the order. Each of these was modeled as a hyperlink on the home page.

Alec then decided to model the full search option as a pop-up search menu that allowed the customer to choose to search the CD catalog based on artist, title, or composer. He further decided that a textbox would be required to allow the customer to type in the name of the artist, title, or composer, depending on the type of search requested. Finally, he chose to use a button to submit the request to the system. After the Submit button was pressed, the system produced a report composed of hyperlinks to the individual information on each CD. A CD report containing the basic information was generated by clicking on the hyperlink associated with the CD. On the basic report, Alec added buttons for choosing to find out additional information on the CD and to add the CD to the shopping cart. If the Detail button was pressed, a detailed report containing the marketing information on the CD was produced. Finally, Alec decided to include a button on this report to add the CD to the shopping cart.

The second basic operation supported on the home page was to allow the user to search the CD catalog by category of music. As with the previous operation, Alec chose to model the category search with a pop-up search menu. In this case, once the customer chose the category, the system would produce the report with the hyperlinks to the individual information on each CD. From that point on, the navigation would be identical to the previous searches.

The third operation supported was to review the contents of the shopping cart. In this case, Alec decided to model the shopping cart as a report that contained three types of hyperlinks: one for removing an individual CD from the shopping cart, one for removing all CDs from the shopping cart, and one for placing the order. The removal hyperlinks would remove the individual CD (or all the CDs) from the shopping cart if the user would confirm the operation. The place order link would send the customer to an order form. Once the customer filled out the order form, the customer would press the order button. The system would then respond with an order confirmation message box.

The fourth operation supported on the home page was to allow the customer to place an order directly. Upon review, Alec decided that the place order and review cart operations were essentially identical. As such, he decided to force the user to have the place order and review shopping cart operations go through the same process.

Alec also envisioned that by using frames, the user would be able to return to the home page from any screen. Documenting these would give the WND too many lines, so Alec simply put a note describing it with the WND.

The Revised WND   Alec then examined the use scenarios to see how well the initial WND enabled different types of users to work through the system. He started with the Browsing Shopping use scenario and followed it through the WND, imagining what would appear on each screen and pretending to navigate through the system. He found the WND to work well, but he noticed a couple of minor issues related to the shopping cart. First, he decided that it would make sense to allow the customer to retrieve the information related to the CDs contained in the shopping cart. As such, he changed the stereotype of the user interface component from Report to HyperLink Rep and added a hyperlink from the Shopping Cart to the Basic Report created by the different search requests. Second, he noticed that the Shopping Cart was using hyperlinks to link to the Removal and Place Order processes. However, in all the other elements of the WND, he was using buttons to model the equivalent ideas. As such, he decided to change the Shopping Cart component to model these connections as buttons. Of course, this forced him to modify the transitions as well.

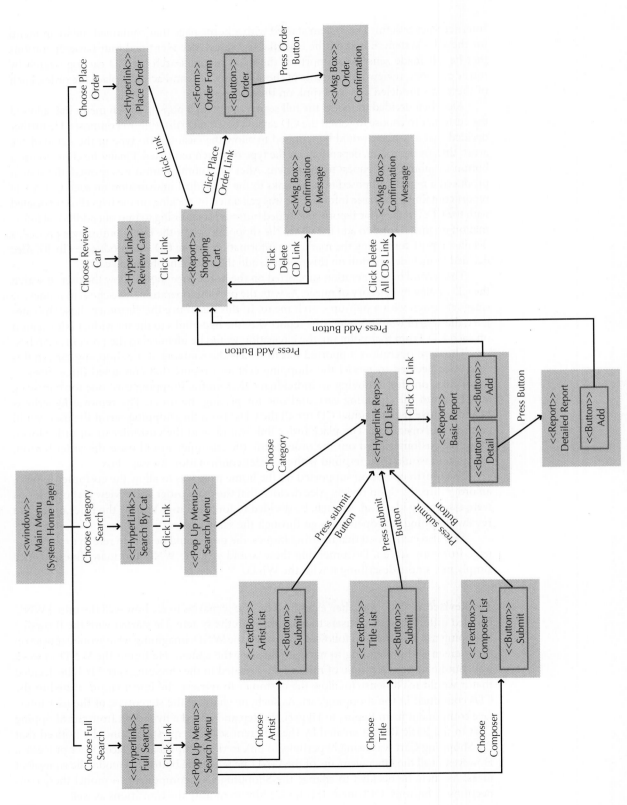

FIGURE 11-19 CD Selections Initial WND for the Web Portion of the Internet Sales System

Alec next explored the Hurry-up Shopper use scenario. In this case, the WND did not work as well. Moving from the home page, to the search page, to the list of matching CDs, to the CD page with price and other information took three mouse clicks. This falls within the three clicks rule, but for someone in a hurry, this may be too many. Alec decided to add a quick-search option to the home page that would enable the user to enter one search criteria (e.g., just artist name or title, rather than a more detailed search as would be possible on the search page) thus, with one click, taking the user to the one CD that matched the criteria or to a list of CDs if there were more than one. This would enable an impatient user to get to the CD of interest in one or two clicks.

Once the CD was displayed on the screen, the Hurry-up Shopper use scenario would suggest that the user would immediately purchase the CD, do a new search, or abandon the Web site and surf elsewhere. This suggested two important changes. First, there had to be an easy way to go to the Place Order screen. As the WND stood (see Figure 11-19), the user had to add the item to the shopping cart and then click on the link on the HTML frame to get to the Place Order screen. Although the ability of users to notice the place order link in the frame would await the interface evaluation stage, Alec suspected, based on past experience, that a significant number of users would not see it. Therefore, he decided to add a button to the Basic Report screen and the Detailed Report screen called Buy (see Figure 11-20).

Second, because the hurry-up shopper might want to search for another CD instead of buying the CD, Alec decided to include the quick-search item from the home page on the frame. This would make all searches immediately available from anywhere in the system. This would mean that all functionality on the home page would now be carried on the frame. Alec updated the note on the bottom attached to the WND to reflect the change.

Finally, upon review of the WND, Alec decided to remodel the Artist List, Title List, and Composer Lists as window stereotypes instead of textbox stereotypes. He then added a Search Item textbox to each of these elements. Figure 11-20 shows the revised WND for the Web portion of the Place Order use case. All changes are highlighted.

Interface Standards Design

Once the WND was complete, Alec moved on to develop the interface standards for the system. The interface metaphor was straightforward: a CD Selections music store. The key interface objects and actions were equally straightforward, as was the use of the CD Selections logo icon (see Figure 11-21).

Interface Template Design

For the interface template, Alec decided on a simple, clean design that had a modern background pattern with the CD Selections logo in the upper-left corner. The template had two navigation areas: one menu across the top for navigation within the entire Web site (e.g., overall Web site home page, store locations) and one menu down the left edge for navigation within the Internet sales system. The left-edge menu contained the links to the top-level operations (see WND in Figure 11-19), as well as the "quick search" option. The center area of the screen is used for displaying forms and reports when the appropriate operation is chosen (see Figure 11-22).

At this point, Alec decided to seek some quick feedback on the interface structure and standards before investing time in prototyping the interface designs. Therefore, he met with Margaret Mooney, the project sponsor, and Chris Campbell, the consultant, to discuss the emerging design. Making changes at this point would be much simpler than after doing the prototype. Margaret and Chris had a few suggestions, so after the meeting Alec made the changes and moved into the design-prototyping step.

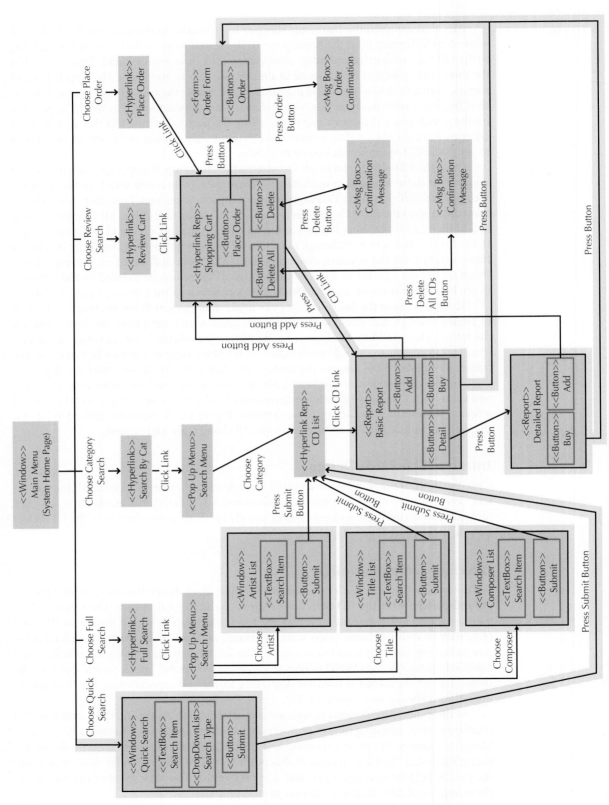

FIGURE 11-20 CD-Selections Revised WND for the Web Portion of the Internet Sales System (Changes Are Highlighted)

Interface Metaphor: A CD Selections Music Store

Interface Objects
- **CD:** All items, whether CD, tape, or DVD, unless it is important to distinguish among them
- **Artist:** Person or group who records the CD
- **Title:** Title or name of CD
- **Composer:** Person or group who wrote the music for the CD (primarily used for classical music)
- **Music Category:** Type of music. Current categories include: Rock, Jazz, Classical, Country, Alternative, Soundtracks, Rap, Folk, Gospel.
- **CD List:** List of CD(s) that match the specified criteria
- **Shopping Cart:** Place to store CDs until they are requested

Interface Actions
- **Search for:** Display a CD list that matches specified criteria
- **Browse:** Display a CD list sorted in order by some criteria
- **Order:** Authorize special order or place hold

Interface Icons
- **CD Selections Logo:** Will be used on all screens

Interface Design Prototyping

Alec decided to develop a hypertext markup language (HTML) prototype of the system. The Internet sales system was new territory for CD Selections and a strategic investment in a new business model, so it was important to make sure that no key issues were overlooked. The HTML prototype would provide the most detailed information and enable interactive evaluation of the interface.

In designing the prototype, Alec started with the home screen and gradually worked his way through all the screens. The process was very iterative, and he made many changes to the screens as he worked. Once he had an initial prototype designed, he posted it on CD Selections intranet and solicited comments from several friends with lots of Web experience. He revised it based on the comments he received. Figure 11-23 presents some screens from the prototype.

FIGURE 11-23

Sample Interfaces from the CD Selections Design Prototype

Interface Evaluation

The next step was interface evaluation. Alec decided on a two-phase evaluation. The first evaluation was to be an interactive evaluation conducted by Margaret, her marketing managers, selected staff members, selected store managers, and Chris. They worked hands-on with the prototype and identified several ways to improve it. Alec modified the HTML prototype to reflect the changes suggested by the group and asked Margaret and Chris to review it again.

The second evaluation was another interactive evaluation, this time by a series of two focus groups of potential customers—one with little Internet experience, the other with extensive Internet experience. Once again, several minor changes were identified. Alec again modified the HTML prototype and asked Margaret and Chris to review it again. Once they were satisfied, the interface design was complete.

Use-Case Name: *Place Order*	ID: *15*	Importance Level: *High*

Primary Actor: *The Browsing Customer*   Use-Case Type: *Detail, Real*

Stakeholders and Interests: *Customer   Wants to search web site to purchase CD*
                                EM Manager   Wants to maximize Customer satisfaction.

Brief Description: *This use case describes how customers can search the web site and place orders.*

Trigger: *Customer visits web site*

Type: *External*

Relationships:
   Association: *Customer*
   Include: *Checkout, Maintain Order*
   Extend:
   Generalization:

Normal Flow of Events:
   1. The Customer visits the Web site.
   2. The System displays the Home Page
      If the Customer wants to do a Full Search, execute S-1: Full Search
      If the Customer wants to Browse by Music Type, execute S-2: Browse by Music Type
      If the Customer wants to see any Special Deals, execute S-3: Special Deals
      If the Customer wants to see the contents of the Shopping Cart, execute S-4: Shopping Cart
      If the Customer wants to Buy Now, execute S-5: Buy Now
   3. The Customer leaves the site.

Subflows:
   S-1: Full Search
      1. The Customer clicks the Full Search hyperlink
      2. The System displays the search type pop-up menu
         If the Customer chooses an Artist search, execute S-1a: Artist List
         If the Customer chooses an Title search, execute S-1a: Title List
         If the Customer chooses an Composer search, execute S-1a: Composer List
   S-1a: Artist List
      1. The System displays the Artist List window in the Center Area of the Home Page.
      2. The Customer enters the Artist Name into the Search Item text box.
      3. The Customer presses the Submit button.
      4. The System executes S-2a: CD List.
   S-2a: CD List
      1. The System displays the CD List hyperlink report.
      2. The Customer chooses a CD to review by clicking the CD link.
      3. The System executes S-2b: Display Basic Report
      4. Iterate over steps 2 and 3.

Alternate/Exceptional Flows:

FIGURE 11-24
The Browsing
Customer Real Use
Case (Partial Listing
Only)

Navigation Design Documentation

The last step that Alec completed was to document the navigation design through the use of real use cases. To accomplish this, Alec gathered together the essential use case (see Figure 5-18), the use scenarios (see Figure 11-6), the window navigation diagram (see Figure 11-20), and the user interface prototype (see Figures 11-22 and 11-23). First, he copied the contents of the essential use case to the real use case. He changed the type from detail, essential to detail, real; the primary actor was specialized to browsing customer instead of simply customer. Second, he wrote the specific set of steps and responses that described the interaction between the browsing customer and system. Figure 11-24 shows a partial listing of the steps in the normal flow of events and subflows sections of the real use case. Last, he repeated the steps for the hurry-up shopper.

YOUR
TURN

11-11 CD Selections Internet Sales System

At this point in time, you should play the role of Alec and complete the design of the human–computer interaction layer. This includes adding use scenarios, updating the WND, updating the user interface template design, updating the user interface prototypes, and adding any additional real use cases required.

SUMMARY

Principles for User Interface Design

The first element of the user interface design is the layout of the screen, form or report, which is usually depicted using rectangular shapes with a top area for navigation, a central area for inputs and outputs, and a status line at the bottom. The design should help the user be aware of content and context, both between different parts of the system as they navigate through it and within any one form or report. All interfaces should be aesthetically pleasing (not works of art) and need to include significant white space, use colors carefully, and be consistent with fonts. Most interfaces should be designed to support both novice/first time users, as well as experienced users. Consistency in design (both within the system and across other systems used by the users) is important for the navigation controls, terminology, and the layout of forms and reports. Finally, all interfaces should attempt to minimize user effort—for example, by requiring no more than three clicks from the main menu to perform an action.

User Interface Design Process

First, analysts develop use scenarios that describe commonly used patterns of actions that the users will perform. Second, they design the interface structure via a WND based on the essential use cases. The WND is then tested with the use scenarios to ensure that it enables users to quickly and smoothly perform these scenarios. Third, analysts define the interface standards in terms of interface metaphor(s), objects, actions, and icons. These elements are drawn together by the design of a basic interface template for each major section of the system. Fourth, the designs of the individual interfaces are prototyped, either through a simple storyboard, an HTML prototype, or a prototype using the development language of the system itself (e.g., Visual Basic). Finally, interface evaluation is conducted using heuristic evaluation, walkthrough evaluation, interactive evaluation, or formal usability testing. This evaluation almost always identifies improvements, so the interfaces are redesigned and evaluated further.

Navigation Design

The fundamental goal of the navigation design is to make the system as simple to use as possible, by preventing the user from making mistakes, simplifying the recovery from mistakes, and using a consistent grammar order (usually object-action order). Command languages, natural languages, and direct manipulation are used in navigation, but the most common approach is menus (menu bar, drop-down menu, pop-up menu, tab menu, button and tool bars, and image maps). Error messages, confirmation messages, acknowledgment messages, delay messages, and help messages are common types of messages. Once the navigation design is agreed upon, it is documented in the form of WNDs and real use cases.

Input Design

The goal of the input mechanism is to simply and easily capture accurate information for the system, typically by using online or batch processing, capturing data at the source, and minimizing keystrokes. Input design includes both the design of input screens and all preprinted forms that are used to collect data before they are entered into the information system. There are many types of inputs, such as text fields, number fields, check boxes, radio buttons, on-screen list boxes, drop-down list boxes, and sliders. Most inputs are validated by using some combination of completeness checks, format checks, range checks, check digits, consistency checks, and database checks.

Output Design

The goal of the output mechanism is to present information to users so they can accurately understand it with the least effort, usually by understanding how reports will be used and designing them to minimize information overload and bias. Output design means designing both screens and reports in other media, such as paper and the Web. There are many types of reports, such as detail reports, summary reports, exception reports, turnaround documents, and graphs.

Nonfunctional Requirements and the Human–Computer Interaction Layer

Nonfunctional requirements can affect the usefulness of the human–computer interaction layer. Because the user/client sees the system as the human–computer interaction layer, not paying sufficient attention to the nonfunctional requirements in the design of this layer can cause the entire system development effort to fail. These requirements include operational, performance, security, and cultural and political issues, and they are intertwined with the design of the data management and physical architecture layers. However, special attention must be paid to cultural and political requirements, such as user interface layout and multilingual capabilities, that address the design of this layer for the system to be effective.

KEY TERMS

Acknowledgment message
Action-object order
Aesthetics
Bar-code reader
Batch processing
Batch report
Bias
Button
Check box
Check digit check
Combo box
Command language
Completeness check
Confirmation message
Consistency
Consistency check
Content awareness
Database check

Data-entry operator
Default value
Delay message
Density
Detail report
Direct manipulation
Drop-down list box
Drop-down menu
Ease of learning
Ease of use
Edit check
Error message
Essential use case
Exception report
Field
Field label
Form
Format check

Grammar order
Graph
Graphical user interface (GUI)
Help message
Heuristic evaluation
Hot key
HTML prototype
Image map
Information load
Input mechanism
Interactive evaluation
Interface action
Interface design prototype
Interface evaluation
Interface icon
Interface metaphor
Interface object
Interface standards

Interface template
Language prototype
Layout
Magnetic stripe readers
Menu
Menu bar
Natural language
Navigation controls
Navigation mechanism
Number box
Object-action order
Online processing
On-screen list box
Optical character recognition
Output mechanism
Pop-up menu
Radio button

Range check
Real-time information
Real-time report
Real use case
Report
Screen
Selection box
Sequence diagrams
Slider
Smart card
Source data automation
State
Stereotype
Storyboard
Summary report
System interface
Tab menu

Text box
Three-clicks rule
Toolbar
Transaction processing
Transition
Turnaround document
Usability testing
Use case
Use scenario
User experience
User interface
Validation
Walkthrough evaluation
White space
Window
Windows layout diagram
Window navigation diagram (WND)

QUESTIONS

1. How can a system be designed to be used by experienced and first-time users?
2. Describe five types of inputs.
3. Why is input validation important?
4. Describe five types of input validation methods.
5. What do you think are three common mistakes that novice analysts make in input design?
6. Compare and contrast check boxes and radio buttons. When would you use one versus the other?
7. Explain three principles in the design of inputs.
8. How can you prevent mistakes?
9. Explain three important user interface design principles.
10. Describe the basic process of user interface design.
11. Why do we prototype the user interface design?
12. Compare and contrast the three types of interface design prototypes.
13. Explain three principles in the design of outputs.
14. Describe five types of outputs.
15. Compare and contrast batch processing and online processing. Describe one application that would use batch processing and one that would use online processing.
16. Why is capturing data at the source important?
17. Describe four devices that can be used for source data automation.
18. When would you use electronic reports rather than paper reports, and vice versa?

19. What do you think are three common mistakes that novice analysts make in output design?
20. How would you improve the form in Figure 11-4?
21. What is white space, and why is it important?
22. Describe four types of navigation controls.
23. Describe three basic principles of navigation design.
24. Explain the differences between object-action order and action-object order.
25. Compare and contrast four types of menus.
26. Why are menus the most commonly used navigation control?
27. Under what circumstances would you use an image map versus a simple list menu?
28. Under what circumstances would you use a drop-down menu versus a tab menu?
29. Describe five types of messages.
30. What are the key factors in designing an error message?
31. What is context-sensitive help? Does your word processor have context-sensitive help?
32. Why is content awareness important?
33. What do you think are three common mistakes that novice analysts make in navigation design?
34. What are some of the nonfunctional requirements that can influence the design of the human computer interaction layer?
35. What are three fundamental parts of most user interfaces?

36. How can different parts of the interface be consistent?
37. Why is consistency in design important? Why can too much consistency cause problems?
38. What are use cases, and why are they important?
39. How do an essential use case and a real use case differ?
40. What is the relationship between real use cases and use scenarios?
41. What is the relationship between essential use cases and use scenarios?
42. Why are interface standards important?
43. Explain the purpose and contents of interface metaphors, interface objects, interface actions, interface icons, and interface templates.
44. Compare and contrast the four types of interface evaluation.
45. Why is it important to perform an interface evaluation before the system is built?
46. Under what conditions is heuristic evaluation justified?

EXERCISES

A. Ask Jeeves (http://www.ask.com) is an Internet search engine that uses natural language. Experiment with it and compare it to search engines that use key words.
B. Draw a WND for Your Turn 11-7 using the opposite grammar order from your original design (if you didn't do it, draw two WNDs, one in each grammar order). Which is best? Why?
C. In Your Turn 11-7, you probably used menus. Design the navigation system again using a command language.
D. Based on the use-case diagram for the health club problem in exercise O in Chapter 5:
    1. Develop two use scenarios.
    2. Develop the interface standards (omitting the interface template).
    3. Draw a WND.
    4. Design a storyboard.
E. Based on your solution to exercise D:
    1. Design an interface template.
    2. Create Windows Layout Diagrams for the interface design.
    3. Develop an HTML prototype.
    4. Develop a real use case.
F. Based on the use-case diagram for Picnics R Us in exercise Q in Chapter 5:
    1. Develop two use scenarios.
    2. Develop the interface standards (omitting the interface template).
    3. Draw a WND.
    4. Design a storyboard.
G. Based on your solution to exercise F:
    1. Design an interface template.
    2. Create Windows Layout Diagrams for the interface design.
    3. Develop an HTML prototype.
    4. Develop a real use case.

H. Based on the use-case diagram for the university library borrowing system in exercise S in Chapter 5:
    1. Develop two use scenarios.
    2. Develop the interface standards (omitting the interface template).
    3. Draw a WND.
    4. Design a storyboard.
I. Based on your solution to exercise H:
    1. Design an interface template.
    2. Create Windows Layout Diagrams for the interface design.
    3. Develop an HTML prototype.
    4. Develop a real use case.
J. Based on the use-case diagram for the Of-the-Month-Club in exercise U in Chapter 5:
    1. Develop two use scenarios.
    2. Develop the interface standards (omitting the interface template).
    3. Draw a WND.
    4. Design a storyboard.
K. Based on your solution to exercise J:
    1. Design an interface template.
    2. Create Windows Layout Diagrams for the interface design.
    3. Develop an HTML prototype.
    4. Develop a real use case.
L. Draw a WND for a Web site that sells some retail products (e.g., books, music, clothes).
M. Describe the primary components of the interface standards for a Web site that sells some retail products (metaphors, objects, actions, icons and template).
N. Develop two use scenarios for a Web site that sells some retail products (e.g., books, music, clothes).

MINICASES

1. Refer to the Professional and Scientific Staff Management (PSSM) Minicase in Chapters 5, 7, 8, and 10.
   a. Develop two use scenarios, interface standards (omitting the interface template), draw a WND, and design a storyboard.
   b. Based on your answers to part a, design an interface template, develop an HTML prototype, and develop a real use case.

2. Tots to Teens is a catalog retailer specializing in children's clothing. A project has been under way to develop a new order entry system for the company's catalog clerks. The old system had a character-based user interface that corresponded to the system's COBOL underpinnings. The new system will feature a graphical user interface more in keeping with up-to-date PC products in use today. The company hopes that this new user interface will help reduce the turnover they have experienced with their order entry clerks. Many newly hired order entry staff found the old system very difficult to learn and were overwhelmed by the numerous mysterious codes that had to be used to communicate with the system.

   A user interface walkthrough evaluation was scheduled for today to give the user a first look at the new system's interface. The project team was careful to invite several key users from the order entry department. In particular, Norma was included because of her years of experience with the order entry system. Norma was known to be an informal leader in the department; her opinion influenced many of her associates. Norma had let it be known that she was less than thrilled with the ideas she had heard for the new system. Due to her experience and good memory, Norma worked very effectively with the character-based system and was able to breeze through even the most convoluted transactions with ease. Norma had trouble suppressing a sneer when she heard talk of such things as "icons" and "buttons" in the new user interface.

   Cindy was also invited to the walkthrough because of her influence in the order entry department. Cindy has been with the department for just one year, but she quickly became known because of her successful organization of a sick child day-care service for the children of the department workers. Sick children are the number-one cause of absenteeism in the department, and many of the workers could not afford to miss workdays. Never one to keep quiet when a situation needed improvement, Cindy has been a vocal supporter of the new system.

   a. Drawing upon the design principles presented in the text, describe the features of the user interface that will be most important to experienced users like Norma.
   b. Drawing upon the design principles presented in the text, describe the features of the user interface that will be most important to novice users like Cindy.

3. The members of a systems development project team have gone out for lunch together and, as often happens, the conversation turns to work. The team has been working on the development of the user interface design, and so far, work has been progressing smoothly. The team should be completing work on the interface prototypes early next week. A combination of storyboards and language prototypes has been used in this project. The storyboards depict the overall structure and flow of the system, but the team went on to develop language prototypes of the actual screens because they felt that the users would gain from looking at the actual screens.

   Robin (the youngest member of the project team): I read an article last night about a really cool way to evaluate a user interface design. It's called usability testing, and it's done by all the major software vendors. I think we should use it to evaluate our interface design.

   Dayita (systems analyst): I've heard of that too, but isn't it really expensive?

   Pankaj (project manager): I'm afraid it is, and I'm not sure if we can justify the expense for this project.

   Robin: But we really need to know if the interface works. I thought this usability testing technique would help us prove we have a good design.

   Man Yee (systems analyst): It would, Robin, but there are other ways too. I suggest we do a thorough walk-through with our users and present the interface to them at a meeting. We can project each interface screen so that the users can see it and give us their reactions. This is probably the most efficient way to get their feedback to our work.

   Dayita: That's true, but I'd sure like to see the users sit down and work with the system. I've always learned a lot by watching what they do, noting where they are stuck, and listening to their comments and feedback.

   Manuel (systems analyst): We've put so much work into this interface design that we really need to review it ourselves. Let's just make a list of the key design principles and ensure that we've followed them consistently. We want to get on with the implementation, you know.

Pankaj: These are all good ideas. It seems like we've all got different views of evaluating the interface design. Let's brainstorm and decide on the technique that is best for our project.

Develop a set of guidelines that can help the project team to select the most appropriate interface evaluation technique for their project.

4. The menu structure for Holiday Travel Vehicle's existing character-based system is shown here. Develop and prototype a new interface design for the system's functions using a graphical user interface. Also, develop a set of real use cases for your new interface. Assume the new system will need to include the same functions as those shown in the menus provided. Include any messages that will be produced as a user interacts with your interface (error, confirmation, status, etc.). Also, prepare a written summary that describes how your interface implements the principles of good interface design as presented in the textbook.

```
┌─────────────────────────────────────────────┐
│           Holiday Travel Vehicles             │
│                                               │
│                  Main Menu                    │
│                                               │
│              1 Sales Invoice                  │
│              2 Vehicle Inventory              │
│              3 Reports                        │
│              4 Sales Staff                    │
│                                               │
│   Type number of menu selection here:____     │
└─────────────────────────────────────────────┘
```

```
┌─────────────────────────────────────────────┐
│           Holiday Travel Vehicles             │
│                                               │
│              Sales Invoice Menu               │
│                                               │
│            1 Create Sales Invoice             │
│            2 Change Sales Invoice             │
│            3 Cancel Sales Invoice             │
│                                               │
│   Type number of menu selection here:____     │
└─────────────────────────────────────────────┘
```

```
┌─────────────────────────────────────────────┐
│           Holiday Travel Vehicles             │
│                                               │
│            Vehicle Inventory Menu             │
│                                               │
│        1 Create Vehicle Inventory Record      │
│        2 Change Vehicle Inventory Record       │
│        3 Delete Vehicle Inventory Record       │
│                                               │
│   Type number of menu selection here:____     │
└─────────────────────────────────────────────┘
```

```
┌─────────────────────────────────────────────┐
│           Holiday Travel Vehicles             │
│                                               │
│                Reports Menu                   │
│                                               │
│          1 Commission Report                  │
│          2 RV Sales by Make Report            │
│          3 Trailer Sales by Make Report       │
│          4 Dealer Options Report              │
│                                               │
│   Type number of menu selection here:____     │
└─────────────────────────────────────────────┘
```

```
┌─────────────────────────────────────────────┐
│           Holiday Travel Vehicles             │
│                                               │
│          Sales Staff Maintenance Menu         │
│                                               │
│          1 Add Salesperson Record             │
│          2 Change Salesperson Record          │
│          3 Delete Salesperson Record          │
│                                               │
│   Type number of menu selection here:____     │
└─────────────────────────────────────────────┘
```

5. One aspect of the new system under development at Holiday Travel Vehicles will be the direct entry of the sales invoice into the computer system by the salesperson as the purchase transaction is being completed. In the current system, the salesperson fills out a paper form (shown on p. 462):

Design and prototype an input screen that will permit the salesperson to enter all the necessary information for the sales invoice. The following information may be helpful in your design process. Assume that Holiday Travel Vehicles sells recreational vehicles and trailers from four different manufacturers. Each manufacturer has a fixed number of names and models of RVs and trailers. For the purposes of your prototype, use this format:

Mfg-A	Name-1 Model-X
Mfg-A	Name-1 Model-Y
Mfg-A	Name-1 Model-Z
Mfg-B	Name-1 Model-X
Mfg-B	Name-1 Model-Y
Mfg-B	Name-2 Model-X
Mfg-B	Name-2 Model-Y
Mfg-B	Name-2 Model-Z
Mfg-C	Name-1 Model-X
Mfg-C	Name-1 Model-Y
Mfg-C	Name-1 Model-Z
Mfg-C	Name-2 Model-X
Mfg-C	Name-3 Model-X

Mfg-D   Name-1 Model-X
Mfg-D   Name-2 Model-X
Mfg-D   Name-2 Model-Y

Also, assume there are ten different dealer options that could be installed on a vehicle at the customer's request. The company currently has ten salespeople on staff.

Holiday Travel Vehicles
Sales Invoice

Invoice #: _____
Invoice Date: _____

Customer Name: _____
Address: _____
City: _____
State: _____
Zip: _____
Phone: _____

New RV/TRAILER
(circle one)   Name: _____
Model: _____
Serial #:_____ Year: _____
Manufacturer: _____

Trade-in RV/TRAILER
(circle one)   Name: _____
Model: _____
Year: _____
Manufacturer: _____

Options: <u>Code</u>   <u>Description</u>   <u>Price</u>

_____   _____   _____
_____   _____   _____
_____   _____   _____
_____   _____   _____

Vehicle Base Cost: _____   _____
Trade-in Allowance: _____   (Salesperson Name)
Total Options: _____
Tax: _____
License Fee: _____   _____
Final Cost: _____   (Customer Signature)

CHAPTER 12

ARCHITECTURE

$\mathbf{A}$n important component of the design of an information system is the design of the physical architecture layer, which describes the system's hardware, software, and network environment. The *physical architecture layer design* flows primarily from the nonfunctional requirements, such as operational, performance, security, cultural, and political requirements. The deliverable from the physical architecture layer design includes the architecture and the hardware and software specification.

OBJECTIVES

- Understand the different physical architecture components.
- Understand server-based, client-based, and client–server physical architectures.
- Be familiar with distributed objects computing.
- Be able to create a network model using a deployment diagram.
- Understand how operational, performance, security, cultural, and political requirements affect the design of the physical architecture layer.
- Be familiar with how to create a hardware and software specification.

CHAPTER OUTLINE

Introduction
Elements of the Physical Architecture Layer
    Architectural Components
    Server-Based Architectures
    Client-Based Architectures
    Client–Server Architectures
    Client–Server Tiers
    Distributed Objects Computing
    Selecting a Physical Architecture
Infrastructure Design
    Deployment Diagram
    Network Model

Nonfunctional Requirements and Physical
    Architecture Layer Design
    Operational Requirements
    Performance Requirements
    Security Requirements
    Cultural and Political Requirements
    Synopsis
Hardware and Software Specification
Applying the Concepts at
   CD Selections
Summary

INTRODUCTION

In today's environment, most information systems are spread across two or more computers. A Web-based system, for example, will run in the browser on a desktop computer but will interact with the Web server (and possibly other computers) over the Internet. A system that operates completely inside a company's network may have a Visual Basic program installed on one computer but interact with a database server elsewhere on the network.

Therefore, an important step of design is the creation of the physical architecture layer design, the plan for how the system will be distributed across the computers and what hardware and software will be used for each computer (e.g., Windows, Linux).

Most systems are built to use the existing hardware and software in the organization, so often the current architecture and hardware and software infrastructure restricts the choice. Other factors, such as corporate standards, existing site-licensing agreements, and product–vendor relationships also can mandate what architecture, hardware, and software the project team must design. However, many organizations now have a variety of infrastructures available or are openly looking for pilot projects to test new architectures, hardware, and software, which enables a project team to select an architecture on the basis of other important factors.

Designing a physical architecture layer can be quite difficult; therefore, many organizations hire expert consultants or assign very experienced analysts to the task.[1] In this chapter, we examine the key factors in physical architecture layer design, but it is important to remember that it takes lots of experience to do it well. The nonfunctional requirements developed during analysis (see Chapter 5) play a key role in physical architecture layer design. These requirements are reexamined and refined into more detailed requirements that influence the system's architecture. In this chapter, we first explain how the designers think about application architectures, and we describe the three primary architectures: server-based, client-based, and client–server. Next, we look at using UML's deployment diagram as a way to model the physical architecture layer. Then we examine how the very general nonfunctional requirements from analysis are refined into more specific requirements and the implications that they have for physical architecture layer design. Finally, we consider how the requirements and architecture can be used to develop the hardware and software specifications that define exactly what hardware and other software (e.g., database systems) are needed to support the information system being developed.

ELEMENTS OF THE PHYSICAL ARCHITECTURE LAYER

The objective of designing the physical architecture layer is to determine what parts of the application software will be assigned to what hardware. In this section we first discuss the major architectural elements to understand how the software can be divided into different parts. Then we briefly discuss the major types of hardware onto which the software can be placed. Although there are numerous ways in which the software components can be placed on the hardware components, there are three principal application architectures in use today: *server-based architectures*, *client-based architectures* and *client–server architectures*. The most common architecture is the client–server architecture, so we focus on it.

Architectural Components

The major *architectural components* of any system are the software and the hardware. The major software components of the system being developed have to be identified and then allocated to the various hardware components on which the system will operate. Each of these components can be combined in a variety of different ways.

All software systems can be divided into four basic functions. The first is *data storage* (associated with the object persistence located on the data management layer—see Chapter 10).

[1] For more information on physical architecture layer design, see Stephen D. Burd, *Systems Architecture,* 4th ed. (Boston: Course Technology, 2003); Irv Englander, *The Architecture of Computer Hardware and Systems Software: An Information Technology Approach,* 3rd ed. (New York: Wiley, 2003); and William Stallings, *Computer Organization & Architecture: Designing for Performance* (Upper Saddle River, NJ: Prentice Hall, 2003).

Most application programs require data to be stored and retrieved, whether the information is a small file such as a memo produced by a word processor or a large database that stores an organization's accounting records. These are the data documented in the structural model (CRC cards and class diagrams). The second function is *data access logic* (associated with the data access and manipulation classes located on the data management layer—see Chapter 10), the processing required to access data, which often means database queries in *SQL* (*structured query language*). The third function is the *application logic* (located on the problem domain layer—see Chapters 5–9), which can be simple or complex, depending on the application. This is the logic documented in the functional (activity diagrams and use cases) and behavioral models (sequence, communication, and behavioral state machines). The fourth function is the *presentation logic* (located on the human–computer interaction layer—see Chapter 11), the presentation of information to the user, and the acceptance of the user's commands (the user interface). These four functions (data storage, data access logic, application logic, and presentation logic) are the basic building blocks of any application.

The three primary hardware components of a system are *client computers, servers,* and the *network* that connects them. Client computers are the input/output devices employed by the user and are usually desktop or laptop computers, but they can also be handheld devices, cell phones, special-purpose terminals, and so on. Servers are typically larger computers that are used to store software and hardware that can be accessed by anyone who has permission. Servers can come in several types: *mainframes* (very large, powerful computers usually costing millions of dollars), *minicomputers* (large computers costing hundreds of thousands of dollars) and *microcomputers* (small desktop computers like the ones we all use to those costing $50,000 or more). The network that connects the computers can vary in speed from a slow cell phone or modem connection that must be dialed, to medium-speed always-on frame relay networks, to fast always-on broadband connections such as cable modem, DSL, or T1 circuits, to high-speed always-on Ethernet, T3, or ATM circuits.[2]

Server-Based Architectures

The very first computing architectures were server-based architectures, with the server (usually a central mainframe computer) performing all four application functions. The clients (usually terminals) enabled users to send and receive messages to and from the server computer. The clients merely captured keystrokes and sent them to the server for processing and accepted instructions from the server on what to display (see Figure 12-1).

Client /(terminal)

Server Host
(mainframe computer)

Presentation logic
Application logic
Data access logic
Data storage

FIGURE 12-1
Server-Based
Architecture

---

[2] For more information on networks, see Alan Dennis, *Networking in the Internet Age* (New York: Wiley, 2002).

Client
(microcomputer)

Server
(microcomputer)

FIGURE 12-2
Client-Based
Architectures

Presentation logic
Application logic
Data access logic

Data storage

This very simple architecture often works very well. Application software is developed and stored on one computer, and all data are on the same computer. There is one point of control, because all messages flow through the one central server. The fundamental problem with server-based networks is that the server must process all messages. As the demands for more and more applications grow, many server computers become overloaded and unable to quickly process all the users' demands. Response time becomes slower, and network managers are required to spend increasingly more money to upgrade the server computer. Unfortunately, upgrades come in large increments and are expensive; it is difficult to upgrade "a little."

Client-Based Architectures

With client-based architectures, the clients are personal computers on a local area network (LAN), and the server computer is a server on the same network. The application software on the client computers is responsible for the presentation logic, the application logic, and the data access logic; the server simply stores the data (see Figure 12-2).

This simple architecture also often works well. However, as the demands for more and more network applications grow, the network circuits can become overloaded. The fundamental problem in client-based networks is that all data on the server must travel to the client for processing. For example, suppose the user wishes to display a list of all employees with company life insurance. All the data in the database must travel from the server where the database is stored over the network to the client, which then examines each record to see if it matches the data requested by the user. This can overload both the network and the power of the client computers.

Client–Server Architectures

Most organizations today are moving to client–server architectures, which attempt to balance the processing between the client and the server by having both do some of the application functions. In these architectures, the client is responsible for the presentation logic, whereas the server is responsible for the data access logic and data storage. The application logic may reside on either the client or the server or be split between both (see Figure 12-3). The client shown in Figure 12-3 can be referred to as a *thick*, or *fat, client* if it contains the bulk of application logic. A current practice is to create client–server architectures using *thin clients* because there is less overhead and maintenance in supporting thin-client applications. For example, many Web-based systems are designed with the Web browser performing presentation, with only minimal application logic using programming languages like Java and the Web server having the application logic, data access logic, and data storage.

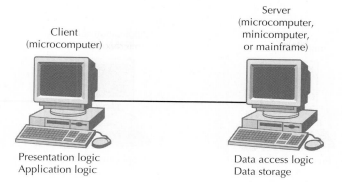

FIGURE 12-3
Client–Server
Architecture

Client–server architectures have four important benefits. First and foremost, they are *scalable.* That means it is easy to increase or decrease the storage and processing capabilities of the servers. If one server becomes overloaded, you simply add another server so that many servers are used to perform the application logic, data access logic, or data storage. The cost to upgrade is much more gradual, and you can upgrade in smaller steps rather than spending hundreds of thousands to upgrade a mainframe server.

Second, client–server architectures can support many different types of clients and servers. It is possible to connect computers that use different operating systems so that users can choose which type of computer they prefer (e.g., combining both Windows computers and Apple Macintoshes on the same network). We are not locked into one vendor, as is often the case with server-based networks. *Middleware* is a type of system software designed to translate between different vendors' software. Middleware is installed on both the client computer and the server computer. The client software communicates with the middleware, which can reformat the message into a standard language that can be understood by the middleware assisting the server software. We describe two approaches to middleware in the distributed objects computing section.

Third, for thin-client server architectures that use Internet standards, it is simple to clearly separate the presentation logic, the application logic, and the data access logic and design so each is somewhat independent. For example, the presentation logic can be designed in HTML or XML to specify how the page will appear on the screen (e.g., the colors, fonts, order of items, specific words used, command buttons, the type of selection lists, etc.; see Chapter 11). Simple program statements are used to link parts of the interface to specific application logic modules that perform various functions. These HTML or XML files defining the interface can be changed without affecting the application logic. Likewise, it is possible to change the application logic without changing the presentation logic or the data, which are stored in databases and accessed using SQL commands.

Finally, because no single server computer supports all the applications, the network is generally more reliable. There is no central point of failure that will halt the entire network if it fails, as there is in server-based computing. If any one server fails in a client–server environment, the network can continue to function using all the other servers (but, of course, any applications that require the failed server will not work).

Client–server architectures also have some critical limitations, the most important of which is its complexity. All applications in client–server computing have two parts, the software on the client and the software on the server. Writing this software is more complicated than writing the traditional all-in-one software used in server-based architectures. Updating the network with a new version of the software is more complicated, too. In server-based architectures, there is one place in which application software is stored; to update the

Client
(microcomputer)

Application server
(microcomputer)

Database server
(microcomputer,
minicomputer,
or mainframe)

Presentation logic

Application logic

Data access logic
Data storage

FIGURE 12-4
A Three-Tiered
Client–Server
Architecture

software, we simply replace it there. With client–server architectures, we must update all clients and all servers.

Much of the debate about server-based versus client–server architectures has centered on cost. One of the great claims of server-based networks in the 1980s was that they provided economies of scale. Manufacturers of big mainframes claimed it was cheaper to provide computer services on one big mainframe than on a set of smaller computers. The personal computer revolution changed this. Since the 1980s, the cost of personal computers has continued to drop, whereas their performance has increased significantly. Today, personal computer hardware is more than 1,000 times cheaper than mainframe hardware for the same amount of computing power.

With cost differences like these, it is easy to see why there has been a sudden rush to microcomputer-based client–server computing. The problem with these cost comparisons is that they ignore the *total cost of ownership,* which includes factors other than obvious hardware and software costs. For example, many cost comparisons overlook the increased complexity associated with developing application software for client–server networks. Most experts believe that it costs four to five times more to develop and maintain application software for client–server computing than it does for server-based computing.

Client–Server Tiers

There are many ways in which the application logic can be partitioned between the client and the server. The example in Figure 12-3 is one of the most common. In this case, the server is responsible for the data, and the client is responsible for the application and presentation. This is called a *two-tiered architecture* because it uses only two sets of computers, clients, and servers.

A *three-tiered architecture* uses three sets of computers (see Figure 12-4). In this case, the software on the client computer is responsible for presentation logic, an application server (or servers) is responsible for the application logic, and a separate database server (or servers) is responsible for the data access logic and data storage.

An *n-tiered architecture* uses more than three sets of computers. In this case, the client is responsible for presentation, database servers are responsible for the data access logic and data storage, and the application logic is spread across two or more different sets of servers. Figure 12-5 shows an example of an *n-tiered* architecture of a software product called Consensus @nyWARE®.[3] Consensus @nyWARE® has four major components. The first is the

---

[3] Consensus @nyWARE® was originally developed by Alan Dennis while at the University of Georgia.

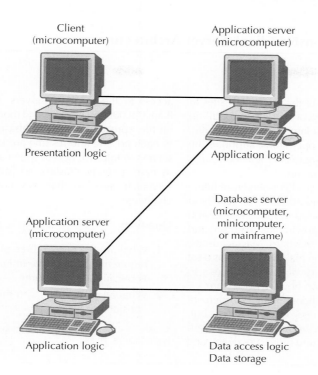

Client
(microcomputer)

Application server
(microcomputer)

Presentation logic

Application logic

Database server
(microcomputer,
minicomputer,
or mainframe)

Application server
(microcomputer)

Application logic

Data access logic
Data storage

FIGURE 12-5
A Four-Tiered
Client–Server
Architecture

Web browser on the client computer employed by a user to access the system and enter commands (presentation logic). The second component is a Web server that responds to the user's requests, either by providing (HTML) pages and graphics (application logic) or by sending the request to the third component (a set of twenty-eight programs written in the C programming language) on another application server that performs various functions (application logic). The fourth component is a database server that stores all the data (data access logic and data storage). Each of these four components is separate, making it easy to spread the different components on different servers and to partition the application logic on two different servers.

The primary advantage of an n-tiered client–server architecture compared with a two-tiered architecture (or a three-tiered with a two-tiered) is that it separates out the processing that occurs to better balance the load on the different servers; it is more scalable. In Figure 12-5, we have three separate servers, a configuration that provides more power than if we had used a two-tiered architecture with only one server. If we discover that the application server is too heavily loaded, we can simply replace it with a more powerful server or just put in several more application servers. Conversely, if we discover the database server is underused, we could store data from another application on it.

There are two primary disadvantages to an n-tiered architecture compared with a two-tiered architecture (or a three-tiered with a two-tiered). First, the configuration puts a greater load on the network. If you compare Figures 12-3, 12-4, and 12-5, you will see that the n-tiered model requires more communication among the servers; it generates more network traffic, so you need a higher-capacity network. Second, it is much more difficult to program and test software in n-tiered architectures than in two-tiered architectures because more devices have to communicate to complete a user's transaction.

CONCEPTS

IN ACTION

12-A A Monster Client–Server Architecture

*E*very spring, Monster.com, one of the largest job sites in the United States with an average of more than 3 million visitors per month, experiences a large increase in traffic. Aaron Braham, vice president of operations, attributes the spike to college students who increase their job-search activities as they approach graduation.

Monster.com uses a three-tier client–server architecture that has 150 Web servers and 30 database servers in its main site in Indianapolis. It plans to move that to 400 over the next year by gradually growing the main site and adding a new site with servers in Maynard, Massachusetts, just in time for the spring rush. The main Web site has a set of load-balancing devices that forward Web requests to the different servers, depending on how busy they are.

Braham says the major challenge is that 90 percent of the traffic is not simple requests for Web pages but, rather, search requests (e.g., the network jobs are available in New Mexico), which require more processing and

access to the database servers. Monster.com has more than 350,000 job postings and more than 3 million resumes on file, spread across its database servers. Several copies of each posting and résumé are kept on several database servers to improve access speed and provide redundancy in case a server crashes, so just keeping the database servers in sync so that they contain correct data is a challenge.

Questions

1. What are the two or three primary nonfunctional requirements that have influenced Monster.com's application architecture?
2. What alternatives do you think Monster.com considered?

Source: "Resume Influx Tests Mettle of Job Sites' Scalability," *Internet Week* (May 29, 2000).

Distributed Objects Computing

From an object-oriented perspective, *distributed objects computing (DOC)* is the next version of client–server computing. DOC represents a software layer that goes between the clients and servers; hence, it is known as middleware. Middleware supports the interaction between objects in a distributed computing environment. Furthermore, middleware treats the actual physical network architecture in a transparent manner. This allows the developer to focus on the development of the application and ignore the peculiarities of a specific distributed environment.

From a practical perspective, DOC allows the developer to simply concentrate on the users, objects, and methods of an application instead of worrying about which server contains which set of objects. The client object simply requests the "network" to locate and execute the server object's method. This basically allows the physical location of the server object to become irrelevant from the client object's perspective. Therefore, because servers are no longer being addressed directly on the client, servers can be added and subtracted from the network without having to update the client code. Only the middleware must be made aware of the new addresses of the server objects. This can greatly reduce the maintenance in a client–server environment. However, because middleware adds an additional layer between the clients and servers, it can reduce the efficiency of the application.

Currently, there are three competing approaches to support DOC: Object Management Group's, Sun's, and Microsoft's. The Object Management Group supports DOC via its *Common Object Request Broker Architecture (CORBA)* standard. Sun supports DOC via its *Enterprise JavaBeans (EJB)* and its *Java 2 Enterprise Edition (J2EE)*. At this point in time, the approaches supported by OMG and Sun seem to be evolving toward one another. Even though Microsoft is involved in OMG, Microsoft has its own competing approach to support DOC: *.net*. At this point in time, all three are competitors to support DOC; it is not clear which approach will win in the marketplace.

	Server-Based	Client-Based	Client–Server
Cost of infrastructure	Very high	Medium	Low
Cost of development	Medium	Low	High
Ease of development	Low	High	Low–medium
Interface capabilities	Low	High	High
Control and security	High	Low	Medium
Scalability	Low	Medium	High

FIGURE 12-6
Characteristics
of Computing
Architectures

Selecting a Physical Architecture

Most systems are built to use the existing infrastructure in the organization, so often the current infrastructure restricts the choice of architecture. For example, if the new system will be built for a mainframe-centric organization, a server-based architecture may be the best option. Other factors such as corporate standards, existing licensing agreements, and product/vendor relationships can also mandate what architecture the project team needs to design. However, many organizations now have a variety of infrastructures available or are openly looking for pilot projects to test new architectures and infrastructures, enabling a project team to select an architecture based on other important factors.

Each of the computing architectures just discussed has its strengths and weaknesses, and no architecture is inherently better than the others. Thus, it is important to understand the strengths and weaknesses of each computing architecture and when to use each. Figure 12-6 presents a summary of the important characteristics of each architecture.

Cost of Infrastructure   One of the strongest driving forces to client–server architectures is cost of infrastructure (the hardware, software, and networks that will support the application system). Simply put, personal computers are more than 1,000 times cheaper than mainframes for the same amount of computing power. The personal computers on our desks today have more processing power, memory, and hard disk space than the typical mainframe of the past, and the cost of the personal computers is a fraction of the cost of the mainframe.

Therefore, the cost of client–server architectures is low compared to server-based architectures that rely on mainframes. Client–server architectures also tend to be cheaper than client-based architectures because they place less of a load on networks and thus require less network capacity.

Cost of Development   The cost of developing systems is an important factor when considering the financial benefits of client–server architectures. Developing application software for client–server computing is extremely complex, and most experts believe that it costs four to five times more to develop and maintain application software for client–server computing than it does for server-based computing. Developing application software for client-based architectures is usually cheaper still, because there are many GUI development tools for simple stand-alone computers that communicate with database servers (e.g., Visual Basic, Access).

The cost differential may change as more companies gain experience with client–server applications; new client–server products are developed and refined; and client–server standards mature. However, given the inherent complexity of client–server software and the need to coordinate the interactions of software on different computers, there is likely to remain a cost difference.

Ease of Development   In most organizations today, there is a huge backlog of mainframe applications, systems that have been approved but that lack the staff to implement them. This backlog signals the difficulty in developing server-based systems. The tools for mainframe-based systems often are not user friendly and require highly specialized skills (e.g., COBOL/CICS)—skills that new graduates often don't have and aren't interested in acquiring. In contrast, client-based and client–server architectures can rely on *graphical user interface (GUI)* development tools that can be intuitive and easy to use. The development of applications for these architectures can be fast and painless. Unfortunately, the applications for client–server can be very complex because they must be built for several layers of hardware (e.g., database servers, Web servers, client workstations) that need to communicate effectively with each other. Project teams often underestimate the effort involved in creating secure, efficient client–server applications.

Interface Capabilities   Typically, server-based applications contain plain, character-based interfaces. For example, think about airline reservation systems such as SABRE, which can be quite difficult to use unless the operator is well trained on the commands and hundreds of codes that are used to navigate through the system. Today, most users of systems expect a GUI or a Web-based interface that they can operate using a mouse and graphical objects (e.g., pushbuttons, drop-down lists, icons, and so on). GUI and Web development tools typically are created to support client-based or client–server applications; rarely can server-based environments support these types of applications.

Control and Security   The server-based architecture was originally developed to control and secure data, and it is much easier to administer because all the data are stored in a single location. In contrast, client–server computing requires a high degree of coordination among many components, and the chance for security holes or control problems is much more likely. Also, the hardware and software used in client–server architecture are still maturing in terms of security. When an organization has a system that absolutely must be secure (e.g., an application used by the Department of Defense), then the project team may be more comfortable with the server-based alternative on highly secure and control-oriented mainframe computers.

Scalability   Scalability refers to the ability to increase or decrease the capacity of the computing infrastructure in response to changing capacity needs. The most scalable architecture is client–server computing because servers can be added to (or removed from) the architecture when processing needs change. Also, the types of hardware that are used in client–server situations (e.g., minicomputers) typically can be upgraded at a pace that most closely matches the growth of the application. In contrast, server-based architectures rely primarily on mainframe hardware that needs to be scaled up in large, expensive increments, and client-based architectures have ceilings above which the application cannot grow because increases in use and data can result in increased network traffic to the extent that performance is unacceptable.

YOUR TURN   **12-1 Course Registration System**

Think about the course registration system in your university. What physical architecture does it use? If you had to create a new system today, would you use the current architecture or change to a different one? Describe the criteria that you would consider when making your decision.

INFRASTRUCTURE DESIGN

In most cases, a system is built for an organization that has a hardware, software, and communications infrastructure already in place. Thus, project teams are usually more concerned with how an existing infrastructure needs to be changed or improved to support the requirements that were identified during analysis, as opposed to how to design and build an infrastructure from scratch. Further, the coordination of infrastructure components is very complex, and it requires highly skilled technical professionals. As a project team, it is best to leave changes to the computing infrastructure to the infrastructure analysts. In this section we summarize key elements of infrastructure design to create a basic understanding of what it includes. We describe UML's deployment diagram and the network model.

Deployment Diagram

Deployment diagrams are used to represent the relationships between the hardware components used in the physical infrastructure of an information system. For example, when designing a distributed information system that will use a wide area network, a deployment diagram can be used to show the communication relationships among the different nodes in the network. They also can be used to represent the software components and how they are deployed over the physical architecture or infrastructure of an information system. In this case, a deployment diagram represents the environment for the execution of the software.

The elements of a deployment diagram include nodes, artifacts, and communication paths (see Figure 12-7). There are also other elements that can be included in this diagram. In our case, we include only the three primary elements and the element that portrays an artifact being deployed onto a node.

A *node* represents any piece of hardware that needs to be included in the model of the physical architecture layer design. For example, nodes typically include client computers, servers, separate networks, or individual network devices. Typically, a node is labeled with its name and, possibly, with a stereotype. The stereotype is modeled as a text item surrounded by "<< >>" symbols. The stereotype represents the type of node being represented on the diagram. For example, typical stereotypes include device, mobile device, database server, Web server, and application server. Finally, there are times that the notation of a node should be extended to better communicate the design of the physical architecture layer. Figure 12-8 includes a set of typical network node symbols that can be used instead of the standard notation.

An *artifact* represents a piece of the information system that is to be deployed onto the physical architecture (see Figure 12-7). Typically, an artifact represents a software component, a subsystem, database table, an entire database, or a layer (data management, human–computer interaction, or problem domain). Artifacts, like nodes, can be labeled with both a name and a stereotype. Stereotypes for artifacts include source file, database table, and executable file.

A *communication path* represents a communication link between the nodes of the physical architecture (see Figure 12-7). Communication paths are stereotyped based on the type of communication link they represent (e.g., LAN, Internet, serial, parallel, or USB) or the protocol that is being supported by the link (e.g., TCP/IP).

Figure 12-9 portrays three different versions of a deployment diagram. Version a uses only the basic standard notation. Version b introduces the idea of deploying an artifact onto a node (see Figure 12-7). In this case, the artifacts represent the different layers of the appointment system described in the earlier chapters. Finally, version c uses the extended notation to represent the same architecture. As you can see, all three versions have their strengths and weaknesses. When comparing version a and version b, the user

A node:

- Is a computational resource, e.g., a client computer, server, separate network, or individual network device.
- Is labeled by its name.
- May contain a stereotype to specifically label the type of node being represented, e.g., device, client workstation, application server, mobile device, etc.

<<stereotype>>
Node Name

An artifact:

- Is a specification of a piece of software or database, e.g., a database or a table or view of a database, a software component or layer.
- Is labeled by its name.
- May contain a stereotype to specifically label the type of artifact, e.g., source file, database table, executable file, etc.

<<stereotype>>
Artifact Name

A node with a deployed artifact:

- Portrays an artifact being placed on a physical node.
- Supports modeling the distribution of the software over a network.

<<stereotype>>
Node Name

<<stereotype>>
Artifact Name

A communication path:

- Represents an association between two nodes.
- Allows nodes to exchange messages.
- May contain a stereotype to specifically label the type of communication path being represented, e.g., lan, Internet, serial, parallel, etc.

<<stereotype>>

FIGURE 12-7   Development Diagram Syntax

can glean more information from version b with little additional effort. However, when comparing version a to version c, the extended node notation enables the user to quickly understand the hardware requirements of the architecture. Finally, when comparing version b to version c, version b supports the software distribution explicitly but forces the

FIGURE 12-8   Extended Node Syntax for Development Diagram

FIGURE 12-9   Three Versions of Appointment System Deployment Diagram

user to rely on the stereotypes to understand the required hardware, whereas version c omits the software distribution information entirely. We recommend that you use the combination of symbols to best portray the physical architecture to the user community.

Network Model

The *network model* is a diagram that shows the major components of the information system (e.g., servers, communication lines, networks) and their geographic locations throughout the organization. There is no one way to depict a network model, and in our experience analysts create their own standards and symbols, using presentation applications (e.g., Microsoft Power Point) or diagramming tools (e.g., Visio). In this text, we use UML's deployment diagram.

The purpose of the network model is twofold: to convey the complexity of the system and to show how the system's software components will fit together. The diagram also helps the project team develop the hardware and software specification that is described later in this chapter.

The components of the network model are the various clients (e.g., personal computers, kiosks), servers (e.g., database, network, communications, printer), network equipment (e.g., wires, dial-up connections, satellite links), and external systems or networks (e.g., Internet service providers) that support the application. *Locations* are the geographic sites related to these components. For example, if a company created an application for users at four of its plants in Canada and eight plants in the United States and it used one external system to provide Internet service, the network model to depict this would contain twelve locations (4 + 8 = 12).

Creating the network model is a top-down exercise whereby we first graphically depict all the locations where the application will reside. This is accomplished by placing symbols that represent the locations for the components on a diagram and then connecting them with lines that are labeled with the approximate amount of data or types of network circuits between the separated components.

Companies seldom build networks to connect distant locations by buying land and laying cable (or sending up their own satellites). Instead, they usually lease services provided by large telecommunications firms such as AT&T, Sprint, and WorldCom/MCI. Figure 12-10 shows a typical network. The "clouds" in the diagram represent the networks at different locations (e.g., Toronto, Atlanta). The lines represent network connections between specific points (e.g., Toronto to Brampton). In other cases, a company may lease connections from many points to many others, and rather than trying to show all the connections, a separate cloud may be drawn to represent this many-to-many type of connection (e.g., the cloud in the center of this Figure 12-10 represents a network of many-to-many connections provided by a telecom firm like Verizon).

This high-level diagram has several purposes. First, it shows the locations of the components needed to support the application; therefore, the project team can get a good understanding of the geographic scope of the new system and how complex and costly the communications infrastructure will be to support. (For example, an application that supports one site will probably have less communications costs as compared to a more-complex application that will be shared all over the world.) The diagram also indicates the external components of the system (e.g., customer systems, supplier systems), which may impact security or global needs (discussed later in this chapter).

The second step to the network model is to create low-level network diagrams for each of the locations shown on the top-level diagram. First, hardware is drawn on the model in a way that depicts how the hardware for the new system will be placed throughout the location. It usually helps to use symbols that resemble the hardware that will be used.

CONCEPTS

IN ACTION

12-B Nortel's Network

Nortel, the giant Canadian manufacturer of network switches, recently implemented a network using its own equipment. Prior to implementing it, Nortel had used a set of more than 100 circuits that was proving to be more and more expensive and unwieldy to manage. Nortel started by identifying the 20 percent of its sites that accounted for 80 percent of the network traffic. These were the first to move to the new network. By the time the switchover is complete, more than 100 sites will be connected.

There are two primary parts to the new network (see Figure 12-7). The largest part uses a public network provided by WorldCom/MCI. Nortel also operates a private network in Ontario. Both parts of the network use 44 Mbps circuits. Nortel is planning to move to 155 Mbps circuits when costs drop.

Question

What are the advantages and disadvantages of Nortel's phased implementation of the new network?

Source: "A Case for ATM," *Network World* (June 2, 1997). Tim Greene, Vol. 14, No. 22, pp. 1-2.

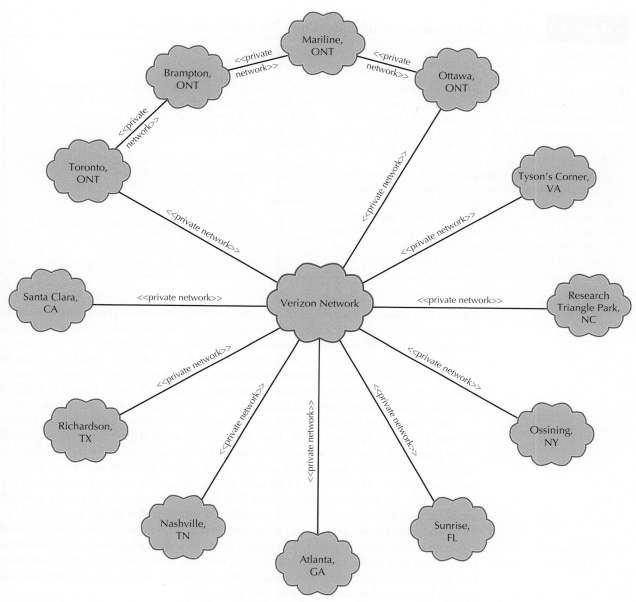

FIGURE 12-10   Deployment Diagram Representation of a Top-Level Network Model

The amount of detail to include on the network model depends on the needs of the project. Some low-level network models contain text descriptions below each of the hardware components that describe in detail the proposed hardware configurations and processing needs; others include only the number of users that are associated with the different parts of the diagram.

Next, lines are drawn connecting the components that will be physically attached to each other. In terms of software, some network models list the required software for each network model component right on the diagram, whereas other times, the software is described in a memo attached to the network model and stored in the project binder.

12-C The Department of Education's Financial Aid Network

Each year, the U.S. Department of Education (DOE) processes more than 10 million student loans, such as Pell grants, guaranteed student loans, and college work-study programs. The DOE has more than 5,000 employees that need access to more than 400 gigabytes of student data.

The financial aid network, located in one Washington, D.C., building, has two major parts (i.e., backbones) that each use high-speed 155 Mbps circuits to connect four LANs and three high-performance database servers (see Figure 12-11). These two backbones are connected into another network backbone that connects the financial aid building using 100 Mbps circuits to five other DOE buildings in Washington with similar networks. The two backbones are also connected to the ten regional DOE offices across the United States and to other U.S. government agencies via a set of lower-speed 1.5 Mbps connections.

Source: Network Computing (May 15, 1996). Maureen Zaprylok, Financial Aid: Department of Education's EdNet, p. 84.

Question

Why does each part of the network have different data rates?

Figure 12-11 shows a deployment diagram that portrays two levels of detail of a low-level network model. Notice, we use both the standard and extended node notation in this figure. Furthermore, in this case, we have included a package (see Chapter 8) to represent a set of connections to the router in the MFA building. By including a package, we show only the detail necessary. The extended notation in many cases aids the user in understanding the topology of the physical architecture layer much better than the standard notation. As such, we recommend using the symbols that get the message across best.

Our experiences have shown that most project teams create a memo for the project files that provides additional detail about the network model. This information is helpful to the people who are responsible for creating the hardware and software specification (described later in this chapter) and who will work more extensively with the infrastructure development. This memo can include special issues that impact communications, requirements for hardware and software that may not be obvious from the network model, or specific hardware or software vendors or products that should be acquired.

Again, the primary purpose of the network model diagram is to present the proposed infrastructure for the new system. The project team can use the diagrams to understand the scope of the system, the complexity of its structure, any important communication issues that may affect development and implementation, and the actual components that will need to be acquired or integrated into the environment.

12-2 Create a Network Model

Suppose that you have just moved into a fraternity or sorority house with twenty other people. The house has a laser printer and a low-cost scanner that its inhabitants are free to use. You have decided to network the house so that everyone can share the printer and scanner and have dial-up access to the university network. Create a high-level network model that describes the locations that will be involved in your work. Next, create a low-level diagram for the house itself.

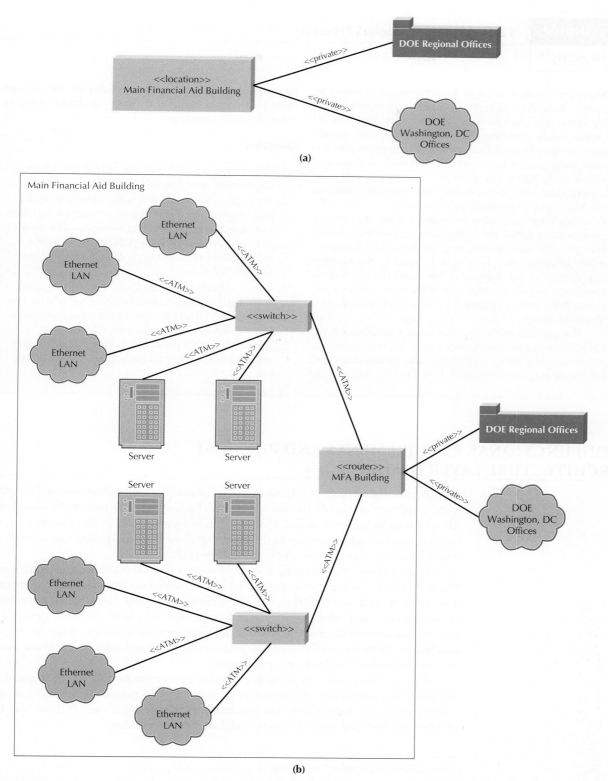

FIGURE 12-11 Deployment Diagram Representation of a Low-Level Network Model

12-D Managing Global Projects

Shamrock Foods is a major food distributor centered in Tralee, Ireland. Originally a dairy cooperative, they branched into various food components [dried milk, cheese solids, flavorings (or flavourings, as they would spell it)]. They have had substantial growth in the past ten years, most coming by way of acquisition of existing companies or facilities. For example, Iowa Soybean in the United States in now a subsidiary of Shamrock Foods, as is a large dairy cooperative in Wisconsin.

They have processing facilities in more than twelve countries and distribution sales in more than thirty countries. With the rapid growth by acquisition, they have generally adopted a hands-off policy, keeping the systems separated and not integrated into unified ERP system. Thus, each acquired company is still largely autonomous, although it reports to Shamrock Foods and is managed by Shamrock Foods.

This separation concept has been a problem for the CFO of Shamrock Foods, Conor Lynch. The board of directors would like some aggregated data for direction and analysis of acquired businesses. Conor has the reports from the various subsidiaries but has to have his staff convert them to a consistent basic currency (generally either to euros or American dollars).

Questions

1. When should a multinational or multisite business consolidate data systems?

2. There are costs associated with consolidating data systems—the various acquired companies had their own functioning accounting systems—with a variety of hardware and software systems. What justification should Conor use to push for a consolidated, unified ERP system?

3. Conor at times has to deal with incomplete and incompatible data. Inventory systems might be first in, first out (FIFO) for some of the subsidiaries and last in, first out (LIFO) or another accounting method for other subsidiaries. How might a multinational CFO deal with incomplete and incompatible data?

NONFUNCTIONAL REQUIREMENTS AND PHYSICAL ARCHITECTURE LAYER DESIGN[4]

The design of the physical architecture layer specifies the overall architecture and the placement of software and hardware that will be used. Each of the architectures discussed before has its strengths and weaknesses. Most organizations are trying to move to client–server architectures for cost reasons, so in the event that there is no compelling reason to choose one architecture over another, cost usually suggests client–server.

Creating a physical architecture layer design begins with the nonfunctional requirements. The first step is to refine the nonfunctional requirements into more detailed requirements that are then used to help select the architecture to be used (sever-based, client-based, or client–server) and what software components will be placed on each device. In a client–server architecture, one also has to decide whether to use a two-tier, three-tier, or n-tier architecture. Then the nonfunctional requirements and the architecture design are used to develop the hardware and software specification.

There are four primary types of nonfunctional requirements that can be important in designing the architecture: operational requirements, performance requirements, security requirements, and cultural/political requirements. We describe each in turn and then explain how they may affect the physical architecture layer design.

---

[4] Nonfunctional requirements were also discussed in Chapters 4, 10, and 11.

Type of Requirement	Definition	Examples
Technical Environment Requirements	Special hardware, software, and network requirements imposed by business requirements	• The system will work over the Web environment with Internet Explorer. • All office locations will have an always-on network connection to enable real-time database updates. • A version of the system will be provided for customers connecting over the Internet via a small screen PDA.
System Integration Requirements	The extent to which the system will operate with other systems	• The system must be able to import and export Excel spreadsheets. • The system will read and write to the main inventory database in the inventory system.
Portability Requirements	The extent to which the system will need to operate in other environments	• The system must be able to work with different operating systems (e.g., Linux and Windows Vista). • The system may need to operate with handheld devices such as a Palm.
Maintainability Requirements	Expected business changes to which the system should be able to adapt	• The system will be able to support more than one manufacturing plant with six months advance notice. • New versions of the system will be released every six months.

FIGURE 12-12   Operational Requirements

Operational Requirements

Operational requirements specify the operating environment(s) in which the system must perform and how those may change over time. This usually refers to operating systems, system software, and information systems with which the system must interact, but it will on occasion also include the physical environment if the environment is important to the application (e.g., it's located on a noisy factory floor, so no audible alerts can be heard). Figure 12-12 summarizes four key operational requirement areas and provides some examples of each.

Technical Environment Requirements   *Technical environment requirements* specify the type of hardware and software system on which the system will work. These requirements usually focus on the operating system software (e.g., Windows, Linux), database system software (e.g., Oracle), and other system software (e.g., Internet Explorer). Occasionally, specific hardware requirements will impose important limitations on the system, such as the need to operate on a PDA or cell phone with a very small display.

System Integration Requirements   *System integration requirements* are those that require the system to operate with other information systems, either inside or outside the company. These typically specify interfaces through which data will be exchanged with other systems.

Portability Requirements   Information systems never remain constant. Business needs change and operating technologies change, so the information systems that support them and run on them must change, too. *Portability requirements* define how the technical operating environments may change over time and how the system must respond (e.g., the

12-E Citywide Broadband

*S*eattle officials might finally be ready to start crunching numbers and nailing down specifics of a long-mulled, but slow-moving, proposal to build a citywide broadband network. But they might yet need to build a case before the city council to justify such a network. The idea has been studied by the city council, a task force, and the mayor's information technology office since 2004. Having estimated last year that it would cost $500 million to build and connect all Seattle to a fiber broadband network, the city is now looking to invite private companies to do that work—perhaps with taxpayer help.

The mayor is asking the city council to free up $185,000 he says his information technology office will use to invite companies to bid on the sweeping and still-undefined job.

Questions

1. How much money (if any) will taxpayers contribute in subsidies or access to public facilities?
2. Should the city start with a small-scale pilot project, as it did with Wi-Fi, before deciding about a citywide wireless network?
3. Is such a network even a feasible, cost-effective idea?

system may run currently on Windows XP, whereas in the future it may be required to deploy the system on Linux). Portability requirements also refer to potential changes in business requirements that will drive technical environment changes. For example, in the future users may want to access a Web site from their cell phones.

Maintainability Requirements *Maintainability requirements* specify the business requirement changes that can be anticipated. Not all changes are predictable, but some are. For example, suppose a small company has only one manufacturing plant but is anticipating the construction of a second plant in the next five years. All information systems must be written to make it easy to track each plant separately, whether for personnel, budgeting, or inventory systems. The maintainability requirements attempt to anticipate future requirements so that the systems designed today will be easy to maintain if and when those future requirements appear. Maintainability requirements may also define the update cycle for the system, such as the frequency with which new versions will be released.

Performance Requirements

Performance requirements focus on performance issues, such as response time, capacity, and reliability. Figure 12-13 summarizes three key performance requirement areas and provides some examples.

Speed Requirements *Speed requirements* are exactly what they say: how fast should the system operate. First and foremost is the *response time* of the system: how long it takes the system to respond to a user request. Although everyone would prefer low response times, with the system responding immediately to each user request, this is not practical. We could design such a system, but it would be expensive. Most users understand that certain parts of a system will respond quickly, whereas others are slower. Those actions that are performed locally on the user's computer must be almost immediate (e.g., typing, dragging and dropping), whereas others that require communicating across a network can have higher response times (e.g., a Web request). In general, response times less than seven seconds are considered acceptable when they require communication over a network.

Type of Requirement	Definition	Examples
Speed Requirements	The time within which the system must perform its functions	• Response time must be less than 7 seconds for any transaction over the network. • The inventory database must be updated in real time. • Orders will be transmitted to the factory floor every 30 minutes.
Capacity Requirements	The total and peak number of users and the volume of data expected	• There will be a maximum of 100–200 simultaneous users at peak use times. • A typical transaction will require the transmission of 10K of data. • The system will store data on approximately 5,000 customers for a total of about 2 meg of data.
Availability and Reliability Requirements	The extent to which the system will be available to the users and the permissible failure rate due to errors	• The system should be available 24/7, with the exception of scheduled maintenance. • Scheduled maintenance shall not exceed one 6-hour period each month. • The system shall have 99% uptime performance.

FIGURE 12-13   Performance Requirements

The second aspect of speed requirements is how long it takes transactions in one part of the system to be reflected in other parts. For example, how soon after an order is placed will the items it contained be shown as no longer available for sale to someone else? If the inventory is not updated immediately, then someone else could place an order for the same item, only to find out later it is out of stock. Or, how soon after an order is placed is it sent to the warehouse to be picked from inventory and shipped? In this case, some time delay might have little impact.

Capacity Requirements   *Capacity requirements* attempt to predict how many users the system will have to support, both in total and simultaneously. Capacity requirements are important in understanding the size of the databases, the processing power needed, and so on. The most important requirement is usually the peak number of simultaneous users because this has a direct impact on the processing power of the computer(s) needed to support the system.

It is often easier to predict the number of users for internal systems designed to support an organization's own employees than it is to predict the number of users for customer-facing systems, especially those on the Web. How *does* weather.com estimate the peak number of users who will simultaneously seek weather information? This is as much an art as a science, so often the team provides a range of estimates, with wider ranges used to signal a less-accurate estimate.

Availability and Reliability Requirements   *Availability and reliability requirements* focus on the extent to which users can assume that the system will be available for them to use. Although some systems are intended to be used only during the forty-hour workweek, some systems are designed to be used by people around the world. For such systems, project team members need to consider how the application can be operated, supported, and maintained *24/7* (i.e., 24 hours a day, 7 days a week). This 24/7 requirement means that users may need help or have questions at any time, and a support desk that is available eight hours a day will not be sufficient support. It is also important to consider what reliability is needed in the system. A system that requires high reliability (e.g., a medical device or telephone switch) needs far greater planning and testing than one that does not have such high-reliability needs (e.g., personnel system, Web catalog).

It is more difficult to predict the peaks and valleys in usage of the system when the system has a global audience. Typically, applications are backed up on weekends or late evenings when users are no longer accessing the system. Such maintenance activities need to be rethought with global initiatives. The development of Web interfaces, in particular, has escalated the need for 24/7 support; by default the Web can be accessed by anyone at any time. For example, the developers of a Web application for U.S. outdoor gear and clothing retailer Orvis were surprised when the first order after going live came from Japan.

Security Requirements[5]

Security is the ability to protect the information system from disruption and data loss, whether caused by an intentional act (e.g., a hacker, a terrorist attack) or a random event (e.g., disk failure, tornado). Security is primarily the responsibility of the operations group—the staff responsible for installing and operating security controls, such as firewalls, intrusion-detection systems, and routine backup and recovery operations. Nonetheless, developers of new systems must ensure that the system's *security requirements* produce reasonable precautions to prevent problems; system developers are responsible for ensuring security within the information systems themselves.

Security is an ever-increasing problem in today's Internet-enabled world. Historically, the greatest security threat has come from inside the organization itself. Ever since the early 1980s when the FBI first began keeping computer crime statistics and security firms began conducting surveys of computer crime, organizational employees have perpetrated the vast majority of computer crimes. For years, 80 percent of unauthorized break-ins, thefts, and sabotage have been committed by insiders, leaving only 20 percent to hackers external to the organizations.

In 2001, that changed. Depending on what survey you read, the percent of incidents attributed to external hackers in 2001 increased to 50 to 70 percent of all incidents, meaning that the greatest risk facing organizations is now from the outside. Although some of this shift may be due to better internal security and better communications with employees to prevent security problems, much of it is simply due to an increase in activity by external hackers.

Developing security requirements usually starts with some assessment of the value of the system and its data. This helps pinpoint extremely important systems so that the operations staff are aware of the risks. Security within systems usually focuses on specifying who can access what data, identifying the need for encryption and authentication, and ensuring the application prevents the spread of viruses (see Figure 12-14).

System Value   The most important computer asset in any organization is not the equipment; it is the organization's data. For example, suppose someone destroyed a mainframe computer worth $10 million. The mainframe could be replaced, simply by buying a new one. It would be expensive, but the problem would be solved in a few weeks. Now suppose someone destroyed all the student records at your university so that no one knew what courses anyone had taken or their grades. The cost would far exceed the cost of replacing a $10 million computer. The lawsuits alone would easily exceed $10 million, and the cost of staff to find and reenter paper records would be enormous and certainly would take more than a few weeks.

---

[5] For more information, see Brett C. Tjaden, *Fundamentals of Secure Computer Systems* (Wilsonville, OR: Franklin, Beedle, and Associates, 2004); for security controls associated with the Sarbanes-Oxley act, see Dennis C. Brewer, *Security Controls for Sarbanes-Oxley Section 404 IT Compliance: Authorization, Authentication, and Access* (Indianapolis, IN: Wiley, 2006).

Type of Requirement	Definition	Examples
System Value Estimates	Estimated business value of the system and its data	• The system is not mission critical but a system outage is estimated to cost $50,000 per hour in lost revenue. • A complete loss of all system data is estimated to cost $20 million.
Access Control Requirements	Limitations on who can access what data	• Only department managers will be able to change inventory items within their own department. • Telephone operators will be able to read and create items in the customer file, but cannot change or delete items.
Encryption and Authentication Requirements	Defines what data will be encrypted where and whether authentication will be needed for user access	• Data will be encrypted from the user's computer to the Web site to provide secure ordering. • Users logging in from outside the office will be required to authenticate.
Virus Control Requirements	Requirements to control the spread of viruses	• All uploaded files will be checked for viruses before being saved in the system.

FIGURE 12-14   Security Requirements

In some cases, the information system itself has value that far exceeds the cost of the equipment as well. For example, for an Internet bank that has no brick and mortar branches, the Web site is a *mission critical system.* If the Web site crashes, the bank cannot conduct business with its customers. A mission critical application is an information system that is literally critical to the survival of the organization. It is an application that cannot be permitted to fail, and if it does fail, the network staff drops everything else to fix it. Mission critical applications are usually clearly identified so their importance is not overlooked.

Even temporary disruptions in service can have significant costs. The cost of disruptions to a company's primary Web site or the LANs and backbones that support telephone sales operations are often measured in the millions. Amazon.com, for example, has revenues of more than $10 million per hour, so if their Web site were unavailable for an hour or even part of an hour, it would cost them millions of dollars in lost revenue. Companies that do less e-business or telephone sales have lower costs, but recent surveys suggest losses of $100,000 to $200,000 per hour are not uncommon for major customer-facing information systems.

CONCEPTS

IN ACTION

12-F Power Outage Costs a Million Dollars

Lithonia Lighting, located just outside of Atlanta, is the world's largest manufacturer of light fixtures, with more than $1 billion in annual sales. One afternoon, the power transformer at its corporate headquarters exploded, leaving the entire office complex, including the corporate data center, without power. The data center's backup power system immediately took over and kept critical parts of the data center operational. However, it was insufficient to power all systems, so the system supporting sales for all of Lithonia Lighting's North American agents, dealers, and distributors had to be turned off.

The transformer was quickly replaced and power was restored. However, the three hour shutdown of the sales system cost $1 million in potential sales lost. Unfortunately, it is not uncommon for the cost of a disruption to be hundreds or thousands of times the cost of the failed components.

Question

What would you recommend to avoid similar losses in the future?

Access Control Requirements   Some of the data stored in the system need to be kept confidential; some data need special controls on who is allowed to change or delete it. Personnel records, for example, should be able to be read only by the personnel department and the employee's supervisor; changes should be permitted to be made only by the personnel department. *Access control requirements* state who can access what data and what type of access is permitted: whether the individual can create, read, update and/or delete the data. The requirements reduce the chance that an authorized user of the system can perform unauthorized actions.

Encryption and Authentication Requirements   One of the best ways to prevent unauthorized access to data is *encryption,* which is a means of disguising information by the use of mathematical algorithms (or formulas). Encryption can be used to protect data stored in databases or data that are in transit over a network from a database to a computer. There are two fundamentally different types of encryption: symmetric and asymmetric. A *symmetric encryption algorithm* [such as Data Encryption Standard (DES) or Advanced Encryption Standard (AES)] is one in which the key used to encrypt a message is the *same* as the one used to decrypt it, which means that it is essential to protect the key and that a separate key must be used for each person or organization with whom the system shares information (or else everyone can read all the data).

An *asymmetric encryption algorithm* (such as *public key encryption*) is one in which the key used to encrypt data (called the *public key*) is different from the one used to decrypt it (called the *private key*). Even if everyone knows the public key, once the data are encrypted, they cannot be decrypted without the private key. Public key encryption greatly reduces the key-management problem. Each user has its public key that is used to encrypt messages sent to it. These public keys are widely publicized (e.g., listed in a telephone book–style directory)—that's why they're called public keys. The private key, in contrast, is kept secret (which is why it's called private).

Public key encryption also permits *authentication* (or digital signatures). When one user sends a message to another, it is difficult to legally prove who actually sent the message. Legal proof is important in many communications, such as bank transfers and buy/sell orders in currency and stock trading, which normally require legal signatures. Public key encryption algorithms are invertible, meaning that text encrypted with either key can be decrypted by the other. Normally, we encrypt with the public key and decrypt with the private key. However, it is possible to do the reverse: encrypt with the private key and decrypt with the public key. Because the private key is secret, only the real user can use it to encrypt a message. Thus, a digital signature or authentication sequence is used as a legal signature on many financial transactions. This signature is usually the name of the signing party plus other unique information from the message (e.g., date, time, or dollar amount). This signature and the other information are encrypted by the sender using the private key. The receiver uses the sender's public key to decrypt the signature block and compares the result to the name and other key contents in the rest of the message to ensure a match.

The only problem with this approach lies in ensuring that the person or organization that sent the document with the correct private key is the actual person or organization. Anyone can post a public key on the Internet, so there is no way of knowing for sure who actually used it. For example, it would be possible for someone other than Organization A in this example to claim to be Organization A when, in fact, he or she is an imposter.

This is where the Internet's public key infrastructure (PKI) becomes important.[6] The PKI is a set of hardware, software, organizations, and polices designed to make public key

---

[6] For more on the PKI, see www.ietf.org/html.charters/pkix-charter.html.

encryption work on the Internet. PKI begins with a *certificate authority (CA)*, which is a trusted organization that can vouch for the authenticity of the person or organization using authentication (e.g., VeriSign). A person wanting to use a CA registers with the CA and must provide some proof of identify. There are several levels of certification, ranging from a simple confirmation from a valid e-mail address to a complete police-style background check with an in-person interview. The CA issues a digital certificate that is the requestor's public key, encrypted using the CA's private key as proof of identify. This certificate is then attached to the user's e-mail or Web transactions in addition to the authentication information. The receiver then verifies the certificate by decrypting it with the CA's public key—and must also contact the CA to ensure that the user's certificate has not been revoked by the CA.

The encryption and authentication requirements state what encryption and authentication requirements are needed for what data. For example, will sensitive data such as customer credit-card numbers be stored in the database in encrypted form, or will encryption be used to take orders over the Internet from the company's Web site? Will users be required to use a digital certificate in addition to a standard password?

Virus Control Requirements   *Virus control requirements* address the single most common security problem: *viruses*. Recent studies have shown that almost 90 percent of organizations suffer a virus infection each year. Viruses cause unwanted events—some harmless (such as nuisance messages), some serious (such as the destruction of data). Any time a system permits data to be imported or uploaded from a user's computer, there is the potential for a virus infection. Many systems require that all information systems that permit the import or upload of user files to check those files for viruses before they are stored in the system.

CONCEPTS   **12-G Securing the Environment**

IN ACTION

Quinnipiac University is a four-year university in Hamden, Connecticut, with about 7,400 students. As a university with residence halls, the IT staff has to support academic functions—but also have to be an Internet Service Provider (ISP) for students. The IT staff can shape much of the academic usage of the Internet, but students living in residence halls can cause havoc. Students (and faculty) inadvertently open the campus to all kind of attacks, such as viruses, malware, worms, spybots and other intrusions through accessing various Web sites. A particularly trying time is after the semester break in late January, when students return to campus and plug in laptops that have been corrupted with other viruses from home networks. These viruses try to infect the entire campus. Quinnipiac University installed an Intrusion Prevention System (IPS) by Tipping Point Technology in August 2006. On a daily basis, this IPS detects and drops thousands of destructive messages and packets. But, the real proof was in late January

2007 when students returned to campus from semester break. In the previous year, the viruses and spyware virtually took the campus network down for three days. But in January 2007, there were no outages, and the network remained strong and functioning at full speed. Brian Kelly, information security officer at Quinnipiac University said, "Without the IPS solution, this campus would have struggled under this barrage of malicious packets and may have shut down. With the IPS system, we were able to function at full speed without any problems."

Questions

1. What might be some of the tangible and intangible costs of having the internet down for three days on a busy college campus?
2. What benefits would an IPS mean to a campus, to end users (e.g., faculty, staff and students)?

Type of Requirement	Definition	Examples
Multilingual Requirements	The language in which the system will need to operate	• The system will operate in English, French, and Spanish.
Customization Requirements	Specification of what aspects of the system can be changed by local users	• Country managers will be able to define new fields in the product database to capture country-specific information. • Country managers will be able to change the format of the telephone number field in the customer database.
Making Unstated Norms Explicit	Explicitly stating assumptions that differ from country to country	• All date fields will be explicitly identified as using the month-day-year format. • All weight fields will be explicitly identified as being stated in kilograms.
Legal Requirements	The laws and regulations that impose requirements on the system	• Personal information about customers cannot be transferred out of European Union countries into the United States. • It is against U.S. federal law to divulge information on who rented what videotape, so access to a customer's rental history is permitted only to regional managers.

FIGURE 12-15 Cultural and Political Requirements

Cultural and Political Requirements

Cultural and political requirements are those specific to the countries in which the system will be used. In today's global business environment, organizations are expanding their systems to reach users around the world. Although this can make great business sense, its impact on application development should not be underestimated. Yet another important part of the design of the system's architecture is understanding the global cultural and political requirements for the system (see Figure 12-15).

Multilingual Requirements   The first and most obvious difference between applications used in one region and those designed for global use is language. Global applications often have *multilingual requirements,* which means that they have to support users who speak different languages and write using non-English letters (e.g., those with accents, Cyrillic, Japanese). One of the most challenging aspects in designing global systems is getting a good translation of the original language messages into a new language. Words often have similar meanings but can convey subtly different meanings when they are translated, so it is important to use translators skilled in translating technical words.

The other challenge is often screen space. In general, English-language messages usually take 20 percent to 30 percent fewer letters than their French or Spanish counterparts. Designing global systems requires allocating more screen space to messages than might be used in the English-language version.

Some systems are designed to handle multiple languages on the fly so that users in different countries can use different languages concurrently; that is, the same system supports several different languages simultaneously (a concurrent multilingual system). Other systems contain separate parts that are written in each language and must be reinstalled before a specific language can be used; that is, each language is provided by a different version of the system so that any one installation will use only one language (i.e., a discrete multilingual system). Either approach can be effective, but this functionality must be designed into the system well in advance of implementation.

CONCEPTS

IN ACTION

12-H Developing Multilingual Systems

I've had the opportunity to develop two multilingual systems. The first was a special-purpose decision support system to help schedule orders in paper mills called BCW-Trim. The system was installed by several dozen paper mills in Canada and the United States, and it was designed to work in either English or French. All messages were stored in separate files (one set English, one set French), with the program written to use variables initialized either to the English or French text. The appropriate language files were included when the system was compiled to produce either the French or English version.

The second program was a groupware system called GroupSystems, for which I designed several modules. The system has been translated into dozens of different languages, including French, Spanish, Portuguese, German, Finnish, and Croatian. This system enables the user to switch between languages at will by storing messages in

simple text files. This design is far more flexible because each individual installation can revise the messages at will. Without this approach, it is unlikely that there would have been sufficient demand to warrant the development of versions to support less commonly used languages (e.g., Croatian).

—*Alan Dennis*

Questions

1. How would you decide how much to support users who speak languages other than English?
2. Would you create multilingual capabilities for any application that would be available to non-English-speaking people? Think about Web sites that exist today.

Customization Requirements   For global applications, the project team will need to give some thought to *customization requirements:* How much of the application will be controlled by a central group, and how much of the application will be managed locally? For example, some companies allow subsidiaries in some countries to customize the application by omitting or adding certain features. This decision has trade-offs between flexibility and control because customization often makes it more difficult for the project team to create and maintain the application. It also means that training can differ between different parts of the organization, and customization can create problems when staff move from one location to another.

Unstated Norms   Many countries have *unstated norms* that are not shared internationally. It is important for the application designer to make these assumptions explicit because they can lead to confusion otherwise. In the United States, the unstated norm for entering a date is the date format MM/DD/YYYY; however, in Canada and most European countries, the unstated norm is DD/MM/YYYY. Another example is the order of a person's name; does the family name come first or last? The answer depends on the locality. When you are designing global systems, it is critical to recognize these unstated norms and make them explicit so that users in different countries do not become confused. Currency is the other item sometimes overlooked in system design; global application systems must specify the currency in which information is being entered and reported.

Legal Requirements   *Legal requirements* are those requirements imposed by laws and government regulations. System developers sometimes forget to think about legal regulations; unfortunately, forgetting comes at some risk because ignorance of the law is no defense. For example, in 1997 a French court convicted the Georgia Institute of Technology of violating French language law. Georgia Tech operated a small campus in France that offered summer programs for American students. The information on the campus Web server was primarily in English because classes are conducted in English, which violated the law requiring French to be the predominant language on all Internet servers in France. By formally considering legal regulations, they are less likely to be overlooked.

Synopsis

In many cases, the technical environment requirements as driven by the business requirements may simply define the physical architecture layer. In this case, the choice is simple: business requirements dominate other considerations. For example, the business requirements may specify that the system needs to work over the Web using the customer's Web browser. In this case, the architecture probably should be a thin client–server. Such business requirements are most likely in systems designed to support external customers. Internal systems may also impose business requirements, but usually they are not as restrictive.

In the event that the technical environment requirements do not require the choice of a specific architecture, then the other nonfunctional requirements become important. Even in cases when the business requirements drive the architecture, it is still important to work through and refine the remaining nonfunctional requirements because they are important in later stages of design and implementation. Figure 12-16 summarizes the relationship between requirements and recommended architectures.

Requirements	Server-Based	Client-Based	Thin Client–Server	Thick Client–Server
Operational Requirements				
System Integration Requirements	✓		✓	✓
Portability Requirements			✓	
Maintainability Requirements	✓		✓	
Performance Requirements				
Speed Requirements			✓	✓
Capacity Requirements			✓	✓
Availability/Reliability Requirements	✓		✓	✓
Security Requirements				
High System Value	✓		✓	
Access Control Requirements	✓			
Encryption/Authentication Requirements			✓	✓
Virus Control Requirements	✓			
Cultural/Political Requirements				
Multilingual Requirements			✓	
Customization Requirements			✓	
Making Unstated Norms Explicit			✓	
Legal Requirements	✓		✓	✓

FIGURE 12-16
Nonfunctional Requirements and Their Implications for Architecture Design

Operational Requirements   System integration requirements may lead to one architecture being chosen over another, depending on the architecture and design of the system(s) with which the system needs to integrate. For example, if the system must integrate with a desktop system (e.g., Excel), this may suggest a thin or thick client–server architecture, whereas if it must integrate with a server-based system, a server-based architecture may be indicated. Systems that have extensive portability requirements tend to be best suited for a thin client–server architecture because it is simpler to write for Web-based standards (e.g., HTML, XML) that extend the reach of the system to other platforms, rather than trying to write and rewrite extensive presentation logic for different platforms in the server-based, client-based, or thick client–server architectures. Systems with extensive maintainability requirements may not be well suited to client-based or thick client–server architectures because of the need to reinstall software on the desktops.

Performance Requirements   Generally speaking, information systems that have high performance requirements are best suited to client–server architectures. Client–server architectures are more scalable, which mean they respond better to changing capacity needs and thus enable the organization to better tune the hardware to the speed requirements of the system. Client–server architectures that have multiple servers in each tier should be more reliable and have greater availability, because if any one server crashes, requests are simply passed to other servers, and users might not even notice (although response time could be worse). In practice, however, reliability and availability depend greatly on the hardware and operating system, and Windows-based computers tend to have lower reliability and availability than Linux or mainframe computers.

Security Requirements   Generally speaking, server-based architectures tend to be more secure because all software is in one location and because mainframe operating systems are more secure than microcomputer operating systems. For this reason, high-value systems are more likely to be found on mainframe computers, even if the mainframe is used as a server in client–server architectures. In today's Internet-dominated world, authentication and encryption tools for Internet-based client–server architectures are more advanced than those for mainframe-based server-based architectures. Viruses are potential problems in all architectures because they easily spread on desktop computers. If a server-based system can reduce the functions needed on desktop systems, then they may be more secure.

Cultural and Political Requirements   As cultural and political requirements become more important, the ability to separate the presentation logic from the application logic and the data becomes important. Such separation makes it easier to develop the presentation logic in different languages while keeping the application logic and data the same. It also makes it easier to customize the presentation logic for different users and to change it to better meet cultural norms. To the extent that the presentation logic provides access to the application and data, it also makes it easier to implement different versions that enable or disable different features required by laws and regulations in different countries. This separation is the easiest in thin client–server architectures, so systems with many cultural and political requirements often use thin client–server architectures. As with system integration requirements, the impact of legal requirements depends on the specific nature of the requirements, but in general, client-based systems tend to be less flexible.

YOUR TURN   12-3 University Course Registration System

Think about the course registration system in your university. First, develop a set of nonfunctional requirements if the system were to be developed today. Consider the operational requirements, performance requirements, security requirements, and cultural and political requirements. Then create an architecture design to satisfy these requirements.

YOUR TURN   12-4 Global e-Learning System

Many multinational organizations provide global Web-based e-learning courses for their employees. First, develop a set of nonfunctional requirements for such a system. Consider the operational requirements, performance requirements, security requirements, and cultural and political requirements. Then create an architecture design to satisfy these requirements.

HARDWARE AND SOFTWARE SPECIFICATION

The time to begin acquiring the hardware and software that will be needed for a future system is during the design of the system. In many cases, the new system will simply run on the existing equipment in the organization. Other times, however, new hardware and software (usually for servers) must be purchased. The *hardware and software specification* is a document that describes what hardware and software are needed to support an application. The actual acquisition of hardware and software should be left to the purchasing department or the area in the organization that handles capital procurement. However, the project team can use the hardware and software specification to communicate the project needs to the appropriate people. There are several steps involved in creating the document. Figure 12-17 shows a sample hardware and software specification.

	Standard Client	Standard Web Server	Standard Application Server	Standard Database Server
Operating System	• Windows • Internet Explorer	• Linux	• Linux	• Linux
Special Software	• Adobe Acrobat Reader • Real Audio	• Apache	• Java	• Oracle
Hardware	• 40 gig disk drive • Pentium • 17 inch Monitor	• 80 gig disk drive • Pentium	• 80 gig disk drive • Pentium	• 200 gig disk drive • RAID • Quad Pentium
Network	• Always-on Broadband preferred • Dial-up at 56 Kbps possible with some performance loss	• Dual 100 Mbps Ethernet	• Dual 100 Mbps Ethernet	• Dual 100 Mbps Ethernet

FIGURE 12-17   Sample Hardware and Software Specification

1. **Functions and Features**   What specific functions and features are needed (e.g., size of monitor, software features)

2. **Performance**   How fast the hardware and software operates (e.g., processor, number of database writes per second)

3. **Legacy Databases and Systems**   How well the hardware and software interacts with legacy systems (e.g., can it write to this database)

4. **Hardware and OS Strategy**   What are the future migration plans (e.g., the goal is to have all of one vendor's equipment)

5. **Cost of Ownership**   What are the costs beyond purchase (e.g., incremental license costs, annual maintenance, training costs, salary costs)

6. **Political Preferences**   People are creatures of habit and are resistant to change, so changes should be minimized

7. **Vendor Performance**   Some vendors have reputations or future prospects that are different from those of a specific hardware or software system they currently sell

FIGURE 12-18
Factors in Hardware and Software Selection

First, we will need to define the software that will run on each component. This usually starts with the operating system (e.g., Windows, Linux) and includes any special-purpose software on the client and servers (e.g., Oracle database). This document should consider any additional costs, such as technical training, maintenance, extended warranties, and licensing agreements (e.g., a site license for a software package). Again, the listed needs are influenced by decisions that are made in the other design activities.

Second, we must create a list of the hardware that is needed to support the future system. The low-level network model provides a good starting point for recording the project's hardware needs because each component on the diagram corresponds to an item on this list. In general, the list can include things like database servers, network servers, peripheral devices (e.g., printers, scanners), backup devices, storage components, and any other hardware component that is needed to support an application. At this time, you also should note the quantity of each item that will be needed.

Third, we must describe, in as much detail as possible, the minimum requirements for each piece of hardware. Typically, the project team must convey requirements like the amount of processing capacity, the amount of storage space, and any special features that should be included. Many organizations have standard lists of approved hardware and software that must be used, so in many cases, this step simply involves selecting items from the lists. Other times, however, the team is operating in new territory and is not constrained by the need to select from an approved list. In these cases, the project team must convey such requirements as the amount of processing capacity, the amount of storage space, and any special features that should be included.

This step becomes easier with experience; however, there are some hints that can help you describe hardware needs (see Figure 12-18). For example, consider the hardware standards within the organization or those recommended by vendors. Talk with experienced

YOUR TURN   **12-5 University Course Registration System**

*D*evelop a hardware and software specification for the university course registration system described in Your Turn 12-3.

YOUR
TURN

12-6 Global e-Learning System

Develop a hardware and software specification for the global e-learning system described in Your Turn 12-4.

YOUR
TURN

12-7 Create a Hardware and Software Specification

You have decided to purchase a computer, printer, and low-cost scanner to support your academic work. Create a hardware and software specification for these components that describes your hardware and software needs.

system developers or other companies with similar systems. Finally, think about the factors that affect hardware performance, such as the response-time expectations of the users, data volumes, software memory requirements, the number of users accessing the system, the number of external connections, and growth projections.

APPLYING THE CONCEPTS AT CD SELECTIONS

Alec realized that the hardware, software, and networks that would support the new application would need to be integrated into the current infrastructure at CD Selections. As such, he had asked Anne to begin by reviewing the high-level nonfunctional requirements developed in the analysis phase (see Figure 4-15) and by conducting a JAD session and a series of interviews with managers in the marketing department and three store managers to refine the nonfunctional requirements into more detail. Figure 12-19 shows some of the results. The clear business need for a Web-based architecture required a thin client–server architecture for the Internet sales portion of the system.

CD Selections had a formal architecture group responsible for managing CD Selections architecture and its hardware and software infrastructure. Therefore, Anne set up a meeting with the development team and the architecture group. During the meeting, she confirmed that CD Selections was still moving toward a target client–server architecture, although the central mainframe still existed as the primary server for many server-based applications.

They discussed the Internet system and decided that it should be built using a three-tier thin client–server architecture. Everyone believed that it was hard to know at this point exactly how much traffic this Web site would get and how much power the system would require, but a client–server architecture would allow CD Selections to easily scale up the system as needed.

By the end of the meeting, it was agreed that a three-tiered client–server architecture was the best configuration for the Internet portion of the Internet sales system (i.e., the Place Order process in Figures 5-17 and 5-19). Customers would use their personal computers running a Web browser as the client. A database server would store the Internet system's databases, whereas an application server would have Web-server software and the application software to run the system.

1. Operational Requirements

Technical Environment | 1.1 The system will work over the Web environment with Internet Explorer and Real Audio.

1.2 Customers will only need Internet Explorer and RA on their desktops.

System Integration | 1.3 The Internet sales system will read information from the main CD information database, which contains basic information about the CD (e.g., title, artist, id number, price, quantity in inventory). The Internet order system will not write information to the main CD information database.

1.4 The Internet sales system will transmit orders for new CDs in the special order system, and will rely on the special order system to complete the special orders generated.

1.5 The Internet sales system will read and write to the main inventory database.

1.6 A new module for the In-store system will be written to manage the "holds" generated by the Internet system. The requirements for this new module will be documented as part of the Internet sales system because they are necessary for the Internet sales system to function.

1.7 A new module will be written to handle the mail order sales. The requirements for this new module will be documented as part of the Internet sales system because they are necessary for the Internet sales system to function.

Portability | 1.8 The system will need to remain current with evolving Web standards, especially those pertaining to music formats.

Maintainability | 1.9 No special maintainability requirements are anticipated.

2. Performance Requirements

Speed | 2.1 Response times must be less than 7 seconds.

2.2 The inventory database must be updated in real time.

2.3 In-store holds must be sent to the store within 5 minutes.

Capacity | 2.4 There will be a maximum of 20–50 simultaneous users at peak use times.

2.5 The system will support streaming audio to up to forty simultaneous users.

2.6 The system will send up to 5K of data to each store daily.

2.7 The in-store hold database will require 10–20K of disk space per store.

Availability and Reliability | 2.8 The system should be available 24/7.

2.9 The system shall have 99 percent uptime performance.

3. Security Requirements

System Value | 3.1 No special system value requirements are anticipated.

Access Control | 3.2 Only store managers will be able to override In-Store Holds.

Encryption/Authentication | 3.3 No special encryption/authentication requirements are anticipated.

Virus Control | 3.4 No special virus control requirements are anticipated.

4. Cultural and Political Requirements

Multilingual | 4.1 No special multilingual requirements are anticipated.

Customization | 4.2 No special customization requirements are anticipated.

Unstated Norms | 4.3 No special unstated norms requirements are anticipated.

Legal | 4.4 No special legal requirements are anticipated.

FIGURE 12-19 Selected Nonfunctional Requirements for the CD Selections Internet Sales System

A separate two-tier client server system would maintain the CD and CD Marketing Material information (i.e., the Maintain CD Information and Maintain CD Marketing Information use cases in Figure 5-19). This system would have an application for the personal computers of the staff working in the Internet sales group that communicate directly

At this point in time, you should play the role of Anne and complete the design of the physical architecture layer. This would include any additional hardware and software that would be required to support the Maintain CD Information and Maintain CD Marketing Information use cases.

with the database server and enable staff to update the information. The database server would have a separate program to enable it to exchange data with CD Selections' distribution system on the company mainframe. Furthermore, the in-store system was currently built using a two-tier client–server architecture, so the portion of the system responsible for the in-store holds would conform to that architecture.

Next, Anne created a network model to show the major components of the Internet sales system (see Figure 12-20). The Internet sales system was on a separate network segment separated from the CD Selections' main network by a firewall that separated the network from the Internet while granting access to the Web and database servers. The Internet sales system had two parts. A firewall was used to connect the Web/Application server to the Internet, and another firewall further protected the Internet sales group's client computers and database server from the Internet. In order to improve response time, a direct connection was made from the Web/Application server to the database server because these would exchange a lot of data. Based on these decisions, she also created a deployment diagram that shows how the problem domain, human–computer interaction, and data management layers would be deployed over the physical architecture layer (see Figure 12-21).

Given that the Web interface could reach a geographically dispersed group, the development team realized that it needed to plan for 24/7 system support. As such, Anne scheduled a meeting to talk with the CD Selections systems operations group and discussed how they might be able to support the Internet system outside of standard working hours.

After examining the network model, the architecture group and the development team decided that the only components that needed to be acquired for the project were a database server, a Web server, and five new client computers for the marketing group, who would maintain the CD marketing materials. They developed a hardware and software specification for these components and handed them off to the purchasing department to start the acquisition process.

SUMMARY

An important component of design is the design of the physical architecture layer, which includes the hardware, software, and communications infrastructure for the new system and the way in which the information system will be distributed over the architecture. The physical architecture layer design is described in a deliverable that contains the network model and the hardware and software specification.

Elements of the Physical Architecture Layer

All software systems can be divided into four basic functions: data storage, data access logic, application logic, and presentation logic. There are three fundamental computing architectures that place these functions on different computers. In server-based architectures, the server performs all the functions. In client-based architectures, the client computers are

FIGURE 12-20 Deployment Diagram of Network Model for the CD Selections Internet Sales System

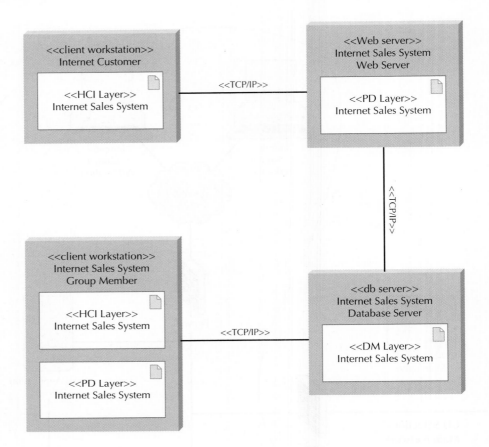

FIGURE 12-21
Deployment Diagram
of Layers for the CD
Selections Internet
Sales System

responsible for presentation logic, application logic, and data access logic, with data stored on a file server. In client–server architectures, the client is responsible for the presentation logic and the server is responsible for the data access logic and data storage. In thin client–server architectures, the server performs the application logic, whereas in thick client–server architectures, the application logic is shared between the servers and clients. In a two-tiered client–server architecture, there are two groups of computers: one client and a set of servers. In a three-tiered client–server architecture, there are three groups of computers: a client, a set of application servers, and a set of database servers. From an object-oriented point of view, the future trend in client–server computing is distributed object computing (DOC). Currently, there are two competing approaches to support DOC: OMG's CORBA and Microsoft's .net.

Each of the computing architectures has its strengths and weaknesses, and no architecture is inherently better than the others. The choice that the project team makes should be based on several criteria, including cost of development, ease of development, need for GUI applications, network capacity, central control and security, and scalability. The project team should also take into consideration the existing architecture and special software requirements of the project.

Infrastructure Design

Deployment diagrams are used to portray the design of the physical architecture layer. The diagrams are composed of nodes, artifacts, and communication links. Also, the diagram can be extended with graphical icons that represent different types of nodes.

The network model is a diagram that shows the technical components of the information system (e.g., servers, personal computers, networks) and their geographic locations throughout the organization. The components of the network model are the various clients (e.g., personal computers, kiosks), servers (e.g., database, network, communications, printer), network equipment (e.g., wire, dial-up connections, satellite links), and external systems or networks (e.g., Internet service provider) that support the application. Creating a network model is a top-down process whereby a high-level diagram is first created to show the geographic sites, or locations, that house the various components of the future system. Next, a low-level diagram is created to describe each location in detail, and it shows the system's hardware components and how they are attached to one another.

Nonfunctional Requirements and Physical Architecture Layer Design

Creating an architecture design begins with the nonfunctional requirements. Operational requirements specify the operating environment(s) in which the system must perform and how those may change over time (i.e., technical environment, system integration, portability, and maintainability). Performance requirements focus on performance issues such as system speed, capacity, and availability and reliability. Security requirements attempt to protect the information system from disruption and data loss (e.g., system value, access control, encryption and authentication, and virus control). Cultural and political requirements are those that are specific to the specific countries in which the system will be used (e.g., language, customization, unstated norms, and legal).

Hardware and Software Specification

The hardware and software specification is a document that describes what hardware and software are needed to support the application. To create a specification document, the hardware that is needed to support the future system is listed and then described in as much detail as possible. Next, the software to run on each hardware component is written down, along with any additional associated costs, such as technical training, maintenance, extended warrantees, and licensing agreements. Although the project team may suggest specific products or vendors, ultimately the hardware and software specification is turned over to the people who are in charge of procurement.

KEY TERMS

.net
24/7
Access control requirements
Application logic
Architectural component
Artifact
Asymmetric encryption algorithm
Authentication
Availability and reliability
   requirements
Capacity requirements
Certificate of authority (CA)
Client-based architecture
Client computer

Client–server architecture
Common Object Request Broker
   Architecture (CORBA)
Communication path
Cultural and political requirements
Customization requirements
Data access logic
Data storage
Deployment diagrams
Distributed objects computing
   (DOC)
Encryption
Enterprise JavaBeans (EJB)
Fat client

Graphical user interface (GUI)
Hardware and software specification
Invertible
Java 2 Enterprise Edition (J2EE)
Legal requirements
Locations
Mainframe
Maintainability requirements
Microcomputer
Middleware
Minicomputer
Mission critical system
Multilingual requirements
Network

Network model
Node
N-tiered architecture
Operational requirements
Performance requirements
Physical architecture layer design
Portability requirements
Presentation logic
Private key
Public key

Public key encryption
Response time
Scalable
Security requirements
Server
Server-based architecture
Speed requirements
SQL (Structured query language)
Symmetric encryption algorithm
System integration requirements

Technical environment
    requirements
Thick client
Thin client
Three-tiered architecture
Total cost of ownership
Two-tiered architecture
Unstated norms
Virus
Virus control requirements

QUESTIONS

1. What is distributed object computing?
2. What are the three competing approaches for supporting distributed object computing?
3. Who is ultimately in charge of acquiring hardware and software for a project?
4. What additional hardware- and software-associated costs may need to be included on the hardware and software specification?
5. What do you think are three common mistakes that novice analysts make in architecture design and hardware and software specification?
6. Are some nonfunctional requirements more important than others in influencing the architecture design and hardware and software specification?
7. What are the three primary hardware components of any physical architecture?
8. Why should the project team consider the existing physical architecture in the organization when designing the physical architecture layer of the new system?
9. Describe the major nonfunctional requirements and how they influence physical architecture layer design.
10. Why is it useful to define the nonfunctional requirements in more detail even if the technical environment requirements dictate a specific architecture?
11. What are the four basic functions of any information system?

12. What do you think are the most important security issues for a system?
13. What are the differences between the top-level network model and the low-level network model?
14. What does the network model communicate to the project team?
15. What six criteria are helpful to use when comparing the appropriateness of computing alternatives?
16. Define scalable. Why is this term important to system developers?
17. Name two examples of a server.
18. What is the biggest problem with client-based computing?
19. What is the biggest problem with server-based computing?
20. Describe the differences among two-tiered, three-tiered, and *n*-tiered architectures.
21. Compare and contrast server-based architectures, client-based architectures, and client–server-based architectures.
22. Describe the major benefits and limitations of thin client–server architectures.
23. Describe the major benefits and limitations of thick client–server architectures.

EXERCISES

A. Using the Web, locate examples of a mainframe component, a minicomputer component, and a microcomputer component. Compare the components in terms of price, speed, available memory, and disk storage. Did you find large differences in prices when the performances of the components are considered?

B. Think about the system that your university currently uses for career services, and suppose that you are in charge of replacing the system with a new one. Describe how you would decide on the computing architecture for the new system using the criteria presented in this chapter. What information will you need to find out before you can make an educated comparison of the alternatives?

C. You have been selected to find the best client–server computing architecture for a Web-based order entry system that is being developed for L.L. Bean. Write a short memo that describes to the project manager your reason for selecting an *n*-tiered architecture over a two-tiered architecture. In the memo, give some idea as to what different components of the architecture you would include.

D. Locate a consumer products company on the Web and read its company description (so that you get a good understanding of the geographic locations of the company). Pretend that the company is about to create a new application to support retail sales over the Web. Create a high-level network model that depicts the locations that would include components that support this application.

E. Using the Web (or past issues of computer industry magazines such as *Computerworld*), locate a system that runs in a client–server environment. Based on your reading, why do you think the company chose that computing environment?

F. Using the Web (or past issues of computer industry magazines such as *Computerworld*), locate a system that runs in a server-based environment. Based on your reading, why do you think the company chose that computing environment?

G. You have been selected to determine whether your firm should invest in CORBA- or .net-based technologies to support distributed objects computing. Write a short memo that describes to the steering committee the advantages and disadvantages of both approaches. (*Note:* You probably should check out the Web sites associated with both.)

H. Suppose that the admissions office in your university has a Web-based application so that students can apply for admission online. Recently, there has been a push to admit more international students into the university. What do you recommend that the application includes to ensure that it supports this global requirement?

I. Create a low-level network diagram for the building that houses the computer labs at your university.

Choose an application (e.g., course registration, student admissions) and include only the components that are relevant to that application.

J. An energy company with headquarters in Dallas, Texas, is thinking about developing a system to track the efficiency of its oil refineries in North America. Each week, the ten refineries—as far as Valdez, Alaska, and as close as San Antonio, Texas—will upload performance data via satellite to the corporate mainframe in Dallas. Production managers at each site will use a personal computer to dial into an Internet service provider and access reports via the Web. Create a high-level network model that depicts the locations that have components supporting this system.

K. Suppose that your mother is a real estate agent, and she has decided to automate her daily tasks using a laptop computer. Consider her potential hardware and software needs, and create a hardware and software specification that describes them. The specification should be developed to help your mother buy her hardware and software on her own.

L. Based on the health club exercises in Chapter 5 (N, O), Chapter 6 (O), Chapter 7 (G, H), Chapter 8 (G, H), Chapter 9 (A), Chapter 10 (I), and Chapter 11 (D, E), suggest a physical architecture design and portray it with a deployment diagram.

M. Based on the Picnics R Us exercises in Chapter 5 (P, Q), Chapter 6 (P), Chapter 7 (I), Chapter 8 (I, J), and Chapter 9 (C), Chapter 10 (J), and Chapter 11 (F, G), suggest a physical architecture design and portray it with a deployment diagram.

N. Based on the Of-the-Month Club exercises in Chapter 5 (T, U), Chapter 6 (R), Chapter 7 (L), Chapter 8 (K, L), Chapter 9 (E), Chapter 10 (K), and Chapter 11 (J, K), suggest a physical architecture design and portray it with a deployment diagram.

O. Based on the library exercises in Chapter 5 (R, S), Chapter 6 (Q), Chapter 7 (K), Chapter 8 (M, N), Chapter 9 (G), Chapter 10 (L), and Chapter 11 (H, I), suggest a physical architecture design and portray it with a deployment diagram.

MINICASES

1. Refer to the Professional and Scientific Staff Management (PSSM) minicase in Chapters 5, 7, 8, 10, and 11. Based on the solutions developed for those problems, suggest a physical architecture design and portray it with a deployment diagram.

2. Refer to the Holiday Travel Vehicles minicase in Chapters 6, 7, 8, 9, 10, and 11. Based on the solutions developed for those problems, suggest a physical architecture design and portray it with a deployment diagram.

3. Clara Silva is a relatively new member of a project team that is developing a retail store management system for a chain of grocery stores. The company's headquarters is in Rio de Janeiro, and the chain has 33 locations throughout Brazil, including multiple stores in several cities.

   The new system will be a networked, client-server architecture. Stores will be linked to one of four regional servers, which will in turn be linked to the corporate headquarters. The regional servers are also interlinked. Each retail store will be outfitted with similar configurations of 6-20 point-of-sale terminals and 10-20 personal computers (based on the size of the store) networked to a local file server. Ms. Silva has been given the task of developing a network model that will document the geographic structure of this system. She has not faced a system of such scope before and is unsure of where to begin.

   a. Prepare a set of instructions for Ms. Silva to follow in developing this network model.
   b. Using a deployment diagram, draw a network model for this organization.
   c. Prepare a set of instructions for Ms. Silva to follow in developing hardware and software specifications.

4. The system development project team at Birdie Masters golf schools has been working on defining the physical architecture design for the system. The major focus of the project is a networked school location operations system, allowing each school location to easily record and retrieve all school location transaction data. Another system element is the use of the Internet to enable current and prospective students to view class offerings at any of the Birdie Masters' locations, schedule lessons and enroll in classes at any Birdie Masters location, and maintain a student progress profile—a confidential analysis of the student's golf skill development.

   The project team has been considering the globalization issues that should be factored into the architecture design. The school's plan for expansion into the golf-crazed Japanese market is moving ahead. The first Japanese school location is tentatively planning to open about six months after the target completion data for the system project. Therefore, it is important that issues related to the international location be addressed now during design.

   Assume that you have been given the responsibility of preparing a summary memo on the globalization issues that should be factored into the design. Prepare this memo discussing the globalization issues that are relevant to Birdie Masters' new system.

PART FOUR

IMPLEMENTATION

During construction, the actual system is built. Building a successful information system requires a set of activities: programming, testing, and documenting the system. In today's global economy, cultural issues also play an important role in managing these activities. Installing an information system requires switching from the current system to the new system. This conversion process can be quite involved; for example, cultural differences among the users, the development team, and the two groups can be quite challenging. Furthermore, not only does conversion involve shutting the old system down and turning the new one on, it also can involve a significant training effort. Finally, operating the system may uncover additional requirements that may have to be addressed by the development team.

CHAPTER 13 ■
DEVELOPMENT

CHAPTER 14 □
INSTALLATION

Programs

Test Plan

Documentation

Conversion Plan

Change Management Plan

Support Plan

Project Assessment

CHAPTER 13

DEVELOPMENT

This chapter discusses the activities needed to successfully build an information system: programming, testing and documenting the system. Programming is time consuming and costly, but except in unusual circumstances, it is the simplest for the systems analyst because it is well understood. For this reason, the system analyst focuses on testing (proving that the system works as designed) and developing documentation.

OBJECTIVES:

- Be familiar with the system construction process.
- Understand different types of tests and when to use them.
- Understand how to develop documentation.

CHAPTER OUTLINE

Introduction
Managing Programming
    Assigning Programmers
    Coordinating Activities
    Managing the Schedule
    Cultural Issues
Designing Tests
    Testing and Object Orientation
    Test Planning
    Unit Tests
    Integration Tests
    System Tests
    Acceptance Tests

Developing Documentation
    Types of Documentation
    Designing Documentation
      Structure
    Writing Documentation Topics
    Identifying Navigation Terms
Applying the Concepts at
  CD Selections
    Managing Programming
    Testing
    Developing User Documentation
Summary

INTRODUCTION

When people first learn about developing information systems, they usually think immediately about writing programs. Programming can be the largest single component of any systems development project in terms of both time and cost. However, it also can be the best understood component and therefore—except in rare circumstances—offers the least problems of all aspects of system development. When projects fail, it is usually not because the programmers were unable to write the programs, but because the analysis, design, installation, and/or project management were done poorly. In this chapter, we focus on the construction and testing of the software and the documentation.

Construction is the development of all parts of the system, including the software itself, documentation, and new operating procedures. Looking back at Figure 1-11, we see that the Construction phase of the Enhance Unified Process deals predominantly with the Implementation and Testing and Configuration and Change Management workflows. Implementation obviously deals with programming. Programming is often seen as the focal point of system development. After all, system development *is* writing programs. It is the reason why we do all the analysis and design. And it's fun. Many beginning programmers see testing and documentation as bothersome afterthoughts. Testing and documentation aren't fun, so they often receive less attention than the creative activity of writing programs.

However, programming and testing are very similar to writing and editing. No professional writer (or student writing an important term paper) would stop after writing the first draft. Rereading, editing, and revising the initial draft into a good paper are the hallmarks of good writing. Likewise, thorough testing is the hallmark of professional software developers. Most professional organizations devote more time and money to testing (and the subsequent revision and retesting) than to writing the programs in the first place.

The reasons are simple economics: downtime and failures caused by software bugs[1] are extremely expensive. Many large organizations estimate the costs of downtime of critical applications at $50,000 to $200,000 *per hour.*[2] One serious bug that causes an hour of downtime can cost more than one year's salary of a programmer—and how often are bugs found and fixed in one hour? Testing is, therefore, a form of insurance. Organizations are willing to spend a lot of time and money to prevent the possibility of major failures after the system is installed.

Therefore, a program is usually not considered finished until the test for that program is passed. For this reason, programming and testing are tightly coupled, and because programming is the primary job of the programmer (not the analyst), testing (not programming) often becomes the focus of the construction stage for the systems analysis team.

The Configuration and Change Management workflow keeps track of the state of the evolving system. The evolving information system comprises a set of artifacts that includes, for example, diagrams, source code, and executables. During the development process, these artifacts are modified. The amount of work, and hence dollars, that goes into the development of the artifacts is substantial. As such, the artifacts themselves should be handled as any expensive asset would be handled—access controls must be put into place to safeguard the artifacts from being stolen or destroyed. Furthermore, because the artifacts are modified on a regular, if not continuous, basis, good version control mechanisms should be established. The *traceability* of the artifacts back through the various artifacts developed, such as data management layer designs, class diagrams, package diagrams, and use-case diagrams, to the specific requirements is also very important. Without this traceability, we will not know which aspects of a system to modify when, not if, the requirements change.

---

[1] When I (Alan Dennis) was an undergraduate, I had the opportunity to hear Admiral Grace Hopper tell how the term bug was introduced. She was working on one of the early Navy computers when suddenly it failed. The computer would not restart properly so she began to search for failed vacuum tubes. She found a moth inside one tube and recorded in the log book that a bug had caused the computer to crash. From then on, every computer crash was jokingly blamed on a bug (as opposed to programmer error), and eventually the term bug entered the general language of computing.

[2] See Billie Shea, "Quality Patrol: Eye on the Enterprise," *Application Development Trends* (November 5, 1998): 31–38.

CONCEPTS

IN ACTION

13-A The Cost of a Bug

My first programming job in 1977 was to convert a set of application systems from one version of COBOL to another version of COBOL for the Government of Prince Edward Island. The testing approach was to first run a set of test data through the old system and then run it through the new system to ensure that the results from the two matched. If they matched, then the last three months of production data were run through both to ensure they, too, matched.

Things went well until I began to convert the gas tax system that kept records on everyone authorized to purchase gasoline without paying tax. The test data ran fine, but the results using the production data were peculiar. The old and new systems matched, but rather than listing several thousand records, the report listed only fifty. I checked the production data file and found it listed only fifty records, not the thousands that were supposed to be there.

The system worked by copying the existing gas tax records file into a new file and making changes in the new file. The old file was then copied to tape backup. There was a bug in the program such that if there were no changes to the file, a new file was created, but no records were copied into it.

I checked the tape backups and found one with the full set of data that were scheduled to be overwritten three days after I discovered the problem. The government was only three days away from losing all gas tax records.

— Alan Dennis

Question

What would have been the cost of this bug if it hadn't been caught?

In this chapter, we discuss three aspects of construction: managing programming, testing, and writing the documentation. Because programming is primarily the job of programmers, not systems analysts, and because this is not a programming book, we devote less time to programming than to testing and documentation. Furthermore, we do not delve into the details of configuration and change management in this chapter (see Chapter 14).[3]

MANAGING PROGRAMMING

In general, systems analysts do not write programs; programmers write programs. Therefore, the primary task of the systems analysts during programming is . . . waiting. However, the project manager is usually very busy managing the programming effort by assigning the programmers, coordinating the activities, and managing the programming schedule.[4]

Assigning Programmers

The first step in programming is assigning modules to the programmers. As discussed in Chapter 9, each module (class, object, or method) should be as separate and distinct as possible from the other modules (i.e., cohesion should be maximized and coupling should

---

[3] A good reference for information on configuration and change management is Jessica Keyes, *Software Configuration Management* (Boca Raton, FL: Auerbach, 2004).

[4] One of the best books on managing programming (even though it was first written over 30 years ago) is that by Frederick P. Brooks, Jr. *The Mythical Man-Month*, 20th Anniversary Edition (Reading, MA: Addison-Wesley, 1995).

CONCEPTS

IN ACTION

13-B Identifying the Best Talent

Quantitative analysis provides the data for making decisions. Some major league baseball teams have used Sabermetrics (and similar data analysis measures) to quantify the value of a player to his salary. For example, is a .250 hitter with a strong base-stealing background getting paid $5 million per year worth more in terms of salary dollars as compared to a .330 hitter with slower speed who is getting paid $15 million per year? Is Alex Rodriguez of the Yankees worth more than Joe Maurer of the Minnesota Twins? Is Randy Johnson of the Arizona Diamondbacks more valuable than Ryan Dempster of the Chicago Cubs? Statistics are gathered and analyzed on all aspects of the game: hitting, fielding, injuries, leadership, coachability, age (and expected lifetime contributions), and much more.

Now take that concept to the business marketplace. Can quantitative measures be placed on IT workers? Is a systems analyst that manages six major projects in a year worth more than an analyst who manages twelve much smaller projects? Is a project leader who is known for

effectively managing teams worth more than one who frequently has disagreements with his or her team? Is a developer who can write 1,000 lines of Java code in a week worth more or less than a developer who writes 500 lines of Visual Basic code? When we discussed tangible and intangible benefits, the authors suggested putting a dollar amount on to intangible benefits if possible.

Questions

1. How could you put quantitative measures on the qualitative efforts of a systems analyst?
2. What might be some of the costs in implementing a statistical analysis of employees?
3. What data might be needed to do such a statistical analysis on employees?
4. Could this system replace the conventional human resource department roles for recruiting and evaluating corporate talent?

be minimized). The project manager first groups together classes that are related so that each programmer is working on related classes. These groups of classes are then assigned to programmers. A good place to start is to look at the package diagrams.

One of the rules of system development is that the more programmers who are involved in a project, the longer the system will take to build. This is because as the size of the programming team increases, the need for coordination increases exponentially, and the more coordination required, the less time programmers can spend actually writing systems. The best size is the smallest possible programming team. When projects are so complex that they require a large team, the best strategy is to try to break the project into a series of smaller parts that can function as independently as possible.

Coordinating Activities

Coordination can be done through both high-tech and low-tech means. The simplest approach is to have a weekly project meeting to discuss any changes to the system that have arisen during the past week—or just any issues that have come up. Regular meetings, even if they are brief, encourage the widespread communication and discussion of issues before they become problems.

Another important way to improve coordination is to create and follow standards that can range from formal rules for naming files, to forms that must be completed when goals are reached, to programming guidelines (see Chapter 3). When a team forms standards and then follows them, the project can be completed faster because task coordination is less complex.

The analysts also must put mechanisms in place to keep the programming effort well organized. Many project teams set up three areas in which programmers can work: a development area, a testing area, and a production area. These areas can be different directories on a server hard disk, different servers, or different physical locations, but the point is that files, data, and programs are separated based on their status of completion. At first, programmers access and build files within the development area and then copy them to the testing area when the programmers are finished. If a program does not pass a test, it is sent back to development. Once all programs, etc., are tested and ready to support the new system, they are copied into the production area—the location where the final system will reside.

Keeping files and programs in different places based on completion status helps manage *change control*, the action of coordinating a system as it changes through construction. Another change control technique is keeping track of which programmer changes which classes and packages by using a *program log*. The log is merely a form on which programmers sign out classes and packages to write and sign in when they are completed. Both the programming areas and program log help the analysts understand exactly who has worked on what and the system's current status. Without these techniques, files can be put into production without the proper testing (e.g., two programmers can start working on the same class or package at the same time).

If a CASE tool is used during the construction step, it can be very helpful for change control because many CASE tools are set up to track the status of programs and help manage programmers as they work. In most cases, maintaining coordination is not conceptually complex. It just requires a lot of attention and discipline to tracking small details.

Managing the Schedule

The time estimates that were produced during project initiation and refined during analysis and design will almost always need to be refined as the project progresses during construction because it is virtually impossible to develop an exact assessment of the project's schedule. As we discussed in Chapter 3, a well-done set of time estimates will usually have a 10 percent margin of error by the time we reach the construction step. It is critical that the time estimates be revised as construction proceeds. If a program module takes longer to develop than expected, then the prudent response is to move the expected completion date later by the same amount.

CONCEPTS   **13-C Finishing the Process**

IN ACTION

As a great analyst, you've planned, analyzed and designed a good solution. Now you need to implement. As part of implementation do you think that training is just a wasted expense?

Stress is common in a help-desk call center. Users of computing services call to get access to locked accounts or get help when technology isn't working as planned, and they frequently are very upset. Employees of the help-desk call center can get stressed out, and this can result in more sick days, less productivity, and higher turnover. Max Productivity Incorporated is a training company that works with people in high-stress jobs. Their training program helps employees in how to relax, how to shake off

tough users—and how to create "win-win" scenarios. They claim to be able to reduce employee turnover by 50 percent, increase productivity by 20 percent, and reduce stress, anger, and depression by 75 percent.

Questions

1. How would you challenge Max Productivity to verify its claims for reducing turnover, increasing productivity, and decreasing stress and anger?

2. How would you conduct a cost–benefit analysis on hiring Max Productivity Incorporated to do ongoing training for your help-desk call-center employees?

PRACTICAL | **13-1 Avoiding Classic Implementation Mistakes**

TIP

*I*n previous chapters, we discussed classic mistakes and how to avoid them. Here, we summarize four classic mistakes in implementation:[5]

1. *Research-oriented development:* Using state-of-the-art technology requires research-oriented development that explores the new technology because "bleeding edge" tools and techniques are not well understood, are not well documented, and do not function exactly as promised.

   Solution: If you use state-of-the-art technology, you need to significantly increase the project's time and cost estimates even if (some experts would say *especially if*) such technologies claim to reduce time and effort.

2. *Using low-cost personnel:* You get what you pay for. The lowest-cost consultant or staff member is significantly less productive than the best staff. Several studies have shown that the best programmers produce software six to eight times faster than the least productive (yet cost only 50 to 100% more).

   Solution: If cost is a critical issue, assign the best, most expensive personnel; never assign entry-level personnel in an attempt to save costs.

3. *Lack of code control:* On large projects, programmers need to coordinate changes to the program source code (so that two programmers don't try to change the same program at the same time and overwrite the other's changes). Although manual procedures appear to work (e.g., sending e-mail notes to others when you work on a program to tell them not to), mistakes are inevitable.

   Solution: Use a source code library that requires programmers to "check out" programs and prohibits others from working on them at the same time.

4. *Inadequate testing:* The number one reason for project failure during implementation is ad hoc testing—where programmers and analysts test the system without formal test plans.

   Solution: Always allocate sufficient time in the project plan for formal testing.

One of the most common causes for schedule problems is scope creep. Scope creep occurs when new requirements are added to the project after the system design was finalized. Scope creep can be very expensive because changes made late in system development can require much of the completed system design (and even programs already written) to be redone. Any proposed change during construction must require the approval of the project manager and should only be done after a quick cost–benefit analysis has been done.

Another common cause is the unnoticed day-by-day slippages in the schedule. One package is a day late here; another one is a day late there. Pretty soon these minor delays add up and the project is noticeably behind schedule. Once again, the key to managing the programming effort is to watch these minor slippages carefully and update the schedule accordingly.

Typically, a project manager will create a risk assessment that tracks potential risks along with an evaluation of their likelihood and potential impact. As the construction step moves to a close, the list of risks will change as some items are removed and others surface. The best project managers, however, work hard to keep risks from having an impact on the schedule and costs associated with the project.

Cultural Issues

One of the major issues facing information systems development organizations is the offshoring of the implementation aspects of information systems development. Conflicts caused by different national and organizational cultures are now becoming a real area of

---

[5] Adapted from Steve McConnell, *Rapid Development*, (Redmond, WA: Microsoft Press, 1996).

CONCEPTS
IN ACTION

13-D Managing a Late Project: When to Say When?

Systems Projects are notorious for being late and over budget. When should management stop a project that is late or costing more than the intended budget? Consider this case.

Valley Enterprises opted to implement Voiceover Internet Protocol (VoIP) service in its Phoenix, Arizona, service area. The company has fifteen locations in the Phoenix area—all with LANs and all with secure Wi-Fi connections. Their current phone system was designed and implemented in the 1950s, when they operated in three locations. As they expanded to additional locations, they generally implemented standard telecommunications' solutions with little thought for compatibility. Over the years, they added phone services as they added new buildings and facilities. The CEO, Doug Wilson, heard of VoIP at a trade show and contacted TMR Telecommunications Consultants for a bid. TMR spent a week with the CIO of Valley Enterprises gathering data and submitted a bid for $50,000 in late 2007. The project was to be started by March 2008 and completed by January 2009. The bid was accepted.

TMR started the project in March 2008. In late July 2008, TMR was bought out by Advanced Communications

of Scottsdale, Arizona. This merger delayed the project initially by over a month. In early September 2008, some of the same personnel from TMR, as well as a new project manager from Advanced Communications, went back to the project.

By March 2009, the project had already cost $150,000 and only eight of the locations had VoIP implemented. Advanced Communications insisted that the LANs were out of date and were unable to carry the expanded load without major upgrades to the bandwidth, the routers, and other telecommunications equipment.

Questions

1. Is it time to end this project? Why or why not?
2. What negotiations should have occurred between TMR and Valley Enterprises prior to December 2008?
3. What should a project manager or project coordinator from Valley Enterprises do when the project first starts to slip?

concern. A simple example that can demonstrate cultural differences with regard to student learning is the idea of plagiarism. What exactly does plagiarism really imply? Different cultures have very different views. In some cultures, one of the highest forms of respect is simply to quote an expert. However, in these same cultures, there is no need to reference the expert. The act of quoting the expert itself is the act of respect. In some cases, actually referencing the expert through the use of quotation marks and a footnote may be viewed as an insult to the expert and the reader because it is obvious to the reader that the writer did not expect the reader to recognize the expert's quote. This expectation was either caused by the reader's own ignorance or the expert's lack of reputation. Either way, the writer would be insulting someone through the use of quotation marks and footnotes. However, in the United States, the opposite is true. If a writer does not use quotation marks and footnotes to appropriately give credit to the source of the quote (or paraphrase), then the writer is guilty of theft.[6] Obviously, in today's global world, plagiarism is not a simple issue.

Another simple example of cultural differences, with regard to student learning, is the idea of students working together to complete homework assignments. Even though we all know that research has shown that students learn better in groups, in the United States, we view students that turn in the same assignment as cheaters.[7] In other cultures, individual performance is not as important as the performance of the group. As such, helping a fellow

[6] A wonderful little book on plagiarism is Richard A. Posner, *The Little Book of Plagiarism* (New York: Pantheon Books, 2007).

[7] In this case, the recent work of Roger Schank is very enlightening. For example, see Roger C. Schank, *Making Minds Less Well Educated Than Our Own* (Mahwah, NJ: Lawrence Erlbaum Associates, 2004).

student to understand the assignment and to perform better in the class would be the expectation. Obviously, this is another example of a substantive cultural difference.

From a project management perspective, information systems development teams can be geographically dispersed and multicultural in their membership. To say the least, this adds a new wrinkle in the management of developing a successful information system. For years, project managers in the United States have had to bring together individuals from very different backgrounds. But, there was always a common spoken and written language, English, and the so-called 'melting pot' idea that guaranteed some level of commonality among the team members.[8] However, in today's "flat world," there is no longer any common culture or common spoken and written language. The common language today tends to be UML, Java, SQL, and Visual Basic, not English. And, as the Internet becomes more prevalent, the only common culture could be the Web.

DESIGNING TESTS[9]

In object-oriented systems, the temptation is to minimize testing. After all, through the use of patterns, frameworks, class libraries, and components, much of the system has been tested previously. Therefore, we should not have to test as much. Right? Wrong! Testing is more critical to object-oriented systems than systems developed in the past. Based on encapsulation (and information hiding), polymorphism (and dynamic binding), inheritance, and reuse, thorough testing is much more difficult and critical. Furthermore, given the global nature of information systems development, testing becomes even more crucial. As such, testing must be done systematically and the results documented so that the project team knows what has and has not been tested.

The purpose of testing is not to demonstrate that the system is free of errors. It is not possible to prove that a system is error free, which is especially true with object-oriented systems. This is similar to theory testing. You cannot prove a theory. If a test fails to find problems with a theory, your confidence in the theory is increased. However, if a test succeeds in finding a problem, then the theory has been falsified. Software testing is similar in that it can only show the existence of errors. As such, the purpose of testing is to uncover as many errors as feasible.[10]

There are four general stages of tests: unit tests, integration tests, system tests, and acceptance tests. Although each application system is different, most errors are found during integration and system testing (see Figure 13-1).

In each of the following sections, we describe the four stages. However, before doing this, we describe the effect that the object-oriented characteristics have on testing and the necessary planning and management activities that must take place to have a successful testing program.

---

[8] Even though individuals who grew up in different areas of the United States (e.g., New York City, Nashville, Minneapolis, Denver, and Los Angeloes) are, in a very real sense, culturally different, the prevalence of the Internet and cable TV has created much more of a shared culture in the United States than in many other parts of the world.

[9] This section is based on material contained in Imran Bashir and Amrit L. Goel, *Testing Object-Oriented Software: Life Cycle Solutions* (New York: Springer Verlag, 1999); Bernd Bruegge and Allen H. Dutoit, *Object-Oriented Software Engineering: Conquering Complex and Changing Systems* (Upper Saddle River, NJ: Prentice Hall, 2000); Philippe Kruchten, *The Rational Unified Process: An Introduction*, 2nd ed. (Boston, MA: Addison-Wesley, 2000); and John D. McGregor and David A. Sykes, *A Practical Guide to Testing Object-Oriented Software* (Boston, MA: Addison-Wesley, 2001). For more information on testing object-oriented software, see Robert V. Binder, *Testing Object-Oriented Systems: Models, Patterns, and Tools* (Reading, MA: Addison-Wesley, 1999); and Shel Sieget, *Object-Oriented Software Testing: A Hierarchical Approach* (New York: Wiley, 1996).

[10] It is not cost-effective to try to get each and every error out of the software. Except in simple examples, it is, in fact, impossible. There are simply too many combinations to check.

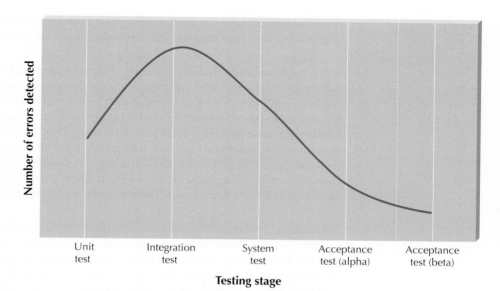

FIGURE 13-1
Error-Discovery
Rates for Different
Stages of Tests

Testing and Object Orientation

Most testing techniques have been developed to support non-object-oriented development. Object-oriented development is still relatively new. As such, most of the testing approaches have had to be adapted to object-oriented systems. The characteristics of object-oriented systems that affect testing the most are encapsulation (and information hiding), polymorphism (and dynamic binding), inheritance, and the use of patterns, class libraries, frameworks, and components. Also, the sheer volume of products that come out of a typical object-oriented development process has increased the importance of testing in object-oriented development.

Encapsulation and Information Hiding   Encapsulation and information hiding allow processes and data to be combined to create holistic entities (i.e., objects). Furthermore, they support hiding everything behind a visible interface. Although this allows the system to be modified and maintained in an effective and efficient manner, it makes testing the system problematic. What do you need to test to build confidence in the system's ability to meet the user's need? You need to test the business process that is represented in the use cases. However, the business process is distributed over a set of collaborating classes and contained in the methods of those classes. The only way to know the effect that a business process has on a system is to look at the state changes that take place in the system. But in object-oriented systems, the instances of the classes hide the data behind a class boundary. How is it possible then to see the impact of a business process?

A second issue raised by encapsulation and information hiding is the definition of a "unit" for unit testing. What is the unit to be tested? Is it the package, class, or method? In traditional approaches, the answer would be the process that is contained in a function. However, the process in object-oriented systems is distributed over a set of classes. Therefore, testing individual methods makes no sense. The answer is the class. This dramatically changes the manner in which unit testing is done.

A third issue raised is the impact on integration testing. In this case, objects can be aggregated to form aggregate objects; for instance, a car has many parts, or they can be grouped together to form collaborations. Furthermore, they can be used in class libraries, frameworks, and components. Based on all of these different ways in which classes can be grouped together, how does one effectively do integration testing?

Polymorphism and Dynamic Binding Polymorphism and dynamic binding dramatically impact both unit and integration testing. Because an individual business process is implemented through a set of methods distributed over a set of objects, as shown before, the unit test makes no sense at the method level. However, with polymorphism and dynamic binding, the same method (a small part of the overall business process) can be implemented in many different objects. Therefore, testing individual implementations of methods makes no sense. Again, the unit that makes sense to test is the class. Furthermore, except for trivial cases, dynamic binding makes it impossible to know which implementation is going to be executed until the system does it. Therefore, integration testing becomes very challenging.

Inheritance When taking into consideration the issues raised about inheritance (see Chapter 9), it should not be a surprise that inheritance impacts the testing of object-oriented systems. Through the use of inheritance, bugs can be propagated instantaneously from a superclass to all its direct and indirect subclasses. However, the tests that are applicable to a superclass are also applicable to all its subclasses. As usual, inheritance is a double-edged sword. Finally, even though we have stated this many times before, inheritance should support only a generalization and specialization type of semantics. Remember, when using inheritance, the principle of substitutability is critical (see Chapter 6). All these issues impact unit and integration testing.

Reuse On the surface, reuse should decrease the amount of testing required. However, each time a class is used in a different context, the class must be tested again. Therefore, anytime a class library, framework, or component is used, unit testing and integration testing are important. In the case of a component, the unit to test is the component itself. Remember that a component has a well-defined API (application program interface) that hides the details of its implementation.

Object-Oriented Development Process and Products In virtually all textbooks, including this one, testing is always covered near the end of system development. This seems to imply that testing is something that takes place only after the programming has ended. However, each and every product[11] that comes out of the object-oriented development process must be tested. For example, it is a lot easier to ensure that the requirements are captured and modeled correctly through testing the use cases. Furthermore, it is a lot cheaper to catch this type of error back in analysis than it is in implementation. Obviously, this is also true for testing collaborations. By the time we have implemented a collaboration as a set of layers and partitions, we could have expended a great deal of time—and time is money—on implementing the wrong thing. So, testing collaborations by role-playing the CRC cards in analysis will actually save the team lots of time and money.

Testing is something that must take place throughout system development, not simply at the end. However, the type of testing that can take place on nonexecutable representations, such as use cases and CRC cards, is different than those on code written in an

---

[11] For example, activity diagrams, use-case descriptions, use-case diagrams, CRC cards, class diagrams, object diagrams, sequence diagrams, communication diagrams, behavioral state machines, package diagrams, contracts, method specifications, use scenarios, window navigation diagrams, storyboards, windows layout diagrams, real use cases, and source code.

Class Test Plan		Page _____ of _____
Class Name:	Version Number:	CRC Card ID:
Tester:	Date Designed:	Date Conducted:

Class Objective:

Associated Contract IDs: _____

Associated Use Case IDs: _____

Associated Superclass(es): _____

Testing Objectives:

Walkthrough Test Requirements:

Invariant-Based Test Requirements:

State-Based Test Requirements:

Contract-Based Test Requirements:

FIGURE 13-2
Class Test Plan

object-oriented programming language. The primary approach to testing nonexecutable representations is some form of an inspection or walkthrough of the representation.[12] Role-playing the CRC cards based on use cases is an example of one type of a walkthrough.

Test Planning

Testing starts with the development of a *test plan,* which defines a series of tests that will be conducted. Since testing takes place through the development of an object-oriented system, a test plan should be developed at the very beginning of system development and continuously updated as the system evolves. The test plan should address all products that are created during the development of the system. For example, tests should be created that can be used to test completeness of a CRC card. Figure 13-2 shows a typical unit test plan form

[12] See Michael Fagan, "Design and Code Inspections to Reduce Errors in Program Development," *IBM Systems Journal,* 15, no. 3 (1976); and Daniel P. Freedman and Gerald M. Weinberg, *Handbook of Walkthrough, Inspections, and Technical Reviews: Evaluating Programs, Projects, and Products,* 3rd ed. (New York: Dorset House Publishing, 1990). Also, Chapter 8 describes the walkthrough process in detail in relation to the verification and validation of the analysis models.

Class Invariant Test Specification		Page _____ of _____
Class Name:	Version Number:	CRC Card ID:
Tester:	Date Designed:	Date Conducted:

Testing Objectives:

Test Cases:

Invariant Description	Original Attribute Value	Event	New Attribute Value	Expected Result	Result P/F
Attribute Name:					
1) _____	_____	_____	_____	_____	_____
2) _____	_____	_____	_____	_____	_____
3) _____	_____	_____	_____	_____	_____
Attribute Name:					
1) _____	_____	_____	_____	_____	_____
2) _____	_____	_____	_____	_____	_____
3) _____	_____	_____	_____	_____	_____
Attribute Name:					
1) _____	_____	_____	_____	_____	_____
2) _____	_____	_____	_____	_____	_____
3) _____	_____	_____	_____	_____	_____

FIGURE 13-3

Class Invariant Test Specification

for a class. For example, in Figure 13-3, a partial list of the invariant test specifications for a class are shown. Each individual test has a specific objective and describes a set of very specific *test cases* to examine. In the case of invariant-based tests, a description of the invariant is given, and the original values of the attribute, the event that will cause the attribute value to change, the actual results observed, the expected results, and whether it passed or failed are shown. *Test specifications* are created for each type of constraint that must be met by the class. Also, similar types of specifications are done for integration, system, and acceptance tests.

Not all classes are likely to be finished at the same time, so the programmer usually writes *stubs* for the unfinished classes to enable the classes around them to be tested. A stub is a placeholder for a class that usually displays a simple test message on the screen or returns some *hardcoded* value[13] when it is selected. For example, consider an application system that provides creating, changing, deleting, finding, and printing functions for some object such as CDs, patients, or employees. Depending on the final design, these different functions could end up in different objects on different layers. Therefore, to test the functionality associated with the classes on the problem-domain layer, a stub would be written for each of the classes on the other layers that interact with the problem-domain classes. These stubs would be the minimal interface necessary to be able to test the problem-domain classes.

---

[13] The word hardcoded means written into the program. For example, suppose you were writing a unit to calculate the net present value of a loan. The stub might be written to always display (or return to the calling module) a value of 100 regardless of the input values.

For example, they would have methods that could receive the messages being sent by the problem-domain layer objects and methods that could send messages to the problem-domain layer objects. Typically, the methods would display a message on the screen notifying the tester that the method was successfully reached (e.g., Delete item from Database method reached). In this way, the problem-domain classes could pass class testing before the classes on the other layers were completed.

Unit Tests

Unit tests focus on a single unit—the class. There are two approaches to unit testing: black-box testing and white-box testing (see Figure 13-4). *Black-box testing* is the most commonly used because each class represents an encapsulated object. Black-box testing is driven by the CRC cards, state chart diagrams, and contracts associated with a class, not by the programmers' interpretation. In this case, the test plan is developed directly from the specification of the class: each item in the specification becomes a test, and several test cases are developed for it. *White-box testing* is based on the method specifications associated with each class. However, white-box testing has had limited impact in object-oriented development. This is due to the rather small size of the individual methods in a class. As such, most approaches to testing classes use black-box testing to assure their correctness.

Class tests should be based on the invariants on the CRC cards, the behavioral state machines associated with each class, and the pre- and postconditions contained on each method's contract. Assuming all the constraints have been captured on the CRC cards and contracts, individual test cases can be developed fairly easily. For example, suppose the CRC card for an order class gave an invariant that the order quantity must be between 10 and 100 cases. The tester would develop a series of test cases to ensure that the quantity is validated before the system accepts it. It is impossible to test every possible combination of input and situation; there are simply too many possible combinations. In this example, the test requires a minimum of three test cases: one with a valid value (e.g., 15), one with a low invalid value (e.g., 7), and one with a high invalid value (e.g., 110). Most tests would also include a test case with a nonnumeric value to ensure the data types were checked (e.g., ABCD). A really good test would include a test case with nonsensical but potentially valid data (e.g. 21.4).

Using a behavioral state machine is a useful way to identify tests for a class. Any class that has a behavioral state machine associated with it will have a potentially complex life cycle. As such, it is possible to create a series of tests to guarantee that each state can be reached.

Tests also can be developed for each contract associated with the class. In the case of a contract, a set of tests for each pre- and postcondition is required. Furthermore, if the class is a subclass of another class, then all the tests associated with the superclass must be executed again. Also, the interactions among the constraints, invariants, and the pre- and postconditions in the subclass and the superclass(es) must be addressed.

Finally, owing to good object-oriented design, to fully test a class, special testing methods may have to be added to the class being tested. For example, how can invariants be tested? The only way to really test them is to have methods that are visible to the outside of the class that can be used to manipulate the values of the class's attributes. However, adding these types of methods to a class does two things. First, they add to the testing requirements because they themselves will have to be tested. Second, if they are not removed from the deployed version of the system, the system will be less efficient, and the advantage of information hiding effectively will be lost. As is readily apparent, testing classes is complex. Therefore, great care must be taken when designing tests for classes.

Stage	Types of Tests	Test Plan Source	When to Use	Notes
Unit Testing	**Black-Box Testing** Treats class as black-box **White-Box Testing** Looks inside the class to test its major elements	CRC Cards Class Diagrams Contracts Method Specifications	For normal unit testing When complexity is high	• Tester focuses on whether the class meets the requirements stated in the specifications. • By looking inside the class to review the code itself the tester may discover errors or assumptions not immediately obvious to someone treating the class as a black box.
Integration Testing	**User Interface Testing** The tester tests each interface function **Use-Case Testing** The tester tests each use case	Interface Design Use Cases	For normal integration testing When the user interface is important	• Testing is done by moving through each and every menu item in the interface either in a top-down or bottom-up manner. • Testing is done by moving through each use case to ensure they work correctly. • Usually combined with user interface testing because it does not test all interfaces.
	Interaction Testing Tests each process in a step-by-step fashion	Class Diagrams Sequence Diagrams Communication Diagrams	When the system performs data processing	• The entire system begins as a set of stubs. Each class is added in turn and the results of the class compared to the correct result from the test data; when a class passes, the next class is added and the test rerun. This is done for each package. Once each package has passed all tests, then the process repeats integrating the packages.
	System Interface Testing Tests the exchange of data with other systems	Use-Case Diagram	When the system exchanges data	• Because data transfers between systems are often automated and not monitored directly by the users it is critical to design tests to ensure they are being done correctly.
System Testing	**Requirements Testing** Tests to whether original business requirements are met	System Design, Unit Tests, and Integration Tests	For normal system testing	• Ensures that changes made as a result of integration testing did not create new errors. • Testers often pretend to be uninformed users and perform improper actions to ensure the system is immune to invalid actions (e.g., adding blank records).
	Usability Testing Tests how convenient the system is to use	Interface Design and Use Cases	When user interface is important	• Often done by analyst with experience in how users think and in good interface design. • Sometimes uses formal usability testing procedures discussed in Chapter 11.
	Security Testing Tests disaster recovery and unauthorized access	Infrastructure Design	When the system is important	• Security testing is a complex task, usually done by an infrastructure analyst assigned to the project. • In extreme cases, a professional firm may be hired.
	Performance Testing Examines the ability to perform under high loads	System Proposal Infrastructure Design	When the system is important	• High volumes of transactions are generated and given to the system. • Often done by using special purpose testing software.
	Documentation Testing Tests the accuracy of the documentation	Help System, Procedures, Tutorials	For normal system testing	• Analysts spot check or check every item on every page in all documentation to ensure the documentation items and examples work properly.
Acceptance Testing	**Alpha Testing** Conducted by users to ensure they accept the system	System Tests	For normal acceptance testing	• Often repeats previous tests but are conducted by users themselves to ensure they accept the system.
	Beta Testing Uses real data, not test data	System Requirements	When the system is important	• Users closely monitor system for errors or useful improvements.

FIGURE 13-4   Types of Tests

Class Case Test Plan		Page _____ of _____
Use Case Name:	Version Number:	Use Case ID:
Tester:	Date Designed:	Date Conducted:

Use Case Objective:

Associated Class IDs: _____

Associated Use Case IDs: _____

Testing Objectives:

Scenario-Based Test Requirements:

FIGURE 13-5
Use-Case Test Plan

Integration Tests

Integration tests assess whether a set of classes that must work together do so without error. They ensure that the interfaces and linkages between different parts of the system work properly. At this point, the classes have passed their individual unit tests, so the focus now is on the flow of control among the classes and on the data exchanged among them. Integration testing follows the same general procedures as unit testing: the tester develops a test plan that has a series of tests, which, in turn, have a test. Integration testing is often done by a set of programmers and/or systems analysts.

From an object-oriented systems perspective, integration testing can be difficult. A single class can be in many different aggregations, due to the way in which objects can be combined to form new objects, class libraries, frameworks, components, and packages. Where is the best place to start the integration? Typically, the answer is to begin with the set of classes, a collaboration, that are used to support the highest-priority use case (see Chapter 5). Also, dynamic binding makes it crucial to design the integration tests carefully to ensure the combinations of methods are tested.

There are four approaches to integration testing: *user interface testing*,[14] *use-case testing*, *interaction testing*, and *system interface testing* (see Figure 13-4). Most projects use all four approaches. However, like unit testing, integration testing must be carefully planned. For example, Figure 13-5 shows a typical form that can be used to plan a use-case test. Because each use case will be supported by a set of classes, the class test plans, and invariant test specifications (see Figures 13-2 and 13-3) can be reused for documenting the details for each test performed for the use case. In the case of use-case testing, only the aspects of the class and class invariants related to the specific use case are included in these use-case context-dependent class test plans and invariant test specifications. In many ways, use-case testing can be viewed as a more rigorous role-playing exercise (see Chapter 5). At this point in the system's development, we have completed the system's specification, and we have completed unit testing. Furthermore, we have the contracts and the methods fully specified for all layers. Therefore, we can more rigorously test whether the use case is fully supported or not.

[14] We describe some of the different types of user interface testing in Chapter 11.

One of the major problems with integration testing and object-oriented systems is the difficulty caused by the interaction of inheritance and dynamic binding. This specific problem has become known as the *yo-yo problem*. The yo-yo problem occurs when the analyst or designer must bounce up and down through the inheritance graph to understand the control flow through the methods being executed. In most cases, this is caused by a rather deep inheritance graph; that is, the subclass has many superclasses above it in the inheritance graph. The yo-yo problem becomes even more of a nightmare in testing object-oriented systems when inheritance conflicts exist and when multiple inheritance is used (See Chapter 9).

System Tests

System tests are usually conducted by the systems analysts to ensure that all classes work together without error. System testing is similar to integration testing but is much broader in scope. Whereas integration testing focuses on whether the classes work together without error, system tests examine how well the system meets business requirements and its usability, security, and performance under heavy load (see Figure 13-4). It also tests the system's documentation.

Acceptance Tests

Acceptance testing is done primarily by the users with support from the project team. The goal is to confirm that system is complete, meets the business needs that prompted the system to be developed, and is acceptable to the users. Acceptance testing is done in two stages: *alpha testing*, in which users test the system using made-up data, and *beta testing*, in which users begin to use the system with real data but are carefully monitored for errors (see Figure 13-4).

DEVELOPING DOCUMENTATION

Like testing, developing documentation of the system must be done throughout system development. There are two fundamentally different types of documentation: system documentation and user documentation. *System documentation* is intended to help programmers and systems analysts understand the application software and enable them to build it or maintain it after the system is installed. System documentation is largely a by-product of the systems analysis and design process and is created as the project unfolds. Each step and phase produces documents that are essential in understanding how the system is or is to be built, and these documents are stored in the project binder(s). In many object-oriented development environments, it is possible to somewhat automate the creation of detailed documentation for classes and methods. For example, in Java, if the programmers use javadoc-style comments, it is possible to create HTML pages that document a class and its methods automatically by using the javadoc utility.[15] Because most programmers look on documentation with much distaste, anything that can make documentation easier to create is useful.

[15] For those that have used Java, javadoc is how the JDK documentation from Sun is created.

User documentation (such as user's manuals, training manuals, and online help systems) is designed to help the user operate the system. Although most project teams expect users to have received training and to have read the user's manuals before operating the system, unfortunately, this is not always the case. It is more common today—especially in the case of commercial software packages for microcomputers—for users to begin using the software without training or reading the user's manuals. In this section, we focus on user documentation.[16]

User documentation is often left until the end of the project, which is a dangerous strategy. Developing good documentation takes longer than many people expect because it requires much more than simply writing a few pages. Producing documentation requires designing the documents (whether on paper or online), writing the text, editing the documents, and testing them. For good-quality documentation, this process usually takes about three hours per page (single-spaced) for paper-based documentation or two hours per screen for online documentation. Thus a "simple" documentation, such as a ten-page user's manual and a set of twenty help screens takes seventy hours. Of course, lower-quality documentation can be produced faster.

The time required to develop and test user documentation should be built into the project plan. Most organizations plan for documentation development to start once the interface design and program specifications are complete. The initial draft of documentation is usually scheduled for completion immediately after the unit tests are complete. This reduces (but doesn't eliminate) the chance that the documentation will need to be changed due to software changes and still leaves enough time for the documentation to be tested and revised before the acceptance tests are started.

Although paper-based manuals are still important, online documentation is becoming more important. Paper-based documentation is simpler to use because it is more familiar to users, especially novices who have less computer experience; online documentation requires the users to learn one more set of commands. Paper-based documentation is also easier to flip through and gain a general understanding of its organization and topics and can be used far away from the computer itself.

There are four key strengths of online documentation that all but guarantee it will be the dominant form for the next century. First, searching for information is often simpler (provided the help search index is well designed) because the user can type in a variety of keywords to view information almost instantaneously, rather than having to search through the index or table of contents in a paper document. Second, the same information can be presented several times in many different formats, so that the user can find and read the information in the most informative way (such redundancy is possible in paper documentation, but the cost and intimidating size of the resulting manual make it impractical). Third, online documentation enables many new ways for the user to interact with the documentation that is not possible in static paper documentation. For example, it is possible to use links, or "tool tips" (i.e., pop-up text; see Chapter 11) to explain unfamiliar terms, and one can write "show-me" routines that demonstrate on the screen exactly what buttons to click and text to type. Finally, online documentation is significantly less expensive to distribute than paper documentation.

Types of Documentation

There are three fundamentally different types of user documentation: reference documents, procedures manuals, and tutorials. *Reference documents* (also called the help system) are designed to be used when the user needs to learn how to perform a specific function (e.g., updating a field, adding a new record). Often people read reference information when they

---

[16] For more information on developing documentation, see Thomas T. Barker, *Writing Software Documentation* (Boston: Allyn and Bacon, 1998).

have tried and failed to perform the function; writing reference documents requires special care because the user is often impatient or frustrated when he or she begins to read them.

Procedures manuals describe how to perform business tasks (e.g., printing a monthly report, taking a customer order). Each item in the procedures manual typically guides the user through a task that requires several functions or steps in the system. Therefore, each entry is typically much longer than an entry in a reference document.

Tutorials—obviously—teach people how to use major components of a system (e.g., an introduction to the basic operations of the system). Each entry in the tutorial is typically longer still than the entries in procedures manuals, and the entries are usually designed to be read in sequence (whereas entries in reference documents and procedures manuals are designed to be read individually).

Regardless of the type of user documentation, the overall process for developing it is similar to the process of developing interfaces (see Chapter 11). The developer first designs the general structure for the documentation and then develops the individual components within it.

Designing Documentation Structure

In this section, we focus on the development of online documentation, because we believe it will become the most common form of user documentation. The general structure used in most online documentation, whether reference documents, procedures manuals, or tutorials, is to develop a set of *documentation navigation controls* that lead the user to *documentation topics*. The documentation topics are the material that user wants to read, whereas the navigation controls are the way in which the user locates and accesses a specific topic.

Designing the structure of the documentation begins by identifying the different types of topics and navigation controls that need to be included. Figure 13-6 shows a commonly used structure for online reference documents (i.e., the help system). The documentation topics generally come from three sources. The first and most obvious source of topics is the set of commands and menus in the user interface. This set of topics is very useful if the user wants to understand how a particular command or menu is used.

However, the users often don't know what commands to look for or where they are in the system's menu structure. Instead, users have tasks they want to perform, and rather than thinking in terms of commands, they think in terms of their business tasks. Therefore, the second and often more useful set of topics focuses on how to perform certain tasks, usually those in the use scenarios, WND, and the real use cases from the user interface design (see Chapter 11). These topics walk the user through the set of steps (often involving several keystrokes or mouse clicks) needed to perform some task.

The third topics are definitions of important terms. These terms are usually the use cases and classes in the system, but sometimes they also include commands.

There are five general types of navigation controls for topics, but not all systems use all five types (see Figure 13-6). The first is the table of contents that organizes the information in a logical form, as though the users were to read the reference documentation from start to finish. The index provides access into the topics based on important keywords, in the same way that the index at the back of a book helps us find topics. Text search provides the ability to search through the topics either for any text the user types or for words that match a developer-specified set of words that is much larger than the set of words in the index. Unlike the index, text search typically provides no organization to the words (other than alphabetical). Some systems provide the ability to use an intelligent agent to help in the search (e.g., the Microsoft Office Assistant—also know as the paper-clip guy). The fifth and final navigation controls to topics are the Weblike links between topics that enable the user to click and move among topics.

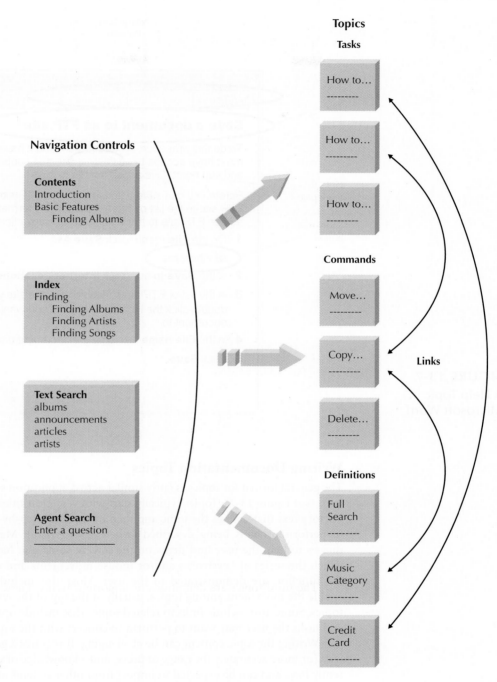

FIGURE 13-6
Organizing Online
Reference Documents

Procedures manuals and tutorials are similar but often simpler in structure. Topics for procedures manuals usually come from the use scenarios, WNDs, and the real use cases developed during interface design and from other basic tasks the users must perform. Topics for tutorials are usually organized around major sections of the system and the level of experience of the user. Most tutorials start with basic, most commonly used commands and them move into more complex and less frequently used commands.

Navigation
Buttons

Title

Links to
other pages

Show me
button

FIGURE 13-7
A Help Topic in
Microsoft Word

Writing Documentation Topics

The general format for topics is fairly similar across application systems and operating systems (see Figure 13-7). Topics typically start with very clear titles, followed by some introductory text that defines the topic and then by detailed, step-by-step instructions on how to perform what is being described (where appropriate). Many topics include screen images to help the user find items on the screen; some also have show-me examples, in which the series of keystrokes and/or mouse movements and clicks needed to perform the function are demonstrated to the user. Most also include navigation controls to enable the movement among topics, usually at the top of the window, plus links to other topics. Some also include links to related topics that include options or other commands and tasks the user may want to perform in concert with the topic being read.

Writing the topic content can be challenging. It requires a good understanding of the user (or more accurately the range of users) and a knowledge of what skills the user(s) currently have and can be expected to import from other systems and tools they are using or have used (including the system the new system is replacing). Topics should always be written from the viewpoint of the user and describe what the user wants to accomplish, not what the system can do. Figure 13-8 provides some general guidelines to improve the quality of documentation text.[17]

---

[17] One of the best books to explain the art of writing is William Strunk and E. B. White, *Elements of Style*, 4th ed. (Needham Heights, MA: Allyn & Bacon, 2000).

Guideline	Before the Guideline	After the Guideline
Use the active voice: The active voice creates more active and readable text by putting the subject at the start of the sentence, the verb in the middle, and the object at the end.	Finding albums is done using the album title, the artist's name, or a song title.	You can find an album by using the album title, the artist's name, or a song title.
Use e-prime style: E-prime style creates more active writing by omitting all forms of the verb to be.	The text you want to copy must be selected before you click on the copy button.	Select the text you want to copy before you click on the copy button.
Use consistent terms: Always use the same term to refer to the same items, rather than switching among synonyms (e.g., change, modify, update).	Select the text you want to copy. Pressing the copy button will copy the marked text to the new location.	Select the text you want to copy. Pressing the copy button will copy the selected text to the new location.
Use simple language: Always use the simplest language possible to accurately convey the meaning. This does not mean you should "dumb down" the text but that you should avoid artificially inflating its complexity. Avoid separating subjects and verbs and try to use the fewest words possible. (When you encounter a complex piece of text, try eliminating words; you may be surprised at how few words are really needed to convey meaning.)	The Georgia Statewide Academic and Medical System (GSAMS) is a cooperative and collaborative distance learning network in the state of Georgia. The organization in Atlanta that administers and manages the technical and overall operations of the currently more than 300 interactive audio and video teleconferencing classrooms throughout Georgia system is the Department of Administrative Service (DOAS). (56 words)	The Department of Administrative Service (DOAS) in Atlanta manages the Georgia Statewide Academic and Medical System (GSAMS), a distance learning network with more than 300 teleconferencing classrooms throughout Georgia. (29 words)
Use friendly language: Too often, documentation is cold and sterile because it is written in a very formal manner. Remember, you are writing for a person, not a computer.	Blank disks have been provided to you by Operations. It is suggested that you ensure your data are not lost by making backup copies of all essential data.	You should make a backup copy of all data that are important to you. If you need more diskettes, contact Operations.
Use parallel grammatical structures: Parallel grammatical structures indicate the similarity among items in list and help the reader understand content.	Opening files Saving a document How to delete files	Opening a file Saving a file Deleting a file
Use steps correctly: Novices often intersperse action and the results of action when describing a step-by-step process. Steps are always actions.	1. Press the customer button. 2. The customer dialogue box will appear. 3. Type the customer ID and press the submit button and the customer record will appear.	1. Press the customer button. 2. Type the customer ID in the customer dialogue box when it appears. 3. Press the submit button to view the customer record for this customer.
Use short paragraphs: Readers of documentation usually quickly scan text to find the information they need, so the text in the middle of long paragraphs is often overlooked. Use separate paragraphs to help readers find information more quickly.		

Source: Adapted from T. T. Barker, *Writing Software Documentation* (Boston: Allyn & Bacon, 1998).

FIGURE 13-8   Guidelines for Crafting Documentation Topics

Identifying Navigation Terms

As we write the documentation topics, we also begin to identify the terms that will be used to help users find topics. The table of contents is usually the most straightforward, because it is developed from the logical structure of the documentation topics, whether reference

13-2 Documentation for an ATM

*S*uppose you are a project manager for a bank developing software for ATMs. Develop an online help system.

topics, procedure topics, or tutorial topics. The items for the index and search engine require more care because they are developed from the major parts of the system and the users' business functions. Every time we write a topic, we must also list the terms that will be used to find the topic. Terms for the index and search engine can come from four distinct sources.

The first source for index terms is set of the commands in the user interface, such as open file, modify customer, and print open orders. All commands contain two parts (action and object). It is important to develop the index for both parts because users could search for information using either part. A user looking for more information about saving files, for example, might search by using the term save or the term files.

The second source is the set of major concepts in the system, which are often use cases and classes. In the case of Appointment system, for example, this might include appointment, symptoms, or patient.

A third source is the set of business tasks the user performs, such as ordering replacement units or making an appointment. Often these will be contained in the command set, but sometimes these require several commands and use terms that do not always appear in the system. Good sources for these terms are the use scenarios and real use cases developed during interface design (see Chapter 11).

A fourth, often controversial, source is the set of synonyms for the three sets of preceding items. Users sometimes don't think in terms of the nicely defined terms used by the system. They may try to find information on how to stop or quit rather than exit or erase rather than delete. Including synonyms in the index increases the complexity and size of the documentation system but can greatly improve the usefulness of the system to the users.

APPLYING THE CONCEPTS AT CD SELECTIONS

Managing Programming

To keep the project on schedule, Anne requested that she be allowed to assign three programmers from the CD Selections programming staff to develop the three major parts of the Internet Sales System. The first was the Web interface, both the client side (browser) and the server side. The second was the client–server-based management system (managing the CD information and marketing materials data bases). The third was the interfaces between the Internet Sales System and CD Selections' existing distribution system and the credit-card clearance center. Programming went smoothly and, despite a few minor problems, according to plan.

Testing

While Anne and the programmers were working, Alec began developing the test plans. The test plans for the three components were similar but slightly more intensive for the

Test Stage	Web Interface	System Management	System Interfaces
Unit tests	Black-box tests	Black-box tests	Black-box tests
Integration tests	User interface tests; use-case tests	User interface tests; use-case tests	System interface tests
System tests	Requirements tests; security tests; performance tests; usability tests	Requirements tests; security tests	Requirements tests; security tests; performance tests
Acceptance tests	Alpha test; beta test	Alpha test; beta test	Alpha test; beta test

Web interface component (see Figure 13-9). Unit testing would use black-box testing based on the CRC cards, class diagrams, and contracts for all components. Figure 13-10 shows part of the class invariant test specification for the order class in the Web interface component.

Integration testing for the Web interface and system management component would be subjected to all user interface and use-case tests to ensure the interface worked properly. The system interface component would undergo system interface tests to ensure that the system performed calculations properly and was capable of exchanging data with the CD Selections' other systems and the credit-card clearance center.

Class Invariant Test Specification		Page 5 of 15
Class Name: Order	**Version Number:** 3	**CRC Card ID:** 15
Tester: Susan Doe	**Date Designed:** 9/9	**Date Conducted:**

Testing Objectives: Ensure that the information entered by the customer on the place order form is valid.

Test Cases:

Invariant Description	Original Attribute Value	Event	New Attribute Value	Expected Result	Result P/F
Attribute Name: CD Number					
1) (1..1)	Null	CreateOrder	Null	F	
2) (unsigned long)	Null	CreateOrder	ABC	F	
3) (unsigned long)	Null	CreateOrder	123	P	
Attribute Name: CD Name					
1) (1..1)	Null	CreateOrder	Null	F	
2) (String)	Null	CreateOrder	ABC	P	
3) (String)	Null	CreateOrder	123	P	
Attribute Name: Artist Name					
1) (1..1)	Null	CreateOrder	Null	F	
2) (String)	Null	CreateOrder	ABC	P	
3) (String)	Null	CreateOrder	123	P	

Systems tests are by definition tests of the entire system—all components together. However, not all parts of the system would receive the same level of testing. Requirements tests would be conducted on all parts of the system to ensure that all requirements were met. Security was a critical issue, so the security of all aspects of the system would be tested. Security tests would be developed by CD Selections' infrastructure team, and once the system passed those tests, an external security consulting firm would be hired to attempt to break into the system.

Performance was an important issue for the parts of the system used by the customer (the Web interface and the system interfaces to the credit-card clearance and inventory systems), but it was not as important for the management component that would be used by staff, not customers. The customer-facing components would undergo rigorous performance testing to see how many transactions (whether searching or purchasing) they could handle before they were unable to provide a response time of two seconds or less. Alec also developed an upgrade plan so that as demand on the system increased, there was a clear plan for when and how to increase the processing capability of the system.

Finally, formal usability tests would be conducted on the Web interface portion of the system with six potential users (both novice and expert Internet users).

Acceptance tests would be conducted in two stages, alpha and beta. Alpha tests would be done during the training of CD Selections' staff. The Internet Sales manager would work together with Alec to develop a series of tests and training exercises to train the Internet Sales group staff on how to use the system. They would then load the real CD data into the system and begin adding marketing materials. These same staff and other CD Selections staff members would also pretend to be customers and test the Web interface.

Beta testing would be done by "going live" with the Web site but only announcing its existence to CD Selections employees. As an incentive to try the Web site (rather than buying from the store in which they worked), employees would be offered triple their normal employee discount for all products ordered from the Web site. The site would also have a prominent button on every screen that would enable employees to e-mail comments to the project team, and the announcement would encourage employees to report problems, suggestions, and compliments to the project team. After one month, assuming all went well, the beta test would be completed, and the Internet Sales site would be linked to the main Web site and advertised to the general public.

Developing User Documentation

While Anne and Alec were busy with the programming and test plans, Brian began the process of developing all necessary documentation. There were three types of documentation (reference documents, procedures manuals, and tutorials) that needed to be produced for the Web interface and the management component. Because the number of CD Selections staff using the system management component would be small, Brian suggested that they produce only the reference documentation (an online help system). After talking with Alec and Anne, the team felt that an intensive training program and a one month beta-test period would be sufficient without tutorials and formal procedures manuals. Likewise, they felt that the process of ordering CDs and the Web interface itself was simple enough to not require a tutorial on the Web—a help system would be sufficient (and a procedures manual didn't make sense).

Brian decided that the reference documents for both the Web interface and system management components would contain help topics for user tasks, commands, and definitions. He also decided that the documentation component would contain four types of

Tasks	Commands	Terms
Find an album	Find	Album
Add an album to my shopping cart	Browse	Artist
Placing an order	Quick search	Music type
How to buy	Full search	Special deals
What's in my shopping cart?		Cart
		Shopping cart

FIGURE 13-11
Sample Help Topics for CD Selections

navigation controls: a table of contents, an index, a find, and links to definitions. He did not feel that the system was complex enough to benefit from a search agent.

After these decisions were discussed with the development team, Brian delegated the development of the reference documents to a technical writer assigned to the project team. Figure 13-11 shows examples of a few of topics that the writer developed. The tasks and commands were taken directly from the interface design. The list of definitions was developed once the tasks and commands were developed, based on the writer's experience in understanding what terms might be confusing to a user.

Once the topic list was developed, the technical writer then began writing the topics themselves and the navigation controls to access. Figure 13-12 shows an example of one topic taken from the task list: how to place an order. This topic presented a brief description of what it was and then led the user through the step-by-step process needed to complete the task. The topic also listed the navigation controls that would be used to find the topic, in terms of the table of contents entries, the index entries, and search entries. It also listed the words in the topic itself that would have links to other topics (e.g., shopping cart).

Help Topic

How to Place an Order

When you are ready to pay for the merchandise you have selected (the items in your shopping cart) you can place your order. There are four steps.

1. Move to the Place order Page

Click on the [Place order] button to move to the place order page.

2. Make sure you are ordering what you want

The place order screen displays all the items in your shopping cart. Read through the list to make sure these are what you want because once you submit your credit card information you cannot change the order.

You can delete an item by

Navigation Controls

Table of Contents list:
  How to Place an Order

Index list:
  Credit Card
  Order
  Pay
  Place order

Search find by:
  Credit Card
  Delete Items
  Order
  Pay
  Place order
  Shopping Cart
  Verify Order

Links:
  Shopping Cart

FIGURE 13-12
Sample Documentation Topic for CD Selections

SUMMARY

Managing Programming

Programming is done by programmers, so systems analysts have few responsibilities during this stage. The project manager, however, is usually very busy. The first task is to assign the programmers to complete the project, ideally the fewest possible because coordination problems increase as the size of the programming team increases. Coordination can be improved by having regular meetings, ensuring that standards are followed, implementing change control, and using CASE tools effectively. One of the key functions of the project manager is to manage the schedule and adjust it for delays. Two common causes of delays are scope creep and minor slippages that go unnoticed. Given today's global development of information systems, different cultural issues need to be taken into consideration.

Designing Tests

Tests must be carefully planned because the cost of fixing one major bug after the system is installed can easily exceed the annual salary of a programmer. A test plan contains several tests that examine different aspects of the system. A test, in turn, specifies several test cases that will be examined by the testers. A unit test examines a class within the system; test cases come from the class specifications or the class code itself. An integration test examines how well several classes work together; test cases come from the interface design, use cases, and the use-case, sequence, and collaboration diagrams. A system test examines the system as a whole and is broader than the unit and integration tests; test cases come from the system design, the infrastructure design, the unit, and integration tests. Acceptance testing is done by the users to determine whether the system is acceptable to them; it draws on the system test plans (alpha testing) and the real work the users perform (beta testing).

Developing Documentation

Documentation, both user documentation and system documentation, is moving away from paper-based documents to online documentation. There are three types of user documentation: Reference documents are designed to be used when the user needs to learn how to perform a specific function (e.g., an online help system), procedures manuals describe how to perform business tasks, and tutorials teach people how to use the system. Documentation navigation controls (e.g., a table of contents, index, find, intelligent agents, or links between pages) enable users to find documentation topics (e.g., how to perform a function, how to use an interface command, an explanation of a term).

KEY TERMS

Acceptance test	Interaction testing	Test plan
Alpha test	Procedures manual	Test specification
Beta test	Program log	Traceability
Black-box testing	Reference document	Tutorial
Change control	Requirements testing	Unit test
Change management	Security testing	Usability testing
Construction	Stub	Use-case testing
Documentation navigation control	System documentation	User documentation
Documentation topic	System interface testing	User interface testing
Hardcoded	System test	White-box testing
Integration test	Test case	Yo-yo problem

QUESTIONS

1. What are the common language or languages used today in information systems development?
2. In *The Mythical Man-Month*, Frederick Brooks argues that adding more programmers to a late project makes it later. Why?
3. What is the primary role of systems analysts during the programming stage?
4. Compare and contrast reference documents, procedures manuals, and tutorials.
5. What are the primary disadvantages of online documentation?
6. Why is online documentation becoming more important?
7. What are the commonly used sources of documentation topics? Which is the most important? Why?
8. What are five types of documentation navigation controls?
9. What are the commonly used sources of documentation navigation controls? Which is the most important? Why?
10. How can different national or organizational cultures affect the management of an information systems development project?
11. What is the purpose of testing?
12. Compare and contrast the terms test, test plan, and test case.
13. What is the primary goal of unit testing?
14. How are the test cases developed for unit tests?
15. What is the primary goal of system testing?
16. How are the test cases developed for system tests?
17. What is the primary goal of integration testing?
18. How are the test cases developed for integration tests?
19. Compare and contrast black-box testing and white-box testing.
20. What are the different types of class tests?
21. What is a stub and why is it used in testing?
22. Describe the yo-yo problem. Why does it make integration testing difficult?
23. What is the primary goal of acceptance testing?
24. How are the test cases developed for acceptance tests?
25. Compare and contrast alpha testing and beta testing.
26. Describe how object orientation impacts testing.
27. Compare and contrast user documentation and system documentation.

EXERCISES

A. If the registration system at your university does not have a good online help system, develop one for one screen of the user interface.
B. Compare and contrast the online help at two different Web sites that enable you to perform some function (e.g., make travel reservations, order books).
C. Examine and prepare a report on the online help system for the calculator program in Windows (or a similar one on the Mac or Unix). (You will probably be surprised at the amount of help for such a simple program.)
D. Create an invariant test specification for the class you chose for Picnics R Us in exercise C in Chapter 9.
E. Create a use-case test plan, including the specific class plans and invariant tests, for a use case from the Picnics R Us exercises in the previous chapters.
F. Create an invariant test specification for the class you chose for the Of-the-Month Club (OTMC) in exercise E in Chapter 9.
G. Create a use-case test plan, including the specific class plans and invariant tests, for a use case from the Of-the-Month Club (OTMC) exercises in the previous chapters.
H. Create an invariant test specification for the class you chose for the health club in exercise A in Chapter 9.
I. Create a use-case test plan, including the specific class plans and invariant tests for a use case from the health club exercises in previous chapters.
J. Create an invariant test specification for the class you chose for the library in exercise G in Chapter 9.
K. Create a use-case test plan, including the specific class plans and invariant tests, for a use case from the library exercises in the previous chapters.
L. Different views of plagiarism and collaborative learning were described as examples of differences among different cultures today. Using the Web, identify other differences that may impact the success of an information systems development team.

MINICASES

1. Rajan Srikant is a project manager on a new systems development project for a major Indian outsourcing firm. This project is Rajan's first, and he has led his team successfully to the programming phase. The project has not always gone smoothly; Rajan himself has made a few mistakes, but he is generally pleased with the progress of his team and the quality of the system being developed. Now that programming has begun, Rajan is hoping for a break from the hectic pace at work.

   Prior to the programming phase, Rajan recognized that the time estimates made earlier in the project were too optimistic. However, he was firmly committed to meeting the project deadline because he wanted his first project to be a success. Anticipating time constraints, Rajan arranged with the human resources department to bring in two new college graduates and two college interns to increase the programming staff. He would have liked to recruit more experienced workers, but he was committed to keeping the already tight project budget under control.

   Rajan lined up the programming assignments, and work on the programs began about two weeks ago. Lately, he has received unfavorable feedback from the programming team leaders. Some bugs have been found when modules from different programmers were integrated. Several programmers have discovered to their dismay that someone had amended their programs without their knowledge.

   a. What problems can you identify in this situation? What advice do you have for the project manager?

   b. Will Rajan be able to achieve his desired goals of being on time and within budget?

2. The systems analysts are developing the test plan for the user interface for the Holiday Travel Vehicles system. As the salespeople are entering a sales invoice into the system, they will be able to either enter an option code into a text box, or select an option code from a drop-down list. A combo box was used to implement this, since it was felt that the salespeople would quickly become familiar with the most common option codes and would prefer entering them directly to speed up the entry process.

   It is now time to develop the test for validating the option code field during data entry. If the customer did not request any dealer-installed options for the vehicle, the salesperson should enter "none"; the field should not be blank. The valid option codes are four-character alphabetic codes and should be matched against a list of valid codes.

   Prepare a test plan for the test of the option code field during data entry.

CHAPTER 14

INSTALLATION

This chapter examines the activities needed to install an information system and successfully convert an organization to using it. It also discusses postimplementation activities, such as system support, system maintenance, and project assessment. Installing the system and making it available for use from a technical perspective is relatively straightforward. However, the training and organizational issues surrounding the installation are more complex and challenging because they focus on people, not computers.

OBJECTIVES

- Be familiar with the system installation process.
- Understand different types of conversion strategies and when to use them.
- Understand several techniques for managing change.
- Be familiar with postinstallation processes.

CHAPTER OUTLINE

INTRODUCTION

It must be remembered that there is nothing more difficult to plan, more doubtful of success, nor more dangerous to manage than the creation of a new system. For the initiator has the animosity of all who would profit by the preservation of the old institution and merely lukewarm defenders in those who would gain by the new.

—Machiavelli, *The Prince*, 1513

533

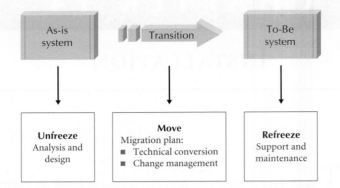

FIGURE 14-1
Implementing Change

Although written almost 500 years ago, Machiavelli's comments are still true today. Managing the change to a new system—whether it is computerized or not—is one of the most difficult tasks in any organization. Because of the challenges involved, most organizations begin developing their conversion and change management plans while the programmers are still developing the software. Leaving conversion and change management planning to the last minute is a recipe for failure.

In many ways, using a computer system or set of work processes is much like driving on a dirt road. Over time, with repeated use, the road begins to develop ruts in the most commonly used parts of the road. Although these ruts show where to drive, they make change difficult. As people use a computer system or set of work processes, those systems or work processes begin to become habits or norms; people learn them and become comfortable with them. These systems or work processes then begin to limit people's activities and make it difficult for them to change because they begin to see their jobs in terms of these processes rather than of the final business goal of serving customers.

One of the earliest models for managing organizational change was developed by Kurt Lewin.[1] Lewin argued that change is a three-step process: unfreeze, move, refreeze (Figure 14-1). First, the project team must *unfreeze* the existing habits and norms (the as-is system) so that change is possible. Most of system development to this point has laid the groundwork for unfreezing. Users are aware of the new system being developed, some have participated in an analysis of the current system (and so are aware of its problems), and some have helped design the new system (and so have some sense of the potential benefits of the new system). These activities have helped to unfreeze the current habits and norms.

The second step is to help the organization move to the new system via a *migration plan*. The migration plan has two major elements. One is technical, which includes how the new system will be installed and how data in the as-is system will be moved into the to-be system; this is discussed in the conversion section of this chapter. The second component is organizational, which includes helping users understand the change and motivating them to adopt it; this is discussed in the change management section of this chapter.

The third step is to *refreeze* the new system as the habitual way of performing the work processes—ensuring that the new system successfully becomes the standard way of performing the business function it supports. This refreezing process is a key goal of the *postimplementation* activities discussed in the final section of this chapter. By providing ongoing support for the new system and immediately beginning to identify improvements for the next version of the system, the organization helps solidify the new system as the new habitual way of doing business. Postimplementation activities include system support,

[1] Kurt Lewin, "Frontiers in Group Dynamics," *Human Relations*, 1, no. 5 (1947): 5–41; and Kurt Lewin, "Group Decision and Social Change," in E. E. Maccoby, T. M. Newcomb, and E. L. Hartley (eds.), *Readings in Social Psychology* (New York: Holt, Rinehart & Winston, 1958), pp. 197–211.

which means providing help desk and telephone support for users with problems; system maintenance, which means fixing bugs and improving the system after it has been installed; and project assessment, which means the process of evaluating the project to identify what went well and what could be improved for the next system development project.

Change management is the most challenging of the three components because it focuses on people, not technology, and because it is the one aspect of the project that is the least controllable by the project team. Change management means winning the hearts and minds of potential users and convincing them that the new system actually provides value.

Maintenance is the most costly aspect of the installation process, because the cost of maintaining systems usually greatly exceeds the initial development costs. It is not unusual for organizations to spend 60 to 80 percent of their total IS development budget on maintenance. Although this may sound surprising initially, think about the software you use. How many software packages do you use that are the very first version? Most commercial software packages become truly useful and enter widespread use only in their second or third version. Maintenance and continual improvement of software is ongoing, whether it is a commercially available package or software developed in-house. Would you buy software if you knew that no new versions were going to be produced? Of course, commercial software is somewhat different from custom in-house software used by only one company, but the fundamental issues remain.

Project assessment is probably the least commonly performed part of system development but is perhaps the one that has the most long-term value to the IS department. Project assessment enables project team members to step back and consider what they did right and what they could have done better. It is an important component in the individual growth and development of each member of the team, because it encourages team members to learn from their successes and failures. It also enables new ideas or new approaches to system development to be recognized, examined, and shared with other project teams to improve their performance.

In this chapter, we describe deploying the new system through the process of transitioning from the old to the new system (i.e., conversion). Next, we describe issues related to managing the changes necessary to adapt to the new business processes. Finally, we describe issues related to placing the system into production (i.e., postimplementation activities). However, before we address these issues, we describe how the cultural issues impact the deployment of a new system.

CULTURAL ISSUES AND INFORMATION TECHNOLOGY[2]

Cultural issues are one of the things that are typically identified as at least partially to blame when there is a failure in an organization. Cultural issues have been studied at both organizational and national levels. In previous chapters we discussed the effect that cultural issues can have on designing the human–computer interaction and physical architecture layers (see Chapters 11 and 12) and the management of programmers. In this chapter, we describe how these issues can impact the successful deployment of an information system that supports a global information supply chain. Cultural issues include things such as power distance, uncertainty avoidance, individualism versus collectivism, masculinity versus femininity, and long- versus short-term orientation.[3]

---

[2] A good summary of cultural issues and information systems is Dorothy E. Leidner and Timothy Kayworth, "A Review of Culture in Information Systems Research: Toward a Theory of Information Technology Culture Conflict," *MIS Quarterly* 30, no. 2 (2006): 357–399.

[3] See Geert Hofstede, *Culture's Consequences: Comparing Values, Behaviors, Institutions and Organizations Across Nations,* 2nd ed. (Thousand Oaks, CA: Sage, 2001).

The first dimension, *power distance*, addresses how power issues are dealt with in the culture. For example, if a superior in an organization has an incorrect belief about an important issue, can a subordinate point out this error? In some cultures, the answer is a resounding no. As such, this dimension could have major ramifications for the successful development and deployment of an information system. For example, it could have affected the determination of the actual requirements used to create the system. If the superior really does not understand the actual problem being solved, then the only way to gather the correct requirements would be to ensure that the subordinates are not negatively impacted when they cooperate. Otherwise, it will be virtually impossible to create a system that will solve the client's problem.

The second dimension, *uncertainty avoidance*, is based on the degree to which the culture depends on rules for direction, how well individuals in the culture handle stress, and the importance of employment stability. Depending on the culture, this dimension can be associated with the power distance dimension. For example, if the culture demands that an employee be subordinate to his or her supervisor, it can influence the power distance dimension, which can influence the manner in which the employee views the stability of his or her employment. As another example, through the use of detailed procedures manuals, a good deal of uncertainty in decision making can be avoided. From an IT perspective, this can influence the importance of the informal structure of an organization. In fact, if the culture really tries to remove any and all uncertainty from decision making, there may be little, if any, informal organizational structures to uncover. However, the opposite could also be true where the informal organization is simply buried very deep.

The third dimension, *individualism* versus *collectivism*, is based on the level of emphasis the culture places on the individual or the collective, or group. The relationship between the individual and the group is important for the success of an information system. In the West, individualism is rewarded. However, in the East, it is believed by focusing on optimizing the group, the individual will be most successful. In other words, it is the group that is the most important. Depending on the culture's orientation, the success of an information system being transitioned into production can depend on whether the focus of the information system will benefit the individual or the group.

The fourth dimension, *masculinity* versus *femininity*, from an information system's perspective, raises the issues of how gender differences are addressed. Some of the differences that could impact the utility of an information system include males' work motivation being based on advancement, earnings, and training, whereas females' work motivations are based on friendly atmosphere, physical conditions, and cooperation.[4] Depending on how the culture views this dimension, different motivations may need to be used to increase the likelihood of the information system being successfully deployed.

The fifth dimension, *long-* versus *short-term orientation*, deals with how the culture views the past and the future. In the East, long-term thinking is highly respected, whereas in the West, short-term profits and the current stock price seem to be the only things that matter. Based on this dimension, all the political concerns raised previously in this text become very important. For example, if the local culture views success only in a short-term manner, then any new information system that is deployed to support one department of an organization may give that department a competitive advantage over other departments in the short run. If only short-run measures are used to judge the success of a department, then it would be in the interest of the other departments

[4] See footnote 3.

to fight the successful deployment of the information system. However, if a longer-run perspective is the norm, then the other departments could be convinced to support the new information system because they could have new supportive information systems in the future.

Obviously, when reviewing these dimensions, they do interact with each other. The most important thing to remember from an IT perspective is that we must be careful not to view the local user community through our eyes; in a global economy, we must take into consideration the local cultural concerns for the information system to be deployed in a successful manner.

CONVERSION[5]

Conversion is the technical process by which a new system replaces an old system. Users are moved from using the as-is business processes and computer programs to the to-be business processes and programs. The migration plan specifies what activities will be performed when and by whom and includes both technical aspects (such as installing hardware and software and converting data from the as-is system to the to-be system) and organizational aspects (such as training and motivating the users to embrace the new system). Conversion refers to the technical aspects of the migration plan.

There are three major steps to the conversion plan before commencement of operations: install hardware, install software, and convert data (Figure 14-2). Although it may be possible to do some of these steps in parallel, usually they must be done sequentially at any one location.

The first step in the conversion plan is to buy and install any needed hardware. In many cases, no new hardware is needed, but sometimes the project requires such new hardware as servers, client computers, printers, and networking equipment. It is critical to work closely with vendors who are supplying needed hardware and software to ensure that the deliveries are coordinated with the conversion schedule so that the equipment is available when it is needed. Nothing can stop a conversion plan in its tracks as easily as the failure of a vendor to deliver needed equipment.

FIGURE 14-2
Elements of a
Migration Plan

---

[5] The material in this section is related to the Enhanced Unified Process's Transition phase and the Deployment workflow (see Figure 1-11).

FIGURE 14-3
Conversion Strategies

Once the hardware is installed, tested, and certified as being operational, the second step is to install the software. This includes the to-be system under development and, sometimes, additional software that must be installed to make the system operational. At this point, the system is usually tested again to ensure that it operates as planned.

The third step is to convert the data from the as-is system to the to-be system. Data conversion is usually the most technically complicated step in the migration plan. Often, separate programs must be written to convert the data from the as-is system to the new formats required in the to-be system and store it in the to-be system files and databases. This process is often complicated by the fact that the files and databases in the to-be system do not exactly match the files and databases in the as-is system (e.g., the to-be system may use several tables in a database to store customer data that were contained in one file in the as-is system). Formal test plans are always required for data conversion efforts (see Chapter 13).

Conversion can be thought of along three dimensions: the style in which the conversion is done (*conversion style*), what location or work groups are converted at what time (*conversion location*), and what modules of the system are converted at what time (*conversion modules*). Figure 14-3 shows the potential relationships among these three dimensions.

Conversion Style

The conversion style is the way in which users are switched between the old and new systems. There are two fundamentally different approaches to the style of conversion: direct conversion and parallel conversion.

Direct Conversion With *direct conversion* (sometimes called cold turkey, big bang, or abrupt cutover), the new system instantly replaces the old system. The new system is turned on and the old system is immediately turned off. This is the approach that we are likely to use when we upgrade commercial software (e.g., Microsoft Word) from one version to another; we simply begin using the new version and stop using the old version.

Direct conversion is the simplest and most straightforward. However, it is the most risky because any problems with the new system that have escaped detection during testing may seriously disrupt the organization.

Parallel Conversion   With *parallel conversion*, the new system is operated side by side with the old system; both systems are used simultaneously. For example, if a new accounting system is installed, the organization enters data into both the old system and the new system and then carefully compares the output from both systems to ensure that the new system is performing correctly. After some time period (often one to two months) of parallel operation and intense comparison between the two systems, the old system is turned off and the organization continues using the new system.

This approach is more likely to catch any major bugs in the new system and prevent the organization from suffering major problems. If problems are discovered in the new system, the system is simply turned off and fixed and then the conversion process starts again. The problem with this approach is the added expense of operating two systems that perform the same function.

Conversion Location

Conversion location refers to those parts of the organization that are converted at what points in time the conversion occurs. Often, parts of the organization are physically located in different offices (e.g., Toronto, Atlanta, Los Angeles). In other cases, location refers to different organizational units located in different parts of the same office complex (e.g., order entry, shipping, purchasing). There are at least three fundamentally different approaches to selecting the way in which different organizational locations are converted: pilot conversion, phased conversion, and simultaneous conversion.

Pilot Conversion   With a *pilot conversion*, one or more locations or units or work groups within a location are selected to be converted first as part of a pilot test. The locations participating in the pilot test are converted (using either direct or parallel conversion). If the system passes the pilot test, then the system is installed at the remaining locations (again using either direct or parallel conversion).

Pilot conversion has the advantage of providing an additional level of testing before the system is widely deployed throughout the organization, so that any problems with the system affect only the pilot locations. However, this type of conversion obviously requires more time before the system is installed at all organizational locations. Also, it means that different organizational units are using different versions of the system and business processes, which may make it difficult for them to exchange data.

Phased Conversion   With *phased conversion*, the system is installed sequentially at different locations. A first set of locations is converted, then a second set, then a third set, and so on, until all locations are converted. Sometimes there is a deliberate delay between the different sets (at least between the first and the second), so that any problems with the system are detected before too much of the organization is affected. In other cases, the sets are converted back to back so that as soon as those converting one location have finished, the project team moves to the next and continues the conversion.

Phased conversion has the same advantages and disadvantages of pilot conversion. In addition, it means that fewer people are required to perform the actual conversion (and any associated user training) than if all locations were converted at once.

Simultaneous Conversion   *Simultaneous conversion*, as the name suggests, means that all locations are converted at the same time. The new system is installed and made ready at all locations; at a preset time, all users begin using the new system. Simultaneous conversion is often used with direct conversion, but it can also be used with parallel conversion.

14-A Too Much Paper (Part 1)

*T*he South Dakota Department of Worker's Compensation was sinking under a load of paper files. As a state agency that oversees that employees are treated fairly when they are injured on the job, the agency had a plethora of paper files and filing cabinets. If a person (or company) called to see the status of an injury claim, the clerk who received the call would have to take a message, get the paper file, review the status, and call the person back. Files were stored in huge filing cabinets and were entered by year and case number (e.g., the 415th person injured in 2008 would be in a file numbered 08-415). But, most people did not remember their file number and would give a name, address, and date of injury. The clerk would look in a spiral notebook for the last name around the date that was given and then find the file number to retrieve the folder. Some folders were small—possibly a minor cut or minor injury that was taken care of quickly and the employee was back to work. Other folders could be very large, with medical reports from several doctors verifying the extent of the injury (such as an arm amputation). A digital solution was suggested; reports could be submitted online using a secure Web site. Medical reports could be submitted electronically, either as a .pdf file or as a faxed digital file. This solution would also mean that the clerk taking the phone call could query the database by the person's name and access the information in a matter of seconds.

Questions

1. The digital solution was going to change the work process of filing injury claims, interacting with people who filed the claims (or companies) wanting to see the status of the claim, and with the process of claims. What might this mean from a work-flow analysis?
2. In many ways, this was a business process reengineering solution. The proposal was to throw out the old process for a completely electronic version. What might a systems analyst do in the data-gathering stage?

Simultaneous conversion eliminates problems with having different organizational units using different systems and processes. However, it also means that the organization must have sufficient staff to perform the conversion and train the users at all locations simultaneously.

Conversion Modules

Although it is natural to assume that systems are usually installed in their entirety, this is not always the case.

Whole-System Conversion A *whole-system conversion*, in which the entire system is installed at one time, is the most common. It is simple and the easiest to understand. However, if the system is large and/or extremely complex (e.g., an enterprise resource-planning system such as SAP or PeopleSoft), the whole system may prove too difficult for users to learn in one conversion step.

Modular Conversion When the *modules*[6] within a system are separate and distinct, organizations sometimes choose to convert to the new system one module at a time—that is, using modular conversion. Modular conversion requires special care in developing the system (and usually adds extra cost). Each module either must be written to work with both the old and new systems or object wrappers (see Chapter 8) must be used to encapsulate the old system from the new. When modules are tightly integrated, this is

---

[6] In this case, a module is typically a component or a package, that is, a set of collaborating classes.

Characteristic	Conversion Style		Conversion Location			Conversion Modules	
	Direct Conversion	Parallel Conversion	Pilot Conversion	Phased Conversion	Simultaneous Conversion	Whole-System Conversion	Modular Conversion
Risk	High	Low	Low	Medium	High	High	Medium
Cost	Low	High	Medium	Medium	High	Medium	High
Time	Short	Long	Medium	Long	Short	Short	Long

FIGURE 14-4   Characteristics of Conversion Strategies

very challenging and, therefore, seldom done. However, when there is only a loose association between modules, module conversion becomes easier. For example, consider a conversion from an old version of Microsoft Office to a new version. It is relatively simple to convert from the old version of Word to the new version without simultaneously having to change from the old to the new version of Microsoft Excel.

Modular conversion reduces the amount of training required to begin using the new system. Users need training only in the new module being implemented. However, modular conversion does take longer and has more steps than does the whole-system process.

Selecting the Appropriate Conversion Strategy

Each of the three dimensions in Figure 14-3 is independent, so that a conversion strategy can be developed to fit in any one of the boxes in this figure. Different boxes can also be mixed and matched into one *conversion strategy*. For example, one commonly used approach is to begin with a pilot conversion of the whole system using parallel conversion in a handful of test locations. Once the system has passed the pilot test at these locations, it is then installed in the remaining locations using phased conversion with direct cutover. There are three important factors to consider in selecting a conversion strategy: *risk*, *cost*, and the *time* required (Figure 14-4).

Risk   After the system has passed a rigorous battery of unit, system, integration, and acceptance testing, it should be bug free … maybe. Because humans make mistakes, nothing built by people is ever perfect. Even after all these tests, there may still be a few undiscovered bugs. The conversion process provides one last step in which to catch these bugs before the system goes live and the bugs have the chance to cause problems.

Parallel conversion is less risky than is direct conversion because it has a greater chance of detecting bugs that have gone undiscovered in testing. Likewise, pilot conversion is less risky than is phased conversion or simultaneous conversion because if bugs do occur, they occur in pilot test locations whose staff are aware that they may encounter bugs. Because potential bugs affect fewer users, there is less risk. Likewise, converting a few modules at a time lowers the probability of a bug because there is more likely to be a bug in the whole system than in any given module.

How important the risk is depends on the system being implemented—the combination of the probability that bugs remain undetected in the system and the potential cost of those undetected bugs. If the system has indeed been subjected to extensive methodical testing, including alpha and beta testing, then the probability of undetected bugs is lower than if the testing was less rigorous. However, there still might have been mistakes made in the analysis process, so that although there might be no software bugs, the software might fail to properly address the business needs.

Assessing the cost of a bug is challenging, but most analysts and senior managers can make a reasonable guess at the relative cost of a bug. For example, the cost of a bug in an automated stock market trading program or a heart-lung machine keeping someone alive is likely to be much greater than a bug in a computer game or word processing program. Therefore, risk is likely to be a very important factor in the conversion process if the system has not been as thoroughly tested as it might have been and/or if the cost of bugs is high. If the system has been thoroughly tested and/or the cost of bugs is not that high, then risk becomes less important to the conversion decision.

Cost   As might be expected, different conversion strategies have different costs. These costs can include things such as salaries for people who work with the system (e.g., users, trainers, system administrators, external consultants), travel expenses, operation expenses, communication costs, and hardware leases. Parallel conversion is more expensive than direct cutover because it requires that two systems (the old and the new) be operated at the same time. Employees must then perform twice the usual work because they have to enter the same data into both the old and the new systems. Parallel conversion also requires the results of the two systems to be completely cross-checked to make sure there are no differences between the two, which entails additional time and cost.

Pilot conversion and phased conversion have somewhat similar costs. Simultaneous conversion has higher costs because more staff are required to support all the locations as they simultaneously switch from the old to the new system. Modular conversion is more expensive than whole-system conversion because it requires more programming. The old system must be updated to work with selected modules in the new system, and modules in the new system must be programmed to work with selected modules in both the old and new systems.

Time   The final factor is the amount of time required to convert between the old and the new system. Direct conversion is the fastest because it is immediate. Parallel conversion takes longer because the full advantages of the new system do not become available until the old system is turned off. Simultaneous conversion is fastest because all locations are converted at the same time. Phased conversion usually takes longer than pilot conversion because once the pilot test is complete, all remaining locations are usually (but not always) converted simultaneously. Phased conversion proceeds in waves, often requiring several months before all locations are converted. Likewise, modular conversion takes longer than whole-system conversion because the models are introduced one after another.

YOUR TURN

14-1 Developing a Conversion Plan

Suppose you are leading the conversion from one word processor to another at your university. Develop a conversion plan (i.e., technical issues only). You have also been asked to develop a conversion plan for the university's new Web-based course registration system. How would the second conversion plan be similar to and different from the one you developed for the word processor?

14-B U.S. Army Installation Support

Throughout the 1960s, 1970s, and 1980s, the U.S. Army automated its installations (army bases, in civilian terms). Automation was usually a local effort at each of the more than 100 bases. Although some bases had developed software together (or borrowed software developed at other bases), each base often had software that performed different functions or performed the same function in different ways. In 1989, the army decided to standardize the software so that the same software would be used everywhere. This would greatly reduce software maintenance and also reduce training when soldiers were transferred between bases.

The software took four years to develop. The system was quite complex, and the project manager was concerned that there was a high risk that not all requirements of all installations had been properly captured. Cost and time were less important because the project had already run four years and cost $100 million.

Therefore, the project manager chose a modular pilot conversion using parallel conversion. The manager selected seven installations, each representing a different type of army installation (e.g., training base, arsenal, depot) and began the conversion. All went well, but several new features were identified that had been overlooked during the analysis, design, and construction. These were added and the pilot testing resumed. Finally, the system was installed in the rest of the army installations using a phased direct conversion of the whole system.

—*Alan Dennis*

Questions

1. Do you think the conversion strategy was appropriate?
2. Regardless of whether you agree, what other conversion strategy could have been used?

CHANGE MANAGEMENT[7]

In the context of a systems development project, change management is the process of helping people to adopt and adapt to the to-be system and its accompanying work processes without undue stress. There are three key roles in any major organizational change. The first is the *sponsor* of the change—the person who wants the change. This person is the business sponsor who first initiated the request for the new system (see Chapter 2). Usually, the sponsor is a senior manager of the part of the organization that must adopt and use the new system. It is critical that the sponsor be active in the change management process because a change that is clearly being driven by the sponsor, not by the project team or the IS organization, has greater legitimacy. The sponsor has direct management authority over those who adopt the system.

The second role is that of the *change agent*—the person(s) leading the change effort. The change agent, charged with actually planning and implementing the change, is usually someone outside of the business unit adopting the system and therefore has no direct management authority over the potential adopters. Because the change agent is an outsider from a different organizational culture, he or she has less credibility than do the sponsor and other members of the business unit. After all, once the system has been installed, the change agent usually leaves and thus has no ongoing impact.

---

[7] The material in this section is related to the Enhanced Unified Process's Transition and Production phases and the Configuration and Change Management workflow (see Figure 1-11). Many books have been written on change management. Some of our favorites are the following: Patrick Connor and Linda Lake, *Managing Organizational Change*, 2nd ed. (Westport, CT: Praeger, 1994); Douglas Smith, *Taking Charge of Change* (Reading, MA: Addison-Wesley, 1996); Daryl Conner, *Managing at the Speed of Change* (New York: Villard Books, 1992); and Mary Lynn Manns and Linda Rising, *Fearless Change: Patterns for Introducing New Ideas* (Boston, MA: Addison-Wesley, 2005).

The third role is that of *potential adopters,* or targets of the change—the people who actually must change. These are the people for whom the new system is designed and who will ultimately choose to use or not use the system.

In the early days of computing, many project teams simply assumed that their job ended when the old system was converted to the new system at a technical level. The philosophy was "build it and they will come." Unfortunately, that happens only in the movies. Resistance to change is common in most organizations. Therefore, the change management plan is an important part of the overall installation plan that glues together the key steps in the change management process. Successful change requires that people want to adopt the change and are able to adopt the change. The change management plan has four basic steps: revising management policies, assessing the cost and benefit models of potential adopters, motivating adoption, and enabling people to adopt through training (see Figure 14-2). However, before we can discuss the change management plan, we must first understand why people resist change.

Understanding Resistance to Change[8]   People resist change—even change for the better—for very rational reasons. What is good for the organization is not necessarily good for the people who work there. For example, consider an order-processing clerk who used to receive orders to be shipped on paper shipping documents but now uses a computer to receive the same information. Rather than typing shipping labels with a typewriter, the clerk now clicks on the print button on the computer and the label is produced automatically. The clerk can now ship many more orders each day, which is a clear benefit to the organization. The clerk, however, probably doesn't really care how many packages are shipped. His or her pay doesn't change; it's just a question of which the clerk prefers to use, a computer or typewriter. Learning to use the new system and work processes—even if the change is minor—requires more effort than continuing to use the existing, well-understood system and work processes.

So why do people accept change? Simply put, every change has a set of costs and benefits associated with it. If the benefits of accepting the change outweigh the costs of the change, then people change. And sometimes the benefit of change is avoidance of the pain that might be experienced if the change were not adopted (e.g., if you don't change, you are fired, so one of the benefits of adopting the change is that you still have a job).

In general, when people are presented with an opportunity for change, they perform a cost–benefit analysis (sometime consciously, sometimes subconsciously) and decide the extent to which they will embrace and adopt the change. They identify the costs of and benefits from the system and decide whether the change is worthwhile. However, it is not that simple, because most costs and benefits are not certain. There is some uncertainty as to whether a certain benefit or cost will actually occur; so both the costs of and benefits from the new system will need to be weighted by the degree of certainty associated with them (Figure 14-5). Unfortunately, most humans tend to overestimate the probability of costs and underestimate the probability of benefits.

There are also costs and, sometimes, benefits associated with the actual *transition process* itself. For example, suppose we found a nicer house or apartment than our current one. Even if we liked it better, we might decide not to move simply because the cost of moving outweighed the benefits from the new house or apartment itself. Likewise, adopting a new computer system might require us to learn new skills, which could be seen as a cost to some people or as a benefit to others, if they perceived that those skills would somehow

[8] This section benefited from conversations with Dr. Robert Briggs, research scientist at the Center for the Management of Information at the University of Arizona.

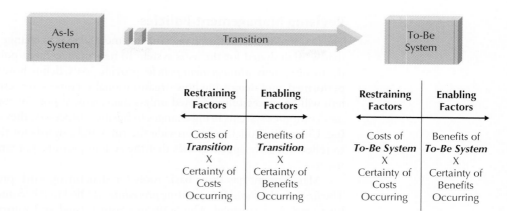

FIGURE 14-5
The Costs and
Benefits of Change

provide other benefits beyond the use of the system itself. Once again, any costs and benefits from the transition process must be weighted by the certainty with which they will occur (see Figure 14-5).

Taken together, these two sets of costs and benefits (and their relative certainties) affect the acceptance of change or resistance to change that project teams encounter when installing new systems in organizations. The first step in change management is to understand the factors that inhibit change—the factors that affect the perception of costs and benefits and certainty that they will be generated by the new system. It is critical to understand that the *real costs* and *real benefits* are far less important than the *perceived costs* and *perceived benefits*. People act on what they believe to be true, not on what is true. Thus, any understanding of how to motivate change must be developed from the viewpoint of the people expected to change, not from the viewpoint of those leading the change.

CONCEPTS **14-C Understanding Resistance to a DSS**

IN ACTION

One of the first commercial software packages I developed was a DSS to help schedule orders in a paper mill. The system was designed to help the person who scheduled orders decide when to schedule particular orders to reduce waste in the mill. This was a very challenging problem—so challenging, in fact, that it usually took the scheduler a year or two to really learn how to do the job well.

The software was tested by a variety of paper mills over the years and always reduced the amount of waste, usually by about 25 percent but sometimes by 75 percent when a scheduler new to the job was doing the scheduling. Although we ended up selling the package to most paper mills that tested it, we usually encountered significant resistance from the person doing the scheduling

(except when the scheduler was new to the job and the package clearly saved a significant amount). At the time, I assumed that the resistance to the system was related to the amount of waste reduced: the less waste reduced, the more resistance because the payback analysis showed it took longer to pay for the software.

—*Alan Dennis*

Questions

1. What is another possible explanation for the different levels of resistance encountered at different mills?

2. How might this be addressed?

Revising Management Policies

The first major step in the change management plan is to change the management policies that were designed for the as-is system to new management policies designed to support the to-be system. *Management policies* provide goals, define how work processes should be performed, and determine how organizational members are rewarded. No computer system will be successfully adopted unless management policies support its adoption. Many new computer systems bring changes to business processes; they enable new ways of working. Unless the policies that provide the rules and rewards for those processes are revised to reflect the new opportunities that the system permits, potential adopters cannot easily use it.

Management has three basic tools for structuring work processes in organizations.[9] The first are the *standard operating procedures* (SOPs) that become the habitual routines for how work is performed. The SOPs are both formal and informal. Formal SOPs define proper behavior. Informal SOPs are the norms that have developed over time for how processes are actually performed. Management must ensure that the formal SOPs are revised to match the to-be system. The informal SOPs will then evolve to refine and fill in details absent in the formal SOPs.

The second aspect of management policy is defining how people assign meaning to events. What does it mean to "be successful" or "do good work"? Policies help people understand meaning by defining *measurements* and *rewards*. Measurements explicitly define meaning because they provide clear and concrete evidence about what is important to the organization. Rewards reinforce measurements because "what gets measured gets done" (an overused but accurate saying). Measurements must be carefully designed to motivate desired behavior. The IBM credit example (Your Turn 4-2) illustrates the problem when flawed measurements drive improper behavior (when the credit analysts became too busy to handle credit requests, they would find nonexistent errors so they could return them unprocessed).

A third aspect of management policy is *resource allocation*. Managers can have clear and immediate impacts on behavior by allocating resources. They can redirect funds and staff from one project to another, create an infrastructure that supports the new system, and invest in training programs. Each of these activities has both a direct and symbolic effect. The direct effect comes from the actual reallocation of resources. The symbolic effect shows that management is serious about its intentions. There is less uncertainty about management's long-term commitment to a new system when potential adopters see resources being committed to support it.

YOUR TURN

14-2 Standard Operating Procedures

*I*dentify and explain three standard operating procedures for the course in which you are using this book. Discuss whether they are formal or informal.

---

[9] This section builds on the work of Anthony Giddons, *The Constitution of Society: Outline of the Theory of Structure* (Berkeley: University of California Press, 1984). A good summary of Giddons's theory that has been revised and adapted for use in understanding information systems is an article by Wanda Orlikowski and Dan Robey: "Information Technology and the Structuring of Organizations," *Information Systems Research* 2, no. 2 (1991): 143–169.

Assessing Costs and Benefits

The next step in developing a change management plan is to develop two clear and concise lists of costs and benefits provided by the new system (and the transition to it) compared with the as-is system. The first list is developed from the perspective of the organization, which should flow easily from the business case developed during the feasibility study and refined over the life of the project (see Chapter 2). This set of organizational costs and benefits should be distributed widely so that everyone expected to adopt the new system should clearly understand why the new system is valuable to the organization.

The second list of costs and benefits is developed from the viewpoints of the different potential adopters expected to change, or stakeholders in the change. For example, one set of potential adopters may be the frontline employees, another may be the first-line supervisors, and yet another might be middle management. Each of these potential adopters, or stakeholders, may have a different set of costs and benefits associated with the change—costs and benefits that can differ widely from those of the organization. In some situations, unions may be key stakeholders that can make or break successful change.

Many systems analysts naturally assume that frontline employees are the ones whose set of costs and benefits are the most likely to diverge from those of the organization and thus are the ones who most resist change. However, they usually bear the brunt of problems with the current system. When problems occur, they often experience them firsthand. Middle managers and first-line supervisors are the most likely to have a divergent set of costs and benefits and, therefore, resist change because new computer systems often change how much power they have. For example, a new computer system may improve the organization's control over a work process (a benefit to the organization) but reduce the decision-making power of middle management (a clear cost to middle managers).

An analysis of the costs and benefits for each set of potential adopters, or stakeholders, will help pinpoint those who will likely support the change and those who may resist the change. The challenge at this point is to try to change the balance of the costs and benefits for those expected to resist the change so that they support it (or at least do not actively resist it). This analysis may uncover some serious problems that have the potential to block the successful adoption of the system. It may be necessary to reexamine the management policies and make significant changes to ensure that the balance of costs and benefits is such that important potential adopters are motivated to adopt the system.

Figure 14-6 summarizes some of the factors that are important to successful change. The first and most important reason is a compelling personal reason to change. All change is made by individuals, not organizations. If there are compelling reasons for the key groups of individual stakeholders to want the change, then the change is more likely to be successful. Factors such as increased salary, reduced unpleasantness, and—depending on the individuals—opportunities for promotion and personal development can be important motivators. However, if the change makes current skills less valuable, individuals may resist the change because they have invested a lot of time and energy in acquiring those skills, and anything that diminishes those skills may be perceived as diminishing the individual (because important skills bring respect and power).

There must also be a compelling reason for the organization to need the change; otherwise, individuals become skeptical that the change is important and are less certain it will, in fact, occur. Probably the hardest organization to change is an organization that has been successful because individuals come to believe that what worked in the past will continue to work. By contrast, in an organization that is on the brink of bankruptcy, it is easier to convince individuals that change is needed. Commitment and support from credible business sponsors and top management are also important in increasing the certainty that the change will occur.

	Factor	Examples	Effects	Actions to Take
Benefits of to-be system	Compelling personal reason(s) for change	Increased pay, fewer unpleasant aspects, opportunity for promotion, most existing skills remain valuable.	If the new system provides clear personal benefits to those who must adopt it, they are more likely to embrace the change.	Perform a cost–benefit analysis from the viewpoint of the stakeholders, make changes where needed, and actively promote the benefits.
Certainty of benefits	Compelling organizational reason(s) for change	Risk of bankruptcy, acquisition, government regulation.	If adopters do not understand why the organization is implementing the change, they are less certain that the change will occur.	Perform a cost–benefit analysis from the viewpoint of the organization and launch a vigorous information campaign to explain the results to everyone.
	Demonstrated top management support	Active involvement, frequent mentions in speeches.	If top management is not seen to actively support the change, there is less certainly that the change will occur.	Encourage top management to participate in the information campaign.
	Committed and involved business sponsor	Active involvement, frequent visits to users and project team, championing.	If the business sponsor (the functional manager who initiated the project) is not seen to actively support the change, there is less certainty that the change will occur.	Encourage the business sponsor to participate in the information campaign and play an active role in the change management plan.
	Credible top management and business sponsor	Management and sponsor who do what they say instead of being members of the "management fad of the month" club.	If the business sponsor and top management have credibility in the eyes of the adopters, the certainty of the claimed benefits is higher.	Ensure that the business sponsor and/or top management has credibility so that such involvement will help; if there is no credibility involvement will have little effect.
Costs of transition	Low personal costs of change	Few new skills needed.	The cost of the change is not borne equally by all stakeholders; the costs are likely to be higher for some.	Perform a cost–benefit analysis from the viewpoint of the stakeholders, make changes where needed, and actively promote the low costs.
Certainty of costs	Clear plan for change	Clear dates and instructions for change, clear expectations.	If there is a clear migration plan, it will likely lower the perceived costs of transition.	Publicize the migration plan.
	Credible change agent	Previous experience with change, does what he/she promises to do.	If the change agent has credibility in the eyes of the adopters, the certainty of the claimed costs is higher.	If the change agent is not credible, then change will be difficult.
	Clear mandate for change agent from sponsor	Open support for change agent when disagreements occur.	If the change agent has a clear mandate from the business sponsor, the certainty of the claimed costs is higher.	The business sponsor must actively demonstrate support for the change agent.

FIGURE 14-6   Major Factors in Successful Change

The likelihood of successful change is increased when the cost of the transition to individuals who must change is low. The need for significantly different new skills or disruptions in operations and work habits may create resistance. A clear migration plan developed by a credible change agent who has support from the business sponsor is an important factor in increasing the certainty about the costs of the transition process.

Motivating Adoption

The single most important factor in motivating a change is providing clear and convincing evidence of the need for change. Simply put, everyone who is expected to adopt the change must be convinced that the benefits from the to-be system outweigh the costs of changing.

There are two basic strategies to motivating adoption: informational and political. Both strategies are often used simultaneously. With an *informational strategy*, the goal is to convince potential adopters that the change is for the better. This strategy works when the cost–benefit set of the target adopters has more benefits than costs. In other words, there really are clear reasons for the potential adopters to welcome the change.

Using this approach, the project team provides clear and convincing evidence of the costs and benefits of moving to the to-be system. The project team writes memos and develops presentations that outline the costs and benefits of adopting the system from the perspective of the organization and from the perspective of the target group of potential adopters. This information is disseminated widely throughout the target group, much like an advertising or public relations campaign. It must emphasize the benefits as well as increase the certainty in the minds of potential adopters that these benefits will actually be achieved. In our experience, it is always easier to sell painkillers than vitamins; that is, it is easier to convince potential adopters that a new system will remove a major problem (or other source of pain) than that it will provide new benefits (e.g., increase sales). Therefore, informational campaigns are more likely to be successful if they stress the reduction or elimination of problems rather than focusing on the provision of new opportunities.

The other strategy for motivating change is a *political strategy*. With a political strategy, organizational power, not information, is used to motivate change. This approach is often used when the cost–benefit set of the target adopters has more costs than benefits. In other words, although the change may benefit the organization, there are no reasons for the potential adopters to welcome the change.

The political strategy is usually beyond the control of the project team. It requires someone in the organization who holds legitimate power over the target group to influence the group to adopt the change. This may be done in a coercive manner (e.g., adopt the system or you're fired) or in a negotiated manner, in which the target group gains benefits in other ways that are linked to the adoption of the system (e.g., linking system adoption to increased training opportunities). Management policies can play a key role in a political strategy by linking salary to certain behaviors desired with the new system.

In general, for any change that has true organizational benefits, about 20 percent to 30 percent of potential adopters will be *ready adopters*. They recognize the benefits, quickly adopt the system, and become proponents of the system. Another 20 percent to 30 percent are *resistant adopters*. They simply refuse to accept the change and they fight against it, either because the new system has more costs than benefits for them personally or because they place such a high cost on the transition process itself that no amount of benefits from the new system can outweigh the change costs. The remaining 40 percent to 60 percent are *reluctant adopters*. They tend to be apathetic and will go with the flow to either support or resist the system, depending on how the project evolves and how their coworkers react to the system. Figure 14-7 illustrates the actors who are involved in the change management process.

The goal of change management is to actively support and encourage the ready adopters and help them win over the reluctant adopters. There is usually little that can be done about the resistant adopters because their set of costs and benefits may be divergent from those of the organization. Unless there are simple steps that can be taken to rebalance their costs and benefits or the organization chooses to adopt a strongly political strategy, it is often best to ignore this small minority of resistant adopters and focus on the larger majority of ready and reluctant adopters.

Sponsor	Change Agent	Potential Adopters
The sponsor wants the change to occur.	The change agent leads the change effort.	Potential adopters are the people who must change. 20–30 percent are ready adopters. 20–30 percent are resistant adopters. 40–60 percent are reluctant adopters.

FIGURE 14-7
Actors in the Change
Management Process

Enabling Adoption: Training

Potential adopters may want to adopt the change, but unless they are capable of adopting it, they won't. Adoption is enabled by providing the skills needed to adopt the change through careful *training*. Training is probably the most self-evident part of any change management initiative. How can an organization expect its staff members to adopt a new system if they are not trained? However, we have found that training is one of the most commonly overlooked parts of the process. Many organizations and project managers simply expect potential adopters to find the system easy to learn. Because the system is presumed to be so simple, it is taken for granted that potential adopters should be able to learn with little effort. Unfortunately, this is usually an overly optimistic assumption.

Every new system requires new skills, either because the basic work processes have changed (sometimes radically in the case of BPR; see Chapter 4) or because the computer system used to support the processes is different. The more radical the changes to the business processes, the more important it is to ensure the organization has the new skills required to operate the new business processes and supporting information systems. In general, there are three ways to get these new skills. One is to hire new employees who have the needed skills that the existing staff does not. Another is to outsource the processes to an organization that has the skills that the existing staff does not. Both these approaches are controversial and are usually considered only in the case of BPR when the new skills needed are likely to be the most different from the set of skills of the current staff. In most cases, organizations choose the third alternative: training existing staff in the new business processes and the to-be system. Every training plan must consider what to train and how to deliver the training.

What to Train   What training should you provide to the system users? It's obvious: how to use the system. The training should cover all the capabilities of the new system so users understand what each module does, right?

Wrong. Training for business systems should focus on helping the users to accomplish their jobs, not on how to use the system. The system is simply a means to an end, not the end in itself. This focus on performing the job (i.e., the business processes), not using the

CONCEPTS IN ACTION

14-D Too Much Paper (Part 2)

Some clerks at the South Dakota Department of Worker's Compensation (see Concepts in Action 14-A) were afraid that the digital solution might not work. What if they could not find an electronic file on the computer? What if a hard drive crashed or files were accidentally deleted? What if they could not retrieve the electronic file?

Question

In terms of organizational feasibility and adoption, what might an analyst do to convince these clerks to adopt the new technology?

system, has two important implications. First, the training must focus on those activities around the system, as well as on the system itself. The training must help the users understand how the computer fits into the bigger picture of their jobs. The use of the system must be put in context of the manual business processes as well as of those that are computerized, and it must also cover the new management policies that were implemented along with the new computer system.

Second, the training should focus on what the user needs to do, not what the system can do. This is a subtle—but very important—distinction. Most systems will provide far more capabilities than the users will need to use (e.g., when was the last time you wrote a macro in Microsoft Word?). Rather than attempting to teach the users all the features of the system, training should instead focus on the much smaller set of activities that users perform on a regular basis and ensure that users are truly expert in those. When the focus is on the 20 percent of functions that the users will use 80 percent of the time (instead of attempting to cover all functions), users become confident about their ability to use the system. Training should mention the other little-used functions but only so that users are aware of their existence and know how to learn about them when their use becomes necessary.

One source of guidance for designing training materials is the use cases. The use cases outline the common activities that users perform and thus can be helpful in understanding the business processes and system functions that are likely to be most important to the users.

How to Train   There are many ways to deliver training. The most commonly used approach is *classroom training,* in which many users are trained at the same time by the same instructor. This has the advantage of training many users at one time with only one instructor and creates a shared experience among the users.

It is also possible to provide *one-on-one training,* in which one trainer works closely with one user at a time. This is obviously more expensive, but the trainer can design the training program to meet the needs of individual users and can better ensure that the users really do understand the material. This approach is typically used only when the users are very important or when there are very few users.

Another approach that is becoming more common is to use some form of *computer-based training (CBT),* in which the training program is delivered via computer, either on CD or over the Web. CBT programs can include text slides, audio, and even video and animation. CBT is typically more costly to develop but is cheaper to deliver because no instructor is needed to actually provide the training.

Figure 14-8 summarizes four important factors to consider in selecting a training method: cost to develop, cost to deliver, impact, and reach. CBT is typically more expensive to develop than one-on-one or classroom training, but it is less expensive to deliver. One-on-one training has the most impact on the user because it can be customized to the user's precise needs, knowledge, and abilities, whereas CBT has the least impact. However, CBT has the greatest reach—the ability to train the most users over the widest distance in the shortest time—because it is much simpler to distribute than classroom and one-on-one training, simply because no instructors are needed.

	One-on-One Training	Classroom Training	Computer-Based Training
Cost to develop	Low–Medium	Medium	High
Cost to deliver	High	Medium	Low
Impact	High	Medium–High	Low–Medium
Reach	Low	Medium	High

FIGURE 14-8
Selecting a
Training Method

YOUR
TURN

14-3 Developing a Training Plan

Suppose you are leading the conversion from one word processor to another in your organization. Develop an outline of topics that would be included in the training. Develop a plan for training delivery.

Figure 14-8 suggests a clear pattern for most organizations. If there are only a few users to train, one-on-one training is the most effective. If there are many users to train, many organizations turn to CBT. We believe that the use of CBT will increase in the future. Quite often, large organizations use a combination of all three methods. Regardless of which approach is used, it is important to leave the users with a set of easily accessible materials that can be referred to long after the training has ended (usually a quick reference guide and a set of manuals, whether on paper or in electronic form).

POSTIMPLEMENTATION ACTIVITIES[10]

The goal of postimplementation activities is the *institutionalization* of the use of the new system—that is to make it the normal, accepted, routine way of performing the business processes. Postimplementation activities attempt to refreeze the organization after the successful transition to the new system. Although the work of the project team naturally winds down after implementation, the business sponsor and sometimes the project manager are actively involved in refreezing. These two—and, ideally, many other stakeholders—actively promote the new system and monitor its adoption and usage. They usually provide a steady flow of information about the system and encourage users to contact them to discuss issues.

In this section, we examine three key postimplementation activities: *system support* (providing assistance in the use of the system), *system maintenance* (continuing to refine and improve the system), and *project assessment* (analyzing the project to understand what activities were done well—and should be repeated—and what activities need improvement in future projects).

System Support

Once the project team has installed the system and performed the change management activities, the system is officially turned over to the *operations group*. This group is responsible for the operation of the system, whereas the project team was responsible for development of the system. Members of the operations group are usually closely involved in the installation activities because they are the ones who must ensure that the system actually works. After the system is installed, the project team leaves but the operations group remains.

Providing system support means helping the users to use the system. Usually, this means providing answers to questions and helping users understand how to perform a certain function; this type of support can be thought of as *on-demand training*.

Online support is the most common form of on-demand training. This includes the documentation and help screens built into the system, as well as separate Web sites that provide answers to *frequently asked questions (FAQs)*, which enable users to find answers

[10] The material in this section is related to the Enhanced Unified Process's Production Phase and the Operations and Support workflow (see Figure 1-11).

TEMPLATE
can be found at
www.wiley.com
/college/dennis

FIGURE 14-9
Elements of a
Problem Report

- Time and date of the report
- Name, e-mail address, and telephone number of the support person taking the report
- Name, e-mail address, and telephone number of the person who reported the problem
- Software and/or hardware causing problem
- Location of the problem
- Description of the problem
- Action taken
- Disposition (problem fixed or forwarded to system maintenance)

without contacting a person. Obviously, the goal of most systems is to provide sufficiently good online support so that the user doesn't need to contact a person, because providing online support is much less expensive than is providing a person to answer questions.

Most organizations provide a *help desk* that provides a place for a user to talk with a person who can answer questions (usually over the phone but sometimes in person). The help desk supports all systems, not just one specific system, so it receives calls about a wide variety of software and hardware. The help desk is operated by *level-1 support* staff, who have very broad computer skills and are able to respond to a wide range of requests, from network problems and hardware problems to problems with commercial software and problems with the business application software developed in-house.

The goal of most help desks is to have the level-1 support staff resolve 80 percent of the help requests they receive on the first call. If the issue cannot be resolved by level 1 support staff, a *problem report* (Figure 14-9) is completed (often using a special computer system designed to track problem reports) and passed to a *level-2 support* staff member.

The level-2 support staff members are people who know the application system well and can provide expert advice. For a new system, they are usually selected during the implementation phase and become familiar with the system as it is being tested. Sometimes the level-2 support staff members participate in training during the change management process to become more knowledgeable about the system, the new business processes, and the users themselves.

The level-2 support staff works with users to resolve problems. Most problems are successfully resolved by the level-2 staff. However, sometimes, particularly in the first few months after the system is installed, the problem turns out to be a bug in the software that must be fixed. In this case, the problem report becomes a *change request* that is passed to the system maintenance group (see the next section).

CONCEPTS
IN ACTION

14-E Too Much Paper (Part 3)

*F*urthermore, the South Dakota Department of Worker's Compensation had legal hurtles to implementing a digital solution to handle workers' compensation claims (see Concepts in Action 14-A and 14-D). One hurtle was that the previous paper method had physical signatures from employees, signing off to indicate that they had received treatment, or from the doctor, signing off on medical treatment performed.

Question

What legal aspect might arise from only having digital signatures or only electronic or paper copies of documents instead of physical documents?

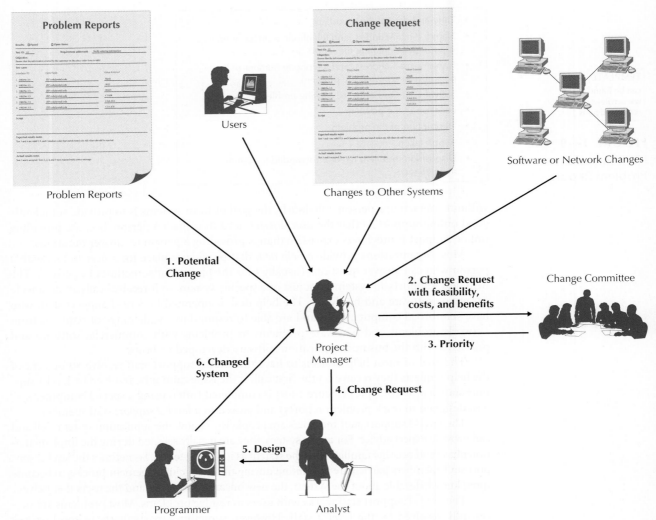

FIGURE 14-10   Processing a Change Request

System Maintenance

System maintenance is the process of refining the system to make sure it continues to meet business needs. Substantially more money and effort is devoted to system maintenance than to the initial development of the system, simply because a system continues to change and evolve as it is used. Most beginning systems analysts and programmers work first on maintenance projects; usually only after they have gained some experience are they assigned to new development projects.

Every system is "owned" by a project manager in the IS group (Figure 14-10). This individual is responsible for coordinating the system's maintenance effort for that system. Whenever a potential change to the system is identified, a change request is prepared and forwarded to the project manager. The change request is a smaller version of the *system request* discussed in Chapters 1 and 2. It describes the change requested and explains why the change is important.

Changes can be small or large. Change requests that are likely to require a significant effort are typically handled in the same manner as system requests: they follow the same process as the project described in this book, starting with project initiation in Chapter 2 and following through installation in this chapter. Minor changes typically follow a smaller version of this same process. There is an initial assessment of feasibility and of costs and benefits, and the change request is prioritized. Then a systems analyst (or a programmer/analyst) performs the analysis, which may include interviewing users, and prepares an initial design before programming begins. The new (or revised) program is then extensively tested before the system is converted from the old system to the revised one.

Change requests typically come from five sources. The most common source is problem reports from the operations group that identify bugs in the system that must be fixed. These are usually given immediate priority because a bug can cause significant problems. Even a minor bug can cause major problems by upsetting users and reducing their acceptance of and confidence in the system.

The second most common source of change requests is enhancement to the system from users. As users work with the system, they often identify minor changes in the design that can make the system easier to use or identify additional functions that are needed. Such enhancements are important in satisfying the users and are often key in ensuring that the system changes as the business requirements change. Enhancements are often given second priority after bug fixes.

A third source of change requests is other system development projects. For example, if the dentist in the appointment problem decided that he or she would like to have a Web-based appointment system that would allow patients to directly interact with the current appointment system, it is likely that other systems, such as billing, would have to be modified to ensure that the two systems would work together. These changes required by the need to integrate two systems are generally rare but are becoming more common as system integration efforts become more common.

A fourth source of change requests is those that occur when underlying software or networks change. For example, new versions of Windows often require an application to change the way the system interacts with Windows or enables application systems to take advantage of new features that improve efficiency. Although users may never see these changes (because most changes are inside the system and do not affect its user interface or functionality), these changes can be among the most challenging to implement because analysts and programmers must learn about the new system characteristics, understand how application systems use (or can use) those characteristics, and then make the needed programming changes.

The fifth source of change requests is senior management. These change requests are often driven by major changes in the organization's strategy or operations. These significant change requests are typically treated as separate projects, but the project manager responsible for the initial system is often placed in charge of the new project.

Project Assessment

The goal of project assessment is to understand what was successful about the system and the project activities (and, therefore, should be continued in the next system or project) and what needs to be improved. Project assessment is not routine in most organizations, except for military organizations, which are accustomed to preparing after-action reports. Nonetheless, assessment can be an important component in organizational learning because it helps organizations and people understand how to improve their work. It is particularly important for junior staff members because it helps promote faster learning. There are two primary parts to project assessment—project team review and system review.

CONCEPTS **14-F Managing a Database Project**

IN ACTION

One of my consulting projects was with a credit-card "bottom feeder" (let's call them Credit Wonder). This company bought credit-card accounts that were written off as uncollectable debts by major banks. Credit Wonder would buy the write-off accounts for 1 or 2 percent of the value. Credit Wonder would then call the owners of the written-off accounts and offer a deal to the credit-card account holders.

Credit Wonder wanted a database for these accounts. Legally, they did own the accounts and could contact the people who had owed the money—but as prescribed in credit law. For example, they could call only during certain hours, could call no more than once a week, and could speak only to the actual account

holder. Any amount collected over the 1 to 2 percent of the original debt would be considered a gain. In their database, they wanted a history of what settlement was offered, the date the account holder was contacted, and additional notes.

Questions

1. How might a systems analyst manage such a system project?
2. Who would they need to interview to get system requirements?
3. How would a database analyst help in structuring the database requirements?

Project Team Review A *project team review* focuses on the way in which the project team carried out its activities. Each project member prepares a short two- to three-page document that reports and analyzes his or her performance. The focus is on performance improvement, not penalties for mistakes made. By explicitly identifying mistakes and understanding their causes, project team members will, it is hoped, be better prepared for the next time they encounter a similar situation—and less likely to repeat the same mistakes. Likewise, by identifying excellent performance, team members will be able to understand why their actions worked well and how to repeat them in future projects.

The documents prepared by each team member are assessed by the project manager, who meets with the team members to help them understand how to improve their performance. The project manager then prepares a summary document that outlines the key learnings from the project. This summary identifies what actions should be taken in future projects to improve performance but is careful not to identify team members who made mistakes. The summary is widely circulated among all project managers to help them understand how to manage their projects better. Often, it is also circulated among regular staff members who did not work on the project so that they, too, can learn from other projects.

System Review The focus of the *system review* is to understand the extent to which the proposed costs and benefits from the new system identified during project initiation were actually recognized from the implemented system. Project team review is usually conducted immediately after the system is installed while key events are still fresh in team members' minds, but system review is often undertaken several months after the system is installed because it often takes a while before the system can be properly assessed.

System review starts with the system request and feasibility analysis prepared at the start of the project. The detailed analyses prepared for the expected business value (both tangible and intangible) as well as the economic feasibility analysis are reexamined and a new analysis is prepared after the system has been installed. The objective is to compare the anticipated business value against the actual realized business value from the system. This helps the organization assess whether the system actually provided the value it was planned to provide. Whether or not the system provides the expected value, future projects can benefit from an improved understanding of the true costs and benefits.

A formal system review also has important behavior implications for project initiation. Because everyone involved with the project knows that all statements about business value and the financial estimates prepared during project initiation will be evaluated at the end of the project, they have an incentive to be conservative in their assessments. No one wants to be the project sponsor or project manager for a project that goes radically over budget or fails to deliver promised benefits.

APPLYING THE CONCEPTS AT CD SELECTIONS

Installation of the Internet Sales System at CD Selections was somewhat simpler than the installation of most systems because the system was entirely new; there was no as-is system for the new system to replace. Also, there was not a large number of staff members who needed to be trained on the operation of the new system.

Conversion

Conversion went smoothly. First, the new hardware was purchased and installed. Then the software was installed on the Web server and on the client computers to be used by the staff of the Internet sales group. There was no data conversion per se, although the system started receiving data downloaded from the distribution system every day, as it would during normal operations.

Alec decided on a direct conversion (because there was no as-is system) in the one location (because there was only one location) of all system modules. The conversion, if we could call it that, went smoothly through the alpha and beta tests and the system was declared technically ready for operation.

Change Management

There were few change management issues because there were no existing staff members who had to change. All new staff were hired, most by internal transfer from other groups within CD Selections. The most likely stakeholders to be concerned by the change would be managers and employees in the traditional retail stores, who might see the Internet sales system as a threat to their stores. Alec therefore developed an information campaign (distributed through the employee newsletter and internal Web site) that discussed the reasons for the change and explained that the Internet sales system was seen as a complement to the existing stores, not as a competitor. The system was instead targeted at Web-based competitors, such as Amazon.com and CDnow.

The new management policies were developed, along with a training plan that encompassed both the manual work procedures and computerized procedures. Alec decided to use classroom training for the Internet sales system personnel because there were not many of them and it was simpler and more cost effective to train them all together in one classroom session.

Postimplementation Activities

Support of the system was turned over to the CD Selections operations group, who had hired four additional support staff members with expertise in networking and Web-based systems. System maintenance began almost immediately, with Alec designated as the project manager responsible for maintenance of this version of the system plus the development of the next version. Alec began the planning to develop the next version of the system.

Project team review uncovered several key lessons learned, mostly involving Web-based programming and the difficulties in linking to existing SQL databases. The project was delivered on budget (see Figure 2-15), with the exception that more was spent on programming than was anticipated.

A preliminary system review was conducted after two months of operations. Sales were $40,000 for the first month and $60,000 for the second, showing a gradual increase (remember that the goal for the first year of operations was $1,000,000). Operating expenses averaged $60,000 per month, a bit higher than the projected average, owing to start-up costs and the initial marketing campaign. Nonetheless, Margaret Mooney, vice president of marketing and the project sponsor, was quite pleased. She approved the feasibility study for the follow-on project to develop the second version of the Internet Sales System, and Alec began the system development process all over again.

SUMMARY

Cultural Issues and Information Technology

Given the global business environment, cultural issues become even more important when deploying information systems today. The cultural dimensions that need to be taken into consideration include power distance, uncertainty avoidance, individualism versus collectivism, masculinity versus femininity, and long- versus short-term orientation. Furthermore, each of these dimensions tends to interact with each other. From an information systems deployment perspective, the most important thing to remember is to take into consideration the local culture before deploying the new system.

Conversion

Conversion, the technical process by which the new system replaces the old system, has three major steps: install hardware, install software, and convert data. Conversion style, the way in which users are switched between the old and new systems, can be via either direct conversion (in which users stop using the old system and immediately begin using the new system) or parallel conversion (in which both systems are operated simultaneously to ensure the new system is operating correctly). Conversion location—what parts of the organization are converted and when—can be via a pilot conversion in one location; via a phased conversion, in which locations are converted in stages over time; or via simultaneous conversion, in which all locations are converted at the same time. The system can be converted module by module or as a whole at one time. Parallel and pilot conversions are less risky because they have a greater chance of detecting bugs before the bugs have widespread effect, but parallel conversion can be expensive.

Change Management

Change management is the process of helping people to adopt and adapt to the to-be system and its accompanying work processes. People resist change for very rational reasons, usually because they perceive the costs to themselves of the new system (and the transition to it) to outweigh the benefits. The first step in the change management plan is to change the management policies (such as standard operating procedures), devise measurements and rewards that support the new system, and allocate resources to support it. The second step is to develop a concise list of costs and benefits to the organization and all relevant stakeholders in the change, which will point out who is likely to support and who is likely to resist the change. The third step is to motivate adoption both by providing information and by using political strategies—using power to induce potential adopters to adopt the new system. Finally, training, whether classroom, one-on-one, or computer based, is essential to enable successful adoption. Training should focus on the primary functions the users will perform and look beyond the system itself to help users integrate the system into their routine work processes.

Postimplementation Activities

System support is performed by the operations group, which provides online and help desk support to the users. System support has both a level 1 support staff, which answers the phone and handles most of the questions, and level 2 support staff, which follows up on challenging problems and sometimes generates change requests for bug fixes. System maintenance responds to change requests (from the system support staff, users, other development project teams, and senior management) to fix bugs and improve the business value of the system. The goal of project assessment is to understand what was successful about the system and the project activities (and, therefore, should be continued in the next system or project) and what needs to be improved. Project team review focuses on the way in which the project team carried out its activities and usually results in documentation of key lessons learned. System review focuses on understanding the extent to which the proposed costs and benefits from the new system that were identified during project initiation were actually recognized from the implemented system.

KEY TERMS

Change agent
Change management
Change request
Classroom training
Collectivism
Computer-based training (CBT)
Conversion
Conversion location
Conversion modules
Conversion strategy
Conversion style
Cost
Direct conversion
Femininity
Frequently asked question
  (FAQ)
Help desk
Individualism
Informational strategy
Institutionalization
Level 1 support
Level 2 support
Long-term orientation

Management policies
Masculinity
Measurements
Migration plan
Modular conversion
Modules
On-demand training
One-on-one training
Online support
Operations group
Parallel conversion
Perceived benefits
Perceived costs
Phased conversion
Pilot conversion
Political strategy
Postimplementation
Potential adopter
Power distance
Problem report
Project assessment
Project team review
Ready adopters

Real benefits
Real costs
Refreeze
Reluctant adopters
Resistant adopters
Resource allocation
Rewards
Risk
Short-term orientation
Simultaneous conversion
Sponsor
Standard operating procedure
  (SOP)
System maintenance
System request
System review
System support
Time
Training
Transition process
Uncertainty avoidance
Unfreeze
Whole-system conversion

QUESTIONS

1. What are the three key roles in any change management initiative?
2. What are the three major elements of management policies that must be considered when implementing a new system?
3. What are the three basic steps in managing organizational change?
4. Explain the three categories of adopters you are likely to encounter in any change management initiative.
5. Some experts argue that change management is more important than any other part of system development. Do you agree or not? Explain.
6. Why do people resist change? Explain the basic model for understanding why people accept or resist change.
7. What are the cultural issues of which developers should be aware?
8. What are the major components of a migration plan?
9. Compare and contrast pilot conversion, phased conversion, and simultaneous conversion.

10. Compare and contrast direct conversion and parallel conversion.
11. Compare and contrast modular conversion and whole-system conversion.
12. Explain the trade-offs among selecting between the types of conversion in questions 9, 10, and 11.
13. In our experience, change management planning often receives less attention than conversion planning. Why do you think this happens?
14. Compare and contrast an information change management strategy with a political change management strategy. Is one better than the other?
15. What is the role of the operations group in system development?
16. Compare and contrast two major ways of providing system support.
17. How should you decide what items to include in your training plan?
18. Compare and contrast three basic approaches to training.
19. What are the major sources of change requests?

20. How is a problem report different from a change request?
21. How is project team review different from system review?
22. Why is project assessment important?

23. What do you think are three common mistakes that novice analysts make in migrating from the as-is to the to-be system?

EXERCISES

A. Consider a major change you have experienced in your life (e.g., taking a new job, starting a new school). Prepare a cost–benefit analysis of the change in terms of both the change and the transition to the change.

B. Suppose you are the project manager for a new library system for your university. The system will improve the way in which students, faculty, and staff can search for books by enabling them to search over the Web, rather than using only the current text-based system available on the computer terminals in the library. Prepare a cost–benefit analysis of the change in terms of both the change and the transition to the change for the major stakeholders.

C. Prepare a plan to motivate the adoption of the system in exercise B.

D. Prepare a training plan that includes both what you would train and how the training would be delivered for the system in exercise B.

E. Suppose you are the project leader for the development of a new airline reservation system that will be used by the airline's in-house reservation agents. The system will replace the current command-driven system designed in the 1970s that uses terminals. The new system uses PCs with a Web-based interface. Develop a migration plan (including both conversion and change management) for your telephone operators.

F. Develop a migration plan (including both conversion and change management) for the independent travel agencies who use the airline reservation system described in exercise E.

G. Suppose you are the project leader for the development of a new Web-based course registration system for your university that replaces an old system in which students had to go to the coliseum at certain times and stand in line to get permission slips for each course they wanted to take. Develop a migration plan (including both technical conversion and change management).

H. Suppose you are leading the installation of a new DSS to help admissions officers manage the admissions process at your university. Develop a change management plan (i.e., organizational aspects only).

I. Suppose you are installing a new accounting package in your small business. What conversion strategy would you use? Develop a conversion plan (i.e., technical aspects only).

J. Suppose you are installing a new room reservation system for your university that tracks which courses are assigned to which rooms. Assume that all the rooms in each building are "owned" by one college or department and only one person in that college or department has permission to assign them. What conversion strategy would you use? Develop a conversion plan (i.e., technical aspects only).

K. Suppose you are installing a new payroll system in a very large multinational corporation. What conversion strategy would you use? Develop a conversion plan (i.e., technical aspects only).

MINICASES

1. Nancy is the IS department head at MOTO Inc., a human resources management firm. The IS staff at MOTO Inc. completed work on a new client management software system about a month ago. Nancy was impressed with the performance of her staff on this project because the firm had not previously undertaken a project of this scale in-house. One of Nancy's weekly tasks is to evaluate and prioritize the change requests that have come in for the various applications used by the firm.

Right now, Nancy has five change requests for the client system on her desk. One request is from a system user who would like some formatting changes made to a daily report produced by the system. Another request is from a user who would like the sequence of menu options changed on one of the system menus to more closely reflect the frequency of use for those options. A third request came in from the billing department. This department performs billing through the use of a billing software package. A major upgrade of this software is being planned, and the interface between the client system and the bill system will need to be changed to accommodate the new software's data structures. The fourth request seems to be a system bug that occurs whenever a client cancels a contract (a rare occurrence, fortunately). The last request came from Susan, the company president. This request confirms the rumor that MOTO Inc. is about to acquire another new business. The new business specializes in the temporary placement of skilled professional and scientific employees and represents a new business area for MOTO Inc. The client management software system will need to be modified to incorporate the special client arrangements that are associated with the acquired firm.

How do you recommend that Nancy prioritize these change requests for the client/management system?

2. San Babila Taxi operates a fleet of taxis in Milan, Italy. The company provides the usual metered taxi services as well as short-term and long-term private car services for companies in Milan. With the increase in the number of cars, drivers, and contracts, the San Babila has experienced difficulty in keeping accurate records of its business activity and the movement of its fleet of cars. It recently purchased a new information system for the business and instructed all its drivers to use the new system. Private car drivers will have to swipe an ID badge through a reader at the start and end of each shift, and key in particulars of the car used and the client(s) served during the shift. Taxi drivers will also have to swipe their ID badge and record particulars of their car and using an in-taxi computer system, record each and every fare they pick up. The in-taxi computer system will also be used for dispatching taxis to customers' requests by phone or through the company's website.

The San Babila office staff were eagerly awaiting the installation of the new system. They agreed the system would reduce problems and errors in billing and would make their work easier. The drivers were less enthusiastic, being unaccustomed to the use of computers and to having their activities closely monitored.

a. Discuss the factors that may inhibit the acceptance of this new system by the drivers.

b. Discuss how an informational strategy could be used to motivate adoption of the new system.

c. Discuss how a political strategy could be used to motivate adoption of the new system.

INDEX

Value-added contracts, 303
Verification
 analysis models and, 273–287
 behavioral models and, 278–280
 functional models and, 275–276
 package diagrams and, 297–299
 structural models and, 276–278
 walkthroughs and, 273–274
View mechanisms, object diagrams and, 221
Virtualization, 394
Virus control requirements, 487
Visibility, class diagrams and, 213
Voiceover Internet Protocol (VolP), 511
Volumetrics, 393

W

Walkthroughs, 273–274, 429
Waterfall development methodology, 8–9
Web-based reports, 443
Web commerce packages, 305
White-box testing, 517, 518
White space, 417

Whole-Part pattern, 333–334
Whole-system conversion, 540–541
Wilson, Carl, 59
Window navigation diagrams (WND), 423–424, 435
Windows layout diagrams, 427
Work breakdown structures, 78–79, 86–90
Workarounds, 301
Workplans, 77–91
 Gantt charts and, 79–80
 Hurricane model and, 82–83
 identifying tasks and, 78–79
 iterative, 87–90
 PERT charts and, 81–82
 project workplan, 79
 refining estimates and, 82–83
 scope management and, 83–85
 task dependencies and, 79
 timeboxing and, 85
 work breakdown structure, 78–79, 86–90

Y

Yo-yo problem, 520